Polymers for Advanced Technologies

Symposium Editor

M. Lewin

School of Applied Science and Technology, Hebrew University and Israel
Fiber Institute, P.O.B. 8001, Jerusalem, Israel 91080

International Advisory Committee

Honorary Chairmen

H. F. Mark *(USA)*   **E. Katzir-Katchalski** *(Israel)*

Chairman

E. M. Pearce *(USA)*

Secretary

G. Kirshenbaum *(USA)*

Members

C. H. Bamford (UK) • **N. M. Bikales** (USA) • **B. Bulkin** (USA) • **E.
Cernia** (Italy) • **N. S. Choi** (Korea) • **J. B. Donnet** (France) • **J. Economy**
(USA) • **D. Fles** (Yugoslavia) • **M. Goodman** (USA) • **G. Hardy** (Hungary) • **R.
Konigsveld** (Netherlands) • **M. Kryszewski** (Poland) • **J. Lenz** (Austria) • **G.
Manecke** (FRG) • **E. Mano** (Brazil) • **R. Marchessault** (Canada) • **N. Igata**
(Japan) • **I. Panayotov** (Bulgaria) • **N. Pino** (Switzerland) • **R. Ranby**
(Sweden) • **I. Sakurada** (Japan) • **C. Simionescu** (Rumania) • **G. Smets**
(Belgium) • **D. Takayanagi** (Japan) • **D. Tanner** (USA)

National Organizing Committee

Chairman

M. Lewin

Secretary

S. Daren

Members

D. Cohn • **M. Friedman** • **L. Goldstein** • **H. Guttman** • **D. Katz** • **S.
Kenig** • **M. Levy** • **S. Margel** • **M. Markis** • **Z. Nir** • **A. Patchornik** • **Z.
Priel** • **A. Silberberg** • **G. Tanny** • **D. Vofsi** •
A. Zlicha

Cover illustration: LCPs with Calamitic and Discotic Mesogens (from
Lecture by Ringsdorf and W. Kreuder, Mainz

International Union of
Pure and Applied Chemistry
Macromolecular Division

In Conjunction with

**Israel Academy of Sciences: Israel Chemical Society • Ministry
of Industry and Trade • Ministry of Science and Development • Ben-
Gurion University of the Negev • The Hebrew University of
Jerusalem • Technion-Israel Institute of Technology • The Weizmann
Institute of Science • Tel-Aviv University**

Invited Lectures and Selected Contributed Papers

Presented at the

International Symposium on Polymers
for Advanced Technologies

**Held in Jerusalem, Israel
August 16-21, 1987**

Volume Edited by

Menachem Lewin

CHEMISTRY

Menachem Lewin
School of Applied Science and Technology
Hebrew University and Israel Fiber Institute
P.O.B. 8001
Jerusalem, Israel 91080

Library of Congress Cataloging-in-Publication Data

Polymers for advanced technologies.

    Includes bibliographies and index.

    I. Polymers and polymerization.   I. Lewin, Menachem, 1918-
QD381.7.P62   1988   547.8'4   88-177
ISBN 0-89573-293-9

Printed in the United States of America.

ISBN 0-89573-293-9 VCH Publishers
ISBN 3-527-26854-5 VCH Verlagsgesellschaft

Distributed in North America by:

VCH Publishers, Inc.
220 East 23rd Street, Suite 909
New York, New York 10010

Distributed Worldwide by:

VCH Verlagsgesellschaft mbH
P.O. Box 1260/1280
D-6940 Weinheim
Federal Republic of Germany

List of Contributors

M.A. Abkowitz
G. Ajroldi
M. Aldissi
J.H. An
B.C. Auman
D.B. Bailey
D.A. Batzel
R. Benrashid
Sylvio Brock
T.B. Brust
A. Buchman
Bernard J. Bulkin
Ivan Cabrera
H.-J. Cantow
S.J. Cau
J.Y. Cavaillé
M.C.O. Chang
Adolphe Chapiro
H. Craubner
R.C. Daley
D. Deveen
S. DiStefano
H. Dodiuk
Manfred Eich
Shi Wen Feng
R. Friedrich
C.W. Fu
J. Fuhrmann
A. Furukawa
Vlodek Gabara
C. Garbuglio
M. Garcia
Umesh Gaur
M. Glotin
P.L. Grady
G. Groeninckx
Sangita S.P. Gupta
S.P. Hersh
M. Hess

J. Hilborn
R. Hosemann
J. Jachowitz
W.J. Jackson Jr.
Michael Jaffe
U. Jahn
S.R. Jin
Wu Jishan
C. Jourdan
Qu Bao Jun
D. Jungbauer
S. Kenig
G. Kennedy
M.E. Kenny
Sung Chul Kim
W.H. Ko
Robert Kosfeld
Valeri Krongauz
Marian Kryszewski
T.K. Kwei
Stephanie L. Kwolek
Jae Heung Lee
R.W. Lenz
Menachem Lewin
Chen Yong Lie
Ling-Yu-He
Jerome B. Lo
H.F. Mark
Wesley Memeger
G.C. Meyer
A. Mey-marom
J. Moacanin
G. Moggi
G. Nagasubramanian
Hermann H. Neidlinger
Naoya Ogata
Eli M. Pearce
J.M. Pearson
V. Percec

A. Peters
M. Pianca
J.F. Pierson
Dusan C. Prevorsek
Béla Pukánszky
Bengt Rånby
L. Rebenfeld
Scott E. Rickert
Helmut Ringsdorf
E. Sacher
Jürg R. Seidel
John R. Schaefgen
Paul Schissel
H.A. Schneider
F. Schosseler
Raymond B. Seymour
I. Shapiro
S. Shkolnik
L.H. Sperling
M. Stolka
H. Sturm
Gwo-Shin Swei
David Tanner
G.C. Tesoro
D.A. Thomas
Ferenc Tüdos
James E. Van Trump
D.S. Varma
I.K. Varma
H.D. Wagner
Joachim H. Wendorff
M.R. Wertheimer
J.M. Widmaier
G. Wis-Surel
J.J. Wortman
C.N. Wu
Hyun-Nam Yoon
Robert J. Young
Andrzej Ziabicki

v

Table of Contents

PREFACE

During the last two decades, significant changes have taken place in the general pattern of industrial developments throughout the world. Novel industrial enterprises have blossomed, based primarily on new scientific and technological achievements. The appearance of these advanced industries represents a growing partnership between scientists, engineers, and industrialists.

Three outstanding characteristics of the new technological trends are

1. The pace of development: The transition from concept, through laboratory and pilot plant, to practical application is highly accelerated.

2. The new developments are interdisciplinary in nature and require close cooperation between several fields of scientific as well as engineering skills. This is perhaps a major reason for the imaginativeness and unexpectedness of many of these developments.

3. These technologies are highly diversified; they branch out into many avenues of human activity and endeavor, meeting ever-increasing challenges of hitherto unsolved problems.

It is not surprising that a considerable part of these new technologies center around polymers. Seventy million tons of polymers, valued at over 140 billion dollars, were produced worldwide in 1986. The bulk of this production was relatively simple commodity polymers. Recently, however, a small but growing proportion of the polymers produced have been specially designed and tailor-made for specific, increasingly sophisticated applications. These spectacular achievements in polymer science led to a recent statement by Herman F. Mark that the time is close when "the potential of polymer chemistry will be limited only by the imagination and ingenuity of scientists."

It appeared to the organizers of the 1987 IUPAC International Symposium on Polymers for Advanced Technologies that the time was ripe for focusing the attention of scientists from academia and industry on recent developments and on several new aspects of polymer research and its impact on the emerging science-based technologies, in order to meet the challenges of the 21st century.

The Symposium took place at the Laromme Hotel in Jerusalem on the 16–21 of August, 1987, with 450 participants from 20 countries. The program consisted of 26 lecture sessions, two poster sessions, and a panel discussion on Strategies for Advanced Polymers in the World.

The Symposium was opened by Professor Menachem Lewin, followed by a welcoming address by the Prime Minister of the State of Israel, the Honorable Yitzhak Shamir, who stated "Science and technology are a striking example of international cooperation in which gifted individuals from many countries reach out to each other across borders and beyond limitations in a joint effort to create knowledge in order to build a new, happier world, a safer future, and a permanent peace for all humanity." The opening lecture was given by Professor Herman F. Mark, who presented an overview on "Polymers for Advanced Technologies." This was followed by the plenary lectures by Professor E. Katchalski-Katzir on "Polymers for Biosystems," by Nobel Laureate Professor W.B. Merrifield on "The Solid Phase Approach to the Chemical Synthesis of Biologically Active Peptides," by Dr. D. Tanner on "Polymers for Advanced Structures," and by Professor A. Chapiro on "Polymers in Radiation-Related Technologies."

The Symposium, organized by Professor Menachem Lewin, was the third IUPAC Macromolecular Symposium held in Israel. The first took place in Rehovot in 1956 and was organized by the late Professor Aharon Katchalski-Katzir. The second symposium was held in Jerusalem in 1975 and was organized by Professor Alex Silberberg.

The Symposium, organized around three major topics (microsymposia) covered a large part of the work in the field: (1) Polymers for Biosystems, (2) Polymers in Radiation-Related Technologies, (3) Polymers for Advanced Structures.

Seventy-five out of the 135 lectures and posters presented at the Symposium are published in the two volumes of the proceedings. They include the plenary and sectional lectures as well as a number of selected contributed presentations by outstanding authors from 20 countries. These represent a wide diversity of subjects, approaches, and experiences and are truly representative of the present state and future trends of the field of polymers for advanced technologies.

One of the volumes, "Polymers for Biosystems," contains 26 chapters and is being published by Hüthig and Wepf Publishers, Basel. The present volume contains 51 chapters and deals with the two other Symposium topics: Polymers in Radiation-Related Technologies and Polymers for Advanced Structures. It is subdivided into nine parts. Part I deals with radiation-sensitive and radiation-modified and cured systems. Silicon-containing resist polymers for bilayer applications and image-selective silylations for single layer schemes are described by *J.M. Pearson. S.P. Hersh, S. Brock, P.L. Grady,* and *J.J. Wortman* compare the ion irradiation method for enhancing permanent conductivity of a number of highly insulating polymers to chemical doping. Modification of polymers by high-energy radiation resulted in highly sophisticated devices applied in fields such as wires, cables, packaging materials, automobile tires; breakthroughs are expected in electronics and biomedical devices (*A. Chapiro*). The state of the art of industrial applications of ultraviolet (uv) lamps and low-energy electron accelerators for instant polymerization of radiation-curable chemicals is critically reviewed by *J.R. Seidel*. Such systems are exemplified by photo-crosslinking of EPDM elastomers in a multi-step reaction with photoinitiated uv crosslinking of polyethylenes and polyesters (*B. Ranby, C.Y. Lie, Q.B. Jun,* and *S.W. Feng*). An electron beam radiation grafting of *p*-styrene sulfonic acid on ethylene-vinyl alcohol is also presented (*A. Mey-Marom* and *S. Shkolnik*). Recent developments in solar energy collection technologies have brought about considerable progress in designing and modifying polymers in order to obtain the desired properties and predicted performances in service. Of particular importance are the optical, thermomechanical, and diffusion barrier behavior properties, as well as environmental aging effects (*H. Neidlinger* and *P. Schissel*).

Part II describes systems of photo-electro conductive and piezo-electric polymers. Photoconductivity of homogeneous and heterogeneous systems with particular emphasis on materials based on poly(N-vinyl carbazole) is discussed by *M. Kryszewski. M.A. Abkowitz* and *M. Stolka* describe the electrophotographic process and discuss the electronic transport phenomena in polymeric photoreceptors with emphasis on polysilylenes. A study on the contact resistance and interfacial properties of *p*-Si/conducting polymer interface for solar cell applications is presented by *G. Nagasubramanian, S.D. Stefano,* and *J. Moacanin*. Phthalocyanine-based Langmuir gas sensors are reviewed by *E. Rickert*.

Three chapters deal with electrical phenomena in poly(vinylidene fluoride) (PVDF) as a piezoelectric and pyroelectric polymer: *J.B. Lando* discusses the relationship between crystallization and the poling behavior of the polymers. He also discloses a way to produce bulk poly-(fluoroacetylene) from PVDF. *J.H. Wendorff* and *O. Jungbauer* report on nonlinear piezoelectricity obtained when uniaxially stretched and poled PVDF was stressed or strained perpendicular to the stretching direction. *J. Fuhrmann, U. Jahn,* and *H. Strum* show that the ferroelectric behavior of PVDF is frequency dependent, due to the dynamic cooperative dipole effect. *M. Aldissi*, in a contributed paper, discusses intrinsic anisotropy in electric and optical properties of conducting polymers obtained by macroscopic and chain orientation of the material. New information is reported on triboelectricity of keratin fibers by *J. Jachowicz, G. Wis-Gurel,* and *M. Garcia*. A comparison of the alpha particle penetration range of several polymers used in the microelectonics industry—poly-*p*-xylylene, polyimide, polyester, and plasma polymerized hexamethyldisiloxane—is reported by *G. Kennedy, E. Sacher, and M.R. Wertheimer*.

Parts III–VIII of the present volume deal with polymers for advanced structures, which are defined by *D. Tanner, V. Gabara,* and *J. Schaefgen* in their overview "as systems of materials which meet unusually high performance requirements" in, for example, aircraft, aerospace, automotive, ballistics, and protective clothing applications. In his overview, *H.F. Mark* discusses recent developments in advanced polymers and regards "competition and cooperation of polymers with metals and ceramics" as a major trend in future developments. This relates to efforts being made to synthesize polymers with properties approaching those of metals and ceramics. In addition, materials composed of polymers combined with metals or ceramics are emerging.

Parts IV and V deal with high-performance polymers. *Dr. Stephanie Kwolek*, who was the first to establish the existence of lyotropic liquid crystalline (LC) solutions of polymers, reviews recent work on aramide LC polymers and fibers and discusses a new vapor phase polycondensation process. *M. Jaffe* and *H.N. Yoon* discuss structure-processing relationships of nematic thermotropic copolyesters. *W.J. Jackson* compares effects of composition and fiber processing conditions on the tensile properties of LC polymers prepared from aromatic copolyesters containing flexible spacers and wholly aromatic copolyesters. *R.W. Lenz* discusses the effect of both main chain and pendant flexible spacers on the LC properties of thermotropic polyesters. *J.H. Wendorff* reports on the effect of local structural variations induced in LC domains by external fields, on the local variation of the refractive index, and on birefringence and its use for reversible information storage.

Materials prepared from polyacrylate and polysiloxane LCPs showing both thermotropic and photochromic mesophases are described by *I. Cabrera*. Novel condensation systems for producing high molecular-weight polyamides and polyesters are presented by *N. Ogata*.

D. Prevorsek discusses factors influencing strength and modulus of high tenacity–high modulus polyethylene fibers. *A. Ziabicki* compares development of structure in processing of polymers built of rigid versus flexible macromolecules. Preparation of single crystal fibers by solid state polymerization of monomeric single crystals of substituted diacetylenes and their properties is presented by *R.J. Young*. *H.D. Wagner* discusses the effect of size on ultimate properties such as strength and lifetime of fibers and whiskers.

Recently, there has been considerable interest in multi-component polymer blends and in their application to the engineering, space, biomedical, electrical, and electronic industries. Understanding the morphology and behavior of various polymer blends and networks is in its beginning stages; nevertheless, the wide possibilities for obtaining materials and devices with improved and novel properties are already widely realized.

Parts VI and VII of this volume contain eight papers on polymer network and blends. *L.H. Sperling* reviews recent work on interpenetrating polymer networks (IPNs), with emphasis on domain size assessment with small angle neutron scattering phase separation and on sound and vibration damping. *S.R. Jin, J.M. Widmayer,* and *G.C. Meyer*, in a contributed paper, report on the formation kinetics of a acrylic network phase in the presence of an already fully-formed polyurethane network, followed by Fourier transform infra red (FTIR) spectroscopy. *Sung Chul Kim* reports on an IPN membrane composed of the hydrophylic polyurethane and the hydrophobic polystyrene for pervaporation of ethanol–water mixtures. Water-swellable ionic polymeric networks composed of polyacrylamide and polyacrylic acid and their swelling kinetics are discussed by *S.J. Candau*.

Four chapters are devoted to polymer blends. *E.M. Pearce, T.K. Kwei*, and *Ling-Yu-He* discuss thermal degradation of blends of poly(vinylacetate) and PVDF. Mechanical spectroscopy is applied for the study of the structure of ethylene–propylene-based polymer blends in a study by *J.F. Pierson*. *R. Kosfeld* describes the thermal behavior of blends of polycarbonate with semiflexible thermotropic copolyesters from poly(ethyleneterephthalate) (PET) and *p*-hydroxybenzoate. *G. Groeninckx* reports on the miscibility behavior, morphology, mechanical properties, and deformation modes of blends of a rubber containing modifier with a random terpolymer mainly based on styrene and maleic anhydride and PVC.

The intensive studies being conducted in many laboratories on composites and adhesives are reflected in the five papers of Part VIII of this book. In a contributed paper by *I.K. Varma*, studies are reported on the thermal behavior of bismaleimide resins with a series of additives. *G. Tesoro* reports on a new system for chemically modifying the surface of Kevlar® fibers to improve composite properties. *F. Tüdos* and *B. Pukanszky* describe the role of the interface in PP composites containing different inorganic fillers. *I. Shapiro* reports on the chain length development in epoxy-amine matrices. *H. Dodiuk, A. Buchman*, and *S. Kenig* describe polyurethane adhesive with silane coupling agents.

Part IX of this volume is devoted to elastomers and several special topics. *C. Garbuglio* reviews recent work on structure and properties of fluoroelastomers—for example, copolymers vinylidene fluoride (VDF)/hexafluoropropene (HFP), and terpolymers VDF-HFP-tetrafluoroethylene. *H.J. Cantow, B.C. Auman, W.A. Jishan, V. Percec*, and *H.A. Schneider* report on a new temperature-resistant thermoplastic elastomer composed of aromatic polyether sulfone with incorporated poly(dimethylsiloxane). *R.B. Seymour* describes the synthesis of block and graft copolymers from macroradicals. *B.J. Bulkin* and *M. Lewin* discuss new developments in Raman spectroscopy of polymers which enable the direct determination of a number of properties. *R. Hosemann* describes microparacrystals and their equilibrium state. *H. Craubner* presents a statistical thermodynamic theory of polymer precipitation.

The Symposium would not have been possible without the sponsorship and financial support of the following institutions: IUPAC, Israel Academy of Sciences and Humanities, Israel Chemical Society, Ministry of Science and Development, Ministry of Industry and Trade, Hebrew University, Ben-Gurion University of the Negev, Tel-Aviv University, The Weizmann Institute of Science, and the Technion-Israel Institute of Technology.

We gratefully acknowledge the financial support of the following companies: Agan (Israel), Arco (USA), Celanese Research Co. (USA), E.I. duPont de Nemours (USA), Gelman Sciences (USA and Israel), Israel Military Industries,

Israel Petrochemical Industry, Instituto Guido Donegani, Montedipe, Montedison SPA, Montedison Composites and Advanced Materials, Montefluos (Italy), Katrit Kfar Aza (Israel), Iscar Blades (Israel), Lidor Chemicals (Israel), Makhteshim Chemical Works (Israel), Monsanto (USA), Nilit (Israel), Rafael (Israel), Rohm and Haas Company (USA), Schering (FRG), The Standard Oil Co. (USA), Tambour (Israel), and the Xerox Research Center (Canada).

I wish to express my gratitude to the Organizing Committee, to the International Advisory Committee, to the sponsoring and supporting institutions and companies, to all the authors and their coworkers, to session chairmen, panel members, discussion speakers, to all participants, and to the editorial staff of VCH Publishers for their cooperation and contributions to the success of this Symposium and to this comprehensive volume. A special vote of thanks is due to the Honorary Chairmen, Professor Herman Mark and Professor Ephraim Katchalski-Katzir, for their valuable advice, support, and contributions.

Menachem Lewin
Hebrew University and Israel Fiber Institute

Part I
Radiation-sensitive, Radiation-modified, and Cured Polymers

1. POLYMERS IN RADIATION-RELATED TECHNOLOGIES

RADIATION MODIFIED POLYMERS

Adolphe Chapiro

Centre National de la Recherche Scientifique
B.P. 28
94320 Thiais, France

High Energy Radiations represent a powerful tool for initiating various chemical reactions and, particularly in the field of polymers, a large number of new materials have been developed, or are in the process of being developed, based on irradiation technology. Such modified polymers exhibit improved properties or even properties not found elsewhere. This fast-growing area of new applications includes up-graded insulators, improved packaging materials, better processable rubbers, hot water pipes, heat-shrinkable tubings, films and moldings, conductive polymers, reliable safety devices based on polymers, membranes, polymers for biomedical applications, and many others.

Several techniques are involved in the production of radiation-modified polymers:
- simple irradiation of existing commercial products;
- irradiation of polymers with special formulations;
- irradiation of polymers with added reactive modifiers;
- irradiation of blends of polymers or of polymers and monomers;
- irradiation of monomers formulated for ultra-fast cure.

In order to review this very broad field, we shall consider the following two areas of interest:
1. Radiation cross-linked polymers;
2. Chemically-modified polymers by radiation methods.

Radiation Cross-linked Polymers

Radiation cross-linking of rubbers and plastics is today a
well-established technology with large volume operations and
a bright future. Polyethylene, silicones and various rubbers
are easily cross-linked under electron bombardment to yield
insoluble materials with improved moduli. Improved mechanical properties in semi-crystalline polymers, such as polyethylene, are particularly obvious at elevated temperatures,
above their crystalline melting point, where these polymers
no longer flow, but exhibit rubber elasticity. For these
applications, it is desirable to use special formulations
with selected stabilizers (antioxidants) which do not interfere with the cross-linking process.

Irradiation of polymers below their glass transition temperature, such as PVC at room temperature, results in very
little cross-linking, owing to the lack of mobility of the
polymeric radicals. The efficiency of the process is highly
increased if the polymer is plasticized, bringing its glass
transition temperature below the processing temperature.
Even better results are obtained by plasticizing the polymer
with a polyfunctional monomer, which acts as a cross-linking
promoter. Such a procedure is widely used for cross-linking
PVC, polyethylene, polypropylene, rubbers, PVDF, and many
others.

Up-graded insulators. Radiation cross-linked polymers are
widely used in the wire and cable industry for the production
of up-graded insulators which can withstand much higher temperatures and are more reliable under normal conditions.
These are finding growing acceptance in the consumer's market for telephones, automobiles, etc. Some of the most reliable wires and cables in an aggressive environment and
suitable for use in the aircraft and space industries, have
been specially tailored for such applications. Their structures are more sophisticated and combine the solvent and
temperature resistance of radiation cross-linked polymers
with multi-layered coating, with the outer layer providing
abrasion resistance and a barrier against oxygen.

Heat-shrinkable devices. Heat-shrinkable devices are based
on the "memory effect" of cross-linked polymers, first described in radiation-treated polyethylene by Charlesby in
1954 [1]. Shaped parts such as tubings, films or moldings
are first cross-linked and thereafter heated above their
transition or melting temperature. The resulting rubbery
material is strained and then cooled so that it retains its

deformed shape. If heated again, the polymer specimen will
return to its initial shape. This property is used commercially on a large scale for packaging films. Shrinkable
tubings are used in the electric and electronic industries
and also for protecting the welded zones in gas and oil pipes.
This last application may involve tubings of very large
dimensions, which brings this technology into the "heavy"
industry class (Fig. 1).

   Sophisticated devices tailored for special uses. Much
more sophisticated devices were developed on the basis of
this "memory" effect. Raychem Corporation is the world lead-
er in this field. Even though the following may seem to be
a "commercial" for Raychem, it actually represents a realistic survey of the new materials and devices which have been
developed by radiation modification of polymers.

Figure 1. Paul Cook, President of Raychem, inspecting a heat-
shrinking corrosion preventing tape on the trans-Alaska
pipeline in 1975.

"Solder sleeves" are small pieces of heat-shrinkable tubings, usually made of PVDF, in the center of which a ring of
low melting metal is secured. Both ends may, or may not, be
fitted with rings of non-cross-linked, and therefore melting,
polyethylene (Fig. 2). When stripped wire ends are introduced
in this device and the whole system is heated, the outer sleeve
shrinks, the solder melts and flows to the termination joints
and the melting polyethylene, if present, fills any residual
void. The results is a strong, highly reliable, soldered connection. Such assembly of multiple wire terminations can be

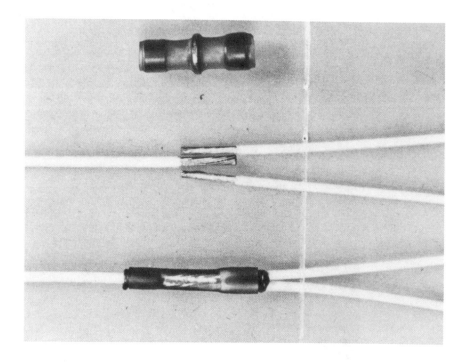

Figure 2. "Solder sleeve" before (top) and after splicing.

carried out in a single and safe operation, as in the case
of plugs with multiple connections (Fig. 3).

 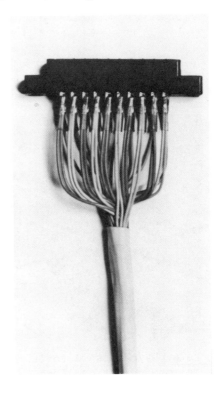

Figure 3. Assembly with Raychem's "Solderpak" on multiple-pin plug.

Based on this same principle, but carrying sophistication a step further, Raychem offers a series of single-step mending devices for damaged telephone lines; these can be easily utilized in the field with a gas or hot air torch or, in the case of a hazardous environment or a lack of space, the device itself may contain a self-regulating electric heating system, based on conductive polymer formulations, as described below (Fig. 4).

Stabilization of molecular structures. Cross-linking also provides a method for stabilizing the molecular packing of a polymer system by preventing any further sliding of the macromolecules. This property is used in the manufacture of self-regulating heating devices which can withstand numerous cycles without losing their characteristic properties. Semi-

Figure 4. Raychem's "Autofit" device for splicing and repairing telecommunication lines.

crystalline polymers loaded with carbon black exhibit a rapid rise of their resistance as the temperature approaches a critical value, characteristic of the particular polymer (Fig. 5). Under constant voltage, the amount of heat evolution will therefore decrease with temperature and become almost zero at the critical temperature. The heater based on such a polymer will reach the critical temperature and remain steady with time afterwards. It is self-regulating on the basis of a very simple principle: if the temperature drops, the resistance decreases and more current goes through the system until the critical temperature is again reached. Raychem has developed heaters based on this principle in which the voltage is supplied to two continuous parallel copper wires separated by a layer of conductive polymer (Fig. 6). The current flows from one wire to the other through the carbon loaded polymer, thereby generating heat. Thermal equilibrium is reached within a few minutes. The heater can be cut to any desired length while still using the same power supply. The total amount of heat evolution is directly proportional to the length of the heater. Such heaters are fully reliable; any local over-heating will stop the current

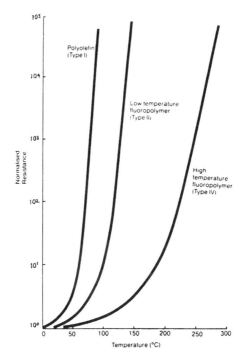

Figure 5. Change of electrical resistance as a function of temperature for various carbon black loaded polymers [2].

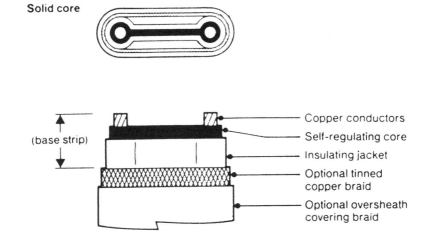

Figure 6. Construction of Raychem's "AutoTrace" heater.

at that particular point without affecting the over-all
operation of the heater. By combining its expertise in the
solid state properties of polymers with its knowledge of
radiation technology, Raychem offers heaters tailored to
stabilize at any desired temperature, which may be as low as
40°C or may exceed 100°C.

"PolySwitch" is a self-resetting circuit breaker (or "fuse")
based on the same principle (Fig. 7). Here the carbon black
loaded polymer protects an electric circuit from excessive
over-current. If the current exceeds a threshold above which
the temperature of the device reaches its critical value, the
resistance suddenly increases dramatically, thereby interrup-
ting the circuit. Upon cooling,the conductive property is
regenerated and, if the overload was only accidental, every-
thing returns to normal. If, on the other hand, the system
suffers a short, the circuit is broken permanently [4].
Figure 8 shows the behavior of the device under multiple faults.
The base resistance slightly increases after each trip, but
eventually reaches a stationary value. The device also trips
if the ambient temperature rises above its critical value.
It thus acts as a sensor for over-temperature protection.

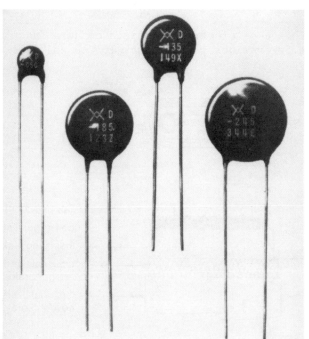

Figure 7. Raychem's self-resetting circuit breakers
"PolySwitch".

We have here two examples of polymeric systems tailored to do a certain job, in which radiation cross-linking provides the necessary reliability and improves the lifetime of the device.

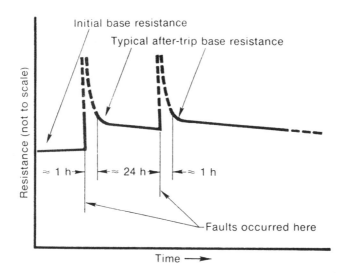

Figure 8. Change of resistance with time of a "PolySwitch" under multiple faults [4].

Gels. Irradiation of polymer solutions under suitable conditions leads to the formation of cross-linked gels. Interesting gels are easily prepared by irradiating water-soluble polymers in aqueous solutions. In such systems, the density of cross-links is determined by the total dose and can be easily controlled. This makes it possible to manufacture gels with any desired water content, from weak, highly swollen specimens to fairly stiff materials [5]. Figure 9 shows that the swelling of a polyvinylpyrollidone gel, which is as high as 10,000% for a dose of 2 kilograys, drops to less than 2000% after 15 kilograys. These doses are very small compared to the doses of 100 to 200 kilograys commonly used for commercial cross-linking in the wire and cable industry. Such gels find increasing uses in the food industry or in biomedical applications as controlled drug-release devices, biocompatible materials, and the like.

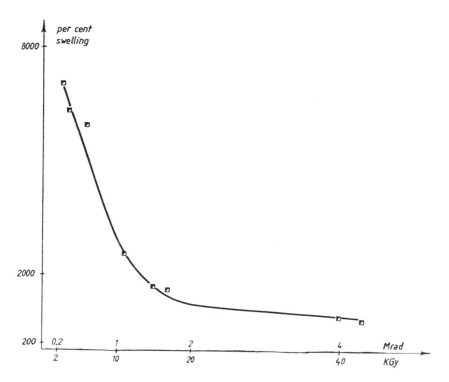

Figure 9. Equilibrium swelling in water of a polyvinylpyrrolidone gel as a function of dose of irradiation [5].

Chemically modified polymers by radiation methods.

Irradiation of polymers in the presence of reactive modifiers results in the formation of new materials in which substantial amounts of the modifier may become chemically bonded
to the polymer. The most obvious reactive chemicals under
free radical attack are halogens and olefins. Very little
work was carried out until now on radiation halogenation of
polymers,and some promising applications may still be found
in this area. On the other hand, vinyl monomers have been
widely used for the production of numerous graft copolymers,
some of which have resulted in commercial applications. This
field is still in its infancy today, however, and since radiation grafting is a very powerful method for producing new
polymers, many more applications are expected to arise in
the future.

Like radiation cross-linking, the process of radiation grafting can be easily carried out with shaped polymeric products such as films, fabrics and tubings. The treatment may either involve the bulk of the article or may be purposely limited to a controlled surface layer. Grafting can also be used to modify polymers in the shape of powders or granules, thereby generating entirely new polymeric materials. A few examples will be considered in the following sections; these were selected from the numerous studies available, either because they are used commercially today, or because their future development looks promising.

Improved cross-linking via grafting. As mentioned earlier, the use of polyfunctional monomers for improving the efficiency of radiation cross-linking is a common procedure for such polymers as PVC, PVDF, and others. The resulting products are graft copolymers in which short, polyfunctional branches form a tight network. Another grafting method was applied to achieve cross-linking of polyaryletherketone polymers [6].

These polymers, with the general formula

$$-C_6H_4-O-C_6H_4-O-C_6H_4-CO-C_6H_4-$$

and trade names "PEEK" from I.C.I., or "STILAN", developed by Raychem, have outstanding mechanical properties and a high glass transition temperature (in the vicinity of 150°C). This makes them suitable as very reliable engineering polymers. For some applications, however, a higher working temperature is desired, particularly in view of their significant creep in the vicinity of Tg. Cross-linking could help improve this behavior, but, owing to the aromatic structure of these polymers and to their high Tg, direct radiation cross-linking is very inefficient. The following procedure was therefore developed:
  a) radiation grafting of acrylonitrile;
  b) heat treatment in the vicinity of 300°C in order to induce the cyclization of the polyacrylonitrile branches, thereby generating cross-links.

The second step is similar to the reaction used for the manufacture of carbon fibers starting from polyacrylonitrile fibers.

The result of such a treatment is a black material which retains the high modulus and tensile strength of the polyaryl polymer, but which has become insoluble in the solvents

of these polymers, such as 98% sulfuric acid, and which exhibits very little creep at 180°C and even at 280°C. In a
typical run, a sample of PEEK was grafted to 6% with acrylonitrile and thereafter heated for one hour to 300°C. Its
tensile modulus was 680 at 180°C (the corresponding value
for PEEK was 330) and 245 at 280°C (not measurable in PEEK).

Figure 10 shows DSC curves of a grafted polyaryletherketone. The upper curve was obtained after grafting, before
the heat treatment. Notice the weak change in slope at ca.
150°C (Tg) and the sharp crystalline melting peak. A second
run on the same sample (lower curve) show a more pronounced
glass transition, but no more melting. During the first run,
the polyacrylonitrile branches underwent cyclization, and
the resulting cross-linked polymer is entirely vitreous at
ordinary temperatures and no longer crystallizes once it is
molten.

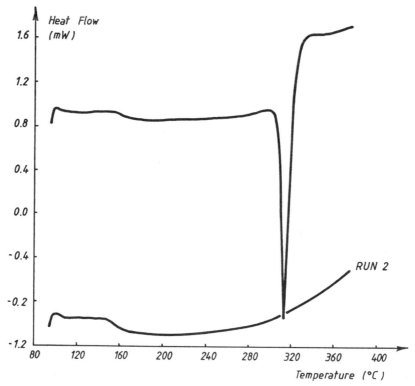

Figure 10. DSC curves of a polyaryletherketone grafted with
acrylonitrile before heat treatment (upper curve) and second
run on same sample (lower curve) [6].

Adhesives based on grafted high-density polyethylene.
This material is manufactured by Compagnie Francaise de
Raffinage, France [7]. A small amount of acrylic acid (1 to
2%) is grafted on high-density polyethylene by an easy and
particularly elegant method. The polymer powder is at first
"activated" for grafting by irradiation in air. This treatment produces stable peroxide groups in the polymer which can
be stored at room temperature. In a second step, the powder
is sprayed with acrylic acid and fed into a special extruder.
The grafting process occurs and the reacted material is
pelletized. It can be then blown into films which exhibit
outstanding adhesive properties with respect to metals and
other surfaces. Aluminium foils laminated with the grafted
film are used for packaging. The ability to seal at moderate
temperature and its chemical inertness make this material
attractive for food wrapping. By mixing the grafted powder
with divided metals and extruding this mixture, a "metal-
plastic alloy" is obtained, which can be extruded or molded
into any desired shape. A similar adhesive based on polypropylene is manufactured by the same Company. It extends
the usability of the product towards higher temperatures.

Membranes. Numerous permselective membranes have been successfully made by radiation grafting of hydrophillic monomers
into polymeric films. This method makes it possible to introduce various functional groups into polymers selected for
their specific properties: chemical inertness, toughness,
temperature resistance, or low cost. One advantage of this
method is that it modifies an existing film, whereas polymers
containing such hydrophilic functional groups often exhibit
very poor film-forming properties. Polymer films used in this
work include polyethylene, polypropylene, PVC and teflon.
The following functional groups were introduced: sulfonic and
carboxylic acids, pyridine, various amines, and pyrrolidone.
Polyfunctional membranes were made by grafting in two successive steps acrylic acid and 4-vinylpyridine. These membranes
exhibited unusual properties [8].

Mosaic membranes were produced, taking advantage of the
possibility of "activating" films with radiation into geo-
metrically well-defined patterns [9]. These membranes were
found to exhibit "negative osmosis" [10]. Figure 11 shows
this effect in the case of the dialysis of NaCl solutions
through a mosaic membrane containing carboxylic and pyridine
groups. Fluxes of water (curve 1) and of salt (curve 2) are
plotted as a function of pH. It clearly appears that, in the
central zone of pH, when both active sites are ionized, water
and salt diffuse in the same direction. Such membranes are

particularly suitable for separating ionic from non-ionic
species by dialysis or piezo-dialysis.

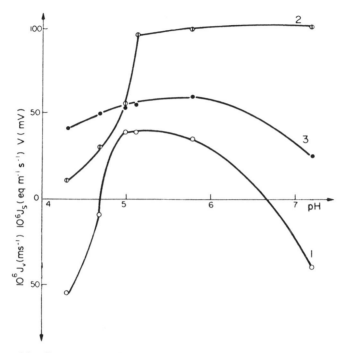

Figure 11. Demonstration of negative osmosis with a radiation
grafted mosaic membrane [10].
                    1. Water flux;
                    2. Flux of NaCl;
                    3. Membrane potential.

Hemocompatible polymers. Various water-soluble polymers
were grafted on non-soluble polymer surfaces using radiation
methods [11]. This technique can be applied to tubes of
various sizes producing a "hydrogel" coating with anti-
thrombogenic properties. Polyvinylpyrrolidone was also
grafted on model prostheses made of polyester fabrics, leading
to materials with improved thrombo-resistance [12]. For this
field of application, the very high flexibility of radiation
grafting methods is of advantage, since it can be applied to
create, in principle, any desired polymeric structure.

References

[1] Charlesby, A. Nucleonics, vol. 12, no. 6, 18 1954.

[2] Mattingley, J; Batiwalla, N. Proceedings of the Design Engineering '85 Conference, Birmingham, England, Oct. 1985.

[3] Sherman, R.D.; Middleman, L.M.; Jacobs, S.M. Polymer Engineering and Science, vol.23, no. 1 January 1983, 36-46.

[4] Carlson, J.R. Machine Design, December 10, 1981.

[5] Chapiro, A.; Legris, C. Europ. Polym. J., vol. 21, no. 1, 49 1985; Radiat. Phys. Chem. vol. 28, no. 2, 143 1986.

[6] Work performed in Raychem's Research Labs in France.

[7] Guimon, C. Radiat. Phys. Chem. vol. 14, 841-846 1979.

[8] Chapiro, A; Bex, G.; Jendrychowska-Bonamour, A.M.; O'Neill, T. Adv. Chem. Serv. vol. 91, 560 1969.

[9] Chaprio, A.; Jendrychowska-Bonamour, A.M.; Misrahi, S. Europ. Polym. J., vol. 12, 773 1976.

[10] Jendrychowska-Bonamour, A.M.; Millequant, J. Europ. Polym. J. vol. 16, 39 1980.

[11] Chapiro, A. Europ. Polym. J., vol. 19, 859 1983.

[12] Unpublished work performed in these laboratories in cooperation with Prof. Menachem Lewin, Jerusalem.

SILICON CONTAINING POLYMERS AND ORGANO-SILICON CHEMISTRIES FOR MICROLITHOGRAPHY

D. B. Bailey, R. C. Daly, T. B. Brust, and J. M. Pearson

Commercial and Information Systems Group,
Research Laboratories, Eastman Kodak Company,
Rochester, NY 14650, USA

Abstract

Multilayer resist technology will be critical in building the new generation of megabit chips. One of the advantages of the concept is the ability to separate functions required of the resist material and optimize them independently in the individual layers. Most multilayer approaches make use of the resistance of silicon-containing materials towards oxygen plasma and reactive ion enhanced etching. Silicon-rich images created in a resist polymer layer can then be transferred through any non-silicon organic underlayer to the substrate using dry etching methods. This article will review the technology with a focus on silicon-containing resist polymers for bilayer applications and on image-selective silylation chemistries for novel single layer schemes.

Introduction

Conventional single layer optical microresist techniques have been the workhorse of the semiconductor industry since the first silicon chips were produced. The shrinkage of device geometries into the sub-micron regime, however, is pushing these solvent-developed, single layer microlithographic processes close to the limits of their imaging capabilities, and the focus has switched to the multilayer approach. It now appears inevitable that some form of multilayer resist processing will be essential in building the new megabit generation of super chips.

Process Overview

A number of multilayer schemes have been described [1]. The structure in Figure 1 illustrates the basic concept. A thick (1-2 μm) polymer coating applied on the wafer serves to smooth over the topographical features encountered on many layers of the device. Dyes may be added to control image deterioration problems which can arise from light reflected from the wafer surface. The planarizing layer is overcoated with a thin (< 0.3 μm) etch barrier layer, generally a silicon-containing inorganic or organic material, and finally a thin (< 0.5 μm) layer of resist. The use of a thin, uniform, smooth imaging layer ensures maximum resolution and linewidth control. After the image is formed in the photoresist layer using conventional microlithographic techniques, it must be transferred through the underlying layers to the wafer surface. This is achieved using new fabrication processes such as plasma and reactive ion enhanced (RIE) etching. These dry etching techniques have the potential for highly selective, anisotropic etching to produce vertical profiles, enabling the accurate transfer of high quality images.

MULTILAYER RESIST SCHEME

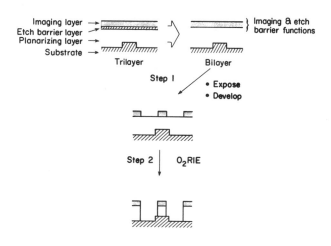

Figure 1.   Schematic of multilayer resist scheme.

Trilayer resist schemes using RIE etch for image transfer are capable of very high resolution. However, this performance is achieved at a high cost in terms of additional process complexity, and recent efforts have been aimed at achieving the same performance with simpler and more manufacturable processes [2].

Combining the etch barrier and imaging functions into one material reduces the system to a bilayer, simplifying the overall process. The relief image generated in the resist

(Figure 1, Step 1) now becomes the etch mask for removal of the underlying polymer using O_2 RIE (Step 2). One way to accomplish this is to incorporate an inorganic element such as silicon into the resist material. In the strongly oxidizing O_2 plasma the organo-silicon is converted into non-volatile SiO_x which forms a thin (100-200 Å) surface skin, significantly reducing the erosion rate of the underlying material [5].

As a general rule, it is necessary to introduce > 10 wt% silicon into the polymer in order to achieve a > 10:1 etch differential relative to the typical organic polymers used for planarizing, e.g., PMMA, novolacs and poly(imides). Several groups [5-8] have pursued the approach of combining imaging chemistry with a polymer containing around 10 wt% silicon to build a bilayer resist.

The structure attained prior to the plasma etching step in a conventional bilayer system consists of a thin silicon-rich image on a relatively thick underlayer. This structure can also be achieved in alternative bilayer and single layer approaches by introducing the silicon imagewise in some post exposure chemical step. The key to this approach is to obtain high selectivity in the additive chemistry, and several schemes for achieving this have recently been described [9-12].

Bilayer Systems

The first organometallic polymers shown to function in a bilayer scheme were the poly(siloxanes). By introducing unsaturation into the polymer, e.g., copolymers of dimethyl-siloxane or methylphenylsiloxane with methylvinylsiloxane, the resins could be imaged in the negative mode with UV and e-beam radiation [13]. Since that publication, a variety of polymers having silicon in the backbone or in pendant groups have been examined directly as resists or as matrices for incorporating other radiation sensitive chemistries.

We have investigated a number of silicon-containing co-polymers for bilayer applications. The copolymers were designed to have high T_g for image stability, to be aqueous base soluble and to function with either positive or negative imaging chemistries. One material which met these requirements was a copolymer of 3-methacryloxypropyltris(trimethyl-siloxy)silane (SMA) with 4-hydroxystyrene (HS), P(SMA-HS).

The silicon content of this methacrylate monomer (26 wt%) is attractive since it allows high silicon loadings to be

achieved in copolymers while keeping the content of the hydrophobic component as low as possible. Copolymers were synthe-
sised by conventional free radical polymerization in toluene
using AIBN as initiator. It was necessary to block the
4-hydroxy group on the styrene monomer, since it interferes
with the polymerization. Several protective groups work
effectively, e.g., acetate, t-butyl carbonate and trimethyl-
silylether. The latter was selected since it could be
readily removed from the copolymer under mild reaction conditions (14% w/v solution of the copolymer in THF with 6% 0.1 N
HCl at 25°C for 15 min) without disturbing the methacrylate
component.

$$CH_2=\overset{\overset{\displaystyle CH_3}{|}}{C}-\overset{\overset{\displaystyle O}{||}}{C}-O\diagdown\diagup\diagdown\;Si\left[OSi(CH_3)_3\right]_3$$

SMA

Copolymerization proceeds as a typical styrene-methacryl-
ate free radical reaction. The reactivity ratios for the
monomer pair, 4-trimethylsiloxystyrene (SHS) [1] and (SMA)
[2], estimated using the Kelen-Tudos method, were $r_1 = 0.62$
and $r_2 = 0.41$ (cf. styrene/methylmethacrylate 0.52 and 0.47,
respectively), indicative of a nearly random copolymer structure. Molecular weight was controlled with a chain transfer
agent, t-butylmercaptan ($C_s = 3.2$). For polymerizations
carried out to high conversions, the transfer agent was added
incrementally throughout the reaction in order to maintain
the molecular weight distribution around 2. A molecular
weight, M_w, range from 10^4 to 10^6 was investigated. Figure 2
shows T_g as a function of composition for both P(SMA-SHS)
and P(SMA-HS).

The O_2 RIE etch characteristics of copolymers containing
from 3 to 15 wt% silicon were very similar to those published
for a number of other silicon-containing polymers [17],
further substantiating the claim that etch resistance is a
function of silicon content and is independent of how the
silicon is present in the matrix. For this study we focused
our efforts on copolymers having > 10 wt% silicon content.
Under the O_2 RIE conditions used in these experiments, the
copolymer with 11 wt% silicon etched at a rate of 100 Å/min,
compared to a value of 2,800 Å/min for novolac and P(HS) and
5,800 Å/min for P(MMA).

One of the primary objectives of the project was to make
the resist aqueous base developable. Introduction of hydrophobic trimethylsilyl groups into a polymer for etch

resistance is in direct conflict with this goal. It was
rationalized that the high solubility of the 4-hydroxystyrene
component in aqueous base media would offset the increasing
hydrophobicity introduced by the methacrylate, and, indeed,
copolymers with a silicon content as high as 15 wt% remain
soluble in aqueous base solvents.

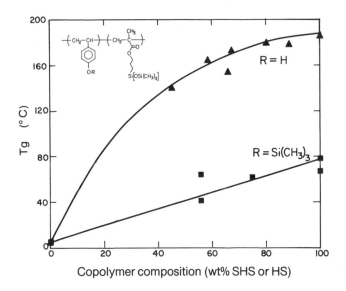

Figure 2. Glass transition temperature, T_g, of copolymers
P(SMA-HS) (▲) and P(SMA-SMS) (■) as a function of copolymer
composition.

The dissolution characteristics of the copolymers were
studied using a laser interferometric technique for monitoring the thickness loss in a thin polymer film as it dissolves
in a solvent [18]. Figure 3 shows the solubility of a series
of copolymers of different composition in aqueous tetramethyl-
ammonium hydroxide (TMAH) solutions. The interferometric
traces were well defined, indicating that the dissolution of
the copolymers is not complicated by anomalous solvent swelling effects. The data shows that for a constant molecular
weight there is a significant decrease in solubility with
increasing content of methacrylate. It was also found that
for the copolymer compositions of interest (around 55 wt%
HS), there is a wide design latitude for controlling solubility by modifying M_w. An activation energy for dissolution
over a rather limited temperature range (18 to 28°C) was
determined to be around 30 kcal/mol, considerably higher than
the value of around 10 kcal/mol measured for a number of

novolac and P(HS) samples under similar experimental conditions. This high activation energy is probably a reflection
of the hydrophobicity introduced into the film by the large
organosilicon sites which inhibit penetration of the aqueous
solvent, retard formation of the phenoxide ion and make solubility and removal of polymer chains from the surface of the
film more difficult.

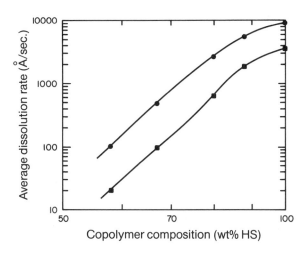

Figure 3. Dissolution rate of P(SMA-HS) copolymers as a
function of composition in aqueous solutions of TMAH at
concentrations of 3% (●) and 2.3% (■) at 21°C.

In addition to using molecular weight to tailor the dissolution response of the copolymers, opportunities exist for
designing improved developers and further optimizing the
development process. In the tetraalkylammonium hydroxide
class of developers it is known that the nature of the alkyl
group determines the partitioning of the active alkaline
species between the aqueous and organic phases [19], and we
are currently investigating the effect of alkyl chain length
on the solubilizing power of these developers.

The copolymer can be utilized for positive or negative
tone imaging by incorporating diazonaphthoquinone (usually
as a sulfonate derivative, NDS) or bis-azide photochemistry
sensitive in the UV (436 to 365 nm) or deep UV (250 nm).
Positive tone imaging was found to be very sensitive to the
water content of the matrix, and although high quality, high
resolution images could be obtained in humidity equilibrated

films, the process was difficult to control. This problem
was not encountered in the negative mode, and the lithographic characteristics of the copolymers were investigated
using 2,6-(4-azidobenzal)-4-methylcyclohexanone (DABMC)
sensitizer at an exposure wavelength of 365 nm. The contrast
curves for a series of copolymers of fixed composition
(56 wt% HS), covering a molecular weight range, M_w 10^4 to

5.10^4, and containing 15 wt% sensitizer are shown in Figure
4. As expected for a cross-linking chemistry, the contrast
increases with increasing molecular weight. For the copolymer with $M_w = 5.10^4$ the sensitivity was around 5 mJ/cm^2 with
a contrast of 4.

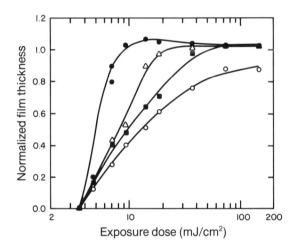

Figure 4.   Contrast curves for a P(SMA-HS/56)-DABMAC resist
as a function of molecular weight, $M_w = 10^4$ (O), 2.10^4 (■),
3.10^4 (Δ) and 5.10^4 (●).

Bilayer imaging studies were carried out using this
particular resist composition coated at 0.5 μm (from chloro-
benzene:cyclohexanone 3:1 v/v) on a 1.2 μm planarizing layer
of hard baked novolac-NDS based Kodak micropositive resist
820 (KMPR 820). Images were developed using a 4.5% aqueous
TMAH solution, and resolution down to 0.5 μm was achieved.
Figure 5 shows an example of a nominal 0.5 μm line/space
pattern created in the photoresist layer (a) and transferred
through the planarizing layer (b) using O$_2$ RIE at a pressure
of 7.6 mtorr at 35 watt power. Exposure was made on an
I-line (365 nm) stepper (courtesy of GCA Corp.) with a NA
of 0.35.

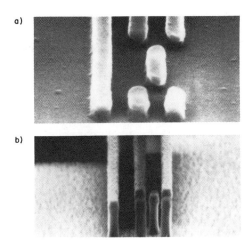

Figure 5. SEM micrograph of 0.5 μm lines and spaces printed
in the P(SMA-HS/56)-DABMAC resist (a), and O_2 RIE etched
through a hard-baked KMPR 820 planarizing layer (b).

Single Layer Systems

The ultimate in process simplification would be to attain
the performance of a trilayer with a single layer coating.
In the schemes described in the literature, etch resistance
has been introduced by a post-exposure chemical treatment
step which introduces its own, as yet unknown, set of process
complications. Nevertheless, the current high level of
activity in this area is a clear recognition of the potential
of single layer technology.

Two basic approaches have been used to design single layer,
organometallic based, plasma developable resist systems. One
approach involves starting with the organometallic species in
the resist film, exposing the film, and using a subsequent
heating step to remove the inorganic component from either
the image or non-image areas. Examples include radiation
induced polymerization of an inorganic containing monomer
within a resist matrix [20], and photochemical cleavage of
silicon-containing side groups from an acrylic copolymer
based resist [21].

The second approach depends on the exposure to modify the
resist, either chemically or physically, so that the inorganic
element can be introduced into selected regions of the film
in a subsequent process step. Imagewise incorporation of
silicon has been reported for both conventional cyclized
poly(isoprene) negative resists [12] and novolac-NDS positive
resists [10,22]. Selective chemical modification of a resist
polymer provides another mechanism for imagewise incorporation
of an inorganic moiety [11]. An example of such an approach
is the conversion of poly(4-t-butoxycarbonyloxystyrene)
P(BOCST) to P(HS) through a photogenerated acid-catalyzed
cleavage of the pendant t-butylcarbonyl groups. The phenolic
groups react preferentially with a silylating reagent such as
hexamethyldisilazane (HMDS), leading to a negative relief
image following O_2 RIE etching. Photogeneration of acid has
also been used to initiate the cationic polymerization of a
monomer such as γ-glycidoxypropyltrimethoxysilane, introduced
following the exposure step, to build a thin relief etch mask
on the surface of a resist film [23].

The silylation approaches are claimed to be capable of
producing and transferring high quality, sub-micron images in
the research environment. However, very little has been
published about the parameters which control the silylation
reaction to define the etch mask and their impact on the
critical plasma image transfer process.

Although both liquid [24] and vapor phase [10] silylation
processing has been disclosed, the vapor method appears to be
favored. We have initiated an investigation of the silylation process, and have studied a number of silicon reagents,
including the silazanes, disilizanes and chlorosilanes. At
this time the factors which control both the reactivity and
the selectivity are not well understood, particularly for
phenolic polymer based photoresists where the photochemistry
only modifies the physical state of exposed areas of the
film.

The chemistry and kinetics of the reaction in 4-hydroxy-
styrene and novolac-NDS positive resists have recently been
studied using a variety of spectroscopic techniques [25]. It
was proposed that the NDS sensitizer molecules act to crosslink the phenolic polymer chains through hydrogen bonding,
and that photodecomposition to produce nitrogen and indene-
carboxylic acid destroys this physical network, increasing
the relaxation of the resist. Selectivity then results from
the more facile diffusion and reaction of the reagent in
exposed areas. Silylation was shown to occur on the phenolic
OH groups, and it was claimed that neither the carboxylic

acid nor the nitrogen photoproducts played any role in the process.

We have studied the gas phase silylation of both novolac-NDS and P(BOCST)-photoacid type resist films using a piezo-electric quartz microbalance technique which allows real-time monitoring of the uptake and incorporation of silicon [26]. A typical mass change curve is shown in Figure 6 for an exposed film of P(BOCST) containing bis(t-butylphenyl)iodonium trifluoromethylsulfonate as the acid precursor. The initial weight loss is due to removal of products of deblocking, iso-butylene and CO_2. Analysis of the remainder of the curve provides information on the silylation process. The characteristic sigmoidal curve indicates that HMDS uptake occurs in stages. The initial rapid mass increase is associated with filling the void space created by the escape of the gaseous products of the deprotection reaction. Annealing of the film removes this component. The second, slower phase results from HMDS saturation of the void volume, and may also reflect a change from reaction- to diffusion-controlled kinetics as the HMDS diffusion to unreacted OH groups becomes the rate limiting step. The final phase shows an accelerated uptake which reaches an equilibrium state whose level is determined by the

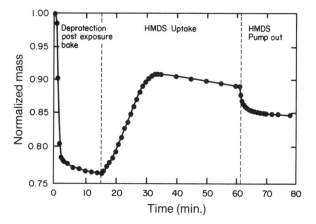

Figure 6. Mass changes measured in an exposed P(BOCST) resist film subjected to a post exposure bake (a), an HMDS vapor treatment (b) and a final reaction chamber pump down (c). The silylation reaction was carried out at an HMDS pressure of 100 Torr at 100°C (from ref. 25 with permission).

exposure dose. When all available OH groups are reacted and the film becomes saturated with HMDS, the uptake stops. Terminating the process by pumping out the reaction chamber results in desorption of significant amounts of unreacted HMDS from the film.

The gas phase functionalization rates are dependant on reaction temperature and reagent pressure. The T_g of the matrix is also an important parameter. At a temperature between 75 and 90°C a dramatic change is noted in the reaction curve profile. Silylation reactions at temperatures below 75°C did not exhibit the usual sigmoidal uptake curve shape, but showed a much earlier and slower approach to equilibrium. It is postulated that the transition results from a change in the reaction mechanism when the temperature falls below the T_g of the matrix. In this particular instance, the T_g values for the polymers P(BOCST) and P(HS) are 125°C and 185°C, respectively, and that for P(SHS), the fully silylated form of P(HS), is 72°C. A combination of the plasticizing influence of the reagent, HMDS, and the decreasing T_g of the silicon-substituted polymer in the reaction zone will clearly play a significant role in determining the rate and temperature sensitivity of the process.

The silylation process in novolac-NDS films proceeds by a similar mechanism. Selective diffusion of HMDS occurs in exposed areas of the film and the diffusion is further enhanced by the reduction in matrix T_g resulting from plasticization and silylation. Figure 7 compares the HMDS uptake of a novolac resin, the resin doped with NDS and an exposed sample of this same film. Clearly, the presence of NDS in the matrix significantly lowers the penetrability of the polymer. On the other hand, either the void volume resulting from the evolution of nitrogen, or the carboxylic acid photoproduct, or both, enhance the uptake of reagent relative to the polymer film alone. The process is temperature dependent, and a discontinuity is found in the silylation reaction around 85°C (cf. the silylation of P(HS)) which further supports a matrix plasticization/T_g effect.

These results are in general agreement with the mechanism proposed from the spectroscopic analysis. They do indicate, however, that at least in the novolac-NDS systems, the nature of the silylating reagent is very important, and that voids formed in the matrix play a role in the silicon uptake. Also, the presence of plasticizing agents and the physical changes in the polymer as a result of increasing degrees of silylation, both of which serve to reduce the T_g in the reaction zone, can dramatically influence the gas phase functionalization process.

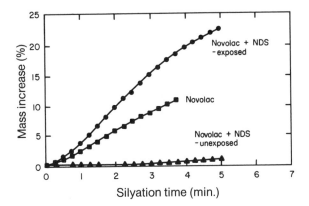

Figure 7. HMDS mass uptake curves for a novolac resin (■),
the resin doped with NDS sensitizer, unexposed (▲) and
exposed (●). The silylation reaction was carried out at an
HMDS pressure of 150 Torr at 120°C.

Hybrid Systems

An intermediate approach which combines features of both
single and bilayer methods has been disclosed [9,12]. It is
a bilayer structure consisting of a thin photoimaging layer
on a planarizing layer which uses a solvent development-gas
phase functionalization-plasma development sequence to produce either positive or negative tone final images.

By introducing silicon following the image formation step,
it is possible to take advantage of existing, conventional
positive photoresist materials. The structure used in these
experiments consisted of a hard baked KMPR 820 (275°C, 90 s)
planarizing layer with a thin (0.3 to 0.5 μm) Kodak micropositive resist 809 (KMPR 809) imaging layer. The relief
image was generated by conventional microlithographic
techniques.

Vapor phase silylation resulted in preferential silicon
incorporation into the resist image. Selectivity is achieved
because of the high permeability and reactivity of the KMPR
809 (low T_g resin) relative to the hard baked KMPR 820. For
some silylating agents it was found that a further enhancement in silicon uptake in the resist image could be achieved

by a short UV flood exposure (unfiltered Hg-Xe lamp) prior to
the silylation step.

Although HMDS is a convenient and effective silylating
agent, image distortion and flow problems were encountered
with the silylated KMPR 809 images. This is shown in Figure
8a. It is interesting to note that the flowed edges of the
lines, which are only 0.1-0.2 μm thick, continue to function
as effective etch masks, Figure 8b, indicating a high silicon
content. This image flow can be readily eliminated by using
multifunctional silylating agents, such as dichlorodimethyl-
silane (DCDMS), which both silylate and cross-link the resin.
The cross-linking reaction provides an additional benefit in
making the images stable up to 160°C, an improvement over the
thermal stability of the original positive resist.

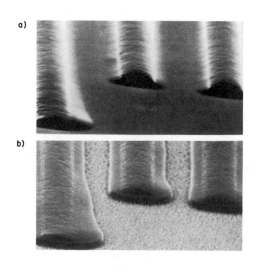

Figure 8. SEM micrograph of resist image flow observed in
1.5 μm lines of KMPR 809 after a 5 s UV flood exposure and
silylation with HMDS at 150 Torr at 90°C for 8 min (a), and
following an O_2 RIE image transfer (b).

The process using DCDMS with a silylation time of a few
minutes at a reagent pressure of around 150 mtorr at 90 to
100°C, can achieve an O_2 RIE selectivity of 10-15:1. An
example of a 0.8 μm line/space pattern over 0.5 μm silicon
oxide steps is shown in the SEM in Figure 9.

Figure 9.   SEM micrograph of 0.8 μm lines and spaces over
0.5 μm silicon oxide steps fabricated using the KMPR 809-
DCDMS procedure.

Summary

There is no doubt that bilayer and single layer microlithographic systems similar to those described here can
achieve sub-micron imagery. It remains to be seen whether
the capabilities that have been demonstrated in the laboratory can be carried over into the more demanding manufacturing environment. For the conventional bilayer approach,
there are a multitude of resist materials which combine
imaging and etch resistance. Many of these are somewhat
exotic materials which may not be commercially viable. For
the more realistic polymers, the trade-offs between sensitivity and etching must be resolved, and the critical process
parameters, particularly those associated with plasma etching, must be better defined. The most realistic approach may
be to design materials which can be more readily assimilated
into existing production lines. In this regard, the silicon-
containing polymers which can be developed in aqueous base
media do not always have desirable etching resistance. The
'hybrid' scheme which uses existing positive resist materials
and introduces the silicon after developing the image, may be
a more acceptable approach.

The process simplicity inherent in the single coating

concept is extremely attractive, and the quality of the images which have been produced is too good to ignore. This approach can also take advantage of existing photoresist materials and process technology. The major issue here is the current lack of understanding of the silylation chemistry and process. There is considerable activity in the area and significant progress has already been made. If its potential can be realized, the single layer approach may well win over bilayer in the technology race.

References

1. B. J. Lin in Introduction to Microlithography, L. F. Thompson, C. G. Willson and M. J. Bowden, Eds., ACS Symposium Series 219, American Chemical Society, Washington, DC, 1983, pp 287-350.
2. L. P. McDonnell-Bushnell, L. V. Gregor and C. F. Lyons, Proceedings of the Kodak Microelectronics Seminar, San Diego, CA, Nov. 1985, pp 59-65.
3. J. A. Mucha and D. W. Hess in Introduction to Microlithography, L. F. Thompson, C. G. Willson and M. J. Bowden, Eds., ACS Symposium Series, 219, American Chemical Society, Washington, DC, 1983, pp 215-285.
4. G. N. Taylor and T. M. Wolf, Polym. Eng. Sci., 1980, 20, 1087.
5. M. Suzuki, K. Saigo, H. Gokan and Y. Ohnishi, J. Electrochem. Soc., 1983, 130, 1962.
6. C. W. Wilkens, Jr., E. Reichmanis, T. M. Wolf and B. C. Smith, J. Vac. Sci. Technol., 1985, 3, 306.
7. S. A. MacDonald, H. Ito and C. G. Willson, Microelectron. Eng., 1983, 1, 269.
8. E. Reichmanis, B. C. Smith, G. Smolinsky and C. W. Wilkens, Jr., J. Electrochem. Soc., 1987, 134, 653.
9. W. C. McColgin, J. Jech,Jr., R. C. Daly and T. B. Brust, 1987 Symposium on VLSI Technology, Karuizawa, Japan, May 1987.
10. F. Coopmans and B. Roland, in Advances in Resist Technology and Processing, C. G. Willson, Ed., Proc. SPIE, 1986, 631, 34.
11. H. Ito, S. A. MacDonald, R. D. Miller and C. G. Willson, US Patent 4,552,833 to IBM, Nov. 1985.
12. L. E. Stillwagon, P. J. Silverman and G. N. Taylor, Proc. PE Regional Technical Conf. on Photopolymers, Ellenville, NY, 1982, 87.
13. M. Hatzakis, J. Paraszak and J. M. Shaw, Proc. Int. Conf. Microlithography and Microcircuit Eng., 1981, 81, 386.

14. D. C. Hofer, R. D. Miller and C. G. Willson, in Advances in Resist Technology, C. G. Willson, Ed., Proc. SPIE, 1984, 469, 16.

15. S. R. Turner, Eastman Kodak Company, unpublished results.

16. A. Tanaka, M. Morita, A. Imamura, T. Tamamura and O. Koyure, Polym. Prepr., Am. Chem. Soc. Div. Polym. Chem., 1984, 25, 309.

17. E. Reichmanis and G. Smolinsky, in Advances in Resist Technology, C. G. Willson, Ed., Proc. SPIE, 1984, 469, 38.

18. F. Rodriguez, P. D. Krasicky and R. J. Groele, Solid State Technol., 1985, 28, 125.

19. R. A. Arcus, Proceedings of the Kodak Microelectronics Seminar, San Diego, CA, Nov. 1985, pp 25-31. 20. G. N. Taylor, M. Y. Hellman, M. D. Feather and W. E. Willenbrock, Polym. Eng. Sci., 1983, 23, 1029.

21. W. H. Meyer, B. J. Curtis and H. R. Brunner, Micro-electron. Eng., 1983, 1, 29.

22. B. Roland and A. Vrancken, European Patent 0184567 to UCB, Oct. 1985.

23. A. Hult, S. A. MacDonald and C. G. Willson, Macro-molecules, 1985, 18, 1804; and US Patent 4,551,418 to IBM, Nov. 1985.

24. K. N. Chiong, B-J. L. Yang and J-M. Yang, US Patent 4,613,398 to IBM, Sept. 1986.

25. R. J. Visser, J. P. W. Schellekens, M. E. Reuhman-Huisken and J. L. van Ijzendoorn, in Advances in Resist Technology and Processing, M. J. Bowden, Ed., Proc. SPIE, 771, 1987, in press.

26. T. B. Brust and S. R. Turner, in Advances in Resist Technology and Processing, M. J. Bowden, Ed., Proc. SPIE, 771, 1987, in press.

POLYMERS IN SOLAR TECHNOLOGIES

Hermann H. Neidlinger and Paul Schissel

Solar Energy Research Institute
1617 Cole Blvd.
Golden, CO 80401

This paper describes some uses of polymeric materials in solar energy collection, identifies limiting problem areas, points out several potential areas of research, and closes with projected areas of future solar application. A key to the successful implementation of the various methods proposed for converting solar energy is the progress made by polymer scientists in developing polymers with the desired properties as well as with resistance to degradation. Polymeric materials offer potentially lower costs, lighter weight, easier processing, and greater design flexibility than materials in current use. However, cost effective design, fabrication, and operation of systems utilizing polymers must be based on a sound understanding of the properties and predicted performance of these materials in service, particularly the optical, thermomechanical, and diffusion barrier behavior and the effects of environmental aging.

Introduction

Solar energy is an appealing alternative energy source and it is attracting attention because it can provide permanent, clean, low cost energy. However, solar energy is difficult to utilize due to its diffuse and intermittent nature. In order for solar radiation to be used on a large scale, it must be collected, stored and/or transmitted to locations where it is needed. In addition, the harvesting and processing of solar energy requires large areas. All the energy needs of the United States in the year 2000 (93

Quads) could be met, if we could only harness and utilize 10% of the total solar energy that falls on about 50,000 square miles [1]. This land area, less than 1.5% of the area of the United States, is equivalent to approximately one-third the area of California or the area of Louisiana or New York. The areas required for capturing solar energy may be in the form of windows, roof top collectors, biological reaction ponds, solar ponds, or reflecting surfaces for focusing mirrors.

In all these cases, polymeric materials are finding wide-spread use not only in such conventional applications as structural materials, adhesives, coatings, moisture bar-riers, electrical and thermal insulation, but also as optical and functional components in solar energy systems [2-4]. Some typical applications of polymers in the major solar technologies are summarized in Table 1.

The controlling factor in selecting polymers for any of these applications is economics, which depends on the opti-mization of the three major parameters: cost, performance, and durability. Some polymers are low in cost and have good physical properties but are not (or have not been demonstra-ted to be) stable in the terrestrial environment. Other polymers have demonstrated resistance to the environment (e.g., silicone, fluorocarbons) but are costly, difficult to process, or do not have appropriate physical and mechanical properties.

Present Applications

Table 1, shows some very general applications of polymers in solar energy systems. Of particular importance in this technology is the use of polymers for optical elements. Various examples of the current development of polymeric optical elements are considered below. Problems such as photodegradation and dirt accumulation, which are common to most optical elements, are considered in a later section.

Solar reflectors. Reflectors are presently receiving considerable attention because they represent the largest economical potential for the application of solar energy. Silver/glass mirrors, while they are able to meet most of the durability and performance requirements for solar energy applications, are heavy and, therefore, expensive in their application. The massive structures which they require cause the heliostat field of a central-receiver solar thermal electric plant to consume nearly half of the capital

Table 1. Typical Applications of Polymers in Solar Energy Systems

Polymer Applications \ Solar Energy Systems	Solar Thermal Conversion	Photo-voltaics	Solar Heating & Cooling of Buildings	Wind	Ocean Thermal	Biological/Chemical
Glazings	X	X	X			X
Encapsulation	X	X	X			
Fresnel Lenses	X	X	X			
Structural Members	X	X	X	X	X	X
Seals/Adhesives	X	X	X	X	X	X
Paints/Coatings	X	X	X	X	X	X
Heat Transfer Loops	X		X	X	X	
Energy Storage	X		X	X		X
Moisture Barriers	X		X		X	X
Thermal & Electrical Insulation	X	X	X	X	X	X
Membranes						X

investment and an even larger fraction of the cost of
systems for process heat production [5].

   Metallized thin polymeric films are one means of making
light-weight mirrors that allow less expensive reflector
designs. These thin, flexible films can be attached with
adhesives to substrates with a single or compound curva-
ture. Studies of aluminized or silvered polymers have
included acrylics, fluorinated polymers, polycarbonate, silicones, and polyesters [6]. Figure 1 shows a parabolic
trough concentrator using a silvered acrylic reflecting film
on a supporting steel substrate.

Figure 1. Parabolic trough collector

   Much current research and development attention is being
directed at the use of polymers as stretched membranes
(Figure 2) for reflectors [7]. This application places
demands on the polymer sheet which may not be compatible
with the properties of any one polymer material; therefore,
it is expected that stretched membrane heliostats will make

use of laminated polymers and composites. For example, one
layer of glass-fiber reinforced structural polyester composite structure might be used for mechanical strength and low
creep characteristics. Other layers, such as polyacryloni-
triles, might be used to isolate the silver from the
weathering effects of atmospheric permeants, while additional layers on the reflecting side of the mirror made of
specially treated acrylics and other materials, have the
ability to screen ultraviolet light and to reject soiling
and abrasion.

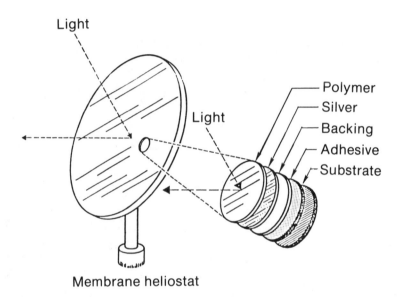

Figure 2.   Schematic of a membrane heliostat.

Solar collectors. Glazings, coverplates, and windows
make up very large areas of transparent material application
in buildings. In the most common usage, glass has been the
material of choice mainly because of its greater resistance
to weathering and scratching. With the increased energy
consciousness of builders, multiple glazings are becoming
the rule, and these offer an opportunity to use intermediate
glazings of polymer materials. However, polymer glazings
must withstand environmental corrosion and, in the case of
flat-plate collectors, also high temperatures. Results on
collector glazings [8] have shown that none of the polymer

materials are completely satisfactory. To date, the more successful glazings have been either polycarbonates, which have great mechanical strength but which have been sensitive to ultraviolet damage, or fluorocarbon polymers, which weather well but are very costly. Other materials that have been used in glazings include polyacrylates and polyethersulfones.

Novel approaches to collector fabrication use honeycomb concepts [9], integral extrusion [10], or laminated thin films [11]. Although suitable polymer materials, which can withstand the operational environment of high temperature solar absorbers, are presently not available, low temperature absorber plates functioning at operating temperatures below $40^{\circ}C$ have been produced of polypropylene with carbon black filler.

Among other polymers which have been used successfully in lower temperature applications as receiver tubes and fluid transfer and containment systems are EPDM, polysulfones, and polyvinylchloride. These materials, while working effectively and having reasonably low cost, have, in some cases, still shown degradation in solar applications at rates higher than expected.

Imaginative applications of polymers to fenestration and solar collectors include the use of solar spectrum transparent conductive coatings on polymer films which reflect selectively long wavelength radiation (heat), and thus greatly increase the efficiency of windows [12]. Polymeric films with reversible thermochromic properties act as automatic window shades which could help control stagnation temperatures [13]. Polymers with photochromic or electrochromic properties have also the potential to perform energy management functions, but research in this area is still in its infancy.

Photovoltaic (PV) encapsulation systems. Polymers can serve several functions in PV encapsulation systems [14]. The core of the encapsulation package is the pottant, which embeds the solar cells and related electrical conductors. Key requirements for a polymer pottant are high transparency in the range of solar cell response, cushioning of the brittle solar cells from thermal and mechanical stresses, electrical insulation to isolate module voltage, resistance to environmental degradation, and cost-effective material and module fabrication processes.

Several polymers can be considered as pottant materials, such as ethylene-vinyl acetate copolymers, ethylene-propylene rubber, and plasticized polyvinylchloride. Hard, durable front cover films to protect the relatively soft pottant include UV-stabilized polyacrylates, polyester, fluorocarbon polymers, and silicon hard-coats.

Luminescent solar concentrators (LSC). LSC devices can be non-tracking and they can concentrate the light input from either direct or diffuse insolation using the principle of light pipe trapping, transmission, and coupling into a photovoltaic cell (PV) [15-17]. Thus, considerable reduction in the area requirements of more expensive PV cells can be achieved. Infrared radiation is not focused on the PV cells with this type of concentrator, and wavelength matching between the solar radiation and spectral response of the PV cell can increase realizable efficiency.

One planar geometry, shown in Figure 3, uses a transparent polymeric host (e.g., polymethylmethacrylate) into which dye molecules (e.g., Rhodamine-6G) are dispersed randomly. Photons with wavelengths in the absorption band of the dye enter the top of the plate, are absorbed, and are re-emitted isotropically at longer wavelengths. A large fraction (70-80%, depending on the refractive index) of the re-emitted photos is trapped by total internal reflection and guided to the edge of the plate where they are collected by PV cells.

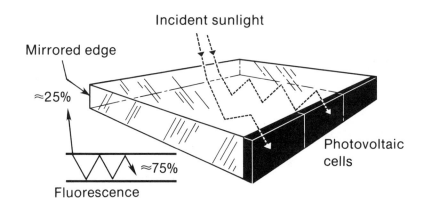

luminescent concentrator

Figure 3. Schematic of a luminescent solar concentrator.

System lifetime is an important unknown factor principally influenced by dye lifetime. Dye-host interactions are also of prime concern. Structure and mobility of the polymer host are factors found to affect collection efficiency [18].

Fresnel lenses. Cast and molded acrylic Fresnel lens concentrators have been studied for both thermal and photovoltaic systems. As in other solar applications of optical polymers, the economic viability of their use depends on a large number of system-related factors, including the performance, cost, and durability of the lenses. Performance requirements include minimum absorption, scattering, and surface reflection.

Other applications. Polymers are also used as structural materials, adhesives, sealants, coatings, moisture barriers, solar pond liners, semi-permeable membranes, electrical and thermal insulation, heat transfer and storage fluids, etc. [2].

Problem and Research Areas

Compared to metals or ceramics, polymers are more permeable, more viscoelastic, and more sensitive to radiation and aging, and their properties are highly process dependent. Used properly in solar energy conversion systems, they can have lower cost, be easier to process, and offer a wider range of properties (and thus design flexibility) than inorganic materials. However, cost effective design, fabrication, and operation of systems utilizing polymers must be based on a sound understanding of the properties and predicted performance of these materials in service, particularly the optical, thermomechanical, and diffusion barrier behavior and the effects of environmental aging.

Optical Properties

Of particular importance in the application of polymers as optical elements in solar systems is the maintenance of their optical performance. The overall hemispherical integrated solar transmittance, is generally easy to achieve with transmittance values as high as 92% regularly available to polymer films. Specular transmittance is more difficult to achieve and maintain. The limitations on specularity in polymers derive from several possible sources [19]. Some are surface specific contributions to light scattering from variations in surface composition and roughness generated during the manufacturing and use of the polymer films. Another possible factor is non-uniformity in the optical refractive index of the polymer film. The refractive index of a polymer depends on the molecular weight of the structural unit and its molar refraction. Fluctuations in bulk composition of a polymer film can cause light scattering by internal optical inhomogeneities which will lead to loss of specularity.

Polymeric protective glazings on silver films can provide high quality, light-weight mirrors. We have recently demonstrated that silvered acrylic polymer films can have an initial specular reflectance of more than 90% into a full-cone acceptance angle of about 1 mrad [20]. Figure 4 shows a characteristic example of the shape of the reflected beam from a silvered acrylic film mounted on a glass substrate. (The short horizontal line in Figure 4 indicates the approximate beam spread (.8 mrad) corresponding to about 90% of the beam, assuming a symmetrical, two-dimensional Gaussian distribution). This performance exceeds our long-range goal of a reflectance greater than 90% with a beam spread of a few mrad. However, the maintenance of such a high specular reflectance in outdoor service environments for many years is a critical issue. Problems which are also common to other solar-related optical elements include photodegradation, dirt retention, cleaning and surface abrasion. A common feature of some of these problems is that the deleterious effects occur at an interface. Solar radiation, atmospheric components, mechanical stress, etc., can have a profound effect on performance by changing surface characteristics. The lifetimes of UV stabilizers can be limited by exudation, permeability can cause harmful reactions at interfaces, and mechanical properties can be influenced by surface crazing. In other applications, the mechanical behavior of the bulk polymer is critical and virtually all applications require that the polymer system withstand multiple environmental stresses simultaneously.

Photodegradation of polymers. Virtually all polymers deteriorate under outdoor weathering exposure and solar radiation, but at greatly varying rates. Polymers in solar equipment must maintain optical, mechanical, and chemical integrity despite prolonged and maximized exposure to solar ultraviolet radiation. Transparency is essential for many of the potentially most cost-effective applications, and conventional approaches to ultraviolet protection such as opaque coatings and fillers are unacceptable.

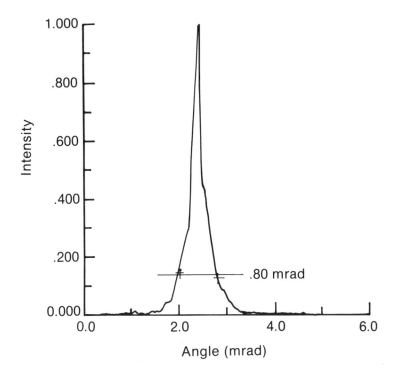

Figure 4. A radial distribution function of the reflected beam from a silvered acrylic film mounted on a glass substrate. (Angle of incidence = 20°, beam diameter = 10 mm, wavelength = 600 nm, unpolarized light.)

Absorption of ultraviolet photon energy by the polymer can cause the breaking of chemical bonds, resulting in embrittlement and increased permeability, or crosslinking which can produce shrinking and cracks. In addition, oxidation often leads to discoloration and reduced transparency, and to the formation of polar groups which could affect electrical properties. These changes obviously limit the performance of solar devices which, for economic reasons, must last satisfactorily for up to 20 years in rather severe environments.

Much work has already been done to unravel the complexities of photooxidative processes and to develop some highly effective light stabilizers which can delay the onset of embrittlement [21]. A long lifetime can best be attained by immobilizing the stabilizer as part of the molecular structure [22,23]. Figure 5 shows some representative changes in

specular transmittance which were observed during exposure
for six weeks in standard accelerated weathering devices
(QUV [24], WOM [25]) for unstabilized polymethylmethacrylate
(PMMA) and PMMA stabilized with a polymeric ortho-hydroxy-
phenylbenzotriazole stabilizer [23].   Optical losses and
mechanical film failure in stabilized films are considerably
reduced in comparison to unstabilized PMMA.   However, very
little is as yet known about the ultimate changes that occur
in polymeric substances after very long periods such as the
20-years lifetime appropriate for solar power plants.

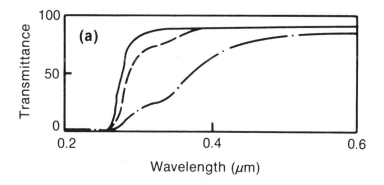

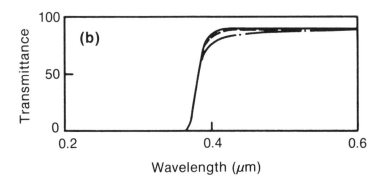

Figure 5.   Transmittance spectra of films made from
(a) unstabilized PMMA and (b) PMMA stabilized with a
polymeric o-hydroxyphenylbenzotriazole (———: Initial;
———: 6 weeks WOM weathering; -·-: 6 weeks QUV weathering).

Surface and interface properties of polymers. Surface phenomena play a significant role in the major problem areas associated with polymers and, therefore, are basic to most of the studies. Surface interactions occur during the fabrication of polymeric materials; the subsequent behavior of a polymer can be critically dependent upon the material against which it is formed. Analytical and experimental studies are needed to improve our understanding of the chemistry, physics, and morphology of surfaces and interfaces between polymers and other materials. Such studies should characterize the interfaces as originally fabricated and after changes caused by typical environmental exposures.

Other examples of surface problems affecting optical elements include abrasion, dust adhesion, and cleaning procedures. The accumulation of airborne particulates is most serious for soft polymers (e.g., silicon rubber) and causes unwanted absorption and scattering of light. An understanding of adhesion mechanisms is required to develop cost-effective cleaning methods and soil-resistant polymeric materials [26,27].

Modifications of polymeric materials, either in bulk or by surface treatment or coating, may result in materials that have better abrasion resistance [28], improved anti-reflective properties [29], UV-screening capability [30], or combinations of several functions.

Adhesive failure of polymer films is of particular concern in solar systems where it can cause corrosion of sensitive optical components, such as reflectors, thin-film electrical conductors, and thin-film photovoltaics. Loss of adhesion may be caused by permeation problems, outgassing and primary bond failure. All polymers are inherently permeable, but to widely varying degrees. Water, oxygen, pollutants, etc., can penetrate polymer films and attack underlying metal components. Furthermore, these penetrants can modify the mechanical and optical properties of the polymer and interfaces between layers.

The permeability of a polymer can be sensitive to how the material is processed and used, and is sometimes enhanced as the material degrades. Although a fundamental understanding of polymer permeability exists [31] and experimental data are available, current information is inadequate to model or control the transport of various species through bulk material and across and along interfaces specifically of interest for solar applications [32].

Some studies and models indicate that the polymer/metal interface morphology, and the changes in the morphology with exposure to the environment, play a key role in corrosion rates [33]. Delamination at the polymer/metal interface could be a likely result of such events.

Thermomechanical Behavior

Requirements for optical performance and mechanical compatibility impose unprecedented requirements for dimensional stability and fatigue response of polymers used, for example, in high concentration reflectors. Moisture, temperature, and UV, separately and in combination can change the volume and thus the stress state of polymers. For example, temperature and humidity cycles in the presence of UV radiation cause microcracks in polycarbonate, while each of the environmental parameters alone does not [34]. Specular reflectance changes in silvered polymer films provide a very sensitive test of the effects of these environmental stresses [20]. An understanding of the relationships between process and environmental effects to mechanical behavior of glassy polymers is essential to permit a reliable design of equipment that uses polymers.

Combined Environmental Effects

Any list of significant effects of the environment on polymers in solar applications will include photodegradation, weathering, permeability, high-temperature performance, delamination/fatigue, dimensional stability, and soiling/cleaning. These effects are not necessarily independent: polymers are expected to suffer more serious degradation during exposure to combined environmental stresses.

The simultaneous and sequential combinations of environmental stresses that alter the properties and affect the performance of polymers need to be identified experimentally. An assessment of the stability of polymers from fundamental rate data or from experimental engineering data requires an understanding of the interactive effects of environmental stresses [6,34].

Improved analytical and effective test methods predicting and verifying the performance of polymers under interactive effects are required. In particular, methods are needed to demonstrate correlations between the result from accelerated and abbreviated tests and real-time behavior.

Emerging Applications

As already mentioned above, for direct conversion of solar energy to heat, collectors in the form of windows, roof-top collectors, biological reaction ponds, solar ponds, or heliostats may be used. However, storage and transmission of the captured thermal energy is limited. More effective methods of energy storage and/or transmission would involve the conversion of the solar energy into some form of chemical energy (e.g., fuel) or into electricity.

Areas of interest include photobiology, photochemistry, and photoelectrochemistry as well as the more mature field of photovoltaics [35]. The natural photosynthesis process, of course, has the advantage of some three billion years of evolutionary development of elegant molecules sequestered in equally elegant macromolecular structures that carry out specific reactions at efficiencies that are optimal for their own survival. Purely synthetic photoconversion systems, on the other hand, offer the prospect of much greater flexibility and higher solar efficiencies, together with the possibility of "tailoring" the chemistry for production of specific fuels and chemicals. The processes of interest [3,35] include not only water photolysis, photofixation of molecular nitrogen, photoreduction of carbon dioxide, but also molecular energy storage in valence isomerization and photoelectrosynthetic reactions at the semiconductor-electrolyte interface. Polymers and molecular assemblies are expected to play a major role in facilitating and stabilizing the one-directional electron transfer necessary for efficient synthetic conversion systems [3,35].

Polymeric materials can provide specific microhetero-geneous reaction environments which facilitate charge separation from the excited state [3,36,37]. The energy migration in synthetic polymers is a model for light-harvesting when the trap of the captured energy is incorporated [38-40]. Synthetic polymers stabilize noble metal colloids used as catalysts for multi-electron processes allowing water photolysis [41,42]. Macrohetero-geneous conversion systems, which can separate the reaction sites and the products, are constructed by utilizing poly-mers as solid phase supports or membranes [43,44]. Stabilization of narrow bandgap semiconductors by polymer coatings is a topic aiming at the utilization of liquid junction semiconductors for solar energy conversion [45-47]. Conductive polymers (e.g., doped polyacetylene) can be employed as p- and n-type semiconductors for the construc-tion of photovoltaic devices made from films [48-52]. Poly-

meric photoredox systems, which can be used for construction
of photogalvanic cells, have also been developed [53,54].

The prospects for the use of polymers in direct solar
conversion processes are exciting and the development of
conversion systems composed of polymers or molecular
assemblies has become an active center of research.
However, much more work and advances in this field (in
particular, a better understanding of energy transfer
processes in polymeric systems) are needed before this
approach can mature into a viable technology. The fact that
this was achieved long ago in nature serves as an
encouragement.

Conclusive Remarks

The technical problems posed by solar technology
development to the polymer R&D community are very similar to
those posed by aerospace and other previously developed
technologies. Frequently, however, we know that the
technical problems can be solved, and it is the economic
constraints that require low cost along with high
performance for solar applications of polymers that provides
the final challenge. One role of the Solar Energy Research
Institute can be to focus the attention of the polymer
community on the most important polymer research and
development areas for solar applications.

References

[1]   The calculation is based on the scenario B described in
      "Energy Projections to the Year 2010"; DOE/PE-0029/2;
      U.S. Department of Energy: Washington, DC, 1983. A
      mean daily direct normal insolation value of 22,932
      kJ/m^2-day, typical for Denver, CO, was chosen from
      Knapp, C.L.; Stoffel, T.L. "Direct Normal Solar
      Radiation Data Manual"; SERI/SP-281-1658; Solar Energy
      Research Institute: Golden, CO, 1982.
[2]   Carrol, W.F.; Schissel, P. In "Polymers in Solar
      Energy Utilization"; American Chemical Society Symp.
      Ser. No. 220; Gebelein, C. G.; Williams, D. J.; Deanin,
      R. D., Eds.; ACS: Washington, DC, 1983, Chapter 1.
[3]   Kaneko, M.; Yamada, A. Adv. Polym. Sci. 1984, 55, 1.
[4]   Rabek, J.F. In "New Trends in the Photochemistry of
      Polymers"; Allen, N.S.; Rabek, J.F., Eds.; Elsevier
      Applied Science Publishers: New York, 1985; Chapter
      16.

[5]   Mavis, C.L. "Status and Recommended Future of Plastic Enclosed Heliostat Development"; SAND 80-8032; Sandia National Laboratories: Albuquerque, NM, 1980.

[6]   Schissel, P.; Czanderna, A.W. Solar Energy Materials 1980, 3, 225.

[7]   Murphy, L.M.; Anderson, J.V.; Short, W.; Wendelin, T. "System Performance and Cost Sensitivity Comparison of Stretched Membrane Heliostat Reflectors with Current Generation Glass/Metal Concepts"; SERI/TR-253-2694; Solar Energy Research Institute: Golden, CO, 1985.

[8]   Clark, E.; Roberts, W.E.; Grimes, J.W.; Embree, E.J. "Solar Energy Systems -- Standard Cover Plates for Flat-Plate Solar Collectors"; NBS Technical Note 1132; National Bureau of Standards, Center for Building Technology: Washington, DC, 1980.

[9]   Lockheed Missiles and Space Co., Inc. "Optimization of Thin Film Transparent Plastic Honeycomb Covered Flat-Plate Solar Collectors"; SAN/1256-78/1; Lockheed: Palo Alto, CA, 1978.

[10] Ramada Energy Systems, Inc. Technical Bulletin, RES TB 180.

[11] Wilhelm, W.G. "Low Cost Solar Energy Collection for Cooling Applications"; BNL51408; Brookhaven National Laboratory: Upton, NY, 1981.

[12] Lampert, C.M. Proc. SPIE 1982, 324, 1.

[13] Suntex Research Associates. "An Energy Efficient Window System: Phase 1 Technical Report"; Suntex: Corte Madera, CA, 1976.

[14] Jet Propulsion Laboratory. "Photovoltaic Module Encapsulation Design and Materials Selection"; LSA Encapsulation Task, Project Report 5101-177.

[15] Batchelder, J.S.; Zewail, A.H.; Cole, T. Appl. Optics 1981, 20, 3733.

[16] Goetzberger, A.; Wittwer, V. Solar Cells 1981, 4, 3.

[17] Hermann, A.M. Solar Energy 1982, 29, 323.

[18] Eisenbach, C.D.; Sah, R.E.; Baur, G. J. Appl. Polym. Sci. 1983, 28, 1819.

[19] Neidlinger, H.H. Solar Energy Materials 1987, in press.

[20] Schissel, P.; Neidlinger, H.H.; Czanderna A.W. Energy 1987, 12, 197.

[21] Carlsson, D.J.; Garton, A.; Wiles, D.M. In "Developments in Polymer Stabilization"; Vol. 1; G. Scott, Ed.; Applied Science: Barking, UK, 1980.

[22] Vogl, O.; Albertson, A.C.; Janovic, Z. Polymer 1985, 26, 1288.

[23] Neidlinger, H.H.; Schissel, P. Polym. Mater. Sci. Eng. 1987, 56, 685.
     Gomez, P.; Neidlinger, H.H.; Polym. Prepr., Am. Chem. Soc., Div. Polym. Chem. 1987, 28 (1), 209.
[24] QUV is a registered trademark of the Q-Panel Company, Cleveland, OH.
[25] Weather-Ometer (WOM) is a registered trademark of the Atlas Electric Devices Company, Chicago, IL.
[26] Hampton, H.L.; Lind, M.A. "The Effects of Noncontact Cleaners on Transparent Solar Materials"; Batelle Pacific Northwest Laboratory: Richland, WA, 1979.
[27] Cuddihy, E.F.; Willis, P.B. "Anti-soiling Technology: Theories of Surface Soiling and Performance of Anti-soiling Surface Coatings"; DOE/JPL-1012-102; Jet Propulsion Laboratory: Pasadena, CA, 1984.
[28] Assink, R.A. Solar Energy Materials 1980, 3, 263.
[29] Jorgensen, G.; Schissel, P. Solar Energy Materials 1985, 12, 491.
[30] Neidlinger, H.H.; Schissel, P. "Polymer Synthesis and Modification Research During FY 1985"; SERI/TR-255-2590; Solar Energy Research Institute: Golden, CO, 1985.
[31] Bixler, M.J.; Sweeting, O.J. In "The Science and Technology of Polymer Films"; Vol. II; Sweeting, O.J. Ed.; Wiley-Interscience: New York, 1971; Chapter 1.
[32] Carmichael, D.C. "Evaluation of Available Encapsulation Materials for Low-Cost Long-Life Silicon Photovoltaic Arrays"; DOE/JPL-954328-28/2; Batelle Columbus Laboratories: Columbus, OH, 1978.
[33] Webb, J.D.; Jorgensen, G.J.; Schissel, P.; Czanderna, A.W.; Smith, D.M.; Chughtai, A.R. In American Chemical Society Symp. Ser. No. 220; ACS: Washington, DC, 1983, Chapter 9.
[34] Blaga, A.; Yamasaki, R.S. J. Mat. Sci. 1976, 11, 1513.
[35] Connolly, J.S., Ed.; "Photochemical Conversion and Storage of Solar Energy"; Academic Press: New York, 1981.
[36] Fendler, J.H. J. Photochem. 1981, 17, 303.
[37] Sassoon, R.E.; Rabani, J. J. Phys. Chem. 1980, 84, 1319.
[38] Tazuke, S.; Inoue, T.; Tanabe, T.; Hirota, S.; Saito, S. J. Polym. Sci., Polym. Lett. 1981, 19, 11.
[39] Hargreaves, J.S.; Webber, S.E. Macromolecules 1985, 18, 734.
[40] Ren, X.X.; Guillet, J.E. Macromolecules 1985, 18, 2012.
[41] Kiwi, J.; Gratzel, M. Nature 1979, 281, 657.
[42] Toshima, N.; Kuriyama, M.; Yamada, Y.; Hirai, H. Chem. Lett. 1981, 793.

[43] Kaneko, M.; Ochiai, M.; Yamada, A.   Makromol. Chem. Rapid Commun. 1982, 3, 299.

[44] Kawai, W. Kobunshi Ronbunshu 1980, 37, 303.

[45] Noufi, R.; Frank, A.J.; Nozik, A.J.   J. Am. Chem. Soc. 1981, 103, 1849.

[46] Frank, A.J.; Honda, K. J. Phys. Chem. 1982, 86, 1933.

[47] Noufi, R.; Tench, D.; Warren, L.F.   J. Electrochem. Soc. 1980, 127, 2310.

[48] Macinnes, D., Jr.; Druy, M.A.; Nigrey, P.J.; Nairns, D.P.; MacDiarmid, A.G.; Heeger, A.J.   J. Chem. Soc., Chem. Commun. 1981, 317.

[49] Chien, J.C.W. J. Poly. Sci., B, 1981, 19, 249.

[50] Pochan, J.M.; Pochan, D.F.; Gibson, H.W. Polymer 1981, 22, 1367.

[51] Metz, P.D.; Teoh, H.; VanderHart, D.L.; Wilhelm, W.G. In American Chemical Society Symp. Ser. No. 220; ACS: Washington, DC, 1983, Chapter 26.

[52] Branston, B.; Duff, J.; Hsiao, C.K.; Loutfy, R.O.   In American Chemical Society Symp. Ser. No. 220; ACS: Washington, DC, 1983, Chapter 27.

[53] Shigehara, K.; Sano, H.; Tsuchida, E.   Makromol. Chem. 1978, 179, 1531.

[54] Kaneko, M.; Yamada, Y. J. Phys. Chem., 1977, 81, 1213.

CONDUCTING POLYMERS BY ION IMPLANTATION

S. P. Hersh, Sylvio Brock, and P. L. Grady

Department of Textile Engineering and Science
N. C. State University, Raleigh, NC 27695 USA

Abstract

The properties of electrically conducting
and semiconducting polymers prepared by implanting ions in insulating polymers is described.
Unlike chemically doped electroactive polymers
which also have conductivities approaching that
of metals, the enhanced conductivity obtained by
ion irradiation is permanent. These conducting
polymers are thus useful for such applications as
large-area photoconductors (electrophotography
and xerography); lightweight, high energy storage
batteries; power transmission lines; shielding
from electromagnetic waves; preventing electrostatic charging of spacecraft; and protective
packaging for delicate items such as precision
bearings, electronic components, and sensitive
instruments. A review of the preparation and
electrical properties of conducting polymers made
by implanting ions in polymers will be presented.
The nature of the interaction between the impinging ions and the substrate and techniques for
assessing those interactions will be described
along with the methods for characterizing the
electrical properties of the implanted polymers.
Possible conduction mechanisms will also be
reviewed and discussed.

1. Introduction

"Ion implantation" is a process by which virtually any
element can be implanted into the near-surface region of
any solid by projecting high velocity ions, commonly with
energies of tens or hundreds of kiloelectron-volts (keV), at
a substrate mounted in a vacuum chamber. This technique
differs fundamentally from diffusion doping since

implantation is not a thermal equilibrium process; rather it
is the movement of the ion into a solid as a consequence of
its accelerating energy and not the result of a concentration
gradient produced by diffusion.

The greatest use of ion implantation has been in the
doping of semiconductor devices which consists of injecting
selected ions to a desired depth into a semiconductor
crystal. Other features of this technique are the ability to
control the average junction depth and to provide a wide
range of density profiles. The utilization of ion
implantation is now extending beyond its original application
to semiconductors to include metals and polymers. In
polymers the major incentive has been to enhance their
electrical conductivity.

Until a decade or so ago synthetic and natural polymers
were known almost exclusively as insulators (conductivity
$\sigma \simeq 1\times10^{-16}-1\times10^{-18}$ S/cm), and hence many of their
applications were plagued with such problems as
triboelectrification, attraction of dust, electrostatic
discharging and the production of radio frequency and other
electromagnetic interference. To overcome some of these
problems, fibers and plastics were frequently rendered
permanently or temporarily more conductive ($\sigma \simeq 1\times10^{-9}-1\times10^{-10}$ S/cm) by adding internal or surface antistatic
agents [1]. Usually these increases in conductivity are
accompanied by an undesirable drop in the mechanical
properties of the polymer. More recently, electroactive
polymers (EAPs) having conductivities in the semiconductor
range ($\sigma \simeq 1\times10^{-5}-1\times10^{-6}$ S/cm) have been produced [2,3]. By
chemically doping these polymers, metallic conductivity is
being approached ($\sigma \sim 1\times10^{3}$ S/cm, about one-tenth that of
mercury) [2]. Motivation for these studies resulted not only
from the desire for a better understanding of the electrical
properties of materials, but also because of the need for
new, better, easier to fabricate, and lower cost materials in
many applications. Examples are large area photoconductors
(electrophotography); lightweight, high energy storage
batteries; power transmission lines; shielding from
electromagnetic waves; preventing electrostatic charging of
spacecraft; and protective packaging for delicate items such
as precision bearings, electronic components, and sensitive
instruments [2,4-8]. Recently ion implantation has been
utilized for creating conductive pathways in circuit
boards [8] and to fabricate microelectronic chips [9,10].
Line widths down to 1 μm could be achieved. It is also
possible to form p/n junctions by implanting polythiophene
with K^{+} ions [11].

Polymers having conductivities above the semiconductor range have been produced by chemically "doping" polymers having conjugated or unsaturated backbones such as polyacetylene, poly(p-phenylene) (PPP), poly(p-phenylene sulfide) (PPS), polypyrole and polythiazyl [(SN)$_x$] [5,12,13]. The most common dopants have been arsenic pentafluoride and iodine (electron acceptors) and sodium naphthalide (electron donor). The increased conductivity achieved by the addition of dopants can be unstable, however, and may decrease with loss of the dopant [12,13].

Another limitation of conventional chemically doped polymers is that many chemical species cannot be used as dopants because they are not volatile enough nor can they be dissolved in suitable solvents. In contrast, virtually any element of the Periodic Table can be used to "dope" a polymer by ion implantation. Also, other polymeric systems that cannot be modified by chemical techniques can be altered by ion implantation.

In an effort to overcome these problems, ion implantation has been utilized to increase the conductivity of polymers [14-18]. Mazurek et al [14] have reported that the enhanced conductivity obtained by implantation of halogens, arsenic, and krypton ions in PPS, unlike that obtained by doping with AsF$_5$, is stable under ambient conditions. For this reason as well as others, interest continues in the study of ion implantation in polymers in spite of the relatively high cost of this method compared with chemical doping.

The response of only a few of the electrically insulating polymers to chemical doping has been examined, and the effect of ion implantation on only a limited number has been reported. Venkatesan et al [17] have implanted five polymer types, including poly(vinyl chloride) (PVC) and poly(methyl methacrylate) (PMMA), with argon ions at an energy of 2 MeV at fluences up to 1×10^{21} ions/m^2 and found that conductivity increases by seventeen orders of magnitude. At lower energies (25 to 260 keV) ion implantation caused an increase of fourteen orders of magnitude in the electrical conductivity of PPS, polyacrylonitrile (PAN) and poly(2,6-dimethyl-phenylene oxide) (PPO) [14,15,19]. Brock et al [20] reported an increase in conductivity of at least 10 orders of magnitude for poly(ethylene terephthalate) (PET) films implanted with 8.3×10^{19} ^{19}F ions/m^2 at 50 keV. Brock [21] has also examined the effect of implanting As$^+$ and Ar$^+$ ions in PET as well as the implantation of F$^+$. He also reported the effect of ion implantation on films of poly(hexamethylene

adipamide) (Nylon 66) (N66); polypropylene (PP); and
polytetrafluoroethylene (PTFE).

The purpose of this paper will be to review the
preparation and properties of electrically conducting and
semiconducting polymers made by implanting ions in insulating
polymers. Possible conducting mechanisms will also be
described.

2. Ion Implantation in Polymers

2.1 General Features

Ion implantation is a technique widely employed in metals,
alloys and semiconductors [19,22,23] but only lately has it
been applied to covalent organic compounds [24,25]. The high
annealing temperatures employed in semiconductors are
impractical for polymers which usually possess lower melting
points (T_m) and much lower glass transition temperatures
(T_g) above which some polymer properties are modified (eg,
modulus) due to initiation of chain mobility [26]. Another
temperature effect appears during ion implantation as a
result of the heating which occurs in the target due to the
implantation current. For this reason, the implantation
current density must be kept at low levels (eg, < 0.50 mA/m^2)
for polymer substrates.

Probably the most important reason that few studies have
been made on ion implantation in polymers is their
crystalline-amorphous heterogeneity which contrasts with the
well-organized structure found in metals and semiconductors.
This morphological inhomogeneity can introduce variations in
the ion-depth profiles and in the ion concentration in the
implanted regions of the polyatomic targets. One more
adverse factor to be considered is that most polymers are
good insulators, which might lead to the buildup of space
charges in the substrates during ion implantation.

The first research dealing with ion implantation in
polymers was reported by Allen et al [16] who found that the
electrical conductivity of polyacetylene $(CH)_x$ films was
enhanced by ion implantation. They used this technique as an
alternative method to adding impurities to the EAPs $(CH)_x$ and
$(SN)_x$. Electroactive polymers are defined as polymers
already having conductivities in the semiconductor range.
These polymers usually are unsaturated and consist of
conjugated carbon chains having a planar conformation over
significant chain lengths or a large content of aromatic

rings in their structures. Earlier the conductivities of the
EAPs had been enhanced by several orders of magnitude by
incorporating oxidizing cations [27] like Ag^+, NO^+ and NO_2^+
or by exposing them to electron acceptor vapors like AsF_5 and
iodine [16] or even sodium naphthalide, an electron donor.
Shirakawa et al [28] reported that such doping of
polyacetylene with fluorine vapor caused physical damage to
the exposed film and decreased its conductivity. Recently,
Rubinstein [29] stated that exposing pressed poly(p-
phenylene) (PPP) pellets and electrochemically formed
polyphenylene films to gaseous AsF_5 for several hours
increases the electrical conductivity of the PPP much more
than it does to the polyphenylene because of the higher
degree of crystallinity of the former.

One advantage of the ion-implantation technique over
chemical doping lies in the fact that implantation can
utilize additives that cannot be applied chemically because
of their low volatility and/or insolubility in suitable
solvents. Another reason is that it is frequently impossible
to modify various polymeric systems by chemical methods.
Another advantage, reported by Weber et al [30], is that the
process of ion implantation appears to "stabilize" the films.
Samples exposed to 25 keV beams of F^+ ions and stored in the
air for over one year [18] maintained the metallic luster and
flexibility of the pristine polymer while part of the film
not exposed to the beam turned brittle and grayish.

Another problem with chemically doped polymers was
reported by Bernett et al [31]. They found that when
antistatic polymer films containing traditional antistatic
agents were used for packaging precision instrument bearings,
the antistatic agents would migrate out of the polymer and
contaminate the bearing surfaces and lubricants. However, no
migration resulted from $(CH)_x$ films implanted with F^+ ions.

The development of a metallic luster usually occurs when
polymers are ion implanted. The color development usually
increases with dose. Iida et al [32] observed that the
polymer turns hard and opaque at high doses, and Hashimoto
et al [33] suggested that this happens due to induced
carbonization and the formation of a highly crosslinked
carbon structure. As cited by Forrest et al [34], the change
in color produced in organic thin films by ion implantation
could convert them into materials useful for optical devices
such as diffraction gratings. An additional benefit of ion
implantation over chemical doping as related by Weber
et al [35] is that the surface of the implanted polymer is
not severely damaged. However, Venkatesan et al [17] found

that the thickness of the film they employed decreased almost 50% at a fluence of $1x10^{17}$ Ar^+/cm^2 and 2 MeV ion-beam energy.

The kinetic energy of the ions fractures the polymeric chains which then undergo molecular rearrangement. This damage has been cited as the major reason for the drastic increase in the polymer's electrical conductivity [14]. Mazurek et al [14] observed increases of up to thirteen orders of magnitude in the conductivity of PPS films after implantation of halogen, krypton and arsenic ions at fluences of $1x10^{16}$ ions/cm^2 with an accelerating energy of 100 keV. Since Kr is a neutral gas, no chemical reactions with the substrate would be expected to occur. The increases in conductivity could therefore be solely attributed to bombardment damage (broken polymer chains) combined with the associated charge formation and crosslinking, rather than reaction with the dopant. As shown in Figure 1, the conductivity increases abruptly with implantation and "plateaus" with a conductivity of $1.5x10^{-5}$ S/cm at a fluence about $1x10^{16}$ ions/cm^2, corresponding to an average impurity concentration of $2.5x10^{21}/cm^3$. A similar enhancement in electrical conductivity was produced in PPS films implanted with As^+ under the same conditions (100 keV) at fluences between $3.2x10^{13}$ and $3.2x10^{16}$ ions/cm^2.

In contrast, Abel et al [15] found that the conductivity of the PPS films implanted with Br^+ at doses greater than $1x10^{16}$ ions/cm^2, instead of reaching a "plateau," continued to increase linearly, suggesting the possibility that the Br^+ ions react chemically with the substrate at these fluences. This increase is also shown in Figure 1.

In the same paper, Abel et al [15] tried to relate the enhancement of the electrical conductivity with the electronegativity of different halogen ions implanted in the PPS films. At a fluence of $1x10^{16}$ ions/cm^2, however, no relationship was observed since they found that fluorine, the most electronegative element, induced the lowest conductivity in the PPS. The order of conductivity was $\sigma(Cl) > \sigma(I) > \sigma(Br) > \sigma(F)$. They suggested that the ordering of the conductivities might be different at higher or lower fluences and that the low relative conductivity of the F^+ implanted samples indicate that bombardment damage dominates over chemical reaction with the substrate.

From Figure 1 it can be observed that the conductivity increases by more than seven orders of magnitude for a two order of magnitude variation in the dose ($1x10^{14}$ – $1x10^{16}$ ions/cm^2).

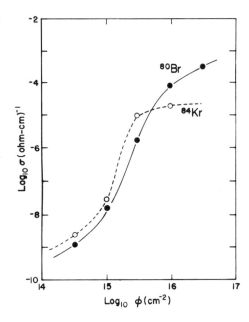

Figure 1. Conductivity σ as a function of fluence φ of PPS films implanted with Br^+ and Kr^+ ions at ion beam energies of 200 keV and 100 keV, respectively. Reproduced with permission [15].

EAPs also display excellent conductivities when chemically doped. Applying this procedure, Ivory et al [36] showed that the electrical conductivity of poly(p-phenylene) films could be adjusted by varying the amount of electron donor or acceptor dopants added to the polymer. These authors reported that the conductivity of less than $1x10^{-14}$ S/cm for the pristine polymer was increased to over $1x10^2$ S/cm via the formation of charge transfer complexes.

Ivory et al [36] obtained the highest conductivity by doping with AsF_5 (an electron acceptor). Doping with alkali-metal compounds (such as sodium naphthalide solution) also provided highly conducting n-type materials having a metallic gold appearance.

It is known that the addition of small concentrations of impurities can greatly modify the electrical conductivity of the doped material via n- or p-type conduction mechanisms. In the n type, the excess of electrons can be thermally excited out of a donor level into the conduction band. In the p type, the localized states created by the extrinsic impurities in the forbidden gap just above the valence band are capable of accepting electrons and, simultaneously, increasing the holes in the valence band.

Recently, Wasserman et al [37] using thermoelectric power (TEP) measurements found drastic differences between PAN and PPO samples implanted with Br^+ at a fluence of $1x10^{16}$ ions/cm^2 at 200 keV ion-beam energy. The PAN exhibited n-type behavior attributed to the formation of a donor band in the mobility gap whereas PPO showed p-type behavior.

Intercalation compounds prepared by introducing chemical dopants between layers of graphite have led to conductivities even higher than shown by some metals [37]. Also, highly oriented pyrolitic graphite (HOPG) intercalated with AsF_5 was reported by Vogel et al [38] as having unusually high electrical conductivity.

2.2   Ion-Substrate Interactions

2.2.1   Penetration and distribution of implanted ions. Ions of virtually any element can be implanted beneath the surface of a solid material after being accelerated to tens or hundreds of keV. The ions reach the target with a high energy which then decreases as a result of collisions with the substrate atoms. The incident ions come to rest at up to hundreds to thousands of nm below the surface as a consequence of losing energy during successive collisions with the atoms in the target. Once the ions come to rest, the implanted impurity can be in a position which serves to alter the electronic properties of the substrate, ie, it beomes a dopant [39]. The energy loss of the ions during their penetration can also cause molecular damage to the substrate which might also modify the electronic properties of the substrate.

In a crystal there are open planes and channels among the rows of atoms which facilitate the penetration of the impinging ions. The resulting ease of ion movement is known as channeling and permits the ions to penetrate deeply before coming to rest in interstitial or substitutional sites. When an impurity atom occupies the position of a lattice atom, the doping is called substitutional. In semiconducters p- and n-type conductivity originates with this kind of doping. When the implanted ions occupy a site between the atoms, the doping is called interstitial. The presence of an ion in this position leads to an additional energy level in the band gap capable of producing the effect of n- or p-type

doping [40]. Wilson and Brewer [39] reported that fluences
between $1x10^{11}$ and $1x10^{16}$ ion/cm^2 are usually required to
dope crystals. At this level, from $1x10^{-4}$ to 10 monolayers
of ions are introduced into the crystal. For covalent
polymers, which are only partially crystalline materials,
higher fluences might be required to obtain adequate
implantation levels.

But even in crystals where channeling is more likely to
take place, not all ions will move freely. Many will hit the
lattice atoms on or close to the exposed surface. These
collisions will dislocate many lattice atoms and can lead to
the ejection of atoms if they recoil with an energy in excess
of the surface binding energy (2-5 eV) [41]. These
collisions that take place in the shallow region near the
surface produce the phenomenon called "sputtering" which is a
removal of the surface layer of atoms. Ion energies ranging
from 4 to 12 keV are employed when it is desired to sputter
surface atoms away rather than implant them. This type of
radiation damage causes the removal of many monolayers only
at fluences of $1x10^{14}$ ions/cm^2 or higher. Dearnaley
et al [23] found that lower doses remove less than one
monolayer of atoms from the target surface. As noted by Liau
and Mayer [42], the sputtering yield is proportional to the
nuclear stopping power of the incident ion in the near-
surface region. Venkatesan et al [17] reported, as
previously mentioned, that the thickness of films used in
their experiments decreased up to 50% at a fluence of $1x10^{17}$
Ar$^+$/cm^2 and 2 MeV beam energy as a consequence of sputtering.

The disordered region produced around the path of the
incident ions reaches its maximum near the end of the
trajectory. At sufficiently high fluences, Elman et al [43]
note that the individual disordered regions start to overlap
until a new crystalline or amorphous layer is obtained. Both
the depth and depth distribution (straggling) of the
implanted ions depend on the mass and energy of the impinging
ion and on the mass of the substrate atoms. The penetration
will be greater when the ion mass is significantly heavier
than the target atoms and, therefore, as pointed out by Mayer
et al [44], the concentration of the ions near the surface
will be smaller, as shown in Figure 2. Here the penetration
depth is reported in units of μg/cm^2 which is the product of
the depth of the particle in the target multiplied by the
density of the target. The effect is magnified as the energy
of the ion increases. For a light ion striking a heavy
target atom, the penetration is reduced as reported by Davies
and Sims [45]. In this system a relatively light ion Na+ was
implanted into "heavy" aluminum at a low energy of 10 keV

giving low penetration and high concentrations near the
surface somewhat similar, but more pronounced, than that
exhibited by the 80 keV curve in Figure 2. This distribution
is frequently described as exhibiting a "truncation effect."
Higher accelerating energies will reduce this effect when the
implantation energy is increased from 10 to 60 keV. The
symmetry of the concentration-depth profiles also increases.

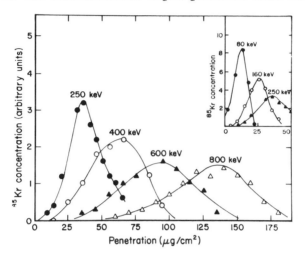

Figure 2. Penetration distribution of [85]Kr implanted in
amorphous Al$_2$O$_3$ at several energies. Reproduced with
permission [75].

A brief quantitative description of each of these
interactions is given in the next section.

2.2.2 Stopping theory. The two main types of interaction
(with consequent energy loss) between incident ions and
target atoms result from nuclear collisions and electronic
interactions. Mayer et al [44] divided these interactions
into three categories based on the energy ranges of the
impinging ion. At low energies nuclear stopping (ie,
screened Coulomb collisions) is the most important process;
at medium energies electronic stopping is the predominant
process and increases linearly with the velocity; and,
finally, at very high energies at which the ion velocity
becomes higher than the velocity of the orbital electrons,
the ion moves as a fully ionized atom. This region is not
useful for ion implantation since the energy is beyond the
energy range of interest in most of these implantations.

Bohr [46] was the first investigator to treat the ion-substrate interaction problem. Later, Lindhard, Scharff, and Schiott [47] developed a unified theory of atomic stopping (commonly referred to as the LSS theory). They expressed the nuclear stopping power $(d\varepsilon/d\rho)_n$ in terms of dimensionless energy (ε) and a dimensionless length (ρ) in which

$$\varepsilon = EaM_2/e^2 Z_1 Z_2 (M_1 + M_2) \quad \text{and} \quad (1)$$

$$\rho = R\pi aNM_1 M_2/(M_1 + M_2) \quad (2)$$

where E = energy of the incoming ion,
       a = the Thomas-Fermi screening radius,
       M = the atomic mass,
       Z = the atomic number, with the subscripts 1 and 2
           denoting the projectile and target atoms,
           respectively,
       N = the atomic density (number of atoms/unit volume) of
           the substrate,
       R = the range (penetration depth), and
       e = electronic charge.

The term a is the measure of the effective radius of the electron clouds surrounding the particles and is calculated by

$$a = 0.8853 \, a_0/(Z_1^{2/3} + Z_2^{2/3})^{1/2} \quad (3)$$

where a_0 = the Bohr radius (0.53Å).

The electronic stopping power $(-d\varepsilon/d\rho)_e$ is given by

$$(-d\varepsilon/d\rho)_e = k\varepsilon^{1/2} \quad (4)$$

where k is a function of Z_1, Z_2, M_1, and M_2.

The nuclear and electronic stopping-power curves derived from the LSS theory are shown as a function of $\varepsilon^{1/2}$ in Figure 3. The slopes of the lines for electronic stopping are given by the value of k which normally ranges from 0.1 to 0.25; the curve shown is for k = 0.15. The significance of the energy regions proposed by Mayer et al [44] described above is evident in Figure 3. Nuclear stopping is clearly more important at low energies and electronic stopping at high energies. The high-energy region in which the nucleus is stripped of its electrons is represented by ε_3. The horizontal line labeled S^0 is an approximation proposed by Nielsen [48].

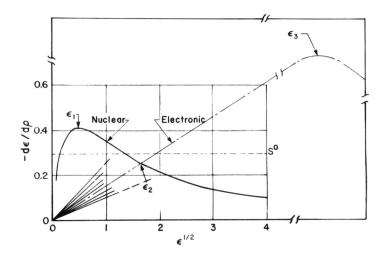

Figure 3. Theoretical nuclear and electronic stopping power curves expressed in terms of the reduced variables ρ and ε in the family of lines for electronic stopping. The dot-dash line is for $k = 0.15$. The S^0 horizontal line represents the constant stopping-power approximation suggested by Nielsen. Reproduced with permission [44].

By integrating the stopping-power curves, the total distance R that the ion moves before reaching the rest position is obtained. If the path wanders through the substrate lattice, the projected range R_p, which gives the actual penetration depth below the surface, is of greater interest than the total path length R. The relationship between R and R_p is also treated in the LSS theory in which it was found that:

$$R/R_p = (1 + bM_2/M_1) \qquad (5)$$

where b equals approximately 1/3 but varies slightly as a function of E and R.

Depth profiles of implanted ions can be calculated for most projectile-substrate combinations from the LSS theory. At low fluence the depth concentration profiles are usually characterized by a Gaussian distribution centered around the average range. At higher fluences and accelerating energies, however, other effects such as sputtering can alter the ultimate depth concentration attainable.

Each distribution can be described by an average projected range (R_p) and a root-mean-square (rms) projected range (ΔR_p). The relative straggling range ($\Delta R_p/R_p$) is another parameter considered. Other useful parameters are the lateral range, the electronic stopping power, and the nuclear stopping power. For amorphous substrates the depth profiles are almost Gaussian in shape. R_p in this case can be accepted as the position of the peak, whereas $2.5 \Delta R_p$ is equal to the full width at half maximum, as cited by Mayer et al [44]. In this situation, the average concentration of implanted ions $\bar{n}$ can be determined since

$$\bar{n} = \phi/2.5 \Delta R_p \qquad (6)$$

where ϕ is the ion fluence (ions/cm^2). Even when the distribution is only approximately Gaussian, Equation 6 is still a valid approximation since the position of the peak (R_p) remains almost the same.

Calculations of projected ranges and electronic stopping powers for monatomic and polymeric substrates are given by Northcliffe and Schilling [49] and Gibbons et al [50]. However, the accuracy of the values tabulated by Gibbons have been questioned, and researchers such as Mazurek et al [14] and Hall et al [51] therefore have preferred, respectively, to use data based on the Brice algorithm [52] and Anderson and Ziegler [53]. More recently, Weber [54] used k values from Equation 4 tabulated by Land and Brennan [55] and LSS theory for his calculations of depth profiles.

R_p and ΔR_p are important parameters, not only for determining the electrical conductivity within the active region of the implanted ions (usually considered to be a layer of thickness $2 \Delta R_p$), but also for calculating the ion concentration $N(x)$ as a function of the depth which can be obtained from the relation:

$$N(x) = \bar{n} \exp[-(x-R_p)^2/2 \Delta R_p^2] \qquad (7)$$

where $\bar{n} = \phi/2.5 \Delta R_p$,
ϕ = the fluence, and
x = the depth measured from the implanted surface.

The depth concentration falls to about $n/10$ at $x = R_p \pm 2 \Delta R_p$ and to $n/100$ at $x = R_p \pm 3 \Delta R_p$ in the absence of channeling. This equation also assumes that no diffusion takes place and does not consider any subsequent annealing effects.

2.3 Ion Implantation in Polymeric Materials

The extent to which the ion implantation technique has been applied to polymeric materials is illustrated in Table 1, which lists the ions, polymers, and implanting conditions that have been reported in the literature. A summary of the results found in these studies is given in the following sections.

Table 1. Research Reporting Electrical Conductivity Enhancement by Ion Implantation of Polymers

Polymer[a]	Implantation Conditions			Refer-ences
	Energy (keV)	Fluence (cm^{-2})	Ion	
$(CH)_x$	25	$3.1 \times 10^{17} - 4 \times 10^{18}$	F^+	24
$(CH)_x$	90	2×10^{16}	Pd^+	16
$(CH)_x$	25-41.6	$3 \times 10^{15} - 3.4 \times 10^{16}$	Br^+	16
$(SN)_x$	25-46	5×10^{15}	Br^+	16
$(CH)_x$	10-40	$1 \times 10^{15} - 3 \times 10^{17}$	F^+, Cl^+ Br^+, I^+	30
$(CH)_x$,HOPG	25	3×10^{17}	F^+	18
$(CH)_x$,COPOL, PPS	25	1×10^{17}	F^+	31
PTCDA,NiPc, NTCDA	2000	$1 \times 10^{14} - 1 \times 10^{17}$	Ar^+	34
$(CH)_x$,HOPG	10-60	$1 \times 10^{14} - 1 \times 10^{18}$	F^+	35
PVC,HPR-204, PMMA,COP,PIQ	2000	$1 \times 10^{14} - 1 \times 10^{17}$	Ar^+	17

Table 1. (continued)

Polymer[a]	Implantation Conditions			Refer-ences
	Energy (keV)	Fluence (cm^{-2})	Ion	
PPS	100	$3.2 \times 10^{14} - 2.2 \times 10^{16}$	$^{75}As, ^{84}K$	14
PPS	200	$3.2 \times 10^{14} - 3.2 \times 10^{16}$	^{80}Br	14
PPS	260	1×10^{16}	^{127}I	14,15
PPS	100	1×10^{16}	^{36}Cl	14,15
PPS	60	1×10^{16}	^{19}F	14,15
PPS	100	$3.2 \times 10^{14} - 3.2 \times 10^{16}$	^{84}K	15
PPS	150; 200	1×10^{16}	^{84}K	15
PPS	100; 150	1×10^{16}	^{80}Br	15
PPO,PPS,PAN	35–230	$3.2 \times 10^{14} - 3.2 \times 10^{16}$	^{80}Br	37
PET	50	$1 \times 10^{14} - 1 \times 10^{16}$	$F^+, Ar^+, As^{+(b)}$	21
N-66,PP,PTFE	50	5×10^{15}	F^+	21

[a] See Glossary.

[b] Implanted at room temperature. F^+ and Ar^+ also implanted at 1×10^{16} ions/cm^2 at 77 K.

3. Characterization of the Implanted Material

3.1 Chemical Composition and Appearance

To characterize the chemical and physical changes produced in the substrate by the penetrating ions and to determine their location, many techniques have been used. These are listed in Table 2. Most of the analytical techniques involve surface spectroscopy since the implanted ions do not penetrate very deeply into polymers unless high accelerating energy is employed. At low fluences, however, many of the analytical techniques are not sensitive enough to detect the ions. Table 3 contains a list of papers in which the analysis of implanted ions and/or the chemical or physical changes induced in polymer substrates by ion implantation have been described.

It is beyond the scope of this review to describe the nature of these techniques and how they are used in the context of ion-implanted polymers. The references should be consulted for details.

3.2 Electrical Properties

Measuring the electrical conductivity of organic polymers has been somewhat difficult because of their high resistivity and the complicated phenomena occurring when attaching electrodes. In many instances poor electrical contact at the interface may introduce a resistivity of the same order as the sample to be measured. It is very difficult to determine whether a contact might be injecting or blocking carriers. The electrode may contribute to the introduction of space charges in the case of injecting electrodes. For those reasons, measuring techniques have been of paramount importance in the determination of the correct resistivity (conductivity) of polymers.

Usually DC potentials are applied in electric circuits designed to measure the resistivity of polymers, but a few experiments have employed AC voltages [37]. For high-resistivity materials, short, wide samples should be used to increase the current flow. Many different electrode systems--such as sandwich type, deposited metal layers, and cells with mercury electrodes--have been applied in resistivity measurements. However, only those employed in conductivity measurements in ion-implanted polymer films will be discussed here.

3.2.1 <u>Low field conductivity</u>. The basic method for
measuring the electrical conductivity of polymers is to
measure the current as a function of an applied potential.
Because of the inherent low conductivity of many polymers,
however, much care has to be exercised in making the
measurements. One big problem, as already mentioned, is that
of attaching or forming suitable electrodes on the polymers.
Three of the common methods are described below.

<u>Four-point probe [16,24,34]</u>: The four-point probe
consists of four equally spaced (or not) metal probes usually
set "in-line" and pressed onto the surface of the implanted
layer. This technique is one of the most preferred because
it is nondestructive and does not require elaborate
preparation. The four-point probe measures all the
electrically active species in implanted layers. A constant
current I is provided by a preset resistor while the voltage
drop ΔV is measured with a high input impedance voltmeter
(electrometer). A switching circuit can be included for
reversing the direction of current flow, changing the current
and voltage probes, and reversing sides of the sample.
Micrometers can be used for moving the probes by known
intervals over the surface of the sample. If the voltage
electrodes are spaced a distance d apart, then $\rho_v = 2\pi d \Delta V/I$,
where ρ_v is the volume resistivity of the sample.

The four-point probe technique was employed by Allen
et al [16] for measuring the electrical conductivity of ion-
implanted polymers. Forrest et al [34] used this method only
for their highly irradiated samples ($\phi > 5 \times 10^{14}$ Ar^+/cm^2).
Other authors did not mention the technique they
used [18,25,30,31,35].

<u>Two-point probe [34,37]</u>: This technique is less suitable
than the four-point probe. The two-point probe depends on
measuring the resistance adjacent to a fine point probe and
requires calibration against known materials [23]. Forrest
et al [34] employed this technique on polymers implanted at
fluences less than 5×10^{14} ions/cm^2. Also, Wasserman
et al [37] used the two-probe technique to measure the
dependence of DC conductivity on temperature and the
thermoelectric power (TEP). They chose this method because
of the very high resistance of their samples at low
temperature.

<u>Planar interdigitated electrodes [14,15]</u>: Mazurek
et al [14,15] argued that the four-point probe technique
produces poor electrical contacts and also that the probes

Table 2. Chemical and Microscopical Techniques
Employed to Analyze Ion-Implanted Polymeric
Substrates

Technique	Abbreviation
Attenuated Total Reflectance,	ATR
Fourier Transmission Infrared, and	FTIR
Infrared Spectroscopy	IR
Auger Electron Spectroscopy	AES
Differential Scanning Calorimetry	DSC
Electron Spin Resonance	ESR
Helium Backscattering	HB
Nuclear Magnetic Resonance	NMR
Optical Reflectance, and	OR
Optical Transmission	OT
Raman Spectroscopy	RS
Rutherford Backscattering	RBS
Scanning Electron Microscopy, and	SEM
Transmission Electron Microscopy	TEM
Ultraviolet and Visible Spectroscopy	UV-Vis
X-Ray Photoelectron Spectroscopy	XPS
(Electron Spectroscopy for Chemical	(ESCA)
Analysis)	
Secondary Ion Mass Spectrometry	SIMS

S.P. Hersh et al.

Table 3. References Reporting the Chemical Analysis
of Ion-Implanted Polymeric Substrates

Authors	Technique Employed[a]	Reference
J. S. Abel et al	AES, FTIR	15
W. N. Allen et al	ESCA, HB	16
L. G. Banks et al	NMR	57
M. K. Bernett et al	ATR, ESCA	31
S. Brock	DSC, ESCA, SEM, SIMS	21
M. S. Dresselhaus et al	ESR, OT	56
S. R. Forrest et al	ESCA, IR, RBS, TEM, UV-Vis	34
H. Mazurek et al	AES, FTIR	14
T. Venkatesan et al	RBS, RS	17
B. Wasserman et al	AES, ESR, OT, UV-Vis	37
D. C. Weber et al	ATR, ESCA, ESR, HB, NMR	24
D. C. Weber et al	ATR, ESCA, ESR, HB, NMR	30
D. C. Weber et al	ESCA, NMR	35
D. C. Weber et al	ESCA, OR	18

[a]See Table 2 for abbreviation code.

can cause physical damage to the implanted layer. Instead, they formed the polymer film to be implanted on a planar, interdigitated set of aluminum electrodes. The electrodes were formed by the photolithographic patterning of a 0.05 µm thick aluminum film evaporated over a 1 µm thick film of thermally grown SiO_2 supported on a silicon substrate. Connections to the electrodes were made with aluminum wires bonded to the contact pads at the base of the electrode pattern. The thin film samples were deposited over the planar interdigitated electrode structure by a modified spin casting technique. Measurements of the inter-electrode resistance were made at room temperature by applying DC voltages between 0.01 and 10 V to the sample and measuring the resulting current with a Keithley 610 C electrometer. A shielded enclosure was used for these measurements.

3.2.2 <u>Conductivity vs temperature</u>. Measurements of conductivity as a function of temperature provide valuable insights into the specific electronic conduction mechanism. For each temperature of the ion-implanted sample, a corresponding conductivity can be obtained from which Arrhenius-type plots can be made. The activation energy of conductivity can be determined from the slope of the straight line. Many examples have been given in the literature for ion-implanted polymers [14,21].

3.2.3 <u>High field conductivity</u>. Another very important technique for characterizing the electrical properties of polymers is to measure the current I as a function of voltage V over a large voltage range using high voltage gradients. The most interesting features of such measurements are illustrated in Figure 4 in which log I is plotted as a function of log V [58]. Three characteristic curves are usually obtained which depend on the nature of the conduction mechanism: (i) if the curve is linear with a slope of one, Ohm's law is obeyed; (ii) if the curve is linear with a slope of two (after a short initial portion with a slope of one), $I \propto V^2$ and Child's law is obeyed; and (iii) if the curve has initially a slope of one followed by a slope of two and then by a steep current rise after which Child's law is again obeyed, a trap-filled-limit (TFL) mechanism is indicated. A more complete discussion of this mechanism is given in section 4.3.

Potentially, this method is a powerful technique for studying polymers, a procedure which has not yet been exploited to any great extent in ion-implanted films. The advantage of the technique arises from the clear distinctions shown between the different mechanisms of charge transport

and the relative ease with which they can be interpreted both
theoretically and experimentally.

3.2.4 <u>Hall effect</u>. The Hall effect is a valuable tool
for measuring the sign and mobilities of the charge carriers
in semiconductors, but it has not usually been employed in
polymers because of their low conductivity and small mobility
of current carriers. However, Dresselhaus et al [56]
succeeded in measuring the mobility of implanted ions from
Hall effect measurements made on a PAN film implanted with
Br^+ at a fluence of $2x10^{16}$ ions/cm^2. The lower limit on the
carrier mobility was found to be $\mu < 1x10^{-3}$ cm^2/V sec.

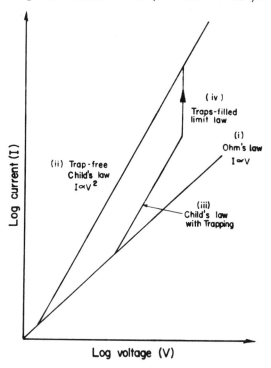

<u>Figure 4.</u> Schematic I-V curves for space-charge-limited
currents. (i) Ohm's law region; (ii) trap-free Child's law;
(iii) Child's law with trapping; (iv) traps-filled limit law
(TFL). Single carrier, single set of traps. Adapted from
[58] with permission.

In general, measurements of Hall potentials are carried
out at rather high temperatures (>300°C) which also
constitutes a major difficulty in applying this technique to
most of the useful polymers.

3.2.5 Thermoelectric power (TEP). The differential
coefficient of thermoelectric power α, also known as the
Seebeck coefficient S, consists of measuring the voltage
difference ΔV formed across opposite faces of a material held
at a constant temperature differential ΔT. Then
α = S = ΔV/ΔT. These values, in combination with
conductivity measurements, provide information concerning
polarity, number, and mobility of charge carriers.

TEP was employed by Grady and Hersh [59] in a study of the
effect of internal additives on the Seebeck coefficient of
nylon 6 and 66. Recently, Wasserman et al [37] applied this
technique to ion-implanted PPO and PAN films and found S to
be slightly dependent on T. Because of the small magnitude
of the Seebeck coefficients (S ≃ μV/K), they suggest that a
high density of carriers having low mobilities ($\mu < 1 \times 10^{-3}$
cm^2/V sec) are present in the implanted samples.

The determination of the dominant charge carrier using the
TEP technique indicated that PAN ion implanted with Br^+ at a
fluence of 2×10^{16} ions/cm^2 exhibits n-type behavior whereas
p-type behavior was observed for PPO and PPS.

3.2.6 Frequency measurements. The frequency dependence
of AC conductivity is a technique which has been used to
confirm the presence of a one-dimensional variable range
hopping mechanism (described in section 4.5) in ion-implanted
polymers since this mechanism is highly frequency dependent
in disordered systems. The frequency dependence of the real
part of the conductivity $\sigma'(\omega)$ follows the relation:

$$\sigma'(\omega) = \sigma_o + A\omega^s \qquad (8)$$

where ω = frequency
   $\sigma'(\omega)$ = real part of conductivity at frequency ω,
     σ_o = conductivity at frequencies near zero,
     A = a value related to the density of pairs of
         localized states, and
     s = a value usually ranging between 0 and 2.

Wasserman et al [37] found s to range between 0 and 1.5
at frequencies below 20 Hz. Above 50 Hz, the frequency
exponent s exceeded two. At low frequencies as ω → 0,
Gogolin [60] reported that $\sigma'(\omega)$ has a $(\omega \ln \omega)^2$ dependence.
This result is in agreement with Mott's qualitative
predictions [61].

3.2.7 <u>Effect of morphology</u>. In addition to effects
related to chemical structure, changes in crystallinity,
orientation, or even tacticity can also affect the electrical
properties of a polymer. For example, a study by Amborsky
[62] on PET revealed that the conductivity and activation
energy of conduction decreased with increasing crystallinity.
Kryszewski [63] also noted that the electrical conductivity
of semiconducting polymers increases as the degree of
crystallinity increases. He suggested that this increase
speaks in favor of an electronic mechanism. Such a rise in
conductivity occurs, however, only if the coplanarity of
the chains remains undisturbed during crystallization since
such disturbances lower the conductivity.

Pressure is another variable that can influence the
conductivity. As pressure increases, conductivity also
increases. Presumably the increasing pressure reduces the
intermolecular distances, which favors a hopping mechanism,
since the jumping of the charge carrier is facilitated by the
narrowing of the potential barriers.

4. Conduction Mechanisms

A recent review of the mechanisms which have been
suggested to explain the electrical conductivity of polymers
has been given by Cotts and Reyes [64]. They pointed out the
need to develop a unified theory of conduction in polymers to
overcome the many diversified approaches reported in the
literature, many applying to each specific polymer studied.
The major features that need to be incorporated into
developing such a theory which would distinguish polymers
from metallic conductors and semiconductors are (1) the
structural disorder in polymers, (2) identification of the
number and sources of the charge carriers, and (3) recognition that conductivity can be limited by the mobility of the
carriers rather than by their number. Those mechanisms which
have been discussed in the literature relating to ion-
implanted polymers will be outlined briefly below,
emphasizing the reasons why ion implantation is thought to be
effective in increasing the electrical conductivity.

4.1 Ionic or Electronic

Charge transport mechanisms can be classified into two
major types: electronic and ionic. Several characteristics
distinguish these mechanisms. It is well known that charge
carriers have greater mobility in electronic conductors.

Electronic conduction does not involve mass transfer while ionic conduction does. The electronic conductors exhibit space charge effects, whereas ionic conductors display electrode and polarization effects. If the conduction mechanism is electronic, then the current levels through the samples do not change when a steady voltage is applied to them. Many other contrasts (eg, photosensitivity) could be cited to distinguish between electronic and ionic conduction.

Mazurek at al [14] rule out the possibility of an ionic conduction mechanism in the ion-implanted PPS films they tested after they observed that when a constant voltage was applied to the samples, the current levels remained constant for a period of several days. Moreover, the conductivities of the ion-implanted polymers stayed stable for several months. In contrast, the conductivity of chemically doped samples dropped after several months (because of loss of dopant).

Another reason favoring the electronic conduction mechanism is that log V vs log I plots displayed in the same research [14] clearly showed a linear (ohmic) portion at the low region of applied voltage, followed by two other regions characteristic of space-charge-limited-current flow. Similar results and conclusions were reached by Brock [21]. In two other experiments [15,34] evidence of an electronic conduction mechanism was also supported.

4.2 Ohm's Law

When Ohm's law is followed, there is a linear relationship between voltage V and current I. If there is no charge carrier trapping and no hindrance to current transport between electrodes, a plot of log I vs log V yields a straight line with slope one. This behavior was observed for voltages between 0.01 and 1 V on As^+ and Kr^+ ions implanted at several fluences and accelerating energies in PPS samples [15] as shown in Figure 5. Brock [21] found his PET samples (implanted with F^+, Ar^+ and As^+) were ohmic up to electric field strengths of 600 kV/m except those implanted at a fluence of ~ 1x16 ions/cm^2 which were ohmic up to an electric field strength of 400 kV/m. At that point the current increased faster than linearly indicative of space-charge-limited-current flow in the presence of carrier traps (see next section). Nylon 66 and PP films implanted with $5x10^{15}$ F^+/cm^2 exhibited almost ohmic behavior--the first had a log V vs log I slope of 0.9 and the latter had a slope of 1.1.

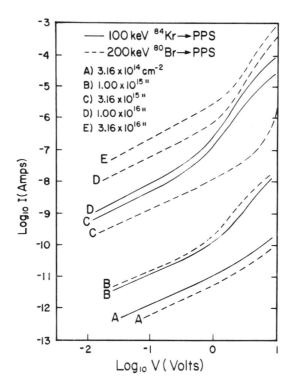

Figure 5. Current as a function of applied voltage for PPS
samples implanted with Br$^+$ and Kr$^+$ at various fluences and
energies. Reproduced with permission [15].

Forrest et al [34] also found their organic thin films
irradiated at fluences higher than 1×10^{14} Ar$^+$/cm^2 to be ohmic
for measured currents in the range of $1 \times 10^{-9} - 1 \times 10^{-4}$ A.

4.3 Space-Charge-Limited Currents (Child's Law)

If the conductor is unable to transport all of the charge
injected by the electrode, a space charge builds up which
limits the current flow. At the point this occurs in
Figure 4, the resistance is no longer ohmic but changes from
$I \propto V$ to $I \propto V^2$, ie, the current rises with the square of the
voltage. This space-charge-limited behavior is known as
Child's law for a trap-free insulator. If there are charge
traps in the insulator, the curve starts out ohmic and then
increases to follow Child's law as the traps fill. Suddenly,
as the last traps are filled, a sharp increase in current
occurs, and the current reaches that of the trap-free, space-
charge-limited conductor.

Well defined changes from ohmic to square law behavior
were observed in a series of current-voltage curves obtained
by Setter [65] for polyethylene at different temperatures.
However, instead of displaying a change to trap-filled-
conductor characteristics at high voltages, the current
leveled off, suggesting electrode saturation occurred.

No high field conductivity measurements have been reported
on ion-implanted polymeric samples. The possible existence
of space-charge-limited currents occurring in samples of this
type at relatively low fields has been suggested, as noted in
the previous section, by Abel et al [15] and Brock [21].

4.4   Thermally Activated Conduction

Several researchers have observed that the temperature
dependence of conductivity of ion-implanted films follows an
Arrhenius activation conduction mechanism typical of
intrinsic and extrinsic semiconductors. Abel et al [15]
observed this type of behavior in PPS samples implanted with
either Br^+ or Kr^+ ions. Their observations are shown in
Figure 6. Based on a plot of the logarithms of conductivity
vs $1/T$, they suggested that a thermally activated mechanism
was responsible for the conduction in PPS samples implanted
with Kr^+ at 1×10^{16} ions/cm^2 and 200 keV over the temperature
range tested (70 to 303 K), whereas for samples implanted
with Br^+ at the same conditions, this conduction mechanism
applied only at $T < 150$ K. The conductivity was consistent
with log $\sigma \propto E_a/kT$ with activation energies E_a for the Kr^+
implanted PPS samples of $0.68 < E_a < 1.5$ eV for $T < 150$ K and
$0.36 < E_a < 0.41$ eV for $T > 150$ K. The PPS films implanted
with Br^+ had an activation energy ranging between 0.54 and
0.68 eV at $T < 150$ K. At $T > 150$ K, the conductivity of the
Br^+ implanted sample was controlled by a hopping mechanism
(described in the next section).

Forrest et al [34] also found an Arrhenius relationship
for PTCDA and NiPc implanted with Ar^+ samples at doses below
1×10^{14} /cm^2.

4.5   Hopping Mechanism

The energetic and spatial distribution of electrons are
affected by disorder in the atomic lattice. With a random
distribution of atoms, the density of electronic energy
states tails into the forbidden zone with the electrons in
these tails being localized. There is then an intermediate
range of electronic energy states in which mobilities are
very low (1×10^{-4} m^2/V sec) and where the overlap is so poor

S.P. Hersh et al.

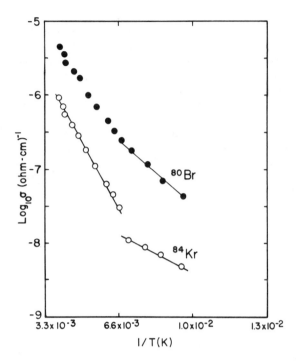

Figure 6. Temperature dependence of conductivity σ of the implanted region as a function of 1/T for Br[+] and Kr[+] ions implanted in PPS at an energy of 200 keV and a fluence of 1×10^{16} ions/cm^2. Reproduced with permission [15].

that it is no longer meaningful to use the band theory [66]. Only when electrons are excited to higher energy states, in which mobilities are also higher, can appreciable conduction occur.

Conduction by way of localized electrons implies discrete jumps across an energy barrier from one site to the next, and this allows the electron either to hop over or tunnel through the potential barrier. The relative importance of these two mechanisms depends on the shape of the barrier and the availability of thermal energy.

Using a model of hopping conductivity between localized states, Mott [61] predicted that the conductivity of amorphous semiconductors as a function of temperature follows the relation:

$$- \log \sigma \propto (1/T)^{1/4} \tag{9}$$

Sheng et al [67] noted that many disordered materials follow such a dependence, but that many others, such as granular metals and some disordered semiconductors, follow the relation:

$$- \log \sigma \propto (1/T)^{1/2} \qquad (10)$$

They demonstrated that the inverse square root dependence of $\log \sigma$ on temperature could be derived from a model, known as the grain model, based on charge transport due to hopping of carriers between isolated conducting islands separated by an insulating matrix.

Other authors [37,56,68] have identified a class of disordered one-dimensional conductors in which charge transport occurs via a variable-range hopping of carriers. In this case the temperature-dependent conductivity follows the relation:

$$\sigma = \sigma_0 \exp(-T_0/T)^{1/2} \qquad (11)$$

where $T_0 = 4\alpha N(E_f)/k$,
$\quad \alpha =$ the coefficient of exponential decay of the localized states,
$N(E_f) =$ the density of states at the Fermi level,
$\quad k =$ the Boltzmann constant, and
$\quad \sigma_0 =$ the conductivity as $T \to \infty$.

As observed by Dresselhaus et al [56], σ_0 is highly fluence dependent.

The same functional form is applicable for the conducting grain model. According to the grain model,

$$T_0 = 4\chi wE/k \qquad (12)$$

where $w =$ the width of the nonconducting barriers between grains,
$\quad \chi = (2m^*\psi/\hbar^2)^{1/2}$,
$\quad m^* =$ the effective mass,
$\quad \psi =$ the barrier height over which the charge hops,
$\quad E =$ the electrostatic energy required to move the charge from an uncharged island to another adjacent initially uncharged island, and
$\quad \hbar = h/2\pi$ where h is Planck's constant.

Thus, conductivity by a hopping mechanism in ion-implanted polymers has been explained either by hopping of carriers between isolated conducting grains (Equations 10,12) or by

the one-dimensional thermally activated hopping in which the
electronic properties are determined mainly by structural
disorder (Equations 9 and 11).

Such hopping has a power law temperature dependence of
$\ln \alpha \propto T^{-m}$ with $1/4 < m < 1/2$ whereas thermally activated
conduction has an Arrhenius activation temperature dependence
of $\log \sigma \propto E_a/kT$, ie, $m = 1$. Thus, by fitting the
exponential values of $\ln \sigma$ and T to the general power law

$$- \ln \sigma \propto T^{-m} \qquad (13)$$

m can be determined, and accordingly, the conduction
mechanism can be established.

As noted in the preceding section, Abel et al [15]
suggested that at temperatures above 150 K, conduction by
hopping applies to PPS films implanted with Br^+ ions at a
fluence of 1×10^{16} ions/cm^2 and accelerating energy of 200
keV. The values obtained for m ($0.2 < m < 0.6$) in the
general power law $\ln \sigma \propto T^{-m}$ were consistent with this type
of conduction mechanism. The discontinuity near 150 K in
Figure 6 was attributed to a phase transition.

The one-dimensional, variable-range hopping model was also
proposed by Wasserman et al [37] for PAN, PPO and PPS films
implanted with Br^+ ions in a fluence range between 3×10^{14} and
3×10^{16} ions/cm^2 and acceleration energies between 50 and 250
keV. Each of the exposed materials becomes highly disordered
due to chain breakage, double bond formation, and generation
of free radicals and free carriers. Thus, they concluded
that the charge transfer in this material occurs by a hopping
mechanism along, between, and across the chain fragments.

Forrest et al [34] found the conducting grain model based
on the Sheng model [67] to apply to high-energy implantations
(eg, 2 MeV). The temperature dependence between 10 and 360 K
of resistivity of PTCDA implanted with Ar^+ at three fluences
is shown in Figure 7. For doses between 5×10^{14} and 1×10^{16}
ions/cm^2, the data fit Equation 11. They noted, based on the
observed temperature dependence ($m = 0.5$), that the data
would also support a mechanism of carrier hopping between
localized states in a one-dimensional conductor [71]. Since
the materials they used, PTCDA, NTCTA, and NiPc, do not link
into long one-dimensional chains, however, they did not
accept the hopping mechanism. At implantation doses less
than 1×10^{14} ions/cm^2 the resistivity followed the Arrhenius
equation (see section 4.4).

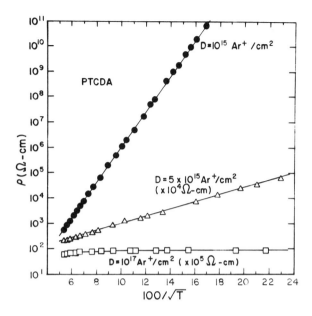

Figure 7. Resistivity ρ vs 1/T for a 3000 Å thick PTCDA film implanted at fluences of 1×10^{15}, 5×10^{15}, and 1×10^{17} ions/cm^2 with 2 MeV Ar$^+$ beams. Reproduced with permission [34].

Brock [21] measured the temperature dependence of PET films implanted with F$^+$, Ar$^+$ and As$^+$ at fluences of 1×10^{16} ions/cm^2 over the range of 80 to 300 K and fit them to the general power law (Equation 13).

The data are shown in Figures 8a and b for the PET film implanted with 8×10^{15} F$^+$/cm^2. The exponent m had values of 1/2 for the F$^+$ and Ar$^+$ implanted films over the range 180 to 300 K and for the As$^+$ implanted film over the range 80 to 170 K. The exponents had values close to zero over the rest of the temperature range. A statistical analysis of the data, however, showed that the conductivity-temperature relationship fit power laws in which m was assigned the value of 1/4, 1/2 or 1 (the Arrhenius law) equally well (except for the As$^+$ implanted sample from 80 K to 170 K in which the fit with m = 1/2 was significantly better. Thus he concluded that it can be difficult to distinguish between σ vs T power laws having exponents between 1/4 and 1. Such a conclusion is perhaps not too dificult to visualize from the data presented in Figures 8a and 8b.

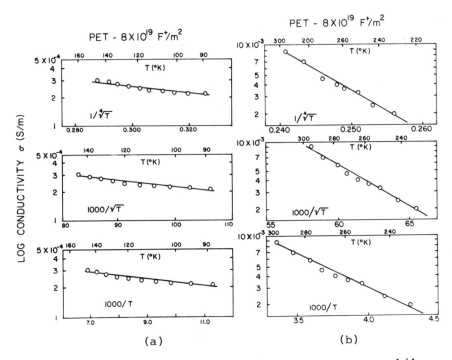

Figure 8. Plots of ln σ vs 1000/T, ln 1000 √T̄ and 1/T$^{1/4}$ for PET films implanted with 8x10^{15} F$^+$ ions/cm^2: (a) temperature range 80-170 K, (b) temperature range 200-300 K [21].

4.6 Quantum Mechanical Tunneling

This type of conduction occurs in lattices characterized by potential wells separated by barriers of the same or different heights and widths. In polymers, which are less ordered structures than crystals, the lowest energy states in a band are localized. The transfer of charges between molecules separated by a potential barrier without actually acquiring the energy to pass over the barrier is known as quantum mechanical tunneling through the barrier from one localized site to another [64,67,69]. The temperature dependence of conductivity by this model is identical to that of Equation 11.

Forrest et al [34] considered the possibility of explaining their measurements made on Ar$^+$ implanted PTCDA and NiPc (see Figure 7 and Section 4.5) on a tunneling conduction mechanism. They concluded, however, that the Sheng model for hopping between isolated, conducting islands was the more likely possibility because they thought the electric fields in their measurements were too low to support tunneling. In addition, a tunneling mechanism should lead to non-ohmic behavior which did not occur.

4.7 Solitons

The term "soliton" was coined to describe a hypothetical pulse-like charge carrier associated with the solution of a wave equation. The soliton is characterized as a nonlinear wave (solitary wave) which emerges unchanged in shape and speed from a collision with a similar pulse. A detailed description of soliton theory is beyond the scope of this paper. Further insight into the nature of the theory can be gained from the review by Scott et al [70].

The existence of solitons is important to explain electrical conductivity in some polymers [71], especially polyenes. In polyacetylene, solitons are considered to be defects or kinks in the alternating carbon-carbon bonds along the chain in which alternations in bond lengths to the left and to the right are out of phase with each other [2].

4.8 Percolation

This mechanism refers to the coalescing of small dispersed conducting elements in an insulating medium to form sufficient interconnections for the system to become conducting [2]. Such a percolation phenomenon exhibits a sharp conductivity threshold transition [72]. In $(CH)_x$ the transition could be the result of a percolation phenomenon in which the $(CH)_x$ chains constitute essentially unconnected metal wires as a result of doping.

Percolation theory is described in detail by Kirkpatrick [72].

4.9 Intercalation and Carbonation

Graphite intercalation compounds are formed by insertion of atomic or molecular layers of a different chemical species between layers in a graphite host material [73]. The intercalation process occurs in highly anisotropic layered structures where the interplanar binding forces are small compared with the intraplanar binding forces. Many types of intercalation agents can be used ranging from simple ionic species such as alkali metals to halogens and to even larger organic molecules [73]. The possibility of utilizing ion implantation to introduce foreign species into graphite hosts

to form intercalation compounds has been considered by
several investigators [73,74].

5.   Summary

5.1   Electrical Measuring Techniques and Conduction
Mechanisms Proposed

Not all the studies of ion implantation in polymers
reported explicitly the measuring technique employed. Only a
few discuss the nature of the conduction mechanism. A
summary indicating the technique used to measure conductivity
and the conduction mechanism suggested for all studies of ion
implantation on polymers is presented in Table 4.

5.2   Common Features

All the studies of ion-implanted polymers have found that
the electrical conductivity is increased by several orders of
magnitude. This enhancement depends on many factors, the
most important of which are ion type and fluence. All
authors also reported that ion implantation causes some
damage to the implanted surface, induces disorder in the
pristine polymer, and leads to molecular rearrangement.

It was commonly reported that polymers change color upon
ion implantation. The change increases with increasing dose,
and the phenomenon has been attributed to beam-induced
carbonization which results in a highly crosslinked carbon
structure [33]. The existence of a crosslinked structure was
observed in several other studies [14,15,17,21,25]. Another
general conclusion is that polymers modified by ion
implantation are stable towards oxidation and retain their
enhanced conductivity after several months.

5.3   Features Common to Halogens

It was suggested [14,15] that the increased conductivity
obtained by implanting halogen ions in PPS films at fluences
up to 1×10^{16} ions/cm^2 results from bond breakage and new
bonding arrangements with the following sequence of
conductivities: $\sigma(Cl) > \sigma(I) > \sigma(Br) > \sigma(F)$. With the
exception of fluorine, these results are consistent with the
results for halogen-implanted polyacetylene [24]. At
fluences above 1×10^{16}, the conductivity of PPS implanted with
Br^+ continued to increase (unlike Kr^+ and As^+ implantations
which plateaued at this fluence). For this reason it was
suggested that the increased conductivity obtained by adding

Table 4. Electrical Conductivity Measurement Techniques and Conduction Mechanisms Reported for Ion-Implanted Polymers

Ion Implanted	Polymer[a]	Implantation Conditions		Measurement Technique[b]	Conduction Mechanism[c]	Refs.
		Energy (keV)	Fluence (cm)$^{-2}$			
F$^+$	(CH)$_x$	25	$3.1 \times 10^{17} - 4 \times 10^{19}$	—	—	24
		10-40	$1 \times 10^{15} - 3 \times 10^{17}$	—	—	30
		25	3×10^{17}	—	—	18
		10-60	$1 \times 10^{14} - 1 \times 10^{18}$	FPP	—	35
F$^+$	(CH)$_x$, PPS, COPOL	25	1×10^{17}	—	—	31
	PPS	60	1×10^{16}	PIE	—	14,15
Br$^+$	(CH)$_x$	25-41.6	$3 \times 10^{15} - 3.4 \times 10^{16}$	FPP	—	16
	(CH)$_x$	10-40	$1 \times 10^{15} - 3 \times 10^{17}$	—	—	30
	(SN)$_x$	25-46	5×10^{15}	FPP	—	16
^{80}Br	PPS	100,150	1×10^{16}	PIE	SCLC	15
	PPS	200	$3.2 \times 10^{14} - 3.2 \times 10^{16}$	PIE	H,T[d]	14,15
Cl$^+$	(CH)$_x$	10-40	$1 \times 10^{15} - 3 \times 10^{17}$	—	—	30
	PPS	100	1×10^{16}	PIE	—	14,15
I$^+$	(CH)$_x$	10-40	$1 \times 10^{15} - 3 \times 10^{17}$	—	—	30
	PPS	260	1×10^{16}	PIE	—	14,15
F$^+$, Ar$^+$	PET	50	1×10^{16}	FECP	SCLC	21

As^+	PET	50	$1x10^{16}$	FECP	SCLC, H[e]	21
F^+,Ar^+,As^+	PET	50	$1x10^{14}$-$5x10^{15}$	FECP	Ohmic	21
F^+	N-66,PTFE,PP	50	$5x10^{15}$	FECP	Ohmic	21
Pd^+	$(CH)_x$	25-90	$3x10^{15}$-$3.4x10^{16}$	FPP	-	16
^{84}Kr	PPS	100	$3.2x10^{14}$-$3.2x10^{16}$	PIE	SCLC	14,15
		150,200	$1x10^{16}$	PIE	T	15
^{75}As	PPS	100	$3.2x10^{14}$-$3.2x10^{16}$	PIE	SCLC	14
Ar^+	PTCDA,NiPc,NTCDA	2000	$1x10^{14}$-$1x10^{17}$	FPP[f]	H[g]	34
	PVC,HPR-204,PMMA COP,PIQ	2000	$1x10^{14}$-$1x10^{17}$	-	H	17
^{14}N,^{80}Br	PAN,PPS	25-230	$3.2x10^{14}$-3.210^{16}	PIE/TPP	H	56
N,As,Br,Kr	PAN	35-230	$3.2x10^{14}$-$3.2x10^{16}$	PIE/TPP	H	37

[a]See Glossary.

[b]FECP: Flat electrodes, conducting Ag paint
FPP: Four-point probe
PIE: Planar interdigitated electrodes
TPP: Two-point probe

[c]H: Hopping
T: Thermally activated

[d]Depending on temperature

[e]80 to 170 K only

[f]TPP for fluences < $1x10^{14}/cm^2$

[g]T for PTCDA and NiPc at fluences < $1x10^{14}/cm^2$

Br^+ above $1x10^{16}$ ions/cm^2 resulted from "specific chemical doping" rather than bombardment damage.

5.4 Features Common to Other Ions

Even knowing that, at least theoretically, any chemical element can be implanted in the polymeric substrate, only a few ions other than halogens have been employed (^{4}He, ^{14}N, ^{40}Ar, ^{75}As, ^{84}Kr, ^{106}Pd, ^{127}I). The choice of arsenic was motivated by its potential effectiveness as an electron donor. Helium, argon, and krypton, since they are inert gases, were generally used because they would not produce any specific chemical bonding with the substrate; they should produce only bombardment-induced disorder. Unexpectedly, however, Mazurek et al [14] found that implanting Kr^+, an inert element, into PPS produces conductivities similar to those obtained when the more reactive As^+ ions are implanted in the same polymer under the same conditions.

Also, the conductivities of polymers [17,21] and organic thin films [34] were increased by implanting them with Ar^+ ions. Hence, again, it was demonstrated that conductivities can be enhanced as a result of mechanical damage produced by bombardment with an ion of an inert element as indicated in Figure 9. The increase in conductivity observed by Brock [21], however, was about one order of magnitude less than that obtained with As^+. Presumably the number of charge carriers is increased as a result of broken bonds [37]. Venkatesan et al [17] also affirm that the enhanced conductivity seems to be a general phenomenon true for many carbon-containing films other than the organic anhydrides and phthalocyanines studies by Forrest et al [34].

5.5 Features Common to Fluence

In many studies the conductivity of several polymers was found to increase upon ion implantation by as many as 13 orders of magnitude when the implantation fluence increased only two orders of magnitude from $1x10^{14}$ to $1x10^{16}$ ions/cm^2. Under and above those critical values the doses do not seem to greatly influence the conductivities. Examples are shown in Figures 1, 9 and 10.

At higher fluences (>$1x10^{16}$/cm^2) Mazurek et al [14], Dresselhaus et al [56], and Wasserman et al [37] state that the conductivity appears to become saturated. Mazurek et al [14] did observe an exception to this general rule,

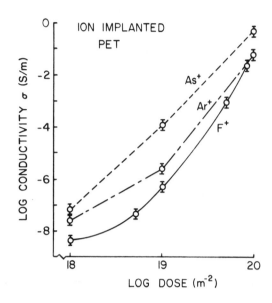

Figure 9. Conductivity as a function of fluence (dose) of
PET films implanted with F$^+$, Ar$^+$, and As$^+$ ions at a beam
energy of 50 keV [21].

however, for PPS film implanted with Br$^+$. As can be seen in
Figure 1, the conductivity continues to rise at a nearly
linear rate at fluences above 1×10^{16} ions/cm^2, an increase
which they attributed to the occurrence of specific chemical
doping following bombardment damage at lower fluences.

It was also noted [17] that extremely high doses
($>1 \times 10^{17}$/cm^2) cause a decrease in film thickness (~50%), at
least for PVC and HPR-204 films implanted with Ar$^+$ at a very
high accelerating energy of 2 MeV.

5.6 Implantation Temperature

Brock [21] reported that the conductivity of PET films
implanted with F$^+$ and Ar$^+$ at fluences near 1×10^{20} ions/m^2 at
77 K was approximately three orders of magnitude less than
similar films implanted at room temperature. The decrease in
conductivity of the samples implanted at liquid nitrogen
temperatures was attributed to the reduced chain damage
expected when the impinging ions have a greater chance of
channeling through the polymer structure. In contrast,
implantation at room temperature would tend to favor higher

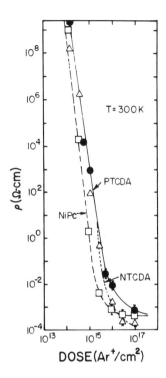

<u>Figure 10.</u> Dependence of room temperature resistivity ρ on dose of Ar^+ ions implanted at 2 MeV in thin films of PTCDA (closed circles), NTCDA (triangles) and NiPc (squares). Reproduced with permission [34].

conductivities since the polymer temperature would probably rise above its T_g (temperature at which polymer chains in noncrystalline regions become mobile) as a result of the heating effects of the ion bombardment. Under these conditions a number of molecular rearrangements would be produced and increase the availability of carriers for conduction.

6.   Glossary (polymer codes)

$(CH)_x$:   Polyacetylene

COP   :   Poly(ethyl acrylate-co-glycidyl acrylate)

COPOL   :   Copolymer of ethylene and methacrylic acid

HOPG   :   Highly oriented pyrolitic graphite

HPR-204:   Hunt positive resist

N-66   :   Poly(hexamethylene adipamide) (nylon 66)

NiPc   :   Nickel phthalocyanine

NTCDA   :   1,4,5,8-napthalenetetracarboxilic dianhydride

PAN   :   Polyacrylonitrile

PET   :   Poly(ethylene terephthalate)

PIQ   :   Polyimide

PMMA   :   Poly(methyl methacrylate)

PP   :   Polypropylene

PPO   :   Poly(2,6-dimethyl phenylene oxide)

PPP   :   Poly(p-phenylene)

PPS   :   Poly(p-phenylene sulfide)

PTCDA   :   3,4,9,10-perylenetetracarboxilic dianhydride

PTFE   :   Polytetrafluoroethylene

PVC   :   Poly(vinyl chloride)

$(SN)_x$:   Poly(sulfur nitride)/(polythiazyl)

References

[1] Hersh, S.P. In "Surface Characteristics of Fibers and Textiles, Part I"; Schick, M.J., Ed.; Dekker: New York, 1975; Chapter 6.

[2] Mort, J. Science 1980, 208, 819.

[3] Mort, J. Adv. Phys. 1980, 29, 367.

[4] Gundlach, R.W. Conference on Electrostatics, Electrostatics Society of America, Troy, NY 1983.

[5] Seymour, R.B. Polym. Plast. Technol. Eng. 1983, 20, 61.

[6] Anonymous. Chem. Eng. News 1983, 61(17), 18.

[7] Verdin, D. J. Electrostatics 1982, 11, 249.

[8] Gabor, A. Business Week 1985, 2890 (April 15), 150F.

[9] Jenekhe, S.A. 193rd ACS National Meeting, I&EC Div., Denver, CO, USA 1987, Paper 84.

[10] Haggin, J. Chem. Eng. News 1987, 65(18), 27.

[11] Usuki, A. 193rd ACS National Meeting, I&EC Div., Denver, CO, USA 1987, Paper 85.

[12] Chance, R.R.; Shacklette, L.W.; Miller, G.G.; Ivory, D.M.; Sowa, J.M.; Elsenbaumer, R.L.; Baughman, R.H. J. Chem. Soc. Chem. Commun. 1980, 348.

[13] Rabolt, J.F.; Clarke, T.C.; Kanazaua, K.K.; Reynolds, J.R.; Street, G.B. J. Chem. Soc. Chem. Commun. 1980, 347.

[14] Mazurek, H.; Day, D.R.; Maby, E.W.; Abel, J.S.; Senturia, S.D.; Dresselhaus, M.S.; Dresselhaus, G. J. Polym. Sci. Polym. Phys. Ed. 1983, 21, 537.

[15] Abel, J.S.; Mazurek, H.; Day, D.R.; Maby, E.W.; Senturia, S.D.; Dresselhaus, G.; Dresselhaus, M.S. In "Metastable Materials Formation by Ion Implantation"; Proceedings of the MRS Symposium on Ion Implantation; Picraux, S.T.; Choyke, W.J., Eds.; North-Holland: New York, 1982; Vol. 7, 173.

[16] Allen, W.N.; Brant, P.; Carosella, C.A.; DeCorpo, J.J.; Ewing, C.T.; Saafeld, F.A.; Weber, D.C. J. Synth. Met. 1980, 1, 151.

[17] Venkatesan, T.; Forrest, S.R.; Kaplan, M.L.; Murray, C.A.; Schmidt, P.H.; Wilkens, B.J. J. Appl. Phys. 1983, 54, 3150.

[18] Weber, D.C.; Brant, P.; Carosella, C.A. In "Metastable Materials Formation by Ion Implantation"; Proceedings of the MRS Symposium on Ion Implantation; Picraux, S.T.; Choyke, W.J., Eds.; North-Holland: New York, 1982; Vol. 7, 167.

[19] Parker, T.; Kelly, R. In "Ion-Implantation in Semiconductors and Other Materials"; Crowder, B.L., Ed.; Plenum Press: New York, 1978; 551.

[20] Brock, S.; Hersh, S.P.; Grady, P.L.; Wortman, J.J. J. Polym. Sci. Polym. Phys. Ed. 1984, 22, 1349.

[21] Brock, S. "Enhancement of Electrical Conductivity of Ion-implanted Polymer Films"; Ph.D. Dissertation, North Carolina State University, Raleigh, NC, 1985.

[22] Stephen, J.; Grimshaw, J.A. In "Ion-Implantation"; Eisen, F.H.; Chadderton, L.D., Eds.; Gordon and Breach Science Publishers Ltd.: London, 1971; 377.

[23] Dearnaley, G.; Freeman, J.H.; Nelson, R.S.; Stephen, J.; "Ion-Implantation"; North-Holland: Amsterdam, 1973.

[24] Weber, D.C.; Brant, P.; Resing, H.A. 183rd ACS National Meeting, Polym. Chem. Div., Las Vegas, NV, USA 1982, Paper 99.

[25] Hall, T.M.; Wagner, A.; Thompson, L.F. J. Appl. Phys. 1983, 53, 3997.

[26] Billmeyer, F.W., Jr. "Textbook of Polymer Science, 2nd. Ed."; Wiley-Interscience: New York, 1971.

[27] Clarke, T.C.; Geiss, R.H.; Kwak, J.K.; Street, G.B. Chem. Commun. 1978, 489.

[28] Shirakawa, H.; Ikeda, S. Polym. J. 1971, 2, 231.

[29] Rubinstein, I. J. Polym. Sci. Polym. Chem. Ed. 1983, 21, 3035.

[30] Weber, D.C.; Brant, P.; Carosella, C.; Banks, L.G. J. Chem. Soc., Chem. Commun. 1981, 522.

[31] Bernett, M.K.; Ravner, H.; Weber, D.C. Memorandum Report (NRL) 1983, 5016.

[32] Iida, Y.; Okabayashi, H.; Suzuki, K. Japanese J. Appl. Phys. 1977, 16, 1313.

[33] Hashimoto, T.; Koguchi, T.; Okuyama, Y.; Yamamoto, K.; Takahata, K.; Kamoshida, M.; Yanagawa, T. IEDM Tech. Dig. 1976, 198.

[34] Forrest, S.R.; Kaplan, M.L.; Schmidt, P.H.; Venkatesan, T.; Lovinger, A.J. Appl. Phys. Lett. 1982, 41, 708.

[35] Weber, D.C.; Brant, P.; Saafeld, F.E.; Carosella, C. U.S. Patent Application Serial No. 389827, Dept. of Navy, 18 June 1982.

[36] Ivory, D.M.; Miller, G.G.; Sowa, J.M.; Schacklette, L.W., Chance, R.R.; Baughman, R.H. J. Chem. Phys. 1979, 71, 1506.

[37] Wasserman, B.; Braunstein, G.; Dresselhaus, M.S.; Wnek, G.E. In "Ion Implantation and Ion Beam Processing of Materials"; Proceedings of the MRS Symposium on Ion Implantation and Ion Beam Processing of Materials; Hubler, G.K.; Holland, O.W.; Clayton, C.R.; White, C.W., Eds.; North-Holland: New York, 1984; Vol. 27, 423.

[38] Vogel, F.L.; Fuzellier, H.; Zeller, C.; McRae, E.J. Bull. Am. Phys. Soc. 1979, 23, 220.

[39] Wilson, R.G.; Brewer, G.R. "Ion Beams"; John Wiley: New York, 1973.

[40] Hirvonen, J.K. In "Treatise on Materials Science and Technology, Ion Implantation"; Hirvonen, J.K., Ed.; Academic Press: New York, 1980; Vol. 18, 1.

[41] Rodriguez, F. "Principles of Polymer Systems, 2nd. Ed."; McGraw-Hill: New York, 1982.

[42] Liau, Z.L.; Mayer, J.W. In "Treatise on Materials Science and Technology, Ion Implantation"; Hirvonen, J.K., Ed.; Academic Press: New York, 1980; Vol. 18, 17.

[43] Elman, B.S.; Dresselhaus, M.S.; Dresselhaus, G.; Maby, E.E.; Mazurek, H. Phys. Rev. B 1981, 24, 1027.

[44] Mayer, J.W.; Eriksson, L.; Davies, J.A. "Ion Implantation in Semiconducers"; Academic Press: New York, 1970.

[45] Davies, J.A.; Sims, G.A. Can. J. Chem. 1961, 39, 601.

[46] Bohr, N. Kgl. Danske Videnskab. Selskab, Mat.-Fys. Medd. 1948, 18, 8.

[47] Lindhard, J.; Scharff, M.; Schiott, H.E. Kgl. Danske Videnskab. Selskab, Mat.-Fys. Medd. 1963, 33, 14.

[48] Nielsen, K.O. In "Electromagnetically Enriched Isotopes and Mass Spectroscopy"; Smith, M.L., Ed.; Academic Press: New York, 1956; 68.

[49] Northcliffe, L.C.; Schilling, R.F. Nucl. Data Tables 1970, A7, 233.

[50] Gibbons, J.F.; Johnson, N.S.; Mylroie, W. "Projected Range Statistics"; Halstead: Stanford, CA, 1975.

[51] Hall, T.M.; Wagner, A.; Thompson, L.F. J. Appl. Phys. 1982, 6, 53.

[52] Brice, D.K. Sandia Laboratories Research Report, SAND75-0622 1977.

[53] Anderson, H.H.; Ziegler, J.F. "Hydrogen Stopping Powers and Ranges in All Elements"; Pergamon Press: New York, 1977.

[54] Weber, D.C. Private Communication 1983.

[55] Land, D.; Brennan, J. Atomic Data and Nuclear Tables 1978, 22, 235.

[56] Dresselhaus, M.S.; Wasserman, B.; Wnek, G.E. In "Ion Implantation and Ion Beam Processing of Materials"; Proceedings of the MRS Symposium on Ion Implantation and Ion Beam Processing of Materials; Hubler, G.K.; Holland, O.W.; Clayton, C.R.; White, C.W., Eds.; North-Holland: New York, 1980; Vol. 27, 413.

[57] Banks, L.G.; Resing, H.A.; Weber, D.C.; Carosella, C.; Myller, G.R.; Brant, P. J. Phys. Chem. Solids 1982, 43, 351.

[58] Seanor, D.A. In "Electrical Properties of Polymers"; Seanor, D.A.; Ed.; Academic Press: New York, 1982; Chapter 1.

[59] Grady, P.L.; Hersh, S.P. Rev. Sci. Instrum. 1975, 46, 20.

[60] Gogolin, A.A. Phys. Reports 1982, 86, 1.
[61] Mott, N.F. Phil. Mag. 1969, 19, 835.
[62] Amborsky, L.E. J. Polym. Sci. 1962, 62, 331.
[63] Kryszewski, M. "Semiconducting Polymers"; Polish
     Scientific Publishers: Warszawa, 1980.
[64] Cotts, D.B.; Reyes, Z. "Electrically Conductive Organic
     Polymers for Advanced Applications"; Noyes Data Corp.:
     Park Ridge, NJ, 1986; Chapters 7,8.
[65] Setter, G. Kolloid Z. 1967, 215, 112.
[66] Blythe, A.R. "Electrical Properties of Polymers";
     Cambridge Univ. Press: Cambridge, Great Britain, 1979.
[67] Sheng, P.; Abeles, B.; Arie, Y. Phys. Rev. Lett. 1973,
     31, 44.
[68] Ambegaokar, V.; Helperin, B.I.; Langer, J.S. Phys.
     Rev. B 1971, 4, 2612.
[69] Zallen, R. "The Physics of Amorphous Solids"; John
     Wiley: New York, 1983.
[70] Scott, A.C.; Chu, F.Y.F.; McLaughin, D.W. Proc. IEEE
     1973, 61, 1443.
[71] Chien, J.C.W. In "Macromolecules, IUPAC"; Benoit, H.;
     Rempp, P., Eds.; Pergamon Press: Oxford, 1982; 233.
[72] Kirkpatrick, S. Rev. Mod. Phys 1973, 45, 574.
[73] Dresselhaus, M.S. In "Theoretical Aspects and New
     Developments in Magnetooptics"; NATO, Adv. Study. Inst.
     Ser. B; 1980; B60, 101.
[74] Dresselhaus, M.S.; Dresselhaus, G. In "Physics and
     Chemistry of Materials with Layered Structures"; Levy,
     F., Ed.; Reidel: Dordrecht, Holland, 1979; Vol. 6, 423.
[75] Jespersgard, P.; Davies, J.A. Can. J. Phys. 1967, 45,
     2983.

ENHANCEMENT OF ELECTRICAL CONDUCTIVITY OF POLYMER FILMS BY ION IMPLANTATION

Sylvio Brock, S. P. Hersh, and P. L. Grady
Department of Textile Engineering and Science

and

J. J. Wortman
Department of Electrical and Computer Engineering
N. C. State University, Raleigh, NC 27695 USA

Abstract

The conductivity of films of nylon 66, poly-propylene, poly(tetrafluoroethylene) (Teflon®) and poly(ethylene terephthalate) implanted with F^+, Ar^+ or As^+ ions increases by 7 to 14 orders of magnitude above that of the pristine films as the ion dose increases from 1×10^{18} to 1×10^{20} ions/m². The conductivity of films implanted with As^+ was approximately one order greater than those implanted with Ar^+ which in turn was approximately one-half order greater than those implanted with F^+. The conductivity of PET films implanted with F^+ and Ar^+ at fluences near 1×10^{20} ions/m² at liquid nitrogen temperatures was approximately 3 orders of magnitude less than similar films implanted at room temperature. Measurements of the temperature dependence of the conductivity for the PET films implanted with F^+ and Ar^+ over the range from 180 K to 300 K and for PET films implanted with As^+ from 80 K to 170 K indicated that conduction takes place most probably by a one-dimensional variable-range hopping mechanism. Techniques such as Secondary Ion Mass Spectroscopy and Scanning Electron Microscopy, among others, were employed to better understand the nature of the interaction of the energetic incident ions with the polymer substrate.

1. Introduction

Until recently, most synthetic and natural polymers were electrical insulators and exhibited many problems such as

producing electrostatic shocks, setting off explosions of
flammable vapors and solids, triboelectrification, soiling,
attraction of dust, etc. By adding internal or surface
antistatic agents, the polymers could be rendered temporarily
or permanently more conductive but not without sacrificing
some of their mechanical properties [1]. During the past
decade or so, electroactive polymers having conductivities in
the semiconductor range have been developed [2]. These
polymers became even more conductive, approaching that of
mercury, by chemically doping with arsenic pentafluoride and
iodine (electron acceptors) and sodium napthalide (electron
donor). However, conductivity of the doped polymers
decreases with loss of the dopant [3,4]. More recently, ion
implantation, a technique well known in the semiconductor
field, has been used to increase the conductivity of
electroactive polymers [5-8].

2. Experimental

Only a few of the traditionally electrically insulating
polymers have been subjected to ion implantation. These
include poly(vinyl chloride), poly(methyl methacrylate),
polyacrylonitrile and poly(ethylene terephthalate)
(PET) [9-11]. The purpose of this paper is to report the
results of ion implantation on four insulating polymers:
PET; poly(hexamethylene adipamide), nylon 66 (N-66);
polypropylene (PP); and polytetrafluoroethylene (Teflon®)
(PTFE).

2.1 Ion Implantation

A general description of the pristine films is given in
Table 1. All samples were cleaned by immersion in carbon
tetrachloride (CCl_4) to remove producer finishes and other
impurities. Thereafter, 3-inch diameter circles were cut
from the cleaned films and mounted in aluminum wafer holders.
Aluminum rings were placed on top of the samples to keep them
flat. These rings also served to reduce heating effects and
minimize carbon buildup.

The holders were then placed at the target site in a
Varian Extrion 400 ion implanter. Implantation was carried
out either at room temperature (RT) or at liquid nitrogen
(LN) temperature (77 K) under a vacuum of 5×10^{-6} torr and 50
keV implantation energy. The PET implantations at LN
temperature were made using a special, almost cubic metallic
holder. Circular film samples, 2 inches in diameter, were
glued to four faces of the holder with a silver conducting

Table 1. Description of Polymer Film Substrates

Films[a]	Manufacturer	Trade Name and/or Type	Density (g/cm^3)	Thickness (inch)	Resistivity $(\Omega-m)$
PET	DuPont	Mylar® 920	1.39	0.001	1×10^{16}[b]
N-66	DuPont	Zytel®	1.15	0.0015	2×10^9[b]
PTFE	DuPont	Teflon®	2.30	0.001	1×10^{17}[b]
PP	Hercules	B-502	0.94	0.00125	$10^{14} - 10^{15}$[c]

[a]PET = Poly(ethylene terephthalate)
  N-66 = Poly(hexamethylene adipamide), Nylon 66
  PTFE = Polytetrafluoroethylene
  PP   = Polypropylene

[b]Values given by the producer

[c]From [22]

paint to reduce the buildup of static charge. The
temperature rise of the samples during implantation could not
be monitored but did not approach the melting point of the
polymer since the samples remained flat after implantation.
The implantation current was kept low (5.4 µA) to minimize
heating effects as well as to reduce carbon buildup during
prolonged implantation.

Three ion species were implanted: F^+, Ar^+, and As^+. F^+
ions were selected for implantation because of their high
electronegativity, Ar^+ because of their chemical inertness,
and As^+ because of their effectiveness as electron donors.
Any increases in electrical conductivity produced by Ar^+
implantation would originate from bombardment-induced
disorder rather than from any specific chemical reaction with
the substrate. The implanted film samples prepared are
listed in Table 2. The films are identified by giving the
polymer code, the ion implanted, and the implantation dose
(in exponential notation), all separated by hyphens.

A knowledge of the mean projected range R_p of the
implanted ions and its statistical standard deviation
(dispersion) ΔR_p is needed in order to calculate the

Table 2. Implantation Conditions, Conductivity, and Linearity of Current I vs Voltage V Curve of Ion-Implanted Films

| Identification Code | Implantation Conditions | | | log σ_v (S/m) | | log σ_b (S/m) | Linearity of I vs V Curve |
	Ion Type	Fluence (ions/m²)	Temp[a]	Mean[b] (S/m)	Standard Deviation	Based on Entire Film	
PET-0	-	0	RT	-	-	-16[c]	-
PET-F-118	F+	1x10¹⁸	RT	-8.33	0.213	-11.01	Linear
PET-F-518	F+	5x10¹⁸	RT	-7.49	0.252	-10.17	Linear
PET-F-119	F+	1x10¹⁹	RT	-6.29	0.248	-8.97	Linear
PET-F-519	F+	5x10¹⁹	RT	-2.98	0.108	-5.66	Linear
PET-F-819	F+	8x10¹⁹	RT	-1.54	0.143	-4.22	Linear up to 3000 V
PET-Ar-118	Ar+	1x10¹⁸	RT	-7.63	0.207	-10.7	Linear above 1000 V
PET-Ar-119	Ar+	1x10¹⁹	RT	-5.58	0.186	-8.65	Linear
PET-Ar-120	Ar+	1x10²⁰	RT	-1.10	0.214	-4.17	Linear up to 3000 V
PET-As-118	As+	1x10¹⁸	RT	-7.20	0.153	-10.6	Linear up to 2000 V
PET-As-119	As+	1x10¹⁹	RT	-3.86	0.130	-7.26	Linear
PET-As-120	As+	1x10²⁰	RT	-0.26	0.084	-3.65	Linear up to 3000 V
PET-F-819-LN	F+	8x10¹⁹	LN	-4.65	0.467	-7.33	Linear
PET-Ar-120-LN	Ar+	1x10²⁰	LN	-3.43	0.127	-6.50	Linear
N66-0	-	0	RT	-	-	-9.3[c]	-
N66-F-519	F+	5x10¹⁹	RT	-5.47	0.200	-8.26	Linear
PP-0	-	0	RT	-	-	-14.7[d]	-
PP-F-519	F+	5x10¹⁹	RT	-6.32	0.266	-9.08	Linear
PTFE-0	-		RT	-	-	-17[c]	-
PTFE-F-519	F+	5x10¹⁹	RT	-7.63	0.461	-10.28	Linear

[a] RT = Room temperature, LN = Liquid nitrogen temperature, 77 K

[b] Four specimens

[c] From producer

[d] From [22]

electrical conductivity of the implanted region. These
values can be calculated theoretically or measured
experimentally by Secondary Ion Mass Spectroscopy (SIMS).
Calculations based on the LSS theory [12] provided by
Weber [13] are reported in Table 3. These data show, as
expected, that ion penetration R_p and ΔR_p are inversely
related to the mass of the implanted ions. If other factors
remain constant, the broader dispersion of the F^+ ions will
produce a lower calculated conductivity. Because of its
higher mass, it also might be expected that implantation with
As^+ produces greater damage to the films. If it is assumed
that polymer damage per se enhances the electrical conductivity, then implantation of As^+ ions would increase polymer
conductivity more than an equivalent dose of lower mass ions.

2.2 Electrical Measurements

Two rectangular specimens were cut from each control and
implanted film; one sample was 22 mm x 11 mm and the other
was 11 mm x 5.5 mm. The length-to-width ratio of the samples
were held constant so that their conductances would be
identical (assuming their conductivities were the same).
Electrodes were then formed on both sides of each end of the
sample strip by painting a band of silver conducting paint
(Type SC 20, Micro-Circuits Co.) on them so that the distance
between electrodes was 1.5 cm for the longer specimens and
0.75 cm for the smaller ones. Test specimens were maintained
at standard textile testing conditions (21°C ± 1°C,
65% RH ± 2% RH) for 24 hours before making the electrical
measurements.

The resistance of the films was measured using the circuit
described by Grady and Hersh [14]. The sample was mounted in
a Keithley Model 6104 shielded test enclosure as illustrated
in Figure 1. The ends of the test specimen were clamped in
custom-built aluminum holders. A dc potential is applied
across the film, and the resulting current is measured with a
Keithley Model 610 C electrometer. Voltages applied ranged
from 500 to 4,500 V in 500 V increments. Higher potentials
were not applied to avoid electrical breakdown in the
connecting cables. The transient currents were permitted to
dissipate after which the current I was measured.
Experimentally it was observed, as expected, that as
conductivity increased with the higher implanted doses, the
transients became smaller and decayed faster. No transients
occurred at fluences above 10^{19} ions/m^2. Regression lines
were calculated from plots of I as a function of applied
voltage V for each sample. The slope of the regression line
is, of course, the electrical resistance.

Table 3. Projected Range R_p and Dispersion ΔR_p of Ions Implanted at an Energy of 50 keV in Polymer Films [13]

	Ion Properties			Polymer							
		Atomic	Ionic	PET		N-66		PTFE		PP	
Element	Mass (amu)	Radius (Å)	Radius (Å)	R_p (Å)	ΔR_p (Å)	R_p (Å)	ΔR_p (Å)	R_p (Å)	ΔR_p (Å)	R_p (Å)	ΔR_p (Å)
F	19	0.70	0.22	1287	264	1403	258	1102	287	1610	266
Ar	40	1.90	1.52	628	107	704	109	500	109	792	111
As	75	1.28	0.70	423	51	485	54	322	50	577	57

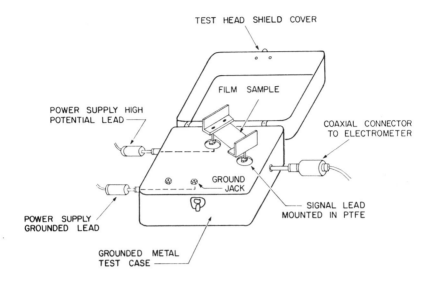

Figure 1. Shielded test enclosure employed in resistance measurements.

The mean volume conductivity within the active region was calculated from the measured resistance by the relation

$$\overline{\sigma}_V = L/2\Delta R_p wR \qquad (1)$$

where L = length of film between electrodes,
     w = film width,
     $2\Delta R_p$ = "effective" thickness of the conductive region, and
     R = film resistance.

Thus, it is assumed that conduction takes place in the implanted region even though its thickness is small compared with the total sample thickness. The "bulk" conductivity σ_b of the sample, ie, the volume conductivity based on the macroscopic thickness of the film, is also reported in Table 2.

2.3 Conductivity vs Temperature

Conductivity as a function of temperature was measured between 80 K and 300 K on the ion-implanted PET samples in order to provide an insight into the conduction mechanism of these films. For these measurements, square samples 10 mm on a side were prepared. Electrodes were painted on the two opposite ends of the ion implanted surface of each sample by applying a 3.5 mm wide band of silver conducting paint.

A bias voltage of 15 V was applied across each sample with an Ambitrol DC power supply model TW 50005, and the resulting current was measured with a Keithley Model 610 C electrometer. The sample, under vacuum, was placed in an Air Products Co. Helitran liquid helium cryostat whose temperature and heating rate were controlled with an AP Model K temperature controller manufactured by Scientific Instruments Co. at a rate usually between 1 and 3 K/min. The system was controlled and monitored by an Adac Computer (LSI 11-23CPV). At a temperature gradient of 3K/min, approximately 1,024 electrometer readings were taken per minute. The conductivity for each measurement was calculated automatically by monitoring the output of the electrometer.

The data obtained were stored on a disk and printed. A continuous conductivity σ vs inverse temperature curve for each sample was plotted with a Hewlett-Packard 7470A plotter along with the results of a least square fit to the power law function

$$\ln \sigma = aT^{-m} \tag{2}$$

where T is the absolute temperature and a and m are constants.

The value of the exponent m of Equation 2 is an important characteristic of the conductivity-temperature relationship and is dependent on the conduction mechanism [2]. For values of m close to 1, the conduction mechanism is thermally activated and follows the Arrhenius equation

$$\ln \sigma/\sigma_0 = - E_a/kT \tag{3}$$

where σ_0 = a temperature-independent constant,
E_a = activation energy for conduction, and
k = Boltzmann's constant (1.38×10^{-23} J/K).

For values of m of 1/2, the conduction mechanism suggested is a one-dimensional variable range hopping model with charge transfer occurring by hopping between, along and across chain fragments [2]. In this case, the temperature-dependent conductivity is given by

$$\ln \sigma/\sigma_0 = - (T_0/T)^{1/2} \tag{4}$$

where $T_0 = 4\alpha N(E_f)/k$,
α = the coefficient of exponential decay of the localized states,
$N(E_f)$ = the density of states at the Fermi level,

k = the Boltzmann constant, and

σ_o = the conductivity as T → ∞.

For a complete understanding of the conduction mechanism, other parameters such as carrier mobility, type and concentration should be known. Values of these parameters can be derived by measuring the Hall effect. An attempt was therefore made to measure the Hall effect on ion-implanted PET films using the procedure described by Buehler and Pearson [15]. However, no Hall effect current could be detected, suggesting that the effective (bulk) conductivity of the samples was too low. Grady and Hersh [16] also reported that the conductivity of antistatic fibers was too low to make Hall effect measurements.

2.4 Secondary Ion Mass Spectroscopy (SIMS)

To understand the influence of the implanted ions on the electrical properties of polymers, it is necessary to determine their locations. This information can be obtained by theoretical and experimental methods. Calculations based on the LSS theory [12] were reported in Table 3. The penetration depth profiles were also measured by SIMS.

Square samples, 10 mm on edge, were prepared for analysis. These samples were mounted on 10 mm square aluminum plates with a silver adhesive and coated with a 200 Å thick gold layer in a Scanning Electron Microscopy (SEM) coating unit (Model E 5100 manufactured by Polaron Instruments Inc.) to minimize charging during ion bombardment. The samples were then fixed to appropriate holders and analyzed in a Cameca IMS 3F ion microanalyzer.

Low primary current (100-180 nA) as well as low doses ($<1 \times 10^{15}$ ions/cm^2) were employed in order to minimize surface damage and prevent carbonization of the carbon-containing substrates. Simko et al [17] reported that these undesirable effects increase noticeably at primary ion fluences above 1×10^{15} ions/cm^2. The implanted films were exposed to a primary beam of Cs$^+$ ions. The dose of the primary ions was approximately 8×10^{14} ions/cm^2.

Once the depth profile has been obtained, the sputtering rate (depth of surface removed per unit of bombardment time at a fixed energy) needs to be obtained in order to determine the actual depth of the implanted ions. Normally, sensitive profilometers are used to determine precisely the depth of the crater created by sputtering away the surface layers during bombardment with the primary ions. Attempts to

measure the crater depth using a Tecnor Instruments
profilometer, however, were unsuccessful because the films
were not completely flat, probably were not uniform in
thickness, and, most seriously, were too soft. Efforts to
measure the depth using incident light interferometer
microscopy were also unsuccessful.

3.   Results and Discussion

3.1   Film Appearance

All the implanted samples, with the exception of PTFE,
acquired a brownish metallic luster varying from a light to a
dark brown hue as the ion fluence increased. The
transparency of the films was also reduced at higher
fluences. The PET-As-120 was the most opaque. All the
changes were reproducible.

The development of color in films upon ion implantation
was attributed by Hashimoto et al [18] and Chapiro [19] to
ion beam-induced lattice disruption and carbonization leading
to a highly cross-linked carbon structure. A recent study
demonstrated that a remarkable decrease in optical
transmission is induced in polymer resists by ion
implantation, making them hard and scratch resistant [18].

3.2   Ambient Conductivity

3.2.1   Effect of polymer type. The results of the
conductivity measurements are reported in Table 2. As shown
in the last column of this table, the current vs voltage
curves for 11 of the 16 implanted films were linear over the
entire voltage range examined (0-4500 V) which indicates that
the conductivity follows Ohm's law. The linearity range of
the other five samples is also reported. The conductivities
of the four polymer types implanted under identical
conditions (50 keV, 5×10^{19} F^+/m^2) are summarized in Table 4.
These data indicate that PET has the highest increase in
electrical conductivity, 13 orders of magnitude. PTFE had
the next greatest increase (9 orders) followed by PP
(8 orders) and N-66 with only 4 orders of magnitude increase.
As shown in Table 4, the electrical conductivities of the
ion-implanted films are much greater than those which can be
achieved by the addition of antistatic agents to fibers [14].

3.2.2   Conductivity as a function of fluence. A log-log
plot of the conductivity of PET samples implanted with F^+,
Ar^+ and As^+ ions as a function of fluence is given in

Table 4. Electrical Conductivity of Polymer Films Obtainable
by Incorporation of Antistatic Agents
and by Ion Implantation

Film	Conductivity, log (S/m)			
	Pristine	Antistatic Agent[a]	Ion Implantation (5x10^19 F^+/m^2)	
			$\log \sigma_v$ [b]	$\log \sigma_b$ [c]
PET	−16	−10.8 to −8.2	−2.98	−5.66
N-66	−9.3	−11.0 to −7.3	−5.47	−8.26
PP	−14.7	−14.64 to −9.97	−6.32	−9.08
PTFE	−17	−	−7.63	−10.28

[a][14]

[b]Based on thickness of conducting layer, $2\Delta R_p$ from Table 3

[c]Based on total film thickness

Figure 2. The 95% confidence intervals shown are for the
means of the conductivities of the four samples measured
(0.208 in log S/m). The conductivity of all three films
increases nearly seven orders of magnitude as the fluence
increases from $1x10^{18}$ ions/m^2 to $1x10^{20}$ ions/m^2. If the
highest of these conductivities is compared with that of the
pristine film, then the increase is even more pronounced and
exceeds 15 orders of magnitude.

3.2.3 Effect of ion type. The order of conductivity
enhancement is $As^+ > Ar^+ > F^+$. As shown in Figure 2,
conductivities of the PET films ion implanted with As^+ are
about one order of magnitude greater than those implanted
with Ar^+. Since Ar is an unreactive noble gas, one would
expect that the enhancement in conductivity of PET is a
consequence of structural changes due to ion implantation-
induced molecular rearrangements rather than to any covalent
bonding. Also, since molecular rearrangements in a target
would increase with increasing mass of the impinging ions and
As^+ ions are more massive than Ar^+ ions (75 amu and 40 amu,
respectively), one would expect an additional increment in
conductivity to be induced by As^+ implantation above that
induced by Ar^+ as indeed was observed. In an analogous

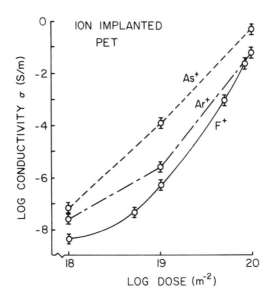

Figure 2. Conductivity as a function of fluence (dose) of PET films implanted with F^+, Ar^+, and As^+ ions at 50 keV.

situation Abel et al [6] and Mazurek et al [5] reported similar relationships for poly(phenylene sulfide) films implanted with As^+ and Kr^+. They suggested that the breakage and reformation of polymer bonds together with the availability of carrier sites facilitated conduction.

Figure 2 also shows that the conductivity of PET films implanted with F^+ ions is nearly 1.5 orders lower than the conductivity of the PET film ion implanted with As^+. This behavior agrees again with the proposed dependence upon the mass of the implanted ion (F^+ = 19 amu, As^+ = 75 amu). However, it is not clear whether the conduction of the F^+ implanted film might surpass that of As^+ implanted films at doses above 10^{20} ions/m^2.

3.2.4 <u>Effect of aging</u>. After measuring the conductivity, one set of implanted films was maintained in a covered container at standard textile testing conditions for up to 15 months. The original conductivities of the films and their conduction after aging are reported in Table 5. The conductivities did not change significantly during this interval. Thus, the increase in conductivity obtained by ion implantation appears to be stable. In contrast, the enhanced conductivity of polymers by chemical doping is relatively unstable [3,4], indicating that the dopant slowly diffuses out of the polymers.

Table 5. Conductivity of the Ion-Implanted Films Before and
After Aging (S/m)

Sample	Aging Times (months)			
	0	3	6	15.6
PET-F-118	4.7×10^{-9}	6.0×10^{-9}	6.8×10^{-9}	–
PET-F-518	32.0×10^{-9}	54.0×10^{-9}	122.0×10^{-9}	61.0×10^{-9}
PET-F-119	510.0×10^{-9}	342.0×10^{-9}	856.0×10^{-9}	–
PET-F-519	1.0×10^{-3}	6.5×10^{-3}	4.4×10^{-3}	3.2×10^{-3}
PET-F-819	37.1×10^{-3}	28.1×10^{-3}	38.1×10^{-3}	37.3×10^{-3}
PET-Ar-118	23.2×10^{-9}	24.6×10^{-9}	27.4×10^{-9}	25.2×10^{-9}
PET-Ar-119	2.6×10^{-6}	1.1×10^{-6}	2.2×10^{-6}	1.8×10^{-6}
PET-Ar-120	77.9×10^{-3}	89.1×10^{-3}	83.4×10^{-3}	79.4×10^{-3}
PET-As-118	63.0×10^{-9}	50.6×10^{-9}	59.1×10^{-9}	56.4×10^{-9}
PET-As-119	138.4×10^{-6}	145.2×10^{-6}	143.8×10^{-6}	139.2×10^{-6}
PET-As-120	546.0×10^{-3}	583.1×10^{-3}	572.2×10^{-3}	–
PET-F-519-LN	22.5×10^{-6}	18.7×10^{-6}	20.4×10^{-6}	19.2×10^{-6}
PET-Ar-120-LN	147.6×10^{-6}	155.4×10^{-6}	151.5×10^{-6}	148.10^{-6}
N66-F-519	3.2×10^{-6}	5.1×10^{-6}	4.4×10^{-6}	–
PP-F-519	0.5×10^{-6}	0.3×10^{-6}	0.7×10^{-6}	–
PTFE-F-519	23.6×10^{-9}	26.8×10^{-9}	24.8×10^{-9}	–

3.2.5 Influence of implantation temperature. Two PET films were implanted at liquid nitrogen temperature (77 K) with an ion beam energy of 50 keV. The samples were identified as PET-F-819-LN and PET-Ar-120-LN. The conductivities of these F^+ and Ar^+ films implanted at 77 K are approximately 1,300 and 210 times lower, respectively, than the films implanted at room temperature as reported in Table 2 and shown in Figure 3. For implantations made at LN temperatures, the conductivity of the Ar$^+$ implanted film is 17 times higher than that of the PET film ion implanted with F$^+$ ions. At implantations conducted at RT, however, the PET film has a conductivity only 2.7 times higher than the conductivity of the F$^+$ implanted film.

These observations may be rationalized as follows. It has been reported that the carbon buildup in the host material during prolonged implantations can be reduced by implanting at low temperatures [7]. Thus, a reduction in conductivity might be expected if the carbon layers do indeed enhance electrical conductivity. In addition, low temperatures restrain chain movements which would decrease chain disruptions and damage, further reducing the conductivity of the samples ion implanted at LN temperatures.

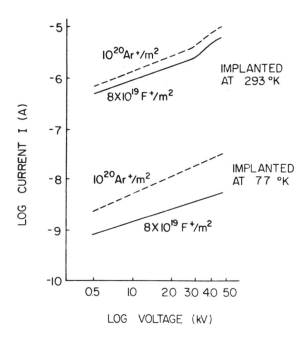

Figure 3. Current-voltage relationship of PET films implanted with 8x10^{19} F$^+$ and 10^{20} Ar$^+$ ions/m^2 at ambient temperature and at 77 K.

3.3   Temperature Dependence of Conduction

3.3.1   Power law dependence and conduction mechanism.   The
conductivity as a function of temperature was measured on the
PET films implanted with F^+, Ar^+ and As^+ at the highest
fluence levels prepared ($\sim 1\times10^{20}$ ions/m^2), As^+ at 1×10^{19}
ions/m^2, and F^+ implanted at 77 K.   The curve obtained for
the PET-F-819 is shown in Figure 4.   Two distinct linear
regions were observed in this film (as well as in PET-Ar-120
and PET-As-120), the first ranging from 80 to 170 K and the
second from 180 to 300 K.   To characterize the data, ten
points were selected from the conductivity-temperature data
and fitted to the power law function (Equation 2) by a
standard least-squares regression analysis.   For curves
having two distinct regions, separate regression lines were
calculated for each region.   Regression lines were also
calculated for Equations 3 and 4.   The calculated regression
parameters are reported in Table 6 together with the
correlation coefficients.   The parameters of the power law
equation based on the ten points were very close to those
calculated automatically by the test instrumentation.

The change in slopes in the σ vs 1/T curves near 180 K
indicates the onset of a transition of some type.

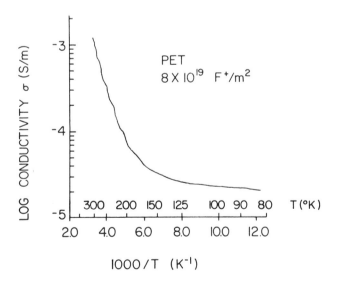

Figure 4.   Conductivity as a function of reciprocal
temperature of PET film implanted with 8×10^{19} F^+ ions/m^2.

Table 6. Power Law (Equation 2), Hopping Model (Equation 4), and Thermal Activation Model (Equation 3) Parameters for Ion-Implanted PET

Film	Temp. Range (K)	Power Law: $\ln \sigma = -aT^{-m}$			Hopping Model: $\sigma = \sigma_0 \exp(-T_0/T)^{1/2}$			Activation Model: $\sigma = \sigma_0 \exp(-E_a/kT)$			log σ @ 21°C, 65% RH (S/m)
		a	m	r_2^a	σ_0(S/m)	T_0(K)	r_4^a	σ_0(S/m)	E_a(eV)	r_3^a	
PET-F-819	80-170	-19.1	0.05	-0.952	0.1×10^{-3}	227	-0.953	0.05×10^{-3}	0.007	-0.941	-1.54
	180-300	-203.6	0.50	-0.996	100	40674	-0.997	0.2	0.14	-0.995	
PET-Ar-120	80-170	-40.9	0.21	-0.965	15×10^{-3}	4733	-0.959	0.7×10^{-3}	0.03	-0.947	-1.10
	180-300	-167.7	0.48	-0.999	60	31117	-0.999	0.2	0.12	-0.997	
PET-As-120	80-170	-111.5	0.49	-0.999	80	13317	-0.999	0.5	0.06	-0.994	-0.26
	180-300	-15.9	0.12	-0.985	150×10^{-3}	924	-0.986	0.6	0.02	-0.990	
PET-As-119	80-200	-17.7	0.05	-0.996	0.6×10^{-3}	296	-0.999	0.28×10^{-3}	0.008	-0.996	-3.86
PET-F-819-LN	80-200	-17.1	0.02	-0.998	0.05×10^{-3}	74	-0.999	0.04×10^{-3}	0.004	-0.996	-4.65
PET-Ar-120-LN	80-200	-11.4	-0.04	—	—	—	—	—	—	—	-3.43

^{a}Correlation coefficient

Abel et al [6] as well as Mazurek et al [5] observed similar behavior in ion-implanted poly(phenylene sulfide) which, however, was not evident in chemically doped films of the same polymer. As shown in Table 6, the slopes in the higher temperature range of 180-300 K for the PET-F-819 and PET-Ar-120 films show a square root power law dependence on temperature (exponent of m = 0.50 and 0.48, respectively); thus they follow Equation 4.

In the lower temperature range 80-170 K, all samples, with the exception of PET-As-120, have a much smaller dependence of conductivity on temperature than they do at higher temperatures. Wasserman et al [11] also found that the conductivity below 10 K was temperature independent but argued that this behavior resulted from a loss in sensitivity of the measuring instrumentation rather than to an intrinsic effect. The shape of the σ vs 1/T curves for PET-As-120 films was sigmoidal having only a small increase in conductivity (<0.01) S/m) from 200 to 300 K. The curve in this range was almost parallel to the 1/T axis indicating a saturation in conductivity at 200 K. The fit of the experimental data between 80-170 K to the $1/T^{1/2}$ model indicates that a hopping conduction mechanism is present in the polymer in this temperature range.

The highest activation energies based on the σ vs 1/T functional dependence shown in Table 6 are 0.12 and 0.14 eV at the higher temperature range for the PET-F-819 and PET-Ar-120, respectively. Since no other similar studies on ion-implanted PET have been reported, it is not possible to compare these values with others. However, the literature does show that the poly(phenylene sulfide) implanted with Br^+ has activation energies between 0.36 and 0.41 eV for T > 150 K [6]. Although these activation energies are about three times those observed here for PET, they are of the same order of magnitude.

The conductivity of PET films implanted with F^+ and Ar^+ at liquid nitrogen temperature was low ($\sim 10^{-4}$ S/m) and varies less than 1×10^{-5} S/m over the temperature range 80-260 K. As shown in Table 6, the m and E_a values for the PET-F-819-LN film are very low (0.02 and 0.004 eV, respectively) confirming the small dependence of conductivity on temperature. As discussed in section 3.2.5, the low values of conductivity of these films implanted at 77 K, even though submitted to high implantation fluences, are consistent with the suggestion that the lower chain mobility during implantation has reduced damage to the polymer structure and thus has not increased the conductivity as much as might be expected at higher implantation temperatures.

An unanticipated observation was that film PET-Ar-120-LN displays a metal-like behavior in which the conductivity decreases with increasing temperature (m has a negative value, -0.04). No explanation for this discrepancy can be offered at this time.

3.3.2 <u>Comparison of temperature-dependent models</u>. From Table 6 it might be concluded that the conductivity-temperature data apparently fit Equations 2, 3 and 4 equally well since the correlation coefficients for each film are high and nearly equal. For this reason, it was decided to determine whether the Arrhenius fit of any of the models provides a significantly better fit of the observed data than do the others. This task was carried out by fitting regression lines to the conductivity data as a function of $1/T$, $1/T^{1/2}$ (Equation 4) and $1/T^{1/4}$, ie, to power laws with m = 1, 1/2 and 1/4, respectively.

The residual mean squares (RMS) of the data fit to the three models and all pairs of their F-ratios are listed in Table 7. The ratio of each pair of residual mean squares, RMS-1, RMS-1/2 and RMS-1/4, which should be distributed as F, were examined to determine whether these ratios are significantly greater than 1. If the ratio of two RMS's are significantly greater than 1, it would indicate that the model with the smaller RMS provided a significantly better fit of the data than did the model having the greater RMS.

For example, for PET-F-819 in the 180-300 K temperature range, the F-ratio of the RMS's comparing the goodness of fit of these three models with each other are 1.52, 1.21 and 1.26, ratios that are likely to occur by chance 28%, 40% and 38% of time, respectively. Since chances of occurrence as low as 5% and 10% are commonly considered to be required before judging two observations to be significantly different from each other, these three models would be judged to be equivalent to each other. At the lower temperature range (80-170 K), the differences between the three models for the PET-F-819 also are not significant. Graphs of these three models with the observed data points for PET-F-819 are shown in Figures 5a and b. These plots clearly verify the conclusion that the data fit all three power laws, ie, ln σ vs $1/T$, $1/T^{1/2}$ and $1/T^{1/4}$, equally well. Thus, in this film it is evident that it is not possible to select a conduction mechanism based on the σ vs T relationship. Perhaps with measurements made over a greater temperature range it might be possible to distinguish between these models.

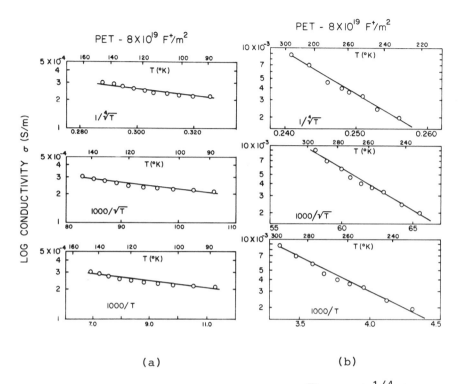

Figure 5. Plots of log σ vs 1000/T, $1000/\sqrt{T}$ and $1/T^{1/4}$ for PET films implanted with 8×10^{19} F$^+$ ions/m^2: (a) temperature range 80-170 K, (b) temperature range 200-300 K.

From the last column in Table 7 (Prob. > F) it is evident that the only film for which one model provides a significantly better fit than the other two is PET-As-120 between 80-170 K for which the $1/T^{1/2}$ model fits the data better than either the 1/T model (99.7% confidence level) or $1/T^{1/4}$ model (99.98% confidence level). The regression lines from these three models for the PET-As-120 data between 80-170 K are shown in Figure 6. Even though significant, a visual inspection of the results suggests that the differences among these three models is trivial indeed.

The only other films indicating a preference among the models (at a 90% probability level or higher) are PET-Ar-120 between 180-300 K and PET-F-819-LN. In the first case, it is suggested that the $1/T^{1/2}$ model fits the experimental data better than does the $1/T^{1/4}$ model. For the PET-F-819-LN film, both the $1/T^{1/2}$ and $1/T^{1/4}$ models fit better than 1/T, but there is no difference between the fits of the $1/T^{1/2}$ and $1/T^{1/4}$ models.

Table 7. Comparison of Regression Models of ln σ vs T^{-m} with m = 1, 1/2 and 1/4 for Ion-Implanted PET Films. The Residual Mean Squares (RMS) all have 8 df.

Film	Temperature Range (K)	m	r^a	RMS x 1000	F-ratio, RMS(m)/RMS(m')		Prob. > F
					m/m'	F	
PET-F-819	180-300	1	-0.995	3.78	1:1/2	1.52	0.284
		1/2	-0.997	2.48	1:1/4	1.21	0.397
		1/4	-0.995	3.13	1/4:1/2	1.26	0.376
	80-170	1	-0.941	1.90	1:1/2	1.23	0.388
		1/2	-0.953	1.55	1:1/4	1.42	0.316
		1/4	-0.961	1.34	1/2:1/4	1.16	0.419
PET-Ar-120	180-300	1	-0.997	4.21	1:1/2	2.43	0.115
		1/2	-0.999	1.73	1/4:1	1.14	0.429
		1/4	-9.999	4.81	1/4:1/2	2.78	0.085
	80-170	1	-0.947	4.13	1:1/2	1.28	0.368
		1/2	-0.959	3.23	1:1/4	1.13	0.434
		1/4	-0.965	3.66	1/4:1/2	1.13	0.434

Sample	Range		[a]				
PET-As-120	180–300	1	-0.990	0.35	1/2:1	1.43	0.312
		1/2	-0.986	0.50	1/4:1	1.03	0.484
		1/4	-0.987	0.36	1/2:1/4	1.39	0.326
	80–170	1	-0.994	0.36	1:1/2	9.00	0.003
		1/2	-0.999	0.04	1/4:1	2.11	0.156
		1/4	-0.999	0.76	1/4:1/2	19.00	0.000
PET-AS-119	80–200	1	-0.998	0.20	1:1/2	1.82	0.208
		1/2	-0.999	0.11	1:1/4	1.43	0.312
		1/4	-0.998	0.14	1/4:1/2	1.27	0.312
PET-F-819-LN	80–200	1	-0.996	0.08	1:1/2	2.67	0.093
		1/2	-0.999	0.03	1:1/4	2.67	0.093
		1/4	-0.999	0.03	1/4:1/2	1.00	0.500

[a]Correlation coefficient

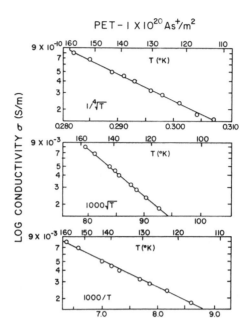

Figure 6. Plots of log σ vs 1000/T, $1000/\sqrt{T}$ and $1/T^{1/4}$ for PET films implanted with 10^{20} As$^+$ ions/m^2 over the temperature range 110-160 K.

3.4   Ion Depth Profiles

The depth profiles of films PET-F-819 and PET-As-120 are shown in Figures 7 and 8. Three characteristic curves are given in each figure. The upper curve gives the count of secondary carbon ions sputtered from the polymer surface as a function of the sputtering time. A drop in the C level occurs near the end of the implanted ion interface (location at which the concentration of the implanted species decays rapidly) which is caused by a decrease in ion yield resulting from the decrease in polymer conductivity at this point.

The Au curve shows how the concentration of the secondary gold ions emitted from the gold coating decreases with depth. The change in concentration of the implanted ions (proportional to the secondary ion count) as a function of the sputtering time (or depth) indicates how the implanted species is distributed in the polymer. The maximum in the implanted-ion curves corresponds, of course, to the region where the concentration of the implanted ion is the highest. The measured depth to this maximum is the projected range R_p described in section 2.1.

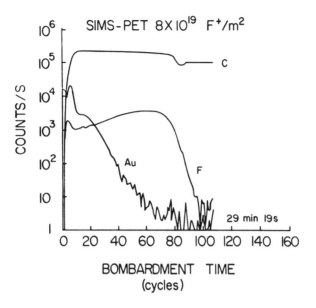

Figure 7. SIMS profile of carbon, fluorine and gold secondary ions emitted by bombardment with Cs+ primary ions from PET film implanted with 8x10^19 F+ ions/m².

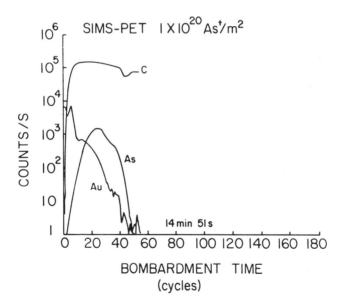

Figure 8. SIMS profile of carbon, arsenic and gold secondary ions emitted by bombardment with Cs+ primary ions from PET film implanted with 10^20 As+ ions/m².

The concentration of F^+ ions in (PET-F-819) increases
slightly with depth before tailing off. This behavior is
characteristic of the "truncation effect" which occurs when a
light projectile is implanted in a heavier substrate. As the
mass of the ionic species increases, the depth profile
acquires a shape closer to the Gaussian distribution. Such a
distribution is evident in Figure 8 (PET-As-120) where the
As^+ implanted species (74.92 amu) is almost four times
heavier than the F^+ (19 amu) implanted in Figure 7. The
curve also suggests that the implanted As^+ ions are buried
closer to the surface and the standard deviation of the
projected range ΔR_p is less than in the F^+ ion-implanted PET
sample. These results confirm theoretical calculations (see
Table 3).

The depth profile of F^+ in PET-819-F-LN (not shown) is
more Gaussian and has a penetration depth approximately 70%
that of the sample implanted at room temperature. The lack
of chain mobility at the LN temperatures at which this sample
was implanted apparently limits both the depth of penetration
and amount of substrate damage. In polymers which usually
consist of two-phase (crystalline-amorphous) systems in which
molecular motions are highly dependent on temperature, it is
not inconceivable that implantation at LN temperature might
behave differently from implantations at RT.

Assuming that, to a first approximation, the R_p of the
implanted ion is proportional to the product of the
sputtering time and the primary ion current, the R_p's of the
F^+ ions implanted in PP, PTFE and N-66 implanted at the same
energy and fluence can be compared. Such calculations
indicate that the depth of the peak is highest in the PP
(187 μC), decreases in N-66 (122 μC), and is lowest in PTFE
(53 μC). Theoretical calculations of the projected range
reported in Table 3 agree with this trend. Unfortunately
this assumption is not likely to be very accurate since the
PTFE film sputtered away at a much faster rate than the PET
and, in fact, the film was almost completely sputtered
through at the end of the analysis. It has been
reported [19] that PTFE is sensitive to radiation and has a
higher tendency to degrade than to cross-link when
irradiated. Until it is possible to measure the depth of the
eroded region, however, (see section 2.4) the exact
sputtering rate remains speculative.

3.5 Cross-linking and Fiber Damage

Degradation and cross-linking are the most important
effects of radiation on polymers. Upon irradiation, PET

has a higher tendency to cross-link than to degrade [19].
Evidence of cross-linking in the ion-implanted PET films was
obtained based on the insolubility of cross-linked polymer in
1,1,1,3,3,3-hexa-2-fluoropropanol, a solvent for PET at RT.
It was not possible, however, to measure the amount of
undissolved polymer gravimetrically because of the small
quantities collected. (The moisture regain of cellulosic
filters is too variable and the solvent extracted too much
material from PTFE filters.) The undissolved residues were
therefore examined by SEM. Polymer residues left on the
filter papers were picked up with double-sided pressure-
sensitive tapes which were then attached to appropriate SEM
stubs. Although it was not possible to weigh the residues of
the implanted samples, it was clear that some undissolved
material was always present on the filters because of its
characteristic brownish color. No undissolved fragments were
detected in the filtered pristine PET.

All the photomicrographs of the residues from the ion-
implanted PET samples show fragments most of which curled
into a cylindrical shape having diameters ranging from 20 to
80 µm. The wall thickness of the cylinders in film PET-F-819
was ~ 0.12 µm, a thickness about 0.5% that of the pristine
film and close to $4\Delta R_p$ calculated for F^+ implanted in PET
(see Table 3).

SEM photomicrographs of the implanted surfaces of films
show that all ions caused some damage. Implantation with F^+,
Ar^+ and As^+ damaged the PET film by forming paths resembling
"river beds" with fragmented segments lying in the beds.
These fragments do not have any preferential shape or size
and might be as small as 0.3 µm by 0.2 µm or as large as 7 µm
by 1 µm. The PET films implanted at 77 K show linear, narrow
striations with no fragments in the beds. Similar formations
were also observed in the Ar^+ samples implanted at 77 K.
These observations suggest that at temperatures where chain
mobility is low, the damage produced is minimal and less
dependent on the ion species implanted.

The earlier speculation (section 3.3.1) that polymer chain
breakage accompanied by the formation of highly disordered
material resulting from ion implantation is one factor which
facilitates electron hopping along, across and between the
fragments is consistent with the results reported here since
the samples implanted at LN temperature display less disorder
and damage and also have lower conductivity as reported in
section 3.2.5.

4. Summary and Conclusions

1. The electrical conductivity of all four polymers evaluated was enhanced by ion implantation indicating that ion beam-induced conductivity is a phenomenon common to many carbon-containing films.

2. Conductivity increases sharply at fluences between 1×10^{18} and 1×10^{20} ions/m^2 in PET implanted with F$^+$, Ar$^+$ and As$^+$. The fluence-dependent conductivity increase in this range was attributed to ion implantation-induced damage which increases with ion mass.

3. The conductivity of PET films implanted with F$^+$ and Ar$^+$ at fluences near 1×10^{20} ions/m^2 at liquid nitrogen temperatures (77 K) was approximately three orders of magnitude less than similar films implanted at RT. The decrease in conductivity of the samples implanted at 77 K was attributed to the reduced damage expected when the impinging ions have a greater chance of channeling through the polymer structure because of reduced chain mobility.

4. The temperature dependence of the conductivity of the three PET films implanted near a fluence of 1×10^{20} ions/m^2 was measured over the range of 80 K < T < 300 K. When fit to the power law ln $\sigma = aT^{-m}$, exponents of 1/2 were obtained for PET-F-819 and for PET-Ar-120 films over the range from 180 to 300 K and for PET-As-120 over the range 80 to 170 K. A value of m = 1/2 suggests a one-dimensional variable range hopping mechanism. The exponents had a value close to zero over the other temperature ranges. Further examination of the conductivity-temperature data, however, showed that the relationship fit power laws in which m was assigned the values of 1/4, 1/2 and 1 equally well (except for the PET-As-120 implanted sample from 80 K to 170 K in which the fit with m = 1/2 was significantly better). Thus, it is difficult to distinguish between conductivity-temperature power laws having exponents between 1/4 and 1.

5. All but five of the ion-implanted samples are ohmic up to electric field strengths of 600 kV/m. One of the anomalous films is ohmic up to an electric field strength of 267 kV/m and three are ohmic up to 400 kv/m, but then the current increases more rapidly than linearly, indicative of space-charge-limited current flow in the presence of carrier traps [20]. One of the films, PET-Ar-118 becomes ohmic above 133 kV/m.

6. The enhanced electrical conductivity of the ion implanted polymers was found to be stable over a period of 15 months. This observation is consistent with other results [21] and suggests that there was no significant diffusion of the implanted ions out of the film.

7. A Gaussian distribution of the implanted species as a function of depth is theoretically expected at the implantation energy employed [12]. This expectation was confirmed experimentally through depth profiles obtained by SIMS. It was observed that the sputtering rate was very high in PTFE, somewhat less in PP, and much less in PET.

8. As a result of the high kinetic energy associated with the incoming ions, polymer chains are broken, rearranged and sometimes cross-linked. Surface damage and film fragments are visible in SEM photomicrographs suggesting that scisson of polymer chains occurred. Based on evidence provided by an SEM examination of insoluble fragments from the implanted films, it was concluded that ion implantation also causes cross-linking.

9. The brownish color of the implanted samples suggested that carbonaceous material is formed in the film during implantation. The color becomes darker as the implantation fluence increases.

5. References

[1] Hersh, S.P. In "Surface Characteristics of Fibers and Textiles, Part I"; Schick, M.J., Ed.; Dekker: New York, 1975; Chapter 6.

[2] Cotts, D.B.; Reyes, Z. "Electrically Conductive Organic Polymers for Advanced Applications"; Noyes Data Corp.: Park Ridge, NJ, 1986.

[3] Chance, R.R.; Shacklette, L.W.; Miller, G.G.; Ivory, D.M.; Sowa, J.M.; Elsenbaumer, R.L.; Baughman, R.H. J. Chem. Soc. Chem. Commun. 1980, 348.

[4] Rabolt, J.F.; Clarke, T.C.; Kanazaua, K.K.; Reynolds, J.R.; Street, G.B. J. Chem. Soc. Chem. Commun. 1980, 347.

[5] Mazurek, H.; Day, D.R.; Maby, E.W.; Abel, J.S.; Senturia, S.D.; Dresselhaus, M.S.; Dresselhaus, G. J. Polym. Sci. Polym. Phys. Ed. 1983, 21, 537.

[6] Abel, J.S.; Mazurek, H.; Day, D.R.; Maby, E.W.; Senturia, S.D.; Dresselhaus, G.; Dresselhaus, M.S. In "Metastable Materials Formation by Ion Implantation"; Proceedings of the MRS Symposium on Ion Implantation;

Picraux, S.T.; Choyke, W.J., Eds.; North-Holland:
New York, 1982; Vol. 7, 173.

[7]   Allen, W.N.; Brant, P.; Carosella, C.A.; DeCorpo, J.J.;
Ewing, C.T.; Saafeld, F.A.; Weber, D.C. J. Synth. Met.
1980, 1, 151.

[8]   Weber, D.C.; Brant, P.; Carosella, C.A. In "Metastable
Materials Formation by Ion Implantation"; Proceedings of
the MRS Symposium on Ion Implantation; Picraux, S.T.;
Choyke, W.J., Eds.; North-Holland: New York, 1982;
Vol. 7, 167.

[9]   Venkatesan, T.; Forrest, S.R.; Kaplan, M.L.; Murray,
C.A.; Schmidt, P.H.; Wilkens, B.J. J. Appl. Phys.
1983, 54, 3150.

[10] Brock, S.; Hersh, S.P.; Grady, P.L.; Wortman, J.J.
J. Polym. Sci. Polym. Phys. Ed. 1984, 22, 1349.

[11] Wasserman, B.; Braunstein, G.; Dresselhaus, M.S.; Wnek,
G.E. In "Ion Implantation and Ion Beam Processing of
Materials"; Proceedings of the MRS Symposium on Ion
Implantation and Ion Beam Processing of Materials;
Hubler, G.K.; Holland, O.W.; Clayton, C.R.; White, C.W.,
Eds.; North-Holland: New York, 1984; Vol. 27, 423.

[12] Lindhard, J.; Scharff, M.; Schiott, H.E. Kgl. Danske
Videnskab. Selskab, Mat.-Fys. Medd. 1963, 33, 14.

[13] Weber, D.C. Private Communications, Sept. 14, 1983,
Oct. 10, 1984.

[14] Grady, P.L.; Hersh, S.P. IEEE Trans. Ind. Appl. 1977,
IA-13, 379.

[15] Buehler, M.G.; Pearson, G.L. Solid State Elec. 1966, 9,
395.

[16] Grady, P.L.; Hersh, S.P. Rev. Sci. Instrum. 1975,
46, 20.

[17] Simko, S.J.; Griffis, D.P.; Murray, R.W.; Linton, R.W.
Anal. Chem. 1985, 57, 137.

[18] Hashimoto, T.; Koguchi, T.; Okuyama, Y.; Yamamoto, K.;
Takahata, K.; Kamoshida, M.; Yanagawa, T. IEDM Tech.
Dig. 1976, 198.

[19] Chapiro, A. "Radiation Chemistry of Polymeric Systems";
Interscience: London, 1962.

[20] Seanor, D.A. In "Electrical Properties of Polymers";
Seanor, D.A.; Ed.; Academic Press: New York, 1982;
Chapter 1.

[21] Weber, D.C.; Brant, P.; Carosella, C.; Banks, L.G.
J. Chem. Soc., Chem. Commun. 1981, 522.

[22] Aggarwal, S.L. In "Polymer Handbook, 2nd Ed.";
Brandrup, J.; Immergut, E.H., Eds.; Interscience:
New York, 1975; V-13.

INDUSTRIAL RADIATION CURING

Jürg R. Seidel

Energy Sciences International
CH-1211 Geneva 13, Switzerland

ABSTRACT

In this paper industrial applications of UV lamps and low energy electron accelerators for the instant polymerization of radiation curable chemicals shall be reviewed under consideration of advantages and limitations of this relatively recent process family.

The motivation to apply these processes are : 1) reduced pollution, because solvents and their recovery systems are not necessary, 2) less energy at higher production rates and the possibility of working only slightly above room temperature which allows to use heat sensitive substrates (polyolefin films) and avoid paper dry-out and rehumidification expenses, 3) the equipments are smaller and safer than thermal curing ovens, 4) new products with improved surfaces can be obtained at more economical speeds.

Certain drawbacks of radiation curing, like the need for special, more reactive, often more expensive and not always available chemicals will be discussed below. An audience like the one attending this Conference may in the future help to overcome problems like poor adhesion on metal surfaces or the lack of FDA approval.

By reviewing proven and recent applications, indications will be given about the most recent conferences and publications dedicated entirely to the field of radiation curing. Where available, statistical values will be presented and commented with the aim of creating additional information and motivation for future suppliers and users of this exciting technology.

SOURCES OF INFORMATION

The easiest way to screen rapidly a large amount of recent documents on applied radiation curing and its suppliers is to attend the three major conferences organized in a two year cycle. Less or more complete proceedings are available for those who could not attend. The events of the last and the next two years are listed in Appendix I, including details on where to obtain the published papers. [3-6]

Vendor information becomes particularly interesting when it can be compared to overall figures in statistics and fore-casts, as they have been presented in Table 1 by Prane [1] for the USA in the period '85 to '90. The proportion 88/12%(A) of UV against EB coatings proves that UV curing is an older, easier, wider applied technology in general, but in particular for clear protective varnishes, for printed and 3-dimensional products. Adhesives in this table (B) show an opposite distribution because their layers are too thick or often hidden (lamination, label complex) and therefore are more easily cured by EB.

Table 1.   Consumption 1985 and Growth Forecast till 1990 for Radiation Curable Chemicals, in 10^3 tons [1]

Application	Comment	UVC	EBC	Avg.Growth/y(%)
Coatings	A)	28 (88%)	3,9 (12%)	7/21
-on Plastic		10	1,7	10/25
Inks	C)	4,3/6,8	0,13/1,3	15/60
Adhesives	B)	0,3 (11%)	2,5 (89%)	24/19
-PSA	D)	-	0,35/1,3	-/30
Electronics and				
Optical Fibers	E)	1,6	0	14
Total (incl.oth.appl.)		48,1(88%)	6,45(12%)	9/23

The '85 figure for EB inks corresponds to the need of a single lithographic web press, 105 cm wide and running at 250m/min for 1600 h/year (C). The 1990 figure for pressure sensitive adhesives (D) is hopefully too small : it is estima-ted that a single one meter wide label machine, running at 200m/min will consume 400-600 tons of PSA per year. Most app-lications in the field of fiber optics (E) will remain with UVC. Up to now one assumes that EB may damage optical fibers and certain electronic components. Strong annual growth rates (based on relatively low 1985 values) are expected for EB curable chemicals; this can also be explained with the high productivity of EB processors.

For the Japanese market only figures for UV cured chemicals are available, but this with a surprising precision [3] : coatings 1420 tons (+18-20 %/year), inks : 2260 tons (+25%/year), adhesives : 220 tons (+8-10 %/year) and printing plates : 2300 tons (+13-15 %/year).

European statistics for radiation curable resins and monomers could not be found. Hopefully the creation of a European radiation curing group will in the future help to correct such a lack of information.

Another Japanese source [3b] gives the distribution of the estimated 25'000 UV systems installed in Japan up to August 1986. From Table 2 we can see that the largest number, i.e. 15'400 systems are used for all kinds of printing applications, from paper to metals and the printing of masks in the production of printed circuit boards. Relatively few UV lamps cure large amounts of resins to produce printing plates, large number of UV lamps cure minute amounts of adhesives -sometimes drops- in the production of electronic components. Contrary to the situation in Europe, mainly in Italy, and in Eastern European countries, only 330 systems are used in the decoration and protection of wood panels, furniture, instruments and building materials.

Table 2. 25'000 UV Systems in Japan (Situation Aug. 86) [3b]

A) For Ink Curing	15400	B) For Adhesives & Coatings	
Paper, Packaging	2250	Printing Plates	1350
Screen Printing	2300	Adh. in Electronics	3550
Business Form Pr.	2830	Plastic Coat.	3230
Seal, Label Pr.	3100	Building Mat. Wood	330
Electron. Masking Pr.	3870	Other	760
Metal Pr., others	1050	Total (Adh. & Coat.)	9620

Another important information at the first CRCA comes from Kumanotani [3]; it shows the worldwide use of EB accelerators for curing at energies up to 300 kV. The values in brackets of Table 3 are the original values indicated in his review paper. They seemed however so low and unfavorable for the USA and Europe that the author added the figures indicated by

Läuppi [7] in the most recent compilation. Pilot line figures
are in some cases too high because units for flue gas desulfu-
rization, low energy crosslinking and discontinued operations
are also included One thing is sure : at the end of 1986, 18
laboratory units were used for R&D and development of chemicals in Europe and almost three times as many production lines
as indicated were in operation (=20).

Table 3.   EB Curing Systems Worldwide 1986 [3][7]
                original figures in (..)

Type, Application	Japan	USA	Europe
R&D, Lab Units	46 (23)	40 (16)	20 (1)
Pilot Lines (a)	12 (2)	45 (7)	35 (10)
Production Lines	12 (10)	65 (23)	25 (7)
Total EBC	70 (35)	150 (46)	80 (18)

    To get a feeling for the growth potential of a new techno-
logy you look at the number of _patents_ being applied for.
About five to ten years later you may expect that several of
them become industrial processes. In Japan [3] 63 patents were
filed for EB applications in 1972 - in 1985/6 1600 were
written per year for the same technology. In 20 months ending
in August 1986 2400 patents were applied for radiation curing,
including UV; more than 50 % for improvements in the electronic industry. A considerable growth can be expected based on
these analyses and forecasts.

    No review papers on Europe and USA were presented at the
last two Radcure Conferences, but information can be picked
from specialty review papers which deal with particular
fields. We shall look at the information made available when
discussing the major fields of application.

INDUSTRIAL EB CURING APPLICATIONS

    Since the development of EB accelerators for industrial cu-
ring processes started one could observe that the beams were
available first. Ideas to use these beams followed second and
were published in a very promising manner by Morganstern,
Bailey and others already twenty years ago. The development of
the chemicals and their coating technology - more difficult

because of higher viscosities – followed much later. There are cases, like metal coatings, where the promises made in 1967 still have not become feasible, mainly because of important shrinkage, poor adhesion and formability of the solvent free coatings.

Builders and vendors of electron accelerators had to learn that their equipment could not be sold as a replacement of a thermal drying system, but that they had to offer and demonstrate the entire process package including the accelerator, safety, coating technology and well performing reasonably priced chemicals in cooperation with other professionals and industries.

Sealing and coating of chipboard panels, wood, doors were for several years the first and major application of EBC in the US, F, B, D, NL. Styrene chemistry was used at the beginning because of its low cost, but reactivity and curing were not very satisfactory. Today most often polyesters are used and the sealer can be UV cured which allows to work in line with intermediate sanding followed by one EB zone (4-5 plants).

The second generation of panel finishing can be considered to be the most ideal EB application : a $90g/m^2$ adhesive layer, followed by the lamination of a $30g/m^2$ decor paper, by one or two layers of varnish at various gloss levels are cured in a single pass. An additional benefit is that the chemicals penetrate from both sides into the decor paper, creating a fiber reinforced material with properties superior to a low pressure laminate [11](2-3 plants).

With the same logic, producers of wood grain printed decor paper impregnate and/or varnish them before selling them to laminators. The disadvantages of melamine impregnation like pollution, enormous ovens and brittleness can be avoided by using EB varnishes with excellent stain and heat resistance, e.g. much better resistance to burn marks from cigarettes (4-5 lines).

Acrylate based radiation cured coatings easily result in high gloss. But in most countries users are looking for a protective coating which is as mat as a natural wood finish. Matting of solventfree varnishes with fillers, like Kaolin, Syloid® etc. is not an easy task because viscosity rises rapidly and the coatability is lost. Other ways have been tried, like adding small amount of water, waxes. In UV curing one can use a dual cure method, which allows to finish cure an inhibited mat surface from curing in air, by a second UV lamp with adapted wave length or by working in an inert atmosphere.

In Japan we observe two coating techniques for building materials with EB curing : concrete roof tiles with glossy enamel-like finishes and gypsum tiles for indoor and dry room applications. Beautiful finishes, also with metallic look are offered but the organic chemistry based varnishes are not sufficiently abrasion resistant to be used for flooring.

In the same field an important contribution was made with chemicals from Bayer, formulated by Wiederhold as first weatherproof coatings on wood cement panels -a replacement for asbestos cement as an outdoor building material with a ten year warranty (2 plants).

Another product used in the chipboard panel industry is the caul stock sheet, an EB cured release paper replacing Al foil as a separation sheet between panels in the glueing press (3 producers).

Still another release paper produced by EB is the casting paper for the artificial leather industry. It must be resistant to hot (TMF) solvents, allow several reuses and release both PUR and PVC layers. With a patented process [12] -envied by other potential users- the casting paper surface allows to obtain a leatherlike structure by curing it through the paper against a structured steel drum in the EB process zone (see also Equipment chapter, Figure 4).

The next important field are the silicone release coatings on calendered or clay coated paper and -thanks to curing near room temperature- the production of release films, e.g. with heat sensitive polyolefins. The major application for these papers is the protection of the PSA layer on label stock. On narrow and relatively slow production units curing is often done by UV - the fastest UV equipment known can produce a one meter wide web at 120m/min using 12 x 1kW UV lamps and developing a considerable heat. This heat is necessary for the hybrid chemistry, otherwise it may be removed by passing the web over a water-cooled support drum [13]. Other applications may be folded release paper for hot-melt packaging, release films for self-adhesive medical supplies, bandages, pace-maker electrodes and reusable closure strips (at least 5 lines).

Printing. For many years UV was the only radiation curing method for printing inks, mainly in screen printing and for sheet-offset printing. Size and equipment cost eliminates EB for low viscosity gravure inks, which need intercolor drying. But in 1980 Album Graphics in Chicago pioneered the use of a single electron accelerator on a web-offset press after the printing of at least four paste-like inks and a protective varnish. In the early eighties Tetrapak helped EB curing to a break-through by equipping new offset presses for the Tetrabrik® laminate with EB accelerators, replacing the expensive gravure process and allowing much nicer print and gloss compared to flexo printing. UVC was considered initially but it could not offer the same productivity, safety and freedom of odor as the EBC. Acrylated offset inks and varnishes are cured at 3Mrad at speeds of 250m/min.

Varnishing. Protective, clear, high-gloss varnishes are in most of the cases (sheet offset, post gravure top coats) cured by UV. There are however certain important exceptions. Three companies produce colored paper and board, with a high-gloss finish obtained without varnishing after the application of a pigmented coating. These products are used for high quality printing or after embossing they find a growing application for book binding.

With a viscosity of 300 to 1000 mPa.s and a curing speed of milliseconds these varnishes do not penetrate into the paper substrate and are often used as a primer for direct vacuum metallization. Metallic looking packaging materials with about 300 Angstroem Al are obtained without the need of laminating an Al-foil of several microns or a metallized polyester film to the substrate.

This technique is used by the largest beer label maker in Europe. He adjusts the gloss level by the applied coating weight : 2g/m^2 result in a dull metallic finish and allow rapid label wash-off during bottle cleaning. 5-6g/m^2 result in a mirror like metal or ink finish when overprinted. In Australia, one of the most performing EB accelerator is used to pre and topcoat label papers and board at a speed of up to 400m/min and a width of 165cm. At a dose of 3Mrad the substrate is barely warmed above 30°C and no moisture is removed during curing -therefore rehumidification is not necessary as in other, thermal curing operations (Figure 1).

After several years of development work, EB has been introduced to transfer metallization to gain in time, quality and productivity. The metallic (Al) layer on a PET or PP film is laminated to a paper, board or textile substrate. After curing -two to three days for solventfree systems without EB -or seconds with EB curing, the laminate is split into a high-gloss product (with optional transferred protective layer) and a demetallized carrier film, which is several times remetallized in vacuum. The economy of this process depends among other criteria from the number of reuses of this film.

Figure 1. Pre- and post metallization varnishing line with a
           200kV, 4 cathode Electrocurtain® accelerator

A particular progress has been made by the largest gift
card producer (Hallmark) with the introduction of <u>selective</u>
<u>metal transfer</u>. After printing several colors by gravure
cylinders, an adhesive pattern is printed on the last press
station before laminating the web to a metallized film.

After instant EB curing through the film, thereby avoiding
the problem of oxygen inhibition, the two webs are split and
metal patterns are transferred on the printed product, e.g. a
metallic birdcage or a brand name etc. It has also been suggested to use this technique to produce cheap PCB's, e.g. for
toys, but it is evident that in both cases it is only economical for rather large volume of production, where it can compete with transfers from hot stamping foils.

Lamination is an ideal process for EB curing because of the adhesive layer hidden for UV light, the speed of curing and the possibility to use heat sensitive substrates. Certain technical products are laminated with EBC (playing cards, film-nonwoven complexes, hot-air balloon skins, etc.) but no application has yet been found in the large field of film/foil lamination for the food packaging area. Lamination adhesives often have too much acrylate or tackyfier odor, they have not yet been accepted by FDA or BGA. In certain cases their green-tack is not sufficient to yield good, instantly cured complexes. Slow progress is made because the few chemical suppliers with interesting adhesives do not want to spend the money for the feeding tests. Extractable and migration tests already give sufficient results (less than 50ppb) to allow their use for "dry" food packaging, but no large user has applied this promising technology yet.

From the many other known or hidden applications let us describe two more before changing to the field of UV curing. A very active group of the host country [15] has developed a process for semi-permeable membranes. A proprietory coating containing non-reactive suspended droplets of controlled size is applied to a non-woven carrier. After two curing steps the non reacted droplets are washed out with solvents leaving a membrane with well defined pore size behind .This product has been presented at Index 1987 in Geneva.

Metallized Products [16] an industrial pioneer in EB applications, has after many contributions to metallization and lamination developed an anti-static coating for packaging materials and films used in the electronic industry. Transparent pouches can be manufactured which no longer destroy sensitive electronic compounds by static electricity discharges, when removing them.

A strong interest can be observed in the curing and cross-linking of magnetic coatings on polyurethane basis. Floppy disks (USA) can be produced with smoother surfaces because they are calendered before EB treatment for abrasion resistance. Development work is observed in abrasive coatings and their substrate impregnation. Finally let's mention a very old but unfortunately only application of curing a flock binders on heat sensitive substrates.

UV CURING APPLICATIONS

The field and volume of industrial UV curing applications
is so large that we first give a listing (Table 4) presented
at a recent Radcure Conference and then look more closely at
the most interesting processes using UV lamps for curing.

Table 4.   Current applications of UV curable systems [6]
                  (underlined items will be discussed below)

Surface Treatment   : Graphic Arts : Inks & Varnishes
                      Wood Finishes : Sealers, Coatings
                      Metal Coatings : Cans, Ends, Sheets
                      Plastic Coatings : Films, Sheets, Bottles
Electronics   : Printed Circuits : neg. Photoresists
                      Sealants : Encapsulation
                      Prot. Coatings : CD, Optical Fibers
                      Patterning : Video Disks
                      Bonding : Conductive Adhesives
Various   : Dental Materials
                      Printing Plates (high volume)
Autom. Industry   : Protective Coatings : Organic Glass
                      Premetallization Coatings : Car Lanterns
                      Laminates : Safety Windshield

No real breakthrough has been realized in metal coatings
with EBC : the Suzuki motorbike and Volkswagen wheel projects
have been discontinued, the Japanese Ellio Sheet project is a
hybrid process where EB brings higher hardness by crosslinking
after coating from solvent base. UV curing is more widely used
mainly in can decoration, preferably in the already shaped
state, because also here adhesion on the metal surface is difficult to be obtained with radiation curable coatings which
shrink instantly during polymerization near room temperature.
The thermal curing of the inside varnish which can not be
acrylate based helps significantly to improve adhesion of the
can decoration.

Considerable improvements are expected from UV initiated
cationic curing epoxies with triallyl-sulfoniumsalt type
photo-initiators which release acids necessary for the polymerization process. Post curing can take place after irradiation. Through ring opening, shrinkage can be avoided. The
slower process and/or a heat shock can reduce stresses and
accelerate postcuring. This process is already used in the
protection of can ends and other areas.

A team at PHILIPS [17] has recently compared cationic UV curing of epoxies with traditional UVC of urethane acrylates as protective layer on compact disks. Adhesion of the protective varnish deposited by spin coating on the sensitive Al reflexion layer is the key problem. But all the known adhesion promotors like acrylic acid, phosphor acrylates and cationic curing tend to corrode the Al mirror. The best alternative to solvent based coatings is sofar a polyether urethane acrylate with benzophenone cured by UV. With the rapid growth of CD's also in car and portable players a radiation cured, solvent and abrasion resistant protection becomes a must and we will favor further development, maybe as far as EB curing.

The automotive industry with its large volume production and very active suppliers or process development departments find regularly attractive applications of UV curing. One example is the production of safety glass or windshield by lamination with or without an organic-film, whereby the adhesive is cured with limited heat development. It will resist to UV light, vibrations and reduce the effect of potential impact encountered on automobiles.

This same industry tries to replace glass and other heavy materials, e.g. by organic glass like polycarbonate or polymethylmethacrylate. But these materials show relative poor resistance against UV exposure and particularly against scratches and abrasion. The introduction of UV absorbers and special silicates [19] brought the looked-for solution, but not without creating certain problems : maintain low coating viscosity, find transparent hard filler, and offer good adhesion. These problems which have now been overcome may allow in the future to create other protective coatings on ceramic, concrete and gypsum tiles with an enamel-like surface without the need to go through a glazing furnace.

Optical fibers. One of the most surprising and fastest growing tecchnologies is the use of optical fibers for telecommunication. Their protection by several UV cured coatings of buffer and jacket layers for improved signal transmission, protection from liquid and gas contaminants, from abrasion and microbending and many other functions can be offered thanks to the efforts of mainly two vendors [20,21]. It all started with a single protective and reflection layer coated by a vertical applicator and cured in a vertically positionned elliptical UV system with the lamps in one focal point, the optical fiber inside of a nitrogen flushed quartz tube in the other focal point.

More recently additional UV cured layers are suggested to build up a strand of optical fibers into a symmetrically circular cable or into ribbon like cable structures. For heavy duty applications UV resins soaked inactive glass filaments are wound aroud the active fibers and create tough "ruggedized" optical cable. Much additional information can be found in references [20] and [21] and other publications of these companies.

UV EQUIPMENT

Sufficient information is offered at every radiation curing conference from the most active UV system suppliers. Most often we have to listen to boring sales talks, but there were exceptions recently. One American supplier complained with humor that there were more than 40 would-be UV lamp and system suppliers fighting like the 40 robbers of Ali-Baba to get more than their 2,5% market share of lamp sales.

A fresh and apparently objective description of pros and cons was presented by Knight [22] a pioneer in UV systems since 1972. From his comparison chart we pick the most important features (in Table 5) to get a short information of the three major lamp types on the market : the medium pressure Mercury arc, the electrodeless and the capillary lamp.

Table 5.   UV Lamp Comparison Chart [22]

Parameter	MPMA	Electrodeless	Capillary
Average lamp cost ¢/W. 1000 h	8-12	10-14	50-60
UV output/WU input	21,4%	15%	15%
Avail. length (m)	up to 2,5	0,1 and 0,2	0,125-0,178
Avail. power (W/cm)	up to 200	120	up to 250
Start time	1-3 min.	10s	instant
Sources	many	very few	very few
Market penetration	95%	few users	few users

This comparison may not be totally objective because the producers of compact special lamps do not aim at the application like graphic arts, three-dimensional objects where large lamps have an advantage.

Two examples of successful application for the compact electrodeless lamps show that they have also found their

market niche. In optical fiber buffer coating lines a 25cm
long highly sophisticated reflector system has become the
standard of the industry. Other applications are coloring of
strands of optical fibers (Figure 2) or coating operations
with two-sided irradiation.

Figure 2. Two sided UV curing in a multi-strand optical fiber
           coloring line

Another novelty with an electrodeless lamp, but in this
case spherical and rotating -the so-called "area exposure
lamp" - is used for uniform static illumination over large
areas in exposure of graphic arts films, printing plates and
printed circuit boards. The spherical bulb system illustrated
in Figure 3 is able to illuminate a surface of almost one m^2
of film at a distance of one meter thanks to a well designed
reflector system, a feature which could not easily be offered
by a one meter wide linear lamp.

ELECTRON BEAM ACCELERATORS

The best way to get an objective listing of available curing equipment is to ask a user of several systems to prepare
a review paper. Karmann [24] accepted this difficult task in
1983 and summarized historical and technical information about
electron beam accelerators used for curing and crosslinking
(Figure 4) on the left side of this chronological chart we
find the oldest industrial systems, based on the point source
(cathode) and scanning principle.

Both High-Voltage Engineering Corp (HVEC) and Radiation
Dynamics Inc. (RDI) use a multistage accelerator. Most of
their accelerators are above the field of low energy (limit
300kV) and require bulky concrete shielding. Ford Motor Co.
produced a certain number of smaller scanners with a cooled
and supported window. All three companies have discontinued
the production of accelerators for EB curing.

Figure 3.   Electrodeless spherical bulb UV system [23]

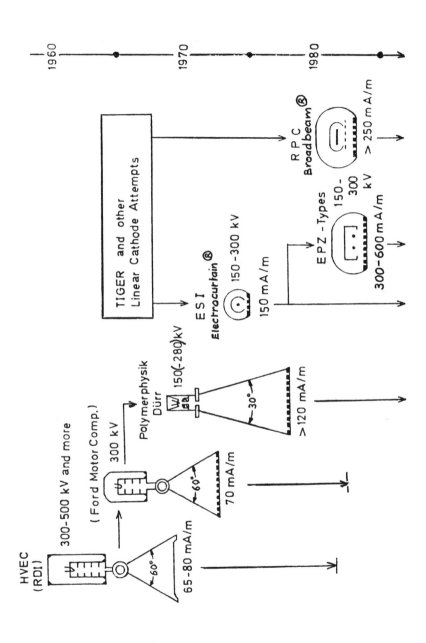

Figure 4.   Electron beam processors for industrial radiation curing in Europe and USA [24]

Polymerphysik (Tuebingen, FRG) continues to offer their selfshielded one-stage accelerator module of the scanner type ESH 150 which is limited in width to 1,2m and to a dose of 1Mrad at 240m/min.

Energy Sciences Inc. separated in 1970 from HVEC to manufacture the Electrocurtain®, a linear cathode accelerator, of compact selfshielded design, which can easily be built into or onto many types of coating or converting machines. It cures the product permanently over the whole width. In the early eighties additional cathodes were introduced to increase the dose delivery up to 1 Mrad at 1500m/min. These EPZ (extended process zone) accelerators are used in a large number for web offset printing and other curing processes with speeds of up to 400m/min, e.g. to cure a varnish requiring a 3Mrad dose.

Radiation Polymer Corp (RPC) is the producer of a multi-cathode, wide window accelerator for relatively high dose delivery, used mainly in the US. An interesting feature is that the electrons are emitted horizontally in 2/3 of the equipments, a fact which makes window maintenance and firing from two sides more easy. The same Company announces since a few years a new accelerator type the WIP (Wire-ion plasma) working with a cold cathode, which is not getting beyond the prototype level(25). Promises of low price, easy curing of three dimensional products apparently without shielding (25, Figure 7) create false hopes and confusion in the market.

Not included in Figure 4 is the major supplier in Japan : Nissin High Voltage; their production equipment is derived from the HVEC scanners, their laboratory equipment looks similar to ESI's and operates in most of the cases with multi-cathodes.

To conclude the EB equipment summary let's look quickly at the origin of the approximately 300 low energy accelerators :

> 215   ESI (USA)
> 25   Polymerphysik (FR Germany)
> 20-25 RPC (USA)
> 35-40 NHV(J), HVEC(USA), others

Recent developments. It has been said before : accelerators exist, but the new coating methods have to be developped to use this new energy source with benefit. A group of processes is influenced by the paper making technology : electrons can not dry water, but highgloss cylinders may be used to improve surface finish. Paper up to a thickness which still allows

penetration is coated on the underside and pressed against a
cylinder which -during its rotation in a shielded treatment
zone- brings it under the beam. There the coating is instantly
"frozen" and shows after leaving the processor the surface
quality copied on the drum surface. The result is similar or
glossier than those obtained by the cast-coat process, but the
energy for drying does not come from a heated cylinder but
from the electron gun and must go through the substrate.

Such an arrangement is shown in Figure 5. The short distance between coating and the possible high speeds reduce the
penetration (wicking) into the substrate, as it happens on the
paper machine with water based coatings. This is not always an
advantage; it is often more difficult to transfer perfectly an
EB varnish.

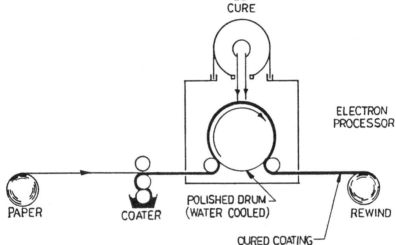

Figure 5.   Curing against a drum for high gloss or
                 patterned finish.

How does an electron accelerator work ? Inside of a vacuum
tube $(10^{-6}$ Torr) we have a tungsten filament (cathode) of a
hairpin or linear shape which is heated above 2000°C and emits
electrons. The negative electrons are attracted by the
anode(+) at the other end of the tube and accelerated to
almost speed of light by a potential difference of at least
150'000 up to 300'000V (low energy field) supplied from a DC
transformer. In multistage scanners this acceleration is done
by steps of 30 to 50'000V. In the stretched cathode accelerator an extraction grid allows, like in a triode radio tube, to
adapt the electron flow to a desired dose. The anode is in all
types identical to a window foil transparent for accelerated
electrons, but separating the vacuum from the normal pressure
treatment zone.

These primary electrons can not go very far, they are slowed down by collisions with molecules. After losses in the window foil and in the air they may be capable to treat a layer of 100 to 300 µm at unit density or a coating or film of 100 to 300 g/m^2 at full dose. This dose has been called Mrad for at least 20 years and 1Mrad equals 10 kJ/kg or 2,4 cal/g of treated matter. This means that a varnish which needs 2Mrad to be cured (or 20 kGray as we should say today) gets an energy of 20 kJ/kg which will increase the material temperature by about 5°C.

We have to overcome two problems : electrons colliding with metal parts emit Bremsstrahlung (X-rays) with an efficiency which goes up with high atomic number (Z) metals, getting close to a real X-ray machine using gold or tungsten as a target. This problem is overcome by shielding the treatment zone with lead. Accelerated electrons also ionize air and ozone is formed. In order to avoid O_3 formation and inhibition of the radical chain polymerization by O_2 and O_3 the treatment zone in EB curing is very often flushed with N_2 to reach a level of 200 to 500 ppm O_2 only.

CHEMICALS FOR RADIATION CURING

This subject could fill and has already filled several books. Conference proceedings, patent flow have been mentionned and are selectively cited in the References and in Appendix I. For the promotion of radiation curing processes rather than telling you how to produce the chemicals, let's look at a few areas where chemists and formulaters have not yet made sufficient progress to allow the replacement of polluting, solvent-based raw materials with EB curing. [6]

Chip and fiberboard panels : acrylated polyesters are in several cases still too expensive

Ceramic, concrete, gypsum tiles : further development to approach glass like, abrasion resistant finishes with room temperature curing.

Mat finishes on wood, flooring and films : additional work on matting agents allowing coatability and offer good long-time abrasion resistance.

Metal coatings for sheets : improved adhesion on Al, TPS,
coils, cans   formability, resistance to steri-
                           lization conditions, replacement
                           of polluting 2K systems, aim
                           future application for inside
                           varnishing. White base coat for
                           UV or EB curing.

Enamelled wire coatings   : enamels resisting to 180°C with-
                           out price for imides, reduce mul-
                           tipass to one or two, avoid air
                           pollution, odor problems.

Silicones for release   : additional(traditional) suppliers
coatings   of adjustable, universally appli-
                           cable and non-ageing silicones.

Pressure sensitive adh.   : additional know-how in compoun-
                           ding, coating techniques and vis-
                           cosity control of encouraging raw
                           materials on saturated polyester
                           base (high-volume potential).

Lamination adhesives   : steps toward lower odor, coating
                           weight, FDA/BGA approvals.

This listing mainly seen by an EB promotor shall not give
the impression that working is impossible with radiation
curing and that chemicals are missing almost everywhere. In
many cases the industry only needs some improvements, a second
source to decide for the investment in radiation curing. In
most of the listed fields several high-volume processors will
be installed, in several countries and continents when the
products are ready or give sufficient promise in pilot evaluation.

With positive thinking we estimate that worldwide about
75'000 UV equipments (about three times the Japanese figures)
and more than 300 EB equipments will consume increasing
amounts of specialty chemicals with an estimated annual equipment growth rate of 7-12% for UV and 10-15% for EB equipment.
The much higher US/EB growth rates in Table 1 are due to a
small base and also to the fact that only about 65 accelerators were used for production. Much higher proportions of
future accelerators installations will be used for production.

CONCLUSION

This limited review of industrial radiation curing has attempted to bring technical -not scientific- information and motivation to polymer chemists and potential suppliers of new ideas and materials. Disadvantages and comparison of the two major processes UV and EB have not been developped in length. Other known applications like : crosslinking, degradation and sterilization were not part of this review. May this text initiate discussion, additional reading and future exchange of know-how and experience, and promote industrial radiation curing by newcomers.

REFERENCES

[1]   Prane, J.W., Radcure '86 Baltimore, MD. (A)
       Radiation Curing - Market, Business, Potential
[2]   Weisman, J., Conf. Rad. Curing Asia 1986 (B) p.11-17
       Radiation Curing in the USA -An Overview
[3]   Kumanotani, J., CRCA 86 (Supplement Proc.) p.5-14
       Radiation Curing Overview - Japan
[4]   Seidel, J., CRCA '86, p.18-23
       Radiation Curing Overview - Europe
[5]   Rubner, R., Radcure '87 Munich (FRG) (C) p.1-1/6
       Radiation Curing in Electronics
[6]   Decker, Ch., do p.1-7/20
       UV Curing Chemistry, Past, Present, and Future
[7]   Läuppi, U.V., IMRAP '87, Ottawa, Canada
       Low Energy, Self-Shielded EB Accelerators
[8]   Quintal, B.S., S.V. Nablo, Radcure '86 Baltimore (A)
       Integration of EB Equipment in Converting Operations
[9]   Tripp E.P., Radcure '87 Munich (C) p.10-9/22
       Recent Developments in EBC
[10]   Ramler, W.J. et al., CRCA '86, Tokyo (Suppl. Proceed)
       p.20-38
       Electron Processors in Novel non-web Applications
[11]   French, D., B.S. Quintal, Radcure '83, Lausanne, AFP FC
       83-277, Low Energy Electron Curing Systems for High
       Speed Wood Panel Finishing
[12]   EUR Pat. 0 106 695A to Scott Paper Co. (USA)
[13]   "Goldschmidt informs..." brochure containing several pa-
       pers on radiation curables silicones, Essen, April '87
[14]   EUR Pat. Appl. 0 130 659 to Elitine/Hallmark (USA)
[15]   Tanny, G.B. et al, Proc. Radcure '86 Baltimore,p.18/1-21
       UV/EB Polymerized Microporous Membranes and Coatings
[16]   Keough, A.H. Proceedings Radcure '84 Atlanta
       Functional Coated Products from EBC
[17]   Lamberts, J.J.M. et al., Preprint Radcure '87 Munich
       UV Curable Acrylates and Epoxies as Protective Coatings
       for CD.

[18] CH Pat. 652 413 (1985) to Chevreux,P.,Battelle, Geneva
      UVC in Safety Glass Lamination
[19] EUR Pat. 069 133 (1985) to Nguyen, V.T. et al.,Battelle
      Geneva
[20] Reese, J.E., Proc. Radcure '86 Baltimore, p. 3/11-22
      UV Curable Resins for Cabling Optical Fibers.
[21] Stowe, R.W., Proc. CRCA; Tokyo Oct.'86, p. 127-137
      Improvements in Efficiency of UV Curing Systems for
      Opt. Fiber Buffer Coatings.
[22] Knight, R.E., Preprint Radcure '87 Munich
      UV Lamps - Reflector Developments, a Review Paper
[23] Levine, L.S., M.G. Ury, Proc. Radcure '86 Baltimore,
      p. 1/1-16
      Microwave-powered Lamp Technology for Photoimaging
[24] Karmann, W., Radcure '83 Lausanne, Reprint AFP FC 83-269
      (EB) Radiation Curing Equipment
[25] Ramler, W.J. et al., CRCA '86 Tokyo, Suppl. p.20-38
      Electron Processors in Novel Non-Web Applications

APPENDIX I

Major Radiation Curing Conferences 1986-1989

(A) : Sept. 86 : Baltimore (USA) Radcure '86, 1000 att.
      Assoc. Fin. Process (SME), Techn. Activities Dept.
      Dearborn, MI 48121 (USA)
(B) : Oct. 86 : Tokyo(Japan) Conf. Rad. Curing Asia, 300 att.
      Prof. Y. Jabata, Dept. of Mech. Engng.,
      University of Tokyo, Tokyo 113, (Japan)
(C) : May 87 : Munich (FRG, Europe) Radcure '87, 600 att.
      AFP/SME (as above)
(D) : April 88 : New Orleans (USA), Radtech '88
      Radtech Int., Northbrook, IL 60062, (founded by former
      AFP members).
(E) : Oct. 88 : Tokyo (Japan) 2nd CRCA
      As in '86 with : UV and EB Surface Fin. Res. Assoc.
(F) : April '89 : Florence (Italy) Rad. Curing Conf. Europe
      (organizer open).

Photo Crosslinking of EPDM Elastomers. Photocrosslinkable Compositions.

J. Hilborn and **B. Rånby**

Department of Polymer Technology
The Royal Institute of Technology, Stockholm, Sweden

Abstract

Mechanism studies with NMR and ESR show that the main photocrosslinking reaction in EPDM is a combination of allylic radicals. Hence, two initiator radicals are needed for every crosslink to be formed. To increase the rate of crosslinking a multifunctional compound was added.

The efficiency and mechanism of photocrosslinking of EPDM elastomers containing different multifunctional crosslinking agents were studied by infrared spectroscopy, swelling and extraction. The different crosslinking agents were trimethylolpropane triacrylate, pentaeythritol tri-/tetra-allylether, triallylcyanurate, dilimonenedimercaptane and dodecyl-bis-maleimide. The efficiency of photocrosslinking and the depth of penetration of the crosslinking reaction are shown to be dependent of the mechanism, e. g. a multistep reaction or a combination of parallel reactions and of the conditions, as type of photoinitiator.

The multistep reaction gives a faster crosslinking than crosslinking through combination of radicals. A photoinitiator which shows a decrease in absorption when it is consumed does give a homogenous density of crosslinking when up to 10 mm layers are crosslinked.

These compositions were shown to be useful as thicker rubber coatings on metal using SiO_2 as reinforcing filler.

Introduction

Naturally occurring rubber from rubber trees has been cured and used since Goodyear in 1839-1842 found that it could be cured using sulphur for crosslinking. Synthetic rubber has been produced since 1940. Crosslinking with sulphur is still the most widely used technique but other methods for special applications are now available (1). Usually there is a need to shape the rubber first and then cure the rubber later. This means that the reaction has to be started in some way, usually by heating.

The crosslinking of rubber prevents the rubber material from flowing and must consequently be formed after the shaping of the material unless the shaping is done by cutting or sawing.

In our study EPDM rubber has been crosslinked by irradiation with ultraviolet light.

Photocrosslinking of EPDM elastomers with ultraviolet light has three main advantages over conventional curing:

1. The saving of energy.

2. The achievement of crosslinking where heat cannot be applied.

3. The health aspect. Considerably less volatile products are formed in photocrosslinking than in conventional curing.

Commercial interest around 1980 for a photocrosslinkable elastomer resulted in a number of patents for photocrosslinking of thin films of commercial elastomers. Thicker layers were not discussed, probably because of a lack of knowledge of the chemistry behind this process. A few papers have been published concerning the photocrosslinking of EPM or EPDM elastomers (2-7).

Our study began with a careful characterization and an investigation of the properties of the starting materials to be able to measure changes thereof. A detailed analysis was made of the crosslinking reaction, including a determination of the sites of the free radicals involved in the crosslinking reaction on the material. The new knowledge together with the available knowledge in photochemistry and polymerization was used to formulate more rapid systems by addition of a small amount of crosslinking agent together with an appropriate photoinitiator.

For practical use elastomers must usually be reinforced with a filler as rubber. Especially EPDM, which was used in our study, has very low strength without reinforcements.

A filler is therefore added to the EPDM when it is applied on metal and cured with UV-light to reach the aim of this work.

UV-curable rubbers can be used as coatings if a light pigment or filler is incorporated onto the material. In the conventional curing of rubber coatings with sulphur or peroxide, energy has to be applied both to the elastomer and to the substrate in order to activate the initiator which typically has a concentration of 1-2%. When the elastomer is photocrosslinked the energy can be applied selectively to the initiator if the components and lamp are appropriately chosen. Examples of such applications are coatings inside containers or tubings which must otherwise be heated under steam pressure.

Primary Photochemical Processes

Photocrosslinking can in general occur only if a photosensitive compound is present. This photosensitive compound can be excited by ultraviolet irradiation,

Table 1. Classification of primary photochemical processes.

Unimolecular process		Bimolecular process	
Photophysical	Photochemical	Photophysical	Photochemical
Luminescence i. e. fluorescence phosphorescence Vibrational energy decay	Dissociation Isomerization	Physical quenching Intermolecular energy transfer	Dimerization Hydrogen abstraction Electron transfer etc.

converted to its first singlet state (S_1) and in most examples by intersystem crossing to its first excited triplet state (T_1). The S_1 or the T_1 state may lose energy by either physical decay or chemical change (8). In either case the process may be first order or second order. The principal pathways are shown in Table 1.

Of these possibilities, photochemical dimerization was one of the first crosslinking reactions commercially used by incorporating pendant cinnamate groups in a prepolymer. These groups can dimerize by cyclobutane ring formation to give crosslinks upon ultraviolet irradiation (9,10,11). In our work we used the photochemical dissociation of photofragmentating initiators and photochemical hydrogen abstraction of hydrogen abstracting photoinitiators. These reactions occur mainly from the triplet state.

The choice of photoinitiator for crosslinking of mm layers of elastomers requires that:

1. The absorption spectrum of the absorbing species must harmonize with the spectral distribution from the lamp used, taking into account the competition from other components, and allowing for the depth of penetration required.

2. The radicals produced must be effective in initiating polymerization of the crosslinking agent or the rubber.

3. The byproducts should preferably not compete for the radiation with the photoinitiator - in other words the chromophore should be the most photoactive component.

4. The initiator must be soluble in the system.

5. The "in cage" reactions between the initiator radicals formed should preferably restore the photoinitiator.

The most successful compounds for photoinitiation are aromatic carbonyl compounds (12). They have a high rate constant for ISC and the formation of the triplet is therefore favoured (13).

Unimolecular photoinitiators. Benzoin ethers (14,15,16,17) are known to be efficient radical producing photoinitiators with varying reactivities for different R_1 groups.

$$[1]$$

They suffer from poor storage stability, however, although large alkyl groups decrease this problem (18).

The poor stability is associated with the α-hydrogen atom (19). The stability may be improved by replacing this α-hydrogen by an alkoxy group, e. g. in αα-dimethoxy α-phenyl acetophenone which has not only improved stability but also improved reactivity (15) compared to other benzoin derivatives.

$$[2]$$

The improved reactivity is due partly to the extra electron-donating alkoxy group, and partly to the well established secondary fragmentation (20) of the substituted benzoyl radical following the initial α cleavage, yielding a highly reactive methyl radical.

Hydroxy alkylphenones have been claimed (21) to have excellent pot life and are also efficient initiators for the crosslinking of rubber.

$$[3]$$

The radicals can also be produced by β-cleavage (14) as in

$$Ph\text{-}O\text{-}PhCO\text{-}CCl_3 \longrightarrow Ph\text{-}O\text{-}PhCO\text{-}\dot{C}Cl_2 + \dot{C}l$$

This is indeed an efficient initiator in rubber but is not very useful in elastomers for the coating of metal because of the formation of HCl which causes corrosion of the metal substrate.

Acylphosphine oxides have recently been developed (22,23,24,25) as useful photoinitiators for the polymerization of acrylates and unsaturated polyester-styrene compositions.

$$[4]$$

The phosphinyl radical is also capable of adding to carbon-carbon double bonds (26).

The high viscosity of the elastomer leads to special problems in the choice of photoinitiator. "In cage" reactions with the benzoyl-1-cyclohexanol is an inefficient use of the initiator radicals (27):

$$[5]$$

Somewhat better as initiator is the $\alpha\alpha$-dimethyl-α-phenyl-acetophenone where disproportionation "in cage" cannot occur. However, the methyl radical formed after β-scission may recombine "in cage" with the benzoyl radical which lowers the efficiency of this photoinitiator compared with acylphosphine oxides.

$$(Ph\overset{\cdot}{C}O + \overset{\cdot}{C}H_3 \longrightarrow PhCOCH_3)_{in\ cage} \qquad [6]$$

The radicals from acylphosphine oxide can also recombine in cage but only to restore the original structure of the photoinitiator which may be used again.

Bimolecular photoinitiators. The excited photoinitiator abstracts a hydrogen from another molecule. Example benzophenone:

$$Ph_2C = O \longrightarrow Ph_2C = O^*(S) \longrightarrow Ph_2C = O^*(T) \qquad [7]$$

This n, π^* state provides the carbonyl with a biradical character

$$Ph_2\overset{\cdot}{C} - \overset{\cdot}{O} \qquad [8]$$

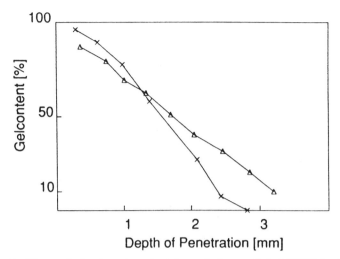

Figure 1. The variation in gel content in a photocrosslinked EPDM rubber using BP as photoinitiator as a function of depth below the surface of the sample, (△)=1% BP (×)=2% BP.

Aromatic ketones are able to abstract hydrogens that are loosely bonded e. g. from thiols (28) and amines (29). This makes benzophenone useful in the crosslinking of elastomers where it may abstract allylic hydrogen atoms.

Benzophenone gives a profile of the crosslinking reaction through the rubber sample as in Figure 1 which shows a decrease of reaction as the depth increases. This profile may be due to the competition for light in the elastomer.

Aromatic ketones with electron-donating substituents frequently have a lowest excited state which is π, π^* rather than n, π^* (30). Since a large extinction coefficient is favoured by two participating orbitals occupying the same region in space, the absorption will be large in this case. The penetration of light into such a system is limited and therefore these aromatic ketones are not useful in thicker rubber coatings.

The source of irradiation. The photoinitiator does not function unless it is irradiated with UV-light of sufficient intensity of the right wavelengths i. e. the emission from the lamp (Figure 2) has to be considered.

In our experiments a high pressure mercury lamp doped with gallium and lead iodides was used. This lamp has an emission useful for UV-curing with peaks around 300 nm (several), around 366 nm and also at 403, 405 nm and 417 nm which gives the light its typical blue colour.

Several other dopants are available (31). An interesting lamp is one containing lead iodide which has an intensity between 350-400 nm 1. 75 times greater than of Hg-lamps.

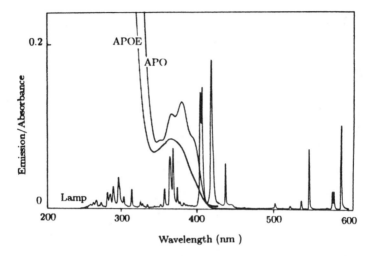

Figure 2. Ultraviolet spectra of 2,4,6-trimethylbenzoyl diphenyl phosphine oxide (0. 31 mM in EtOH), 2,4,6-trimethylbenzoyl phenylphosphinio ethyl ester (0. 37 mM in EtOH) and the emission from the high pressure Hg-lamp.

Table 2.   Some properties of photoinitiators used for crosslinking of EPDM elastomers.

Type of photoinitiator	Depth of reaction	Overlap with lamp emission	Inactivation by recombination "in cage"	Sensitive to visible light
Hydroxyalkylacetophenones	poor	medium	yes	no
αα-Dimethoxy-α-phenylacetophenone	medium	poor	some	no
Acylphosphineoxides	good	good	no	yes
Benzophenone	medium	medium	no	no

The ability to use different photoinitiators for UV crosslinking of elastomers is summarized in Table 2.

Secondary Photochemical Processes

Once the photoinitiator radicals have been produced and escaped from the cage they react with the surroundings. The radicals can add to double bonds, abstract hydrogen atoms, combine or disproportionate with other radicals.

The photoinitiator benzoyl-1-cyclohexanol was mixed into four different types of EPDM elastomers and irradiated for different periods of time (32). The irradiated samples showed different degree of crosslinking as shown in Figure 3.

Can the reason for this behaviour be found in the molecular structure of the elastomers?

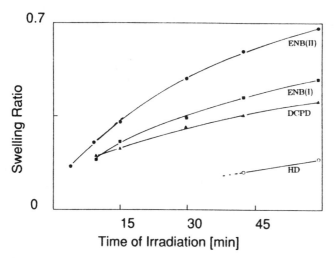

Figure 3. *Four different EPDM elastomers containing benzoyl-1-cyclohexanol have been photocrosslinked by UV-irradiation for different periods of time to prod- ucts of different swelling ratios in cyclohexane (25° C). ENB = ethylidenenorbornene, DCPD = dicyclopentadiene, HD = hexadiene are the dienes in the elastomer samples studied.*

The molecular weight as given in table 3 has only got a small influence on the swelling ratio as can be calculated from the simplified Flory equation (32):

$$q_m^{5/3} \simeq (\bar{v}M_c)(1 - 2M_c/M)^{-1}(1/2 - \chi)/V_1 \qquad [9]$$

q_m=Ratio of swelled polymer to nonswelled.

M_c=Average molecular weight between crosslinks.

χ=Interaction parameter with solvent.

V_1=Molar volume of polymer.

The ENB (containing ethylidenenorbornene) shows a higher reactivity than the others but it also has a higher diene content as shown by ^{1}H NMR. The investigated properties are summarized in Table 3.

It is concluded that the difference in reactivity must be due to the type of diene.

The peak positions as derived from ^{13}C NMR in solution can be used to interpret the solid state spectra shown in Figure 4.

During UV irradiation of the ENB elastomer containing benzoyl-1-cyclohexanol,

Table 3. Some properties (32) of the four types of EPDM elastomers used for photocrosslinking.

Measured property	ENB(I)	ENB(II)	DCPD	HD
$\overline{M}_n$	91000	53000	40000	50000
$\overline{M}_w$	210000	180000	120000	150000
Mol % Diene	2. 6	10	1. 5	1. 6
Mol % Ethylene	62	59	71	66
Mol % Propylene	36	32	28	32
T_g K	227		222	217
Diene isomer	Entgegen	Entgegen	Endo	Trans
IR C=C at (cm^{-1})	1688, 808	1688, 808	948, 699	966, 889
Olfin ^{1}H peak at (ppm)	5. 2 4. 9	5. 2 4. 9	4. 6	4. 7

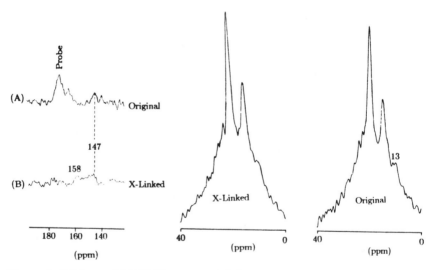

Figure 4. Solid state ^{13}C NMR spectra of (A) noncrosslinked elastomer and on (B) photocrosslinked rubber.

the volatile products were sampled for GC-MS analysis. Cyclohexanol and benzaldehyde were found as initiator fragments. On the basis of the solid state spectra and on the GC-MS the following mechanism is proposed. The benzoyl-1-cyclohexanol cleaves upon UV irradiation to form the benzoyl radical and the hydroxycyclohexyl radical. The benzoyl radical abstracts an allylic hydrogen atom from the elastomer.

[10]

It may also add to the double bond:

$$[11]$$

The hydroxycyclohexyl radical with low reactivity acts as a terminator and disproportionates with any radical to form cyclohexanol.

$$[12]$$

As the concentration of crosslinks is rather low ($M_c = \sim 10^{-5}$) in the systems with low diene content, the crosslinks are difficult to detect by direct methods. However, if the mechanism is known the structure may be determined.

As the benzoyl radical is formed from almost all photoinitiators of interest for the crosslinking of EPDM elastomers it was specially studied (33) with model compounds of the dienes investigated here (27). The three most common types of EPDM elastomers contain 1,4-hexadiene, dicyclopentadiene or 2-ethylidene-5-norbornene, respectively.

The upper double bond in the figure is used for the copolymerization of the diene with ethylene and propylene. Radicals are conveniently studied by electron spin resonance. Even though electron spin resonance is a very sensitive method for analysing unpaired electrons (34) we did not succeed in direct identification of the radicals formed on the elastomers during the crosslinking reaction. The following model compounds for the dienes were therefore used.

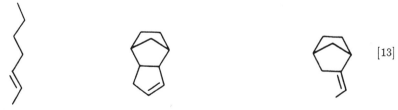

$$[13]$$

2-Heptene   Dihydrodicyclopentadiene   Ethylidenenorbornane

To form the benzoyl radical selectively benzil was irradiated in a solution of the model compound with a 308 nm laser which may induce cleavage of benzil from a higher excited triplet state (35).

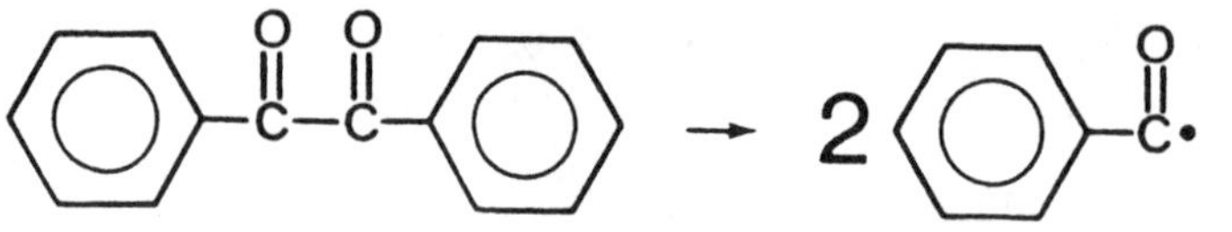

[14]

Only in the case of trans-2-heptene a radical was recorded in detectable concentrations, 2-benzoyl-3-heptenyl formed by addition of benzoyl radicals to the double bond:

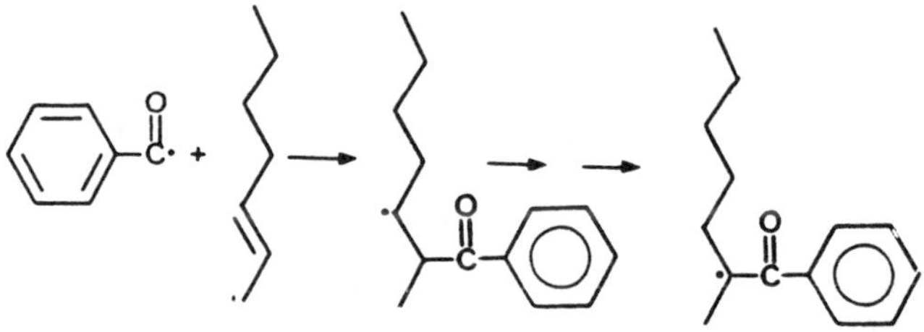

[15]

and rearranged to the 2-benzoyl-2-heptenyl radical. As shown in reference (32), the trans form of the hexene diene is the most common. This may now serve as an explanation of the low reactivity in photocrosslinking found in EPDM elastomers containing hexadiene monomers as compared to those containing dicyclopentadiene or ethylidenenorbornene. The benzoyl radical is "lost" by the formation of the relatively stable 2-benzoyl-2-heptenyl radical.

Spin Trapping. For the other model compounds the reactions were to fast too be detected by direct measurement. The technique of spin trapping first described in 1967 (36) and later in 1968 (37,38,39) was therefore used.

Radicals formed in EPDM rubber crosslinked with benzoylperoxide by heating have recently also been detected in this way (40).

The principle of spin trapping is the addition of the radical to a compound which forms a relatively stable radical which is accumulated in the system to detectable concentrations:

RADICAL· + SPIN TRAP ⟶   ADDUCT·

The spin trap pentamethylnitrosobenzene described by Doba et al. (41) was shown to be useful in our system.

The photoinitiator benzoyl-1-cyclohexanol was dissolved together with pentamethyl nitrosobenzene in the model compounds and irradiated with UV-light. Abstraction of allylic hydrogen atoms was shown to be the main reaction for all compounds and the following radicals could be detected:

Table 4. Some properties of crosslinking agents used in UV crosslinking of EPDM rubbers.

Compound[*]	Reaction Mechanism	Solubility	Other Features	V_2^{**}
TMPTA	Homopolymerization	Poor	Skin irritant	0. 24
PETAE	Allylic coupling	Medium		0. 15
TAC	Good			0. 23
DSH	Alternating addition	Good	Strong odour	0. 53
MI	Alternating coplymerization	Poor	UV absorbing Possibly toxic	0. 47

* Abbreviations are given in the text. ** Swelling ratio.

$C_4H_9CH=CH-CH_2$, $C_3H_7CH-CH=CHCH_3$

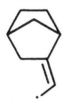  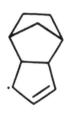

If a less stable radical is formed in the system it is rapidly converted to some of these more stable radicals. As these radicals cannot add to nonpolarized 1,2-disubstituted double bonds, the main mode of crosslinking must be a combination of allylic radicals. Hence, two initiator radicals are needed for each crosslink to be formed.

If the rate of the crosslinking reaction is to be increased, this crosslinking mechanism must be altered to give a better usage of the initiator.

Crosslinking Agents

By mixing the rubber with soluble UV-transparent multifunctional monomers, the utilization of photoinitiator radicals is improved and the crosslinking reaction enhanced (42). The efficiency of enhancing the crosslinking reaction by these components is dependent on the reaction mechanism and on the solubility of the crosslinking agent. The properties of these components in the UV curing of elastomers are summarized in table 4.

The crosslinking agents were:

TMPTA   Trimethylolpropane triacrylate $CH_3CH_2C(CH_2OCOCH=CH_2)_3$
PETAE   Pentaerythritol tri- and $HO-C(CH_2OCH_2CH=CH_2)_3$
         tetra-allylether $C(CH_2OCH_2CH=CH_2)_4$
TAC   Triallylcyanurate $C_3N_3(OCH_2CH=CH_2)_3$

DSH   Dilimonenedimercaptane

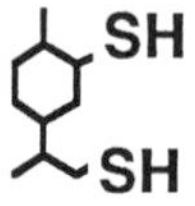

MI   Dodecyl-bis-maleimid

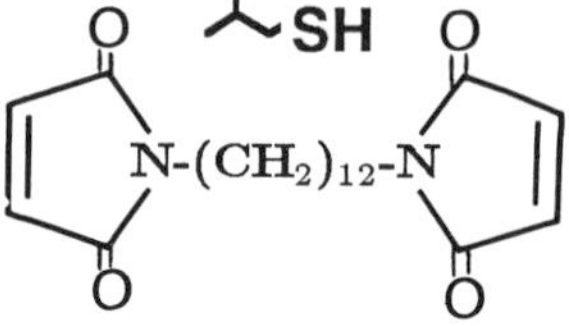

It was found that the system with trimethylolpropane triacrylate crosslinked efficiently to give a swelling ratio of 0.24 (polymer/swelled sample). As the EPDM-trimethylolpropane triacrylate rubber showed a large elongation at break (43) it was believed that trimethylolpropane triacrylate forms long crosslinks between elastomer chains rather than an interpenetrating network which was given as a possibility in our work (42).

To increase the solubility and decrease the amount of homopolymerization, PETAE was used. However, the result was a low swelling ratio of 0.15 . It was suggested that this is due to intramolecular reactions in the crosslinking agent and allylic inhibition of polymerization.

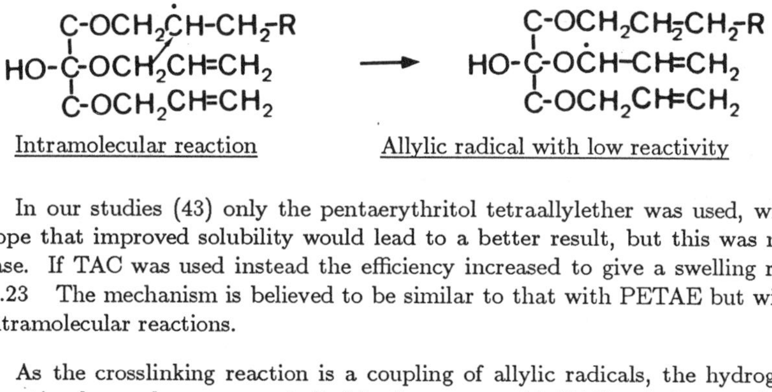

$$
\begin{array}{c}
\text{C-OCH}_2\dot{\text{C}}\text{H-CH}_2\text{-R} \\
\text{HO-C-OCH}_2\text{CH=CH}_2 \\
\text{C-OCH}_2\text{CH=CH}_2
\end{array}
\longrightarrow
\begin{array}{c}
\text{C-OCH}_2\text{CH}_2\text{CH}_2\text{-R} \\
\text{HO-C-O}\dot{\text{C}}\text{H-CH=CH}_2 \\
\text{C-OCH}_2\text{CH=CH}_2
\end{array}
\qquad [16]
$$

Intramolecular reaction   Allylic radical with low reactivity

In our studies (43) only the pentaerythritol tetraallylether was used, with the hope that improved solubility would lead to a better result, but this was not the case. If TAC was used instead the efficiency increased to give a swelling ratio of 0.23   The mechanism is believed to be similar to that with PETAE but with less intramolecular reactions.

As the crosslinking reaction is a coupling of allylic radicals, the hydrogen abstracting benzophenone was studied in the system with TAC and shown to be more efficient due to a deeper penetration of the crosslinking reaction.

To avoid the unwanted homopolymerization and intramolecular reactions completely, a thiol-ene system was formulated. The hydrogen atom on the thiol group is easily abstracted to form a thiyl radical:

$$-SH + R\cdot \longrightarrow -S\cdot + RH \qquad [17]$$

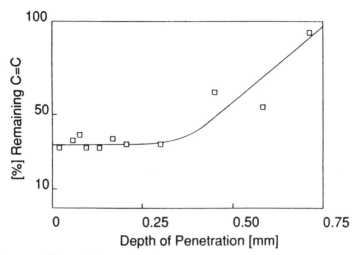

Figure 5. The variation in number of dodecyl-bis-maleimide double bonds in the photocrosslinked EPDM rubber as a function of depth from the surface into the sample.

The thiyl radical adds to the elastomer double bond to form a macro radical:

$$-S \cdot + C = C \longrightarrow -S - C - C \cdot \qquad [18]$$

When the elastomer radical abstracts a thiol hydrogen atom, an alternation reaction is established:

$$-S - C - C \cdot + - SH \longrightarrow -S \cdot + - S - C - CH \qquad [19]$$

This system is efficiently crosslinked by UV irradiation to give a swelling ratio of 0.54   A drawback of this formulation is the strong odour of the thiol.

The possibility of dimerization as a crosslinking reaction was pointed out (42) when dodecyl-bis-maleimide (MI) was incorporated into the rubber, subsequent ultraviolet irradiation showed an exceptional behaviour. The number of double bonds in the dodecyl-bis-maleimidedecreased to a large extent at the surface of the rubber sample. The double bond consumption was shown to decrease with the depth into the rubber samples as shown in Figure 5.

N-substituted maleimides are reported (44) to copolymerize with cyclohexene to form an almost perfect alternating copolymer. Homopolymerization of N-substituted maleimides only occurs by anionic mechanism. As the concentration of rubber double bonds is rather low (9 C=C per 1000 C-atoms, paper 1) the copolymerization of a rubber double bond with a maleimide double bond cannot be responsible for the large consumption of dodecyl-bis-maleimidedouble bonds.

The maleimide may undergo several reactions:

A photochemical 2+2 addition:

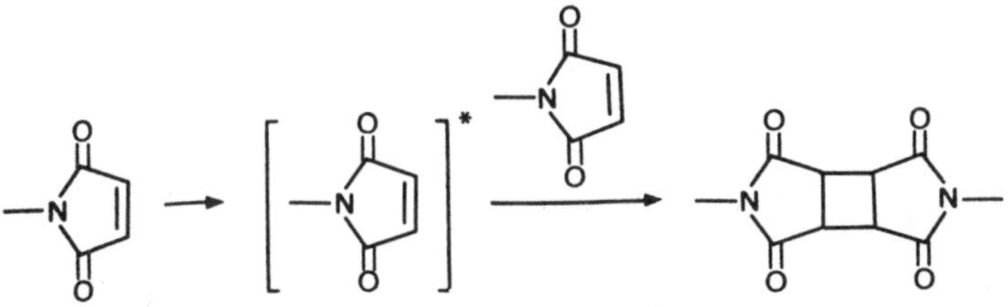

[20]

which means that one maleimid group is excited by UV light and combines with another maleimide group to form a cyclobutane structure. This may also occur with the rubber double bond.

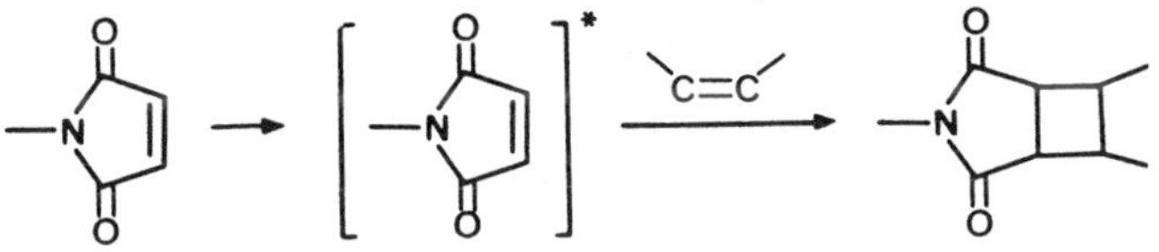

[21]

As the dodecyl-bis-maleimide has a long alkyl chain intramolecular hydrogen abstraction can occur to give a ring expansion.

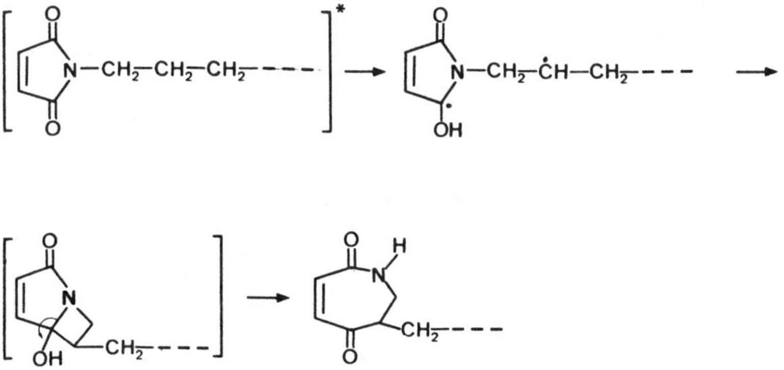

[22]

If these suggestions are correct no photoinitiator would be needed. This was indeed shown to be the case for thin layers (42).

Using the electron poor double bond of maleimide in MI, it was possible to formulate a system of alterating copolymerization (44) or possibly of an allylic hydrogen abstraction on the elastomer followed by addition to the maleimide (45). Even though the dodecyl-bis-maleimidesolubility was poor, the efficiency was high to give a swelling ratio of 0.47.

Rubber Application

Is it possible to use this crosslinking reaction for rubber in practice ?

To answer this question the study (43) was carried out with each formulation containing SiO_2 as reinforcing filler. To improve the adhesion between the elastomer and SiO_2, a coupling agent, γ-mercaptopropyltrimethoxysilane, was added. The formulations were made using the same crosslinking agents as described (42) with the modification of PETAE where only the tetraallylether was used here.

As photoinitiator, αα-dimethoxy α-phenylacetophenone, was used which gives two reactive radicals as earlier described (20). For the maleimide formulation 2,4,6-trimethyl diphenyl phosphineoxide was used.

The trimethylolpropane triacrylate system and the dodecyl-bis-maleimidesystem gave satisfactory results in our testing of adhesion, tensile strength and swelling ratio.

The thiol-ene system was unexpectedly weak in tensile strength and adhesion. This was believed to be due to a thiol reaction with double bonds which in the other systems could add to the coupling agent. This problem may possibly be overcome by the use of another coupling agent with a pendant vinyl group instead of the thiol.

Finally the trimethylolpropane triacrylate formulation was adhered to the inside of a steel tube and crosslinked with UV-light   to show how it is possible to use this system in practice.

Conclusions

Photocrosslinking of mm thick EPDM rubber layers can be achieved by the appropriate choice of photoinitiator and crosslinking agent.

The photoinitiator must have a low absorption of its chromophore and must bleach upon irradiation allowing for the depth of penetration required.

Due to the viscosity in a rubber system the "in cage" reactions of the photoinitiator are favoured and should therefore be kept to a minimum.

By the addition of a multifunctional agent the rate of crosslinking can be increased. For an efficient use of the these agents the crosslinking reaction should not be a coupling but a chain reaction, preferentially to form an alternating polymer consisting of crosslinking agent and rubber. This can be achieved by using a multifunctional thiol which acts as an active chain transfer agent or by the use of an electron poor double bond in the crosslinking agent which will add to the electron rich rubber double bond with preference.

Acknowledgments

The financial support from the National Board for Technical Development (STU), is greatfully acknowledged.

References

(1) F. R. Eirich, "Science and Technology of Rubber", p292-335, Academic Press, New York (1978)

(2) A. M. Kokurina, S. D. Stolyarova, T. Jurre, Deposited Doc. , p112(1977)

(3) J. A. Bousquet, J. P. Fouassier and J. Faure, *Polym. Bull.* , **1** 233(1978)

(3) J. A. Bousquet, J. Faure and J. P. Fouassier, *J. Polym. Sci: Polym. Chem. Ed.* , **17** 1685(1979)

(4) J.A. Bousquet, J.B. Donnet, J. Faure and J.P. Fouassier, *J. Polym. Sci: Polym. Chem. Ed.*, **18** 765(1980)

(5) J.A. Bousquet, B. Haidar, J.P. Fouassier and A. Vidal, *Eur. Polym. J.*, **19** 135(1983)

(6) J.A. Bousquet and J.P. Fouassier, *J. Photochem.*, **20** 53(1982)

(7) B. Haidar and J.B. Donnet, *J. Appl. Polym. Sci.*, **31** 385(1986)

(8) J.G. Calvert and J.N. Pitts, "Photochemistry", John Wiley & Sons, New York (1967)

(9) L.M. Minsk, J.G. Smith, W.P. Van Deusen and J.F. Wright, *J. Appl. Polym. Sci.*, **2** 302(1959)

(10) M. Hepher, *J. Photogr. Sci.*, **12** 181(1964)

(11) G.A. Delzenne, "*Encyclopedia of Polymer Science and Technology*", Suppl. 1 p401, Wiley-Interscience, New York(1976)

(12) A. Ledwith, *J. Oil. Colour Chem. Assoc.*, **59** 157(1976)

(13) P.J. Wagner, "Topics in Current Chemistry 66: Triplet States III", p1-52. Springer Verlag, Berlin (1976)

(14) H.-G. Heine, *Tetrahedron Lett.* 4755(1972)

(15) H.-G. Heine and H.J. Traenchner, *Prog. Org. Coatings*, **3** 115(1975)

(16) H.-G. Heine and H.J. Rosenkranz and H. Rudolph, *Angew. Chem. Internat. Edit.*, **11** 924(1972)

(17) H. Rudolph, H.J. Rosenkranz and H.-G. Heine, *Appl. Polym. Symp.*, **26** 157(1975)

(18) S.P. Pappas, "UV Curing Science and Technology", (Ed. Pappas), p1-22. Technology Marketing Corporation, Stamford, Conn. US (1978)

(19) G. Berner, R. Kirchmeyer and G. Rist, *J. Oil Colour Chem. Assoc.*, **61** 105(1978)

(20) M.R. Sandner and C.L. Osborn, *Tetrahedron Lett.*, 415(1974)

(21) C.P. Herz and J. Eichler, *Farbe Lack*, **85** 933(1979)

(22) H.G. Hageman, W.J. de Klein and E.A. Geizen, Eur. Pat., 37152(1981)

(23) P. Lechtken, I. Buethe and A. Hesse, Ger. Off., 2 830 927(1980)

(24) P. Lechtken, I. Buethe, M. Jacobi and W. Trimborn, Ger. Off., 2 909 994(1980)

(25) A. Henne, A. Hesse, M. Jacobi, G. Wallbillick and B. Bronstert, Eur. Pat. 62 839(1980)

(26) W.G. Bentrude, J.K. Kochi (ed), "Free Radicals", Vol. II, John Wiley, New York, p595(1973)

(27) J. Hilborn and B. Rånby, "Photocrosslinking of EPDM Elastomers. Elastomer Characterization and Mechanism of Crosslinking.", To be published.

(28) C.R. Morgan and A.D. Ketley, *ACS Div. Org. Coat. Plast. Chem.*, **33** 281(1973)

(29) S.G. Cohen, A. Parola and G.H. Parsons, *Chem. Rev.*, **73** 141(1973)

(30) A. Beckett and G. Porter, *Trans. Faraday Soc.*, **59** 2038(1963)

(31) R. Phillips,"Sources and Applications of Ultraviolet Radiation", Chapt. 9, Academic Press, London (1978)

(32) P.J. Flory, "Principles of Polymer Characterization", p580, Cornell Univ. Press, USA (1953)

(33) J. Hilborn and B. Rånby, "Photocrosslinking of EPDM Elastomers. Studies on Model Compounds by Electron Spin Resonance." *Macromolecules*, To be published.

(34) C.D. Pool, "Electron Spin Resonance", Chapt.14, Interscience Publishers, John Wiley & Sons, New York (1967)

(35) W.G. McGimpsey and J.C. Scaiano, A Two-Photon Study of the "Reluctant" Norrish Type I reaction of Benzil, submitted for publication. Personal communication.

(36) A. Mackor, T.A.J.W. Wajer, T.J. de Boer, *Tetrahedron Lett.*, 385(1967)

(37) C. Lagercrantz and S. Forshult, *Nature*, **218** 1247(1968)

(38) G. R. Chalfont, H.J. Perkin and A. Horsfield, *J. Amer. Chem. Soc.*, **90** 7141(1968)

(39) E.G. Janzen and B.J. Blackburn, *J. Amer. Chem. Soc.*, **90** 5909(1968)

(40) T. Miyake, S. Moriuchi, Y. Mashuda et.al., *Nippon Gomu Kyokashi*, **10** 568(1986)

(41) T. Doba, T. Ichitawa and H. Yoshida, *Bull. Chem. Jpn.*, **50** 3124(1977)

(42) J. Hilborn and B. Rånby, "Photocrosslinking of EPDM Elastomers. Photocrosslinkable Compositions.", To be published.

(43) J. Hilborn and B. Rånby, "Photocrosslinking of EPDM Elastomers. A New Method for Rapid Curing at Room Temperature." *J. Rubb. Chem. Tech.*, To be published.

(44) E. Wipfelder and H. Heusinger, *J. Polym. Sci.: Polym. Chem. Ed.*, **16** 1779(1978)

(45) P. Kovacic and P.W. Hein, *Rubb. Chem. Tech.*, **35** 528(1962)

PHOTOINITIATED CROSSLINKING OF POLYETHYLENES

AND POLYESTERS

Bengt Rånby, Chen Yong Lie[*], Qu Bao Jun[**], and
Shi Wen Feng[**]

Department of Polymer Technology,
The Royal Institute of Technology,
S-100 44 Stockholm, Sweden

Various types of commercial polyethylenes (HDPE,
LDPE, LLDPE) have been mixed in a Brabender Plas-
ticorder with a small amount of UV-absorbing initiator (0.5 to 1.0 % benzophenone or derivatives)
and a small amount of a multifunctional allylether
(0.5 to 1.0 % triallylcyanurate or alifatic allylether). Plates of thickness from 0.1 to 5.0 mm were
pressed near the melting point and irradiated with
UV-light from a high pressure mercury lamp (HPM 15
of 2 kW) at ambient temperatures from 50 to 180°C.
The degree of crosslinking was measured by extraction with boiling xylene and given as gel content
(in % of original weight). Under optimal conditions
gel contents of 70-80 % for LDPE and 80-90 % for
HDPE were obtained, for irradiation times of about
10 sec. near the melting point of the samples. –
Unsaturated polyesters of low molecular mass (1200-
1500) have been modified by acrylation at the endgroups, mixed with 10-30 % di- or triacrylate monomers and about 1 % of a photofragmenting initiator
and cured by UV irradiation for 10-20 sec. at room
temperature. With 50 % glassfibre mats the modified
polyesters were cured to laminates of high modulus
(~ 10,000 MPa) and high tensile strength (~ 250 MPa).

---

[*] Visiting scientist from Zhongshan University, Guangzhou,
P.R. of China.

[**] Visiting scientists from China University of Science and
Technology (CUST), Hafei, Anhui, P.R. of China.

Photoinitiated crosslinking of polyethylene using UV ir-radiation was first reported by Oster et al more than 30 years ago [1, 2]. A few other studies followed [3, 4, 5, 6, 7], but no commercial breakthrough in applications has occurred. The UV light as radiation source has been considered having sev-eral disadvantages in practical applications: Only thin speci-mens of polyethylene ($\leq$ 0.3 mm) could be treated due to poor penetration of the UV light and long irradiation times (sev-eral min) were required to obtain more complete crosslinking due to low yields and low crosslinking efficiency. Therefore, most of the crosslinking studies of polyethylene have been carried out using high energy radiation, ie, ^{60}Co γ-rays or electron beams (EB) from accelerators [8, 9], or chemical initiation methods [10, 11]. Electron beam crosslinking (EBC) and chemically initiated crosslinking (CIC) have been developed into commercial processes [12].

In our recent work we have found that UV initiation for crosslinking of polyethylene has several advantages compared with γ-rays and electron beams: UV-based processes can be optimized to give high yields of crosslinking at short ir-radiation times (10-20 sec), UV light sources are inexpensive and readily available, and the radiation protection for UV irradiation is easily arranged, eg, with thin aluminium film. The research projects on polyethylenes and polyesters reported in this paper, are part of a general research program on the photochemistry of polymers in this department, supported by the National Swedish Board for Technical Development (STU). Other projects in this program are photoinitiated crosslink-ing of ethylene-propylene-diene elastomers (EPDM) and photo-initiated surface grafting of vinyl monomers onto polyolefins and polyesters which both will be reported separately at this symposium by Jöns Hilborn (orally) and Klas Allmér (as a pos-ter), respectively.

The first report on our studies of UV-initiated crosslink-ing of polyethylenes was given in 1985 [13], and the first communication on photocrosslinking of unsaturated polyesters was presented in 1987 [14].

Part I.   PHOTOCROSSLINKING OF POLYETHYLENES

Experimental

A selection of commercial samples of polyethylene are used in our studies (Table 1).

Table 1

Polyethylene samples HDPE and LDPE[xxx)]	Density	Melt Index (MI)	$\overline{M}_n$	$\overline{M}_w$
Lupolen 5270 Z (BASF)	0.949-0.953	1.0-2.3 (190/21.6)	2.7×10^4	8.7×10^5
Hostalen 412 (Hoechst)	(HD)	< 0.1	9.8×10^4	1.5×10^6
DFDS 47 (with 0.2 % Santonox R)	0.922	~ 2	3.2×10^{4} [*)]	–
S-118975 (no anti-oxidant	0.922	~ 2	3.2×10^{4} [*)]	–
DFDS 6600 (no anti-oxidant)	(LD)	0.2-0.3	$4.2 - 4.4 \times 10^{4}$ [*)]	–
News 8572[xx)] (with anti-oxidant)	(LD)	~ 2	3.2×10^{4} [*)]	–

[*)] These $\overline{M}_n$-values of molar mass are calculated from the measured MI-values as $(\overline{M}_n)^{1/2} = 188 - 30 \log(MI)$

[**)] This is a linear low density polyethylene (LLDPE), ie, a copolymer of ethylene with an olefin comonomer.

[***)] All low density polyethylene samples were obtained from Unifos Kemi AB (now Neste Polyethylene Co).

The following four photosensitizers were used in the experiments: Benzophenone (BP), 2-Chlorobenzophenone (2-CBP), 4-Chlorobenzophenone (4-CBP) and 4,4'-Dichlorobenzophenone (4,4'-DCBP). BP was supplied by KEBO AB, Sweden, 4-CBP by Fluka AG, Switzerland, 2-CBP and 4,4'-DCBP by Novakemi AB, Belgium. All sensitizers were used as received without further purification.

As crosslinker triallylcyanurate (TAC) from Novakemi AB, Switzerland, was used, and a few alifatic allyl ethers which will be reported in later papers.

Sample Preparation. 50 g samples of polyethylene powder or granulate with the desired amounts of sensitizer and cross-linker (usually 0.25 to 0.50 g of each) were mixed in a Brabender for 8-10 min at 140 to 200°C. The compound samples were hotpressed to sheets at 140 to 180°C for 3 to 4 min in a Carver press.

UV Irradiation. The irradiation with UV light of the polyethylene sheets of a thickness from 0.1 to 5 mm was carried out in a UV-Cure equipment constructed in this laboratory as shown schematically in Figure 1. The irradiation box B is ventilated with a fan at the top V to cool the UV lamp L which is mounted in a holder H which can be moved up and down along two stands P to change the distance from lamp to samples. The irradiation cavity C can be moved in and out of the box along a track T_1 to start and stop the irradiation. The temperature of the sample S on a metal plate can be measured (±2°C and regulated up to 250°C by a temperature detector T, and a heating wire W connected to a temperature controller. The cavity is covered at the top by a fused quartz plate Q transparent to the UV light. To create an inert atmosphere nitrogen gas is introduced into the cavity through a small hole I and blown out through thin slits under the quartz plate Q. The UV lamp used in most experiments is a PHILIPS HPM-15 high pressure mercury lamp (doped with lead/gallium iodide additives) operated at 1 or 2 kW at a distance of 100 to 200 mm from the sample.

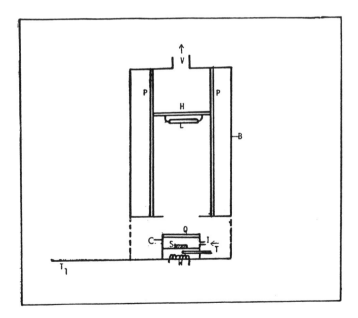

Figure 1. Schematic diagram of the irradiation equipment UV-CURE.

Measurement of Gel Content.   Samples after irradiation are cut into thin slices and put in a basket made from 200 mesh stainless steel net. The basket is immersed in boiling xylene (containing 0.2 % TINUVIN 144 antioxidant) for 48 hours (solvent renewed after 24 hours) with pure nitrogen bubbling through to prevent oxidation. After extraction the basket is washed with pure acetone and dried in vacuum over night at 65-70°C. The gel content is taken as the weight percent of insoluble polymer network in the sample.

Density Measurement.   The density of the samples is measured using a density gradient column according to ASTM D 1505-68. Ethanol/water mixtures are used to form the density gradient in the column and all measurements are made at 23°C.

Results and Discussion

High Density Polyethylenes.   In preliminary experiments 4-chlorobenzophenone (4-CBP) was found to be the most efficient photosensitizer for crosslinking both without and with triallylcyanurate (TAC) added. The dependence of gel content on the concentration of 4-CBP added to HDPE is shown in Figure 2.

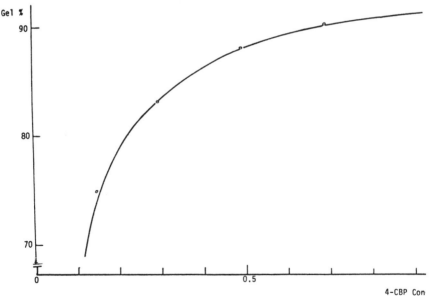

Fig.2 Dependence of gel content on concentration of 4-CBP
TAC 1% N_2: 1 1/min, T=155°C, irr. time: 15 sec.

After 15 sec of irradiation time at 155°C a gel content of
about 90 % is obtained with 0.7-1.0 % 4-CBP and 1 % TAC. A high
molecular mass of PE favours crosslinking as shown in Figure 3.

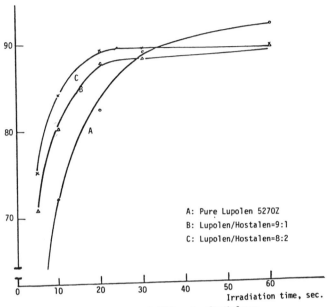

Fig.3 Crosslinking of different polyethylenes

The pure Hostalen sample has too high melt viscosity to be
handled in our process (cf Table 1). Blends of Hostalen and
Lupolen with 1 % 4-CBP and 1 % TAC were therefore prepared and
UV-irradiated. Pure Lupolen required about 30 sec irradiation
to reach 90 % gel content while the 9:1 and 8:2 blends with
Hostalen reached that gel content in 15 to 20 sec. The effect
of molar mass is most pronounced at short irradiation times
which is of importance for commercial applications.

The sample temperature during UV irradiation has a strong
effect on the crosslinking reaction as shown in Figure 4. Below
the melting point gel contents of about 70 % are obtained and
above MP about 90 % under otherwise the same conditions as in
Fig. 2 and 3 (4-CBP and TAC 1 %, temp. 155°C, PE blends 8:2).

The addition of TAC as crosslinker is very important for the
initial rate of the crosslinking reaction (Figure 5). With 1 %
4-CBP and without TAC it takes about 30-40 sec at 155°C to
reach 25 % gel content. With 1 % TAC added 85-90 % gel content
is reached in 10-15 sec. An increase of the TAC content from
1 to 2 % has hardly any effect.

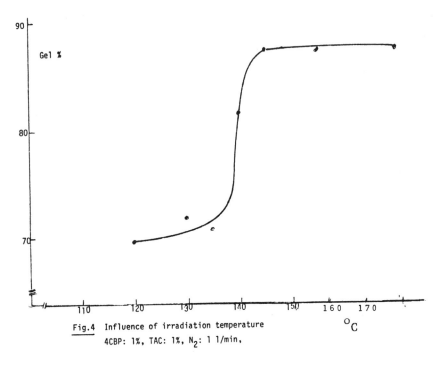

Fig.4   Influence of irradiation temperature

4CBP: 1%, TAC: 1%, N_2: 1 l/min,

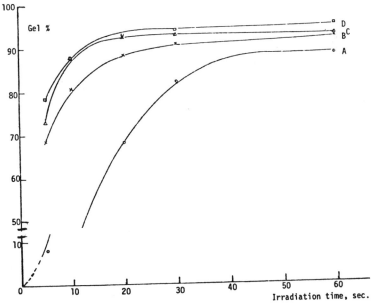

Fig. 5 Effect of TAC on crosslinking rate

4-CBP: 1%, N_2: 1 l/min, T=155°C

TAC : A--0%, B--0.5%, C--1%, D--2%

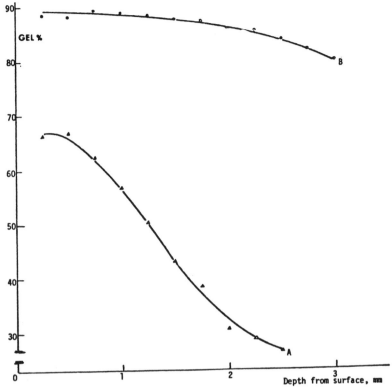

Fig. 6 Effect of TAC on homogeneity of crosslinked PE
4-CBP : 1%, N_2: 1 l/min, T=155°C   TAC : A--0%, B--1%

The addition of TAC as crosslinker has a most pronounced effect on the depth of penetration of the crosslinking reaction (Figure 6). Sheets of 5 mm thickness with 1 % 4-CBP and without TAC (curve A) and with 1 % TAC (curve B) were irradiated for 15 sec at 155°C. The irradiated sheets were cut into 0.25 mm thick slices using a microtome and the gel content of each slice measured by extraction. The homogeneity of the crosslinking is greatly improved by addition of TAC.

Stability of Crosslinked PE in Hot Water. A possible future use of photocrosslinked polyethylene is for hot water pipes. To test the stability in boiling water, PE samples with 1 % 4-CBP and 2 % TAC were pressed to 0.30-0.35 mm thick sheets and UV-irradiated to a gel content around 90 % and then immersed in boiling distilled water. Samples were taken out at intervals (after 7, 16, 30, 60 and 90 days) and dried over silica gel in a desiccator to remove adsorbed water. Gel content and IR spectra were recorded and the results given in Table 2.

Bengt Rånby et al.

Table 2

Time im-mersed in boiling water days	Sample 1			Sample 2		
	Gel content	-C-O-C- bonds	C₃N₃ groups	Gel content	-C-O-C- bonds	C₃N₃ groups
	%	%	%	%	%	%
0	88.8	100	100	88.8	100	100
7	89.7	93.5	98.5	90.3	95.3	98.9
16	87.4	91.7	96.7	88.6	92.5	95.5
30	91.2	88.0	92.3	87.6	90.7	92.0
60	90.5	85.6	90.1	88.1	88.8	90.9
90	-	83.3	(95.6)	-	86.4	(96.6)

The data of gel content show no deterioration of the net-work structure of the PE on prolonged boiling water treatment, indicating stability against hydrolysis. The IR adsorption spectra, however, reveal a steady decrease in both ether bond (1140 cm^{-1}) and triazine ring (820 cm^{-1}) content in the samples. There is no change in OH group content (~3600 cm^{-1}). A possible interpretation is that part of the cyanurate groups are isomerized to isocyanurate as Gillham |15| has reported on heat treatment of poly(TAC).

Low Density Polyethylenes. Low density polyethylenes have a more irregular and branched structure and are usually less stable to oxidation and degradation than high density poly-ethylenes which are essentially linear polymers. It is there-fore of great interest to compare the crosslinking behaviour of the two types of polyethylenes. In a first series of exper-iments, a conventional LDPE (DFDS 6600 in Table 1) was photo-crosslinked with 1 % of the four different sensitizers added (BP, 2-CBP, 4-CBP and 4,4'-CBP) but without TAC added as crosslinker. The samples were pressed to 2 mm thick sheets and irradiated at 140°C with a 2 kW HPM-15 UV lamp at 100 mm dis-tance. Gel content measurements show (Figure 7) that the crosslinking reaction for LDPE is slower than for HDPE (comp. Figure 5) and that 2-CBP is a less efficient sensitizer than the other three. The gel content for photocrosslinked LDPE has an upper limit of 65-70 % compared with about 90 % for HDPE. Without sensitizer added, the gel formation is insignifi-cant (Figure 7).

The sensitizer and the crosslinker in combination enhance the rate of photocrosslinking strongly as shown in Figure 8 for 0.5 mm thick samples containing 1 % TAC (A), 1 % sensitizer (B) and 1 % of both additives (C). 4,4'-CBP (Δ and x) is more efficient than 4-CBP (o and ●).

The rate of gel formation is a function of sample thickness which is shown for sheets containing 2 % 4-CBP and 2 % TAC, irradiated increasing length of time (Figure 9). To reach 20 % gel content requires irradiation times of 16, 23, 32 and 37 sec for samples of 0,5, 2, 3 and 5 mm thickness, respectively.

The presence of an antioxidant in the polyethylene sample slows down the rate of the crosslinking reaction and requires larger amounts of sensitizer (Figure 10). DFDS 47 contains 0,2 % SANTONOX R and S-118975 no antioxidant. The antioxidant apparently consumes some of the sensitizer in the photocrosslinking process.

The temperature of the specimen is important for the photocrosslinking reaction (Figure 11) both for LDPE and HDPE, obviously an effect of chain segment mobility. For HDPE a sharp increase in gel content is recorded at the melting point of the polymer (135-140°C) as previously shown (Figure 4). For LDPE the increase in gel content is gradual through the whole melting range (Figure 12) for all sensitizers from room temperature to 100-110°C where complete melting occurs. The LDPE reaches a level of gel content (70 to 75 %) which is 15 to 20 % lower than that of HDPE (Figure 11). Again the 4,4'-DCBP is a somewhat more efficient sensitizer than 4-CBP.

The mechanism of the photoinitiated crosslinking reaction for polyethylene is still not resolved. Our preliminary analysis has related the basic reaction to the double bonds and to hydrogen abstraction, the probable formation of allyl radicals on the chains or at chain ends and crosslinking by combination of allyl radicals or addition to other radicals, eg, polyethylene chain radicals. Our IR spectra before and after photocrosslinking have clearly shown that vinyl, vinylidene and vinylene groups of the polyethylene are involved in the reaction. This is well in line with the results on photocrosslinking of EPDM elastomers by Hilborn and Rånby |16|. Further work on the mechanism is in progress.

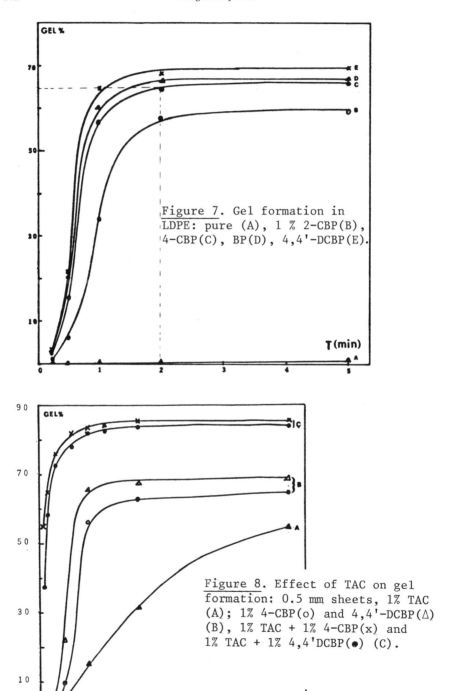

Figure 7. Gel formation in LDPE: pure (A), 1 % 2-CBP(B), 4-CBP(C), BP(D), 4,4'-DCBP(E).

Figure 8. Effect of TAC on gel formation: 0.5 mm sheets, 1% TAC (A); 1% 4-CBP(o) and 4,4'-DCBP(Δ) (B), 1% TAC + 1% 4-CBP(x) and 1% TAC + 1% 4,4'DCBP(●) (C).

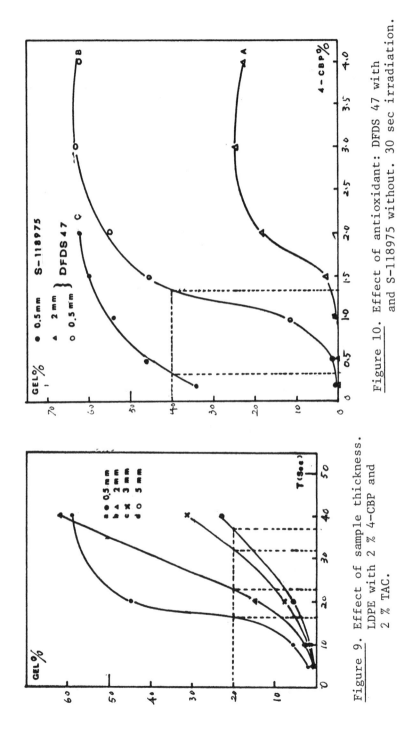

Figure 10. Effect of antioxidant: DFDS 47 with and S-118975 without. 30 sec irradiation.

Figure 9. Effect of sample thickness. LDPE with 2 % 4-CBP and 2 % TAC.

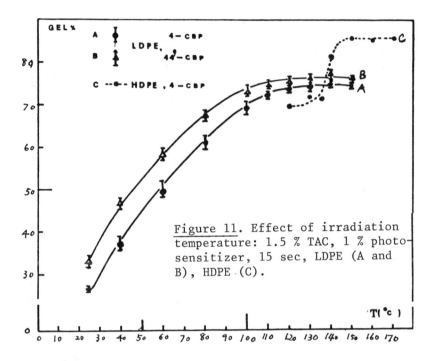

Figure 11. Effect of irradiation temperature: 1.5 % TAC, 1 % photosensitizer, 15 sec, LDPE (A and B), HDPE (C).

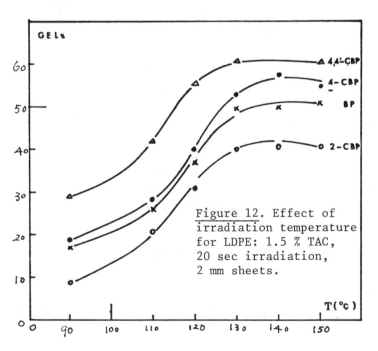

Figure 12. Effect of irradiation temperature for LDPE: 1.5 % TAC, 20 sec irradiation, 2 mm sheets.

Part II. PHOTOINITIATED CROSSLINKING OF POLYESTERS

The commercial glass fibre reinforced polyesters are of various types but most of them are crosslinked at elevated temperatures (>100°C) using a free radical initiator and styrene as added monomer [17]. Styrene is volatile and causes environmental problems, both during curing in the plants and during enduses of the resulting laminates due to unreacted residues of styrene [18]. Therefore, we have studied the possible use of non-volatile monomers for photoinitiated curing of polyesters at low temperatures. Such systems are described in this paper, involving synthesis of modified unsaturated polyesters with reactive endgroups, addition of polyfunctional acrylate monomers and a photofragmenting initiator, and photocrosslinking of the polyester compound as matrix in glassfibre mats using UV irradiation. The properties of the cured polyesters have been measured.

Experimental

Materials. Fumaric acid (FA), phtalic anhydride (PA), propylene glycol (PG), and neopentyl glycol (NPG) as monomers and glycidyl acrylate as modifier are used for synthesis and endcapping of the unsaturated polyester, respectively. For the curing trimethylolpropane trimethacrylate (TMPTMA) or n-hexanediol diacrylate (HDDA) and benzoyldimethylketone (BDK) are added as reactive diluent monomers and photoinitiator, respectively. As reinforcing material, glass fibre mats of E-type glass weighing 450 g/m^2 are used.

Preparation of Polyester. The unsaturated polyesters are prepared by a two-step condensation reaction under nitrogen with toluene added as azeotrop at the end of each step. In the first step, the temperature is raised to 160°C. In the second step, the temperature is raised to 200°C and the reaction vessel evacuated below 40 mm Hg. The excess of acid monomer regulates the molecular weight of the polymers which then are terminated with carboxyl groups.

The molar mass of the polyester is determined by titration of the carboxyl groups and acetylation of the residual hydroxyl groups. The $\overline{M}_n$-values are calculated by dividing the amount by the sum of acid and hydroxyl endgroups. The carboxyl polyester endgroups are reacted with an excess of glycidyl acrylate which is slowly added as a toluene solution during vigorous stirring at about 70°C. The toluene is removed by evacuation below 30 mm Hg and the endcapped polyester is stored under refrigeration until used.

Lamination. A given amount of multifunctional acrylate ester and photoinitiator are added to molten acrylate modified unsaturated polyester using a glass rod for stirring. To obtain a 2 mm thick laminate with about 50 % glass fibre, four layers of glassfibre mats are used. The resin is uniformly cast on the preweighed glassfibre mats in an open mold (88x88 mm) with a Mylar film at the bottom and heated on an electric plate to 100°C. When the glassfibre mats are completely impregnated, another Mylar film is applied as cover, and the air bubbles still present are squeezed out with a roller.

Photocuring. The laminates are UV-irradiated at room temperature under nitrogen atmosphere in the UV-CURE irradiator, described previously in this paper (Figure 1), at a distance of 100 mm from the HPM-15 lamp operated at 2 kW. Each side of the laminate is irradiated separately. After curing, the laminate is separated from the mold and the Mylar films are stripped off. The cured laminates are cut into 88x12 mm strips using a diamond-tooled circular cutting sawblade. Some of the strips are cut with V-shaped notches from both sides. The distance between the notch bottoms is 5.7 mm.

Measurements of Laminate Properties. The glassfibre/resin ratio is analyzed by ashing in the samples in a muffle furnace at a final temperature of about 700°C for half an hour.

The water absorption is measured by weighing after immersion in distilled water at 23±1°C for 24 hours.

For tensile testing the specimen is clamped into a 100 KN INSTRON universal tester with a jaw separation of 50 mm and extended at a rate of 0.5 mm/min to failure.

For bending testing of flexural modulus (FM) the specimen is loaded with a three-point system in an INSTRON 1122 instrument using a 50 mm span with the rate of crosshead motion 0.2 mm/min.

The Charpy impact strength is measured with a 1.67 kg pendulum of 100 kgcm capacity to break unnotched specimens.

Results and Discussion

We have found that benzoyldimethylketone (BDK) is a most efficient initiator in our system of polyester compound/glassfibre/Mylar film. BDK has a broad absorption band at 315 to 390 mm which is in the same range as the UV spectrum of the HPM-15 lamp (Figure 13). The Mylar film has practically no

absorption in this UV range and has very low permeability for
oxygen. Therefore, when the air bubbles are squeezed out, the
Mylar film protects the laminate from atmospheric oxygen and
does not interfere with the UV light during photocuring.

The tensile strength (TS) and tensile modulus (TM) of the
cured polyester laminates with polyesters of different molar
mass ($\overline{M}_n$) are shown in Table 3. Both TS and TM decrease with
increasing $\overline{M}_n$. This is interpreted as due to the relative
amounts of acrylate endgroups which are more reactive than
the maleic/fumaric double bonds along the polyester chains.
In the continued studies we have, therefore, used polyesters
in the low $\overline{M}_n$ range of 1000 to 1300.

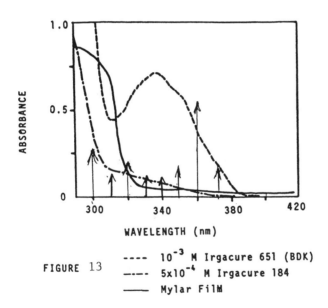

FIGURE 13

---- 10^{-3} M Irgacure 651 (BDK)

—·—· 5×10^{-4} M Irgacure 184

—— Mylar Film

The UV curing is a fast reaction. Glass fibre laminates
containing 50 % glass fibres, 35 % unsaturated polyester and
15 % HDDA to which 1 % BDK is added, reaches "level-off"
properties already after 10 sec UV irradiation (Figure 14).

To reach sufficiently low viscosity of the polyester compound for lamination, it is necessary to add a diluent monomer, preferentially multifunctional and reactive to increase
the rate of UV-curing. The effect of TMPTMA concentration
from 10 to 60 % in the resin with 38.8 % glassfibre is shown
in Figure 15. TS and TM reach a high level already with 20 %

Table 3 Effect of Molecular Weight of the Unsaturated Polyester on Tensile Properties of the Laminate

Glass Fiber Content[*](%)	42.5		
TMPTMA Content[**](%)	20		
Molecular Weight (Mn)	1180	2158	3400
TS (10^2 MPa)	1.62	1.53	1.43
TM (10^3 MPa)	9.30	8.85	7.13

[*] : based on the total weight of laminate.

[**] : based on the weight of the resin mixture.

TMPTMA in the resin while impact strength increases more gradually and reaches a high level at about 40 % TMPTMA.

The glassfibre content, varied from 24.2 to 65.4 % of the laminate, has also a strong effect on the mechanical properties (Table 4). In this case the resin contains 70 % polyester and 30 % HDDA to which 1 % BDK is added as photoinitiator. Tensile strength (TS), tensile modulus (TM) and flexural modulus (FM) all initially increase with increasing glassfibre content, reach a maximum level at about 55 % glass fibre content and then decrease. The mechanical properties of the photocured polyester laminates compares well with those of conventional laminates cured thermally with styrene as reactive monomer. – The photocured polyester laminates are well crosslinked as indicated from their low water absorption (0-12 %) after 24 hours immersion in distilled water.

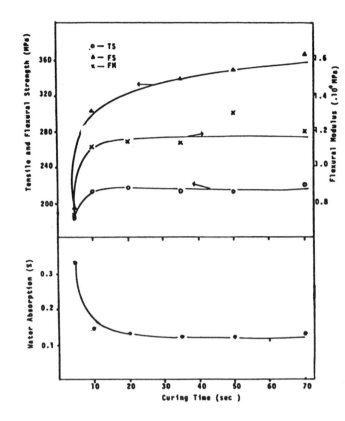

Figure 14. Effect of curing time on mechanical strength and water absorption.

Table 4 Effect of Glass Fiber on Mechanical Properties of the Laminate**

Glass Fiber Content (%)*	24.2	31.2	42.0	47.7	54.9	65.4
TS (.10^{2}MPa)	0.75	1.20	1.52	1.73	2.32	1.82
TM (.10^{4}MPa)	0.42	0.53	0.77	0.77	0.98	0.74
FS (.10^{2}MPa)	1.55	2.10	2.72	2.98	3.86	1.49
FM (.10^{4}MPa)	0.85	0.90	1.07	1.30	1.72	0.38

* based on the total of glass fiber plastic laminate.
** with 30% of HDDA and 1%of BDK, both based on the weight of the resin mixture.

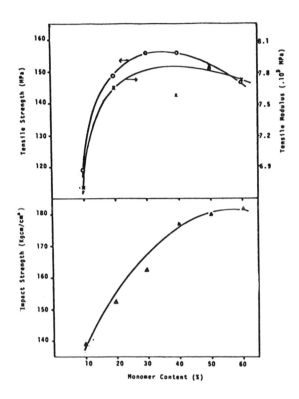

Figure 15. Effect of added monomer on tensile
strength and modulus and impact
strength of polyester laminate.

Acknowledgements

Our thanks are due to the National Swedish Board for Technical Development (STU) for financial support throughout this work and to Professor Jan Bäcklund for good advise and help with the measurements of mechanical properties.

References

[1] Oster, G. J. Polymer Sci. 1956, 22, 185; ibid. Chem. Ed.
     1962, B2, 1181.
[2] Oster, G.; Oster, G.K.; Moroson, H. J. Polymer Sci.
     1959, 34, 671.
[3] Wilski, H. Angew. Chem. 1959, 71, 612.
[4] Chien, Pao-Kung et al. Collected Papers, Inst. of
     Applied Chemistry (PR of China), 1960, 4, 145.
[5] Chien, Pao-Kung et al. Sci. Sinica 1962, 11, 903.
[6] Charlesby, A.; Grace, C.S.; Pilkington, F.B. Proc. Royal
     Soc. 1962, A268, 205.
[7] Labana, S.S. (Ed) Ultraviolet Light Induced Reactions
     in Polymers, ACS Symp. Ser. 1976, Vol. 25.
[8] Lawton, E.J.; Balwit, J.S.; Powell, R.S. J. Polymer Sci.
     1958, 32, 257.
[9] Lu, Xi-Ci; Chen, Dong-Lin Chinese J. Applied Chem. 1984,
     2, 87.
[10] de Boer, J.; Pennings, A.J. Makromol. Chem., Rapid
     Commun, 1981, 2, 749.
[11] de Boer, J.; Pennings, A.J. Polymer 1982, 23, 1944.
[12] Bousquet, J.A.; Haldar, B.; Fouassier, J.P.; Vidal, A.
     Europ. Polymer J. 1983, 19(2), 135.
[13] Rånby, B. "Photochemistry of Polymers - New Trends and
     Possibilities" in "Polymer Science in the Next Decade"
     (ed. O. Vogl). Presented at the International Symposium
     Honoring Professor Herman F. Mark on his 90th Birthday,
     May 1985.
[14] Rånby, B.; Shi Wen Feng ACS Natl. Meeting, Denver,
     Colorado, April 1987. Polymer Preprints 1987, 28:1, 297.
[15] Gillham, J.K.; Mentzer, C.C. J. Appl. Polymer Sci. 1973,
     17, 1113.
[16] Hilborn, J.; Rånby, B. Photocrosslinking of EPDM Elasto-
     mers. Photocrosslinkable Compositions. Oral Presentation
     at this Symposium.
[17] Parkyn, B.; Lamb, F.; Clifton, B.V. Polyesters, Vol. 2;
     Unsaturated Polyester and Polyester Plasticizers. Iliffe
     Books Ltd, London, 1967.
[18] Encyclopedia of Polymer Science and Technology, Vol. 11,
     p. 129-168. Interscience Publ., New York and London, 1969.

RADIATION INDUCED GRAFTING OF STYRENE SULFONIC ACID

K-SALT ON ETHYLENE-VINYL ALCOHOL COPOLYMER

A. Mey-Marom and S. Shkolnik

Department of Radiation Chemistry
Soreq Nuclear Research Center
Yavne 70600, Israel

Introduction

Radiation induced grafting has been used frequently to impart new properties to given polymers [1]. Many studies have been published on the grafting of weak acids such as acrylic acid (AA) on polyethylene (PE) but very few on the grafting of strong acids, e.g. styrene sulfonic acid.

An indirect method for introducing a sulfonic group onto PE has been reported by Chen et al [2]. In their work styrene was radiolytically grafted on PE; the benzene ring was then chlorosulfonated and finally hydrolyzed to form the styrene sulfonic acid graft copolymer. Indirect methods are used since attempts to graft directly vinyl monomers bearing sulfonic groups yield very low degrees of grafting. Shkolnik and Behar [3] concluded that the highly ionized SO_3^- groups with their hydration spheres are incompatible with PE and cannot diffuse into the PE bulk, thus preventing the polymerization from taking place. They reported a two-stage method of grafting where styrene sulfonic acid K-salt (SSKS) was directly grafted on PE which had been previously hydrophylized by radiation grafting of AA.

Hydrophylic monomers such as Nylon-4 and Nylon-6 have been used as substrates for radiation grafting of styrene sulfonic acid Na-salt by Lai et al [4] and Yamakita et al [5]. They applied a ^{60}Co γ-ray direct irradiation method and studied the characteristics of the obtained membranes for desalination or separation purposes by reverse osmosis.

In the present study a copolymer of ethylene and vinyl alcohol was used as the substrate for the preirradiation grafting of SSKS. The synthesis and several characteristics of the obtained membranes were studied. The influence of the

substrate's orientation   on   the   synthesis   and   membrane
features was emphasized.

Experimental

Materials. Monooriented   (F   type,   20   microns)   and
biooriented (XL type,   15   microns)   films of ethylene-vinyl
alcohol copolymer   (32:68   mol %) EVAL-EF grade from Kuraray
Inc., Japan   were   used   as received. 4-Vinylbenzenesulfonic
acid K-salt >97%, 4-tert-butylcatechol purum (both from
Fluka AG) and   potassium chloride, analytical reagent (BDH)
were used with no further purification.

Graft polymerization. Film   samples   of 40x80 or 80x100mm
were preirradiated with   a   High Voltage electron beam (EB)
accelerator at   520   kV,   4   mA,   1-5   passes (1-5 Mrad) and
placed in   SSKS   solution   at   50-80°C   for   2-40   min under
CO_2 flushing. After   grafting,   the   samples were heated in
distilled water   at   60°C   for   2   hours   in order to remove
traces of   monomer   or   homopolymer. The   washed films were
dried in   vacuum   oven   4   hours   at   50°C   and weighed. The
percent of grafting %Gr (add-on) was calculated as follows

$$\%Gr = \frac{(w_g - w_i)}{w_i} \times 100$$

where w_i and w_g represent   the weights of the initial and
grafted films,   respectively.

Viscosity measurements.   The   grafting   solutions   were
collected after   use   and   diluted with water by a factor of
10. An   Ostwald-Fenske   capillary   viscosimeter   was used to
determine the viscosity of the diluted solutions.

Thermal analysis. A Du Pont 990 Thermal Analyzer was used
for DSC   measurements   under   conditions   of   50   ml/min N_2
purge flow,   10°C/min   heating   rate   (up   to   300°C) and
calibration with indium.

Tensile strength   and   elongation measurements. The films
were cut   to   standard   "dogbone"   samples   and tested on an
Instron tensile   machine Model D-1024-16 at a strain rate of
5mm/min.

Measurement of water uptake and change in area by water
swelling. The   grafted films were swelled in distilled water
for 24   hours,   wiped and weighed as quickly as possible.This

film, referred to as the wet film, was then dried in a vacuum oven, and weighed again. The area of both the wet and dry films was measured after a similar treatment. The water uptake WU was calculated as following:

$$WU = \frac{W_{wet} - W_{dry}}{W_{dry}} \times 100$$

where W_{wet} and W_{dry} are the weight of wet and dry films, respectively.

Electric resistance measurements. The grafted films were conditioned 2 hours in distilled water at $70^{\circ}C$ and 48 hours in 3N KCl. The electric resistance was measured in a 3N KCl solution at RT by using a Hewlett-Packard 4328 A milliohmmeter. The specific electric resistance was calculated according to the following equation:

$$\rho = R_m \times S/l$$

where ρ is the specific electric resistance, R_m is the electric resistance of membrane (the measured resistance minus electrolyte resistance, 0.295 ohm) ,S is the membrane area in contact with electrolyte (0.95 cm^2) and l is the membrane thickness.

Grafting Reaction Characteristics

At identical conditions the monoriented grafted films showed higher add-on of SSKS than the bioriented ones. This phenomenon may be attributed to the higher hydrophylicity of the mono- vs bioriented EVAL films as it is reflected in the higher moisture capacity (8 vs 5.9%) and higher water transmission rate (10 vs 2 g/m^2 hr).

Chapiro [6] has found a minimum (4%) water content in PVA films which is required in order to achieve any grafting, as most common monomers are unable to diffuse into the dry PVA. However, when the packing of the polymer chains is loosened through the swelling action of water, the monomer can penetrate more readily into the film and graft copolymerization is favored. It may be assummed that these conclusions are valid also in the case of Eval films.

Dimensional changes were observed in the grafted films. It was found that the grafting reaction induces an increase of the monoriented film's area whereas the bioriented film

shrank. It is supposed that biorientation of the films
generates a greater degree of macromolecular alignment in
comparison with monoorientation. In this case the growth of
grafted SSKS chains would be hindered in the machine and
transverse directions. Actually it was found that the bioriented film increased in thickness as a result of grafting
More detailed results of dimensional changes will be
presented later.

Preirradiation dose. Figures 1 and 2 show the degree of
grafting and time curves, for various preirradiation doses
on monoriented and bioriented Eval films. It can be seen
that in all cases the degree of grafting is higher when the
preirradiation dose is increased and this behavior is
independent of film orientation. However, in the case of the
monoriented film the maximum add-on is almost achieved at 40
minutes of grafting. The logarithmic relationship between
the rate of grafting (in the low range of grafting times)
and the preirradiation dose is shown in Figure 3. The
dependence of the grafting rate on the preirradiation dose
was calculated from this figure and found to be 0.47 and 0.7
for the monoriented and the bioriented films, respectively.

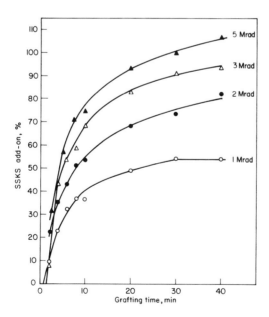

Figure 1. Plot of SSKS add-on vs grafting time for
monoriented EVAL film preirradiated at various doses.
Grafting conditions: 20% SSKS/H_2O, 70°C, purging gas CO_2.

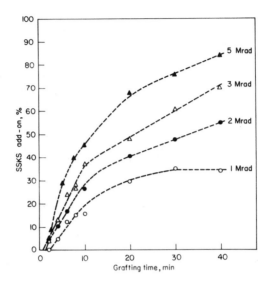

<u>Figure 2</u>. Plot of SSKS add-on vs grafting time for
bioriented EVAL film preirradiated at various doses.
Grafting conditions as in Figure 1.

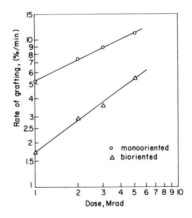

<u>Figure 3</u>. Logarithmic plots of grafting rate vs
preirradiation dose for mono¬ and bioriented EVAL films.
Grafting conditions as in Figure 1.

These values indicate that the grafting rate on bioriented
films grows faster when the preirradiation dose is increased
This phenomenon is unexpected as both films are supposed to

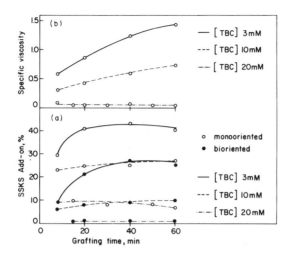

Figure 4. SSKS add-on (a) and specific viscosity of grafting solution (b) vs concentration of tert-butylcatechol. Grafting conditions: EB preirradiation, 520 kV, 4 mA (3 Mrad); 20% SSKS/H_2O; 70°C; 8 min; purging gas CO_2.

have the same chemical structure. Further studies are required to clarify this point.

Inhibition of homopolymerization. The grafting of SSKS is accompanied by homopolymerization in the grafting solution. Already after 8 min of grafting in a 20% SSKS/H_2O solution at 70°C the specific viscosity grows up to a 2.75 value. In order to diminish this undesirable side reaction, tert-butyl catechol (TBC), a common inhibitor for styrene, was introduced into the grafting solution. The influence of TBC on SSKS grafting add-on and solution viscosity was determined. The results are presented in Figure 4. They show that at 10mM TBC the specific viscosity is reduced five-fold to a 0.5 value and yet the add-on values are 10 and 27% for bi- and monooriented films, respectively. Decreasing TBC concentration to a 3mM value raises the SSKS add-on to 25 and 40%, yet the specific viscosity is 1 (at 25 min of grafting). The use of this inhibitor has to be optimized and a search has to be made for other more efficient ones.

Membrane Characterization

ΔH Fusion of grafted EVAL films. Heat of fusion of semicrystalline polymers is assumed to be proportional to

the crystalline content [8]. In order to assess the
influence of SSKS grafting on EVAL copolymer crystallinity,
DSC of grafted EVAL samples was performed. The thermograms
of the monooriented films are presented in Figure 5. It may
be seen that the untreated film has a melting point of
178°C. When SSKS is loaded on the film by grafting there
is no significant change in the melting point. However the
area of the melting peaks decreases gradually with the
increase of SSKS add-on. This phenomena indicates, as
expected, that the grafting process induce damage to the
crystallinity of EVAL polymer.

A similar phenomenon was observed for the bioriented film.

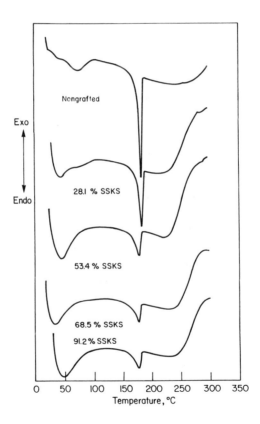

Figure 5. DSC thermograms of monooriented EVAL films grafted
with various amounts of SSKS. Experimental conditions:
purging gas 50 ml/min N_2, heating rate 10°C/min.
Grafting conditions: EB preirradiation, 520 kV, 4 mA (3 Mrad)
20% $SSKS/H_2O$; 70°C; 2-40 min; purging gas CO_2.

Mechanical properties. The tensile strength in the
machine direction of SSKS grafted films is shown in Figures
6 (monooriented) and 7 (bioriented). Grafting of small
amounts of SSKS (10-20%) reduces considerably the tensile
strength of EVAL films. In the case of monooriented film
~20% $SSKS_2$ add-on reduces the tensile strength from 720 to
400 kg/cm^2; further loading with SSKS does not essentially
change the strength. For the bioriented film, an initial
relatively high deterioration of strength is observed at 10%
SSKS add-on and then a further gradual decrease proportional
to the degree of grafting. The maximum strength decrease
observed for the bioriented film is from 1720 to 800
kg/cm^2. This shows, that the bioriented film even with a
grafting degree of 70% is stronger than the virgin
monooriented film.

The yield strength of the untreated monooriented film is
approximately the same as that of the bioriented one.
However, the bioriented film exhibits a strengthening beyond
the yield point which is reflected in a much higher break
strength when compared to monooriented film. The deformations
which occur at this stage are attributed to the actual
displacement of the macromolecular chains with respect to
each other [9]. Thus in the case of bioriented film which is
supposed to have a higher degree of molecular order than the
monooriented one, it may be expected that molecular slippage
will be more remarkable. Introduction of grafted SSKS

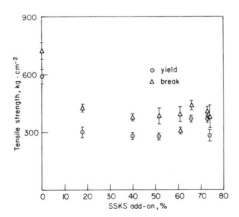

Figure 6. Tensile strength (machine direction) vs SSKS
add-on for monooriented EVAL film. Grafting conditions: EB
preirradiation, 520 kV, 4 mA (5 Mrad); 20% $SSKS/H_2O$;
$80°C$; 2-40 min; purging gas CO_2.

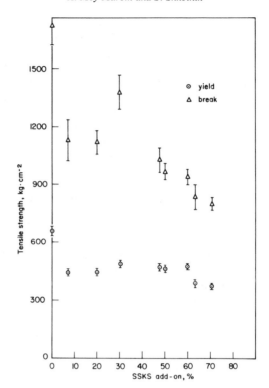

Figure 7. Tensile strength (machine direction) vs SSKS add-on for bioriented EVAL film. Grafting conditions as in Figure 6.

macromolecules will reduce the "contact area" between EVAL chains, as well as the molecular slippage. Such a phenomenon will be reflected in a reduced break strength of bioriented EVAL films, as was actually found.

Elongation in the machine direction of monooriented grafted films is deteriorated from 240 to 160%. The bioriented film's elongation is improved from 80 to 140%. Overall it is seen that at 70-80% SSKS add-on, the elongation of monooriented films is slightly higher than that of the bioriented films.

Sorption of water. The water absorption of polymeric membranes is of importance for different uses of the membranes [10]. The water uptake and change of area (ΔS) of SSKS grafted mono- and bioriented EVAL films is presented in Figure 8. The absorption and dimensional change behavior of the monooriented grafted films may be divided into two parts. First, from 0 to 17% add-on SSKS, WU increases from 8 to 117% and ΔS increases up to 22%. Second, from 17 to 90% add-on SSKS, WU increases from 117 to 190%, while the change of ΔS is more pronounced (from 22 to 110%). Thus, two assumptions are reasonable concerning the swelling behavior of monooriented films grafted with 17-90% SSKS: a) the absorbed water molecules cause extension of polymeric chains in the machine and transverse directions; b) the grafted film acts as an "expandable water reservoir" whose surface area increases in proportion to the amount of absorbed water.

The water absorption and change of area of bioriented SSKS grafted films are linearly dependent on the degree of grafting. However, the actual values are much lower than those obtained for monooriented films under similar conditions. These results indicate a better structural stability of bi- versus monooriented films, a phenomenon which was already observed during grafting experiments. The structural stability of the bioriented grafted films when exposed to water absorption may be attributed to the extension of the polymeric chains when the film was drawn in

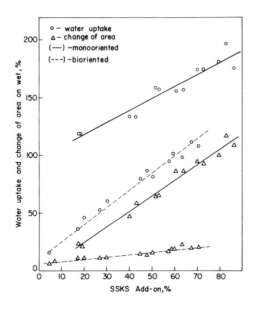

Figure 8. Water uptake and change of area of EVAL films grafted with various amounts of SSKS. Grafting conditions as in Figure 6.

two directions; thus the potential for further expansion of
the polymeric chains and, hence, increased water uptake are
diminished.

It has to be mentioned that the monooriented EVAL films
grafted with SSKS have higher WU values than the maximum of
83% in membranes of PE grafted by a two stage method with
102% polyacrylic acid and 49% SSKS [3]. This feature may be
an advantage of EVAL/SSKS membranes for uses where high
water uptake is required.

Specific electric resistance. In order to evaluate the
potential of SSKS/EVAL membranes for separation uses in
electrochemical cells, the specific electric resistance ρ
was determined. The results are presented in Figure 9. Two
main characteristics are observed in the results: a) the
grafting with   SSKS induces a very steep decrease in ρ of

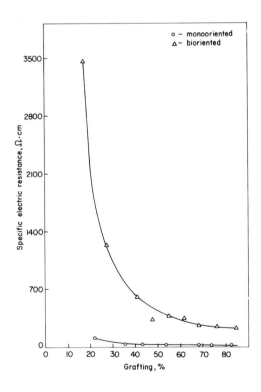

Figure 9. Specific electric resistance of EVAL films grafted
with various amounts of SSKS. Grafting conditions: EB
preirradiation, 520   kV,   4 mA (2 Mrad);   20% SSKS/H_2O;
70°C, 2-40 min, purging gas CO_2.

EVAL films, and b) the ρ values obtained for the
monooriented films are generally lower than for the
bioriented films with the same SSKS add-on. These phenomena
are expected since the ρ actually reflects the transport of
electrolyte mass through the membrane under electrodriving
conditions. The monooriented film, with less alignment of
the macromolecular chains, would be more permeable to
electrolyte than the bioriented one.

The specific electric resistance of several radiation
grafted membranes is presented in the following table:

Table 1. Specific Electric Resistance of Several Radiation
Grafted Polymeric Membranes.

Polymer[a]	Grafted Monomer[b]	ρ, $\Omega \cdot$ cm	Ref.
LDPE	62% MAc	4-10	11
LDPE	100% MAc	7	12
LDPE	212% AAm	27	13
LDPE	100% MAc	180	14
PE	120% MTFA	100	15
PVF_2	82% ($VPY+CH_3J$)	130-160	16
PVF_2	70% AAc-K-salt	30-40	16
EVAL	73% SSKS	15	This work

[a] LDPE = Low density polyethylene
    PE = Polyethylene
    PVF_2 = Polyvinylidene fluoride

[b] MAc = Methacrylic acid
    AAm = Acrylamide
    MTFA = Methyl α,α,β-trifluoroacrylate
    VPY = Vinyl pyridine
    AAc = Acrylic acid

The data presented in the table show that the performance
of the SSKS/EVAL membrane is in the lower range of ρ
compared with other radiation grafted polymeric membranes.

Conclusions

1. Copolymer of ethylene and vinyl alcohol is readily grafted with p-styrenesulfonic acid K-salt by the EB preirradiation method. The grafted copolymers obtained have a SO_3K content of up to 2.5 and 2.1 meq/g for mono- and bioriented films, respectively.

2. The degree of grafting is directly related to the irradiation dose in the range of 1-5 Mrad.

3. 4-tert-Butylcatechol is an effective inhibitor of SSKS homopolymerization. However, the grafting degree is also reduced. Thus, the conditions under which the inhibitor is used have to be optimized.

4. The effects of SSKS grafting on mono- versus bioriented ethylene and vinyl alcohol copolymer films are:
   a. The bioriented film is loaded with less SSKS than the monooriented one, under the same conditions. It is assumed that this phenomenon is linked to a greater degree of macromolecular chains alignment in bi- vs monooriented film.
   b. In monooriented film the changes in area and thickness are nearly of the same magnitude, whereas in the bioriented film most of the change is seen in the thickness.
   c. The H fusion is reduced by the grafting process, but is not influenced by the film's orientation. It may be assumed that the film orientation has no influence on the crystallinity of the films.
   d. Tensile strength in the machine direction is diminished by the grafting process: from 720 to 400 kg/cm^2 for mononooriented films and from 1720 to 800 kg/cm^2 for bioriented ones. Though the damage of the bioriented films is higher than that of the monooriented ones, their ultimate strength is still higher.
   e. Water uptake increases linearly with SSKS add-on and is in the ranges of 20-100% and 120-190% for bi- and monooriented films, respectively. Swelling of the membranes expands their area; this phenomenon is more remarkable in the monooriented films.

   f. The specific electric resistance is reduced by SSKS add-on. From a grafting degree of about 70%, no change in the ρ (15 $\Omega \cdot cm$) of the monooriented films is observed. The ρ of the bioriented films reveals a leveling off at 85% SSKS add-on when its value is 115 $\Omega \cdot cm$

Acknowledgments

The authors wish to acknowledge the technical assistance of Ms. C. Barash in performing the grafting experiments. They also wish to thank Kuraray Inc., Japan and Cotrimex Co. Tel-Aviv for the kind supply of EVAL films.

References

[1] See, for example, Chapiro A. "Radiation Chemistry of Polymeric Systems", Interscience Publishers: London, 1962.

[2] Chen, W.K.W.; Mesrobian, R.B.; Ballantine, D.S.; Metz, D.J.; Glines, A. J. Polym. Sci. 1957, 23, 903.

[3] Shkolnik, S.; Behar D. J. Appl. Polym. Sci. 1982, 27, 2189.

[4] Lai J.Y.; Chang T.C.; Wu Z.J.; Hsieh T.S. J. Appl. Polym. Sci. 1986, 32, 4709.

[5] Hiromi Yamakita; Kiyoshi Hayakawa J. Appl. Polym. Sci. 1979, 23, 303.

[6] Ref. 1, p. 675.

[7] Shkolnik S.; Mey-Marom A. unpublished results.

[8] Wendlandt, W.Wm. "Thermal Methods of Analysis"; John-Wiley & Sons: New York, 1974; p.304.

[9] Kinney, G. F. "Engineering Properties and Applications" John-Wiley & Sons: New York, 1957; pp. 182-184.

[10] Haruvy, Y.; Rajbenbach, L.A.; Jagur-Grodzinski, J. J. Appl. Polym. Sci. 1986, 32, 4649.

[11] Pekala W.; Achmatowicz, T.; Kroh, J. Radiat. Phys. Chem. 1986, 28(2), 173.

[12] Ging-Ho Hsiue; Wen-Kuei Huang J. Appl. Polym. Sci. 1985, 30, 1023.

[13] Hegazy, E.A.; El-Dessouki, M.M.; El-Sharabasy, S.A. Radiat. Phys. Chem. 1986, 27(5), 323.

[14] Hidechi Omichi; Chandbury, D.; Stannett, V.T. J. Appl. Polym. Sci. 1986, 32, 4827.

[15] Hidechi Omichi; Jiro Okamoto J. Appl. Polym. Sci. 1985 30, 1277.

[16] Ellinghorst, E.; Niemoller, A.; Vierkotten, D. Radiat. Phys. Chem. 1983, 22(3-5), 635.

Part II
Photo/electro-conductive and Piezo-electric Polymers

HOMOGENEOUS AND HETEROGENEOUS PHOTOCON-
DUCTING POLYMERIC SYSTEMS

Marian Kryszewski

Center of Molecular and Macromolecular
Studies, Polish Academy of Sciences
90-362 Łódź, Poland

Abstract

Photoconductivity of various polymers
received considerable attention with
regard to electrophotography and related
processes. Structure-property-relat-
ionship in photoconductivity has been
investigated only in some cases. This
work reviews some studies directed towards structural modifications of polymers with specific photoconducting properties, e.g., polypeptides with carbazole chromophores. New intense UV light
sources also make it possible to study the
photoconductivity of some known polymers,
which are not considered as photoconductors,e.g. polyamides, polyimides, polyethylene terephtalate. Blending is a
known method which is widely used for
modification of polymer properties. We
have supplied it to change the properties of the blend of poly /N-vinyl car-
bazole/ /PVK/ with polycarbonate /PC/.
This system is not miscible on a molecular scale. We have carried out photoconductivity studies of this system,inclu-
ding carrier generation, trapping and
recombination,as well as simultaneous
thermoluminescence and thermally stimulated currents. They have shown that
it is possible to obtain information on
photophysical behaviour /different trapping processes/ in relation to specific
relaxation processes in such heterogene-

ous systems. The knowledge gained can
establish a base from which photoconducting processes of this and similar heterogeneous systems can be modified according to specific applications.

Introduction

A large variety of polymeric systems has been
studied and evaluated with regard to photoconductive and photosensitive properties. It is due to the
increasing demand for such materials in electrophotography and related processes. Since 1957 when the
photoconductive properties of poly/N-vinylcarbazole/
/PVK/ were first reported, the interest in photoconductive polymers is concentrated around PVK and
other polymer structures containing carbazole as
the active chromophore. Through incorporation of
carbazole moiety in other polymer chains, many possibilities exist for tailoring appropriate chemical
structures to gain enhanced photoconductivity. PVK
and related systems can be appreciably modified by
introduction of different dopants.
    The aim of this work is not to discuss the basic
concepts of photoconductivity of polymers and to
analyse the experimental techniques commonly employed in this context along with a review of original and patent literature.
    The large number of reviews and books dealing
with this subject /see, e.g. [1-7] / which have appeared in recent years indicate that photoconduction
in polymers holds considerable interest. It appears
that there are still many possibilities to synthesize new photoconducting polymeric systems but there
are also many difficulties in establishing a realistic quantitative description of photoconduction
in macromolecular systems. Important progress has
already been made regarding new possibilities opened by general theories of charge transport in organic solids and their various applications.
    This paper attempts to highlight some aspects of
the progress in characterization and preparation of
selected photoconducting systems. Emphasis will be
on discussion of structure-photoconductivity relationships taking into account also such materials
which are not considered as typical photoconductors.
In addition, some aspects of the photoconductivity
in heterogeneous polymeric systems /polymer blends/

will be discussed in view of the efforts made to obtain photoconductivity in systems with improved
mechanical properties.

In this article I have omitted tables and figures
which can be found in the cited papers. I present
some results and a personal view on some questions
which seem to be of interest concerning the progress
in this area.

We will also exclude from our discussion the
characterization of large number of materials which
are photosensitive and became crosslinked.

Influence of chemical constitution

and structure on photoconductivity

of polymers based on carbazole

Photoconductivity of carbazole-based polymers is
still the subject of many works. In spite of their
wide practical application in electrophotography,
there are many unsolved problems related to elucida-
tion of different aspects of photoconducting behaviour of these materials.

The irradiation of poly/N-vinyl carbazole/ /PVK/
films with strong UV light sources, e.g. mercury
lamps,causes the appearance of new bands in its electronic and fluorescence spectrum. This is due to
photooxidation /evidenced by the photooxidative
behaviour of model substances/. The same effect was
confirmed by IR spectra, and the presence of carbonyl
groups was shown [8]. Even if the concentration of
photooxidation products is very low, they act as an
efficient electron accepting impurity and enhance
the yield of photocarrier generation. The presence
of impurities, however, influences many electrophotographic characteristics of PVK and the purification of commercial PVK would be advised [9] as well
as such preparation of PVK which leads to a high
extent of syndiotactic diades [10]. Such PVK exhibits much higher electrophotographic sensitivity
after sensitization, e.g., with crystal violet.

It is interesting to compare PVK films obtained
in different experimental conditions. Using the
glow discharge method [11] one can obtain PVK films
which are photoconducting in UV and visible regions
of the spectrum /conventional PVK is not photocondu-

ctive in visible range/. This difference in the behaviour of glow discharge PVK is due to creation of
various chemical structures, stable radicals, and
higher concentration of vinyl groups, as well as
crosslinking /insolubility/. Nespurek and Cimrova
[12] have analysed in detail the intrinsic and extrinsic photogeneration of charge carriers in amorphous
and partially crystalline PVK as well as in films
of PVK obtained by plasma polymerization or electrochemical polymerization. It was shown that the polymer disorder influences the initial separation distance r_o of carriers. There is a distribution of electron-hole separation distances, and the distribution
parameter α for intrinsic photogeneration was found
to be different for crystalline and amorphous PVK
films. Different values of primary quantum yield for
the formation of electron-hole pairs were also established for crystalline and amorphous materials.
The presence of oxygen as well as of other acceptors
causes an increase of the quantum efficiency. A similar effect was observed for films treated by ano-
dic oxidation or ion implantation. These results
show that molecular structure and organization
/crystallinity/ play an important role in photoconducting behaviour of PVK.

Knowledge of the exact mechanism of recombination
of electron-hole pairs created in PVK and other amorphous photoconducting solids under illumination is
of basic importance; thus many studies have been
devoted to this problem. In molecular crystals or
organic liquids, geminate recombination /GR/ of coul-
ombically bound pairs is a diffusive process which
is completed within 10-100ps. In PVK, however, the
life time of hole-electron pairs was shown to be up
to 100s [13]. This difference is due to the disorder
inherent to a polymer or any organic glass. The
kinetics of geminate pair recombination in PVK has
been recently studied by monitoring the temporal
decay of the delayed fluorescence by Bässler et.al.
[14]. They have shown that for 20 μs<t<10s the recombination rate obeys a $R(t) \approx t^{-s}$ law, s being of the
order of unity. A comparison with Monte Carlo simulation indicates that intrinsic energetic disorder
of PVK controls the short time behaviour and for
the long time behaviour trap-to-trap recombination
should be taken into consideration, contrary to
what is expected for a bimolecular excitation reaction. It was found that the fluorescence intensity

is temperature independent for 20K<<T<<80 and decreases by a factor of 4 as T is raised to 294K,
whereas diffusion-controlled bimolecular reactions
should follow an exp $\left(- T/T_0\right)^2$ temperature dependence characteristic of electronic transport in
energetically random media [15-18]. These conclusions are important because they show that amorphous
structure of the photoconducting materials plays
an important role in their photoconductivity. There
are, however, investigations of electronic transport
in glassy silicon-backbone polymers which are in
some disagreement with above-mentioned conclusions
on the recombination mechanism. They will be discussed later considering double layer systems.

It seems interesting to note a novel attempt to
use PVK for the studies of photochemical phenomena
in membranes. A very thin PVK film was mounted between two electrolite solutions /these conditions
ensure the lack of photoinjection and reflection
usually associated with metal electrodes/ [19] . For
the film thickness 10 μm and cross-section 1 cm^2,
the dark current of $6 \cdot 10^{-15}$ A was observed. Irradiation λ>330 nm of the surface with positive bias
yields higher photocurrents /by $\approx$ 10%/ then that
observed for a negative bias. Doping of these thin
films with dyes or electron acceptors /e.g. with
crystal violet/ results in an increase of the photocurrent intensity/I_{ph}/ by six orders of magnitude as
compared with the dark current. For undoped membrane this ratio is 10^3. This photocurrent increases
linearly with applied field and with dye concentration at first but subsequently decreases with the
amount of dye being introduced.

In common with other photovoltaic cells, there is
a linear relation-ship between photovoltage and
logarithm of light intensity for open circuit while
under short-circuit condition I_{ph} is a linear function of light intensity. Finally, one shall note
the quasi linear relation between film thicknes
and I_{ph}. All these observations, as well as the day
concentration dependence, which has a maximum for
equal relative concentration of dye and carbazole
unit, can be explained taking into consideration
the penetration of dye molecules between carbazole
rings. In that way one can elucidate the sequence
of charge migration events between one helix and
adjacent helix of PVK. This model was confirmed by
the kinetics studies of charge transfer which was

shown to be dependent on the intensity and wavelength of irradiation as well as on the rate of redox step. The excess dye molecules are ineffective in photoelectron transfer /they may disturb the polymer conformation and reduce the effective light intensity/. Such a system is pH sensitive because of the excited dye reaction with O_2 molecules. The created superoxide radical can reduce the carbazole radical cation when it is not protonated and transfered to the electrolyte solution. The presence of acid is important because of further redox reaction.

These remarks lead us to a consideration of some features of doped PVK systems. Owing to a variety of dopants used and large number of papers dealing with these systems, they can not be treated here in detail. It is reasonable, however, to mention that dopants are used as charge transfer agents in order to shift the photoconductivity of PVK into visible region. An inherent defect in the sensitization of PVK is that the sensitizing material is dispersed throughout the polymeric matrix /molecularly dispersed molecules or their aggregates/. In such systems there is a possibility for time-dependent phenomena which are related to formation of a metastable solid solution. Alternatively, evaporation of sensitizers onto the polymer surface causes their removal by abrasion. It was suggested that one can use the systems in which the sensitizing molecule is chemically bound to the PVK in a controlled manner [20]. These molecules can be attached preferentially to the surface of a film which eliminates the time-dependent phenomena.

The most efficient photoreceptors used in electrophotography today are layered devices in which photogeneration function is separated from the charge transfer function [21,22]. Such design provides maximum design flexibility and an opportunity for an independent optimization of each layer. The carrier generation layer /CGL/ is optimized primarily for spectral response /including IR/ and the carrier supply efficiency. As a matter of fact, the CGL can be only thick enough as to absorb most of the incident light. The CGL materials are highly absorbing and the useful thickness of the CGL is typically 0.3-2 μm. The mechanical properties /usually poor/ are unimportant since they are thin. The charge transport layer CTL is characteristic of flexibility, toughness, mechanical integrity

and resistance to the environment. These CTL layers
are organic glasses which can transport charges, e.g.
PVK or solid solutions of discrete transport molecules dispersed in appropriate polymers, e.g. polycarbonate or polyester. Many papers have been dealing with charge carrier transport, which is chromophore concentration, temperature and electric field
dependent. Carrier mobility u is a tunable factor.
    Photoconductivity of double-layer polymer-poly-
mer systems doped with charge-transfer complex has
been examined. Using polycarbonate and PVK doped
with charge-transfer agent for carrier generation,
it was shown that the photoconductivity threshold
and long wave absorption band of the charge-transfer complex are correlated [23]. Several xerographic characteristics have been examined for a double
layer polymeric receptor comprising CTL of poly
(-9-vinyl-anthracene) /40μm/ and CGL of PVK /1μm/.
It has been shown that these systems exhibit a good
charge retention and that the potential barrier at
the interface of two layers is sufficient to stop
carriers flow in the dark. This is to be regarded
as a very advantageous feature since the electrostatic pattern on the photoreceptor can be retained
for a considerable time. The quantum efficiency is
more or less constant over the visible spectral
range. This system is characteristic of a specific
dependence of log/exposure/ v s. log/intensity/. It
is not straight line parallel to the log/intensity/
axis . It seems to be suitable for short exposure
time photocopier systems and, due to high charge
retention, it seems to be good for systems with time
lag between exposure and development.
    In view of the particularly effective influence
of the carbazole moiety on photoconductivity of
organic molecules many attempts have been made to
synthesize several novel polymer systems containing
carbazole. It is not possible to discuss here all
of the chemically modified poly/N-vinylcarbazole/
systems. They have been discussed to large extent
in a review paper by Biswas and Uryu [30], taking
into consideration the methods of preparation as
well as some characteristics of photoconducting
properties. Usually these systems are copolymers of
N-vinylcarbazole with various comonomers; see, e.g.,
[24-29] . It is not easy to make a general correlation between chemical constitution and photoconducting properties. The physical structure of these

materials is not known; usually they are amorphous solids but some of them can crystallize. A number of these systems exhibit photoconductivity without dopants but some of them require dopants.

It seems interesting to mention some of them in which the photoconductivity-structure relationship is clearly shown. For a random coil, amorphous isotropic polymer, such as PVK, imperfect tacticity, conformational distribution as well as the distribution of chain lengths causes many structural defects in spatial arrangements of chains and side chains. In that case, the relative position of the carbazole unit is not regular. For theoretical analysis of photoconductivity which requires averages of the angular distribution function this problem is very complex. The effect of chain length on photoconductivity was discussed and it was shown that higher molecular weight materials exhibit better conductivity [31]. Other molecular parameters have been investigated only in a few cases. It was expected that a regular structure with respect to spatial arrangement of carbazole units will enhance the photoconductivity because carrier transport occurs between the nearest chromophores /electronic properties of carbazole units are anisotropic/.

This concept of molecular order has been shown to be valid for many conducting polymers, e.g. oriented doped polyacetylenes. It shall be noted that the stretching results in enhancement of conductivity in the orientation direction without any appreciable decrease of that in perpendicular direction. This fact can be interpreted in terms of a cooperative dependence of the conductivity on the long range order. For polydiacetylene and poly-p-phenylene single crystals, the anisotropy of conductivity was clearly shown which is easy to be elucidated as well as for strong CT-complexes which crystallize in stacked structures of various types. Charge carrier mobility of 1,3-diphenyl-5-/p-chlorophenyl/-2-pyrazoline was clearly shown to increase by three orders of magnitude in single crystals as compared with the glassy state [32]. Thus it was obvious to study the influence of orientation on conductivity as well as to examine liquid crystalline materials. The stretching of PVK above glass transition temperature produces only small orientation. The order parameters $<P_2>$ was of <0.2 $<P_2> = 3<\cos^2\Theta>-1$ where Θ is the angle between stretching direction

and the vector describing a chain backbone segment/.
The order parameter was determined by dichroic IR
absorption [33] . The ratio of photoconductivity in
the stretching direction and normal to the film
plane is not known. Some interesting information on
the photoconductivity can be obtained from studies
on excimer emission /emission from excited state of
two chromophores in intimate contact/. The emission
anisotropy r as a function of the order parameter
was investigated /film plane excited at 344 nm and
the emission was observed from the thin edge/. The
shape of emission spectra was not affected by the
orientation but it should be noted that r changes
slightly with parameter $<P_2>$. It is not a sufficient
result for a detailed analysis because the theory
requires r to be a function of $(1+B<cos^2\theta>-<cos^4\theta>)^{-1}$,
B being a parameter of the system. The data on
$<cos^4\theta>$ are not available [34] .

Our studies of the conductivity of oriented PVK
films obtained by extrusion of PVK at temperatures
higher then T_g have shown that the dark conductivity
increases four times in the stretching direction
for stretching ratio $\varepsilon/\varepsilon_0=4$. Photoconductivity of
this material was also higher by one order of magni-
tude as compared with non-stretched material. The
same behaviour was found for solvent cast films,
stretched and annealed at high temperature /270 -
330 C/ however, the increase of photoconductivity
was smaller [35] . These results show that morphology and orientation have an influence on photoelectric behaviour of this material. Highly ordered
structures in PVK have been found by other authors
/see, for e.g., reference [36] / .

Another method of obtaining oriented systems is
the use of a liquid crystalline polymer containing
photoconductive chromophores which can be brought
into a stacked array due to the specific structure
of these systems.

The films of poly-[S-(2-9-carbazolyl) ethyl] -
L-cysteine (β-forming poly-α-aminoacid with a carbazole side chain) cast from dimethyl formamide
/DMF/ and 1,2-dichloroethane show high photoconductivity [37] . The photocurrent in the film obtained
from DMF /a good solvent/ is smaller than that in
the film cast from 1,2-dichloroethane /a poor sol-
vent/. It seems that the carbazole rings are to be
stacked in greater order in the latter case.

The photoconductivity of a poly-γ- (β'-N-carba-

zolylethyl - L-glutamate, alone or doped with low
concentrations of 2,3,5-trinitrofluorenone /TNF/,
has also been reported [38]. The system behaves as
a p-type photoconductor at a dopant concentration
of < 30% and as an n-type above 30%. This allows
rectification of the multilayer polymer films, such
as creating a p-n junction by forming an n-type
monomer dopant layer on top of a p-type film by
immersing the latter in a solution /benzene or methyl
ethyl ketone/ of the dopant. At lower concentrations
of the electron acceptor, $I_{ph}^+ > I_{ph}$, similar to the
trend observed in the case of the unmodified polymer.
On the other hand, with increasing TNF concentration
$I_{ph}^- > I_{ph}$. It is known that PVK shows similar behaviour; thus one can conclude that unmodified poly
(γ-β-N- carbazolylethyl)L-glutamate has the properties of a p-type conductor and the well-doped polymer exhibits n-type behaviour.

Because the poly-α-amino acids are forming lyotropic cholesteric phases due to their α-helical
conformation, work is still going on obtaining
such materials. A novel liquid crystalline system
has been obtained having pendent carbazole groups:
poly- N- chlorocarbonylcarbazole lysine. This polymer of high molecular weight in the α-helical conformation exists in concentrated solutions in tetrahydrofurane in a form of cholesteric lyotropic mesophase clearly shown in the micrographs. This structure causes that the emission and absorption oscil-
lators of the carbazole unit are planar; thus such
system in proper homotropic nematic alignment may
exhibit properties of photoelectric liquid crystal.
It should be pointed out that this polymer shows in
dilute solutions very structured excimer emission
[39,40] .

Similar or even more simple oriented structures
can be obtained by doping of commercially available
low-molecular-weight materials with carbazole molecules. They can be oriented by guest-host mechanism
[41] . Due to the fact that carbazole-containing compounds are photoconductive in the UV-range of the
spectrum /c.a. 335 nm/, the host matrix and electrodes have to be transparent in that range. It has
been shown that the appropriate low-molecular-weight
liquid crystals doped with purified carbazole of
the concentration of dopant in the range $10^{-1}-10^{-2}$ M
exhibit photoconductivity. The measurements were made
in thin cell /liquid crystalline film thickness 23um/

consisting of two quartz plates with Indium-Tin
Oxide layers which were subsequently covered with
thin insulating SiO film using an ion gun at 30° to
the plane. It insures the homogeneous orientation
of the nematic phase. The reorientation of the
liquid crystal molecules, due to their positive
dielectric anisotropy, was induced by application
of 10V. The presence of the dopant causes an increase of the steady state conductivity by a factor of
four. Similar experiments carried out with carbazole
doped isotropic paraffin oil have shown no measurable effect. The change of conductivity, upon switching on the light, is fast but its decay is slow.
The reason for it is not clear as yet. The analysis
of the above-mentioned effects is difficult because
the liquid crystalline matrix exhibits its own photoconductivity due to the presence of impurities [42].
Evidently the photoconductivity is not related to
specific phenomena at the electrode. The dependence
of the observed photoconductivity on applied voltage
shows that the critical field is needed which corresponds to the Freederiksz transition being easily
observed in the polarizing microscope. I have discussed these results in some detail because they
show the possibility of doping of liquid crystalline
materials resulting in photoconductivity in analogy
to other photoelectric effects involving such types
of systems [43,44].

These remarks aimed to show that the appropriate
structure and mutual orientation of photoactive
groups in photoconducting polymers leads to an improvement of photoresponse. The real situation is really
very clear because even in the case of simple polymers and copolymers based vinyl carbazole it was
shown that there are some contradictions.
A reinvestigation [45] of photoconductive behaviour of poly β-N-carbazolylethyl vinyl ether has
shown that hole mobility in this material is comparable to that of PVK at similar field strengths. It
shows the difficulties of attempting to predict simple
steric models for design of photoconductive polymers
with high quantum efficiency of carrier generation
and appropriate transport properties.

Influence of chemical constitution and structure on photoconductivity of miscellaneous polymers

Recent years have also witnessed   progress in the studies of photoconductivity of a large number of polymers, including: polyolefins, polystyrene, substituted styrene polymers, vinyl polymers, various unsaturated polymers, polyesters and condensed aromatics. In some papers the studies on photoconductivity were connected with the discussion of photo-cross-linking and photodegradation reactions in these polymers. It is due to the increasing interest in the use of polymers as photoresist materials.

It is not possible to analyse here in detail the results presented in an ample literature concerning this subject. It seems reasonable, however, to mention some of them, taking into consideration photoconductivity-structure relationship and some really new materials.

Photoconduction of polyethylene in the spectral range between the visible and UV involves extrinsic carrier generation [46]. In the vacuum range one has found /from the action spectrum studies/ a threshold energy of 84 eV which corresponds to the band gap in agreement with theoretical predictions. Similar UV-induced photoconductivity was found in polytetrafluoroethylene and electrons or holes are the predominant photocarriers.

Photoconductivity studies of polyvinylidene fluoride [47] with $\lambda < 3000$Å have shown that the photoconductivity is roughly independent of the existence of different crystalline phases in the sample /phases α and δ/ but the β-phase causes a reduction of crystallinity. This may be due to irreversible changes which occur at high fields in addition to structural phase transitions within the crystalline phase /crystalline reorientation, pooling in the amorphous regions, etc./ leading to new structural trap formation. Recent investigations of photoconduction in i-polystyrene [48] did not result in any new observations which could change our knowledge in that respect. All features are well explained in terms of Onsager model of geminate dissociation and recombination of the excited benzene nuclei. In the case of modified PS /by addition of charge transfer agents or by copolymeri-

zation/ the photosensitivity can be shifted to visible range of spectrum. Charge carrier transport
through day-sensitized polystyrene films occurs via
electron hopping through localized states.

Interesting photosensitive polystyrene-based polymers have been described but these materials cannot
be discussed here. The same remark concerns acrylate
polymers which, due to various modifications, exhibit
photoconductivity and can be made photocrosslinkable
/incorporation of pendent photosensitive groups/.

Uniform, structurally pure polyvinyl-chloride/
/PVC/ films exhibit photoconductivity alone or with
iodine doping [49] . Photoconductivity increases with
increasing voltage for both polarities. The increase
of photoresponse in PVC with iodine concentration
is apparently due to charge-transfer interactions
/creation of holes/.

A special interest shall be given to poly/vinyl
alcohol/ films containing Cu^{+2} complexes obtained
by treatment with $Cu_2/NO_3/_2$ and other simple Cu-salts
[50] . These materials exhibit a photoconductivity
which depends on the nature of the salt-anion, its
mole fraction and wavelength of illumination. The
photocurrent is also bias dependent. These features
can be explained by complex formation and free
radical formation. Unpaired electron clouds overlap
among a number of complexes, implying the formation
of a network. Similar, but somehow different, phenomena were observed for this and other polymers containing transition metal salts. Photoconductivity
of these systems is due to the photoredox reactions
of the polymer-metal complex [51] . In that case the
structure of the system related to specific coordination of metal ions is of basic importance.

Acetylene polymers /unmodified polyacetylene
and modified diacetylens/ have been investigated,
with regard to their photoconductivity, to a high
extent. This is due to conductivity of acetylene
polymers when doped with appropriate reducing or
oxidizing agents. Photoconductivity of polyacetylenes
has been deeply discussed in the handbook of conducting polymers [52]; thus it seems not necessary
to give here a condensed review of the photobehav-
iour of these polymers. Polydiacetylenes modified
with highly polarizable carbazolyl groups in the
side chain exhibit a photoconductivity which is
red-shifted as compared to some other polydiacetylenes [53] .

Single crystals of poly-1,6-N-carbazoyl-2,4-hexa-
diene are a very good model for the investigation of
sensitized photoconductivity in polydiacetylene in
the range of 300-400nm. There is a correlation between carrier yield and carbazole absorption which
indicates that the excitation of the carbazole moiety produces carriers at a higher quantum efficiency
than the excitation of the conjugated backbone chain.
Polydiacetylene-toluene sulfonate was a subject of
many photoconductivity investigations because this
polymer exists in perfect single crystals. The
photoconductivity spectrum of this crystal is characteristic of a steep photoconductivity edge at
0.8 eV [54]. It is related to a localized state of
the same energy below the conduction band and the
dominant charge carriers are electrons. The photocurrent dependence on electric field is superlinear. These facts were interpretated using a model
which assumes that current flow can be analysed applying one-and three-dimensional versions of Onsager's theory.

Polyphenlyacetylenes are semiconductors which
can be doped with certain acceptors and easily form
thin films. Doping with iodine enhances the photocurrent, which increases linearly with I_2 concentration /in the range of 0.4-3%/. In undoped polpheny-
lacetylenes dark and photocurrents are controlled by
shallow traps for electrons, but in the doped materials carrier generation results from charge-transfer interactions in the bulk of polymers [55].

Poly/arylene-vinylens/ are an interesting class
of photoconductive polymeric systems [56-58]. The
spectral sensitivity of these polymers can be enhanced by many electron transfer agents. Recently
their electrochemical properties have been investigated. It was shown that these polymers are a novel
class among the main chain conjugated polymers which
can be oxidized to different radical ion states.

Attention has been given also to naphtalene and
anthracene polymers but their photoconductivity in
the undoped and doped state is not particularly different from that of polyphenlacetylenes, thus they
do not need special considerations.

Photoconductivity of polyamides was found already in the early 70's. Nylons exhibit a photoconductivity induced by pulsed UV light. This method was
applied in order to reduce thermal activation in
these ionic macromolecular systems [59]. The dark

conductivity in these polymers is of ionic nature but investigations on drawing of nylon films have shown that stretching has an effect on the photoconductivity. For nylon 6 the photocurrent decreases with the draw ratio. The photocurrent intensity is higher in the stretching direction than in the transverse direction. It is a clear example of the crystallinity and morphology influence on photoconducting behaviour. It is interesting to note that the photoconductivity of nylon films does not depend significantly on introduced dye concentration [60]. The maximum of photocurrent for dyed nylons is found to be proportional to the cube of the power of light quanta being absorbed in such a material. This superlinear relation is connected with the change in the carrier mobility proportional to the light intensity and with the change of carrier concentration proportional to the square of the light intensity.

The photoconduction of polyimides /e.g. Kapton H/ is characteristic of a maximum at 460 nm. It is independent of polarity or electrode material being used. These observations can be elucidated taking into account an intrinsic photogeneration since the photocurrent exhibits a maximum which corresponds to the onset of optical absorption. The generation process of carriers seems to be uniform in the bulk of material. Tricyclic polyimides show enhanced photoconductivity. The photocurrents in the visible range of the spectrum as well as the dark currents show an ohmic relation against the applied electric field. It is interesting to note that there is not an evident structure-photoconductivity relationship because the photoconducting behaviour changes in a different way than, e.g., the thermal stability, which is directly related to the increase of the size of the fused rings.

Polyethylene terephtalate /PET/ is an interesting material from the view point of photoconductivity. Thin films of this polymer exhibit X-ray-induced photoconductivity characteristic of transit time which is shorter with decreasing electric field. Radiation-induced charge carriers are highly localized. Their transport occurs through hopping process. There is a distribution of the energy levels of the localized sites as well as a distribution in the intersite separation. All these processes show that there is a distribution of hopping times.

The transport of charges induced by irradiation can
be explained on the basis of a dispersive transport
model [61], assuming a power law involving a parameter depending on the hopping mechanism on the distribution of traps depth and intersite separations.
This power law of hopping times, however, is not
valid over a broad range and that additional traps
with long hopping time are present in this polymer.
PET also displays the strong field-dependent generation rate and short-time photoconduction is thermally activated.

An interesting feature is that the transient
photoconductivity of amorphous and of semicrystalline PET is indistinguishable. This fact suggests
that the morphology of the polymer does not really
influence the radiation-induced photoconductivity
in opposite to the relations which have been found
for the dark conductivity. It is possible that in
region with long-range order the charge carriers
become localized on a very short time scale and
subsequently the short-range order is limiting the
transport. The radiation-induced photoconductivity
of PET in its crystalline from is considerably reduced by doping this polymer with 2,4,6-trinitro-
fluorenone which acts as a deep trap for the photocarriers. This result seems to be significant, since the production of radiation-hardened good dielectric can be made by such chemical doping of this
and other polymers. The studies on photoconductivity of PET show that not only the chemical constitution but also the physical structure of this semicrystalline material influences its photoconductivity.

Many other results of photoconductivity investigations of various polymers, e.g., aromatic polymers,
ladder or partially ladder polymers, polyquinones,
doped rubbers and different polyquinoxalines, have
been published. The last-mentioned polymers show a
rather high photoconductivity.

These remarks show that the knowledge of photoconducting behaviour of various polymers is already
substantial. The availability of intense light sources offers many possibilities for futher pursuit,
particularly in the field of photoconductivity
studies of polymers which absorb in the UV range
of the spectrum. Further work is needed in order to
obtain materials with enhanced photoactivity, and
on their possible applications.

Recently a new class of photoconducting polymers, polysilylens, has been discovered and investigated in systems consisting of double-layer system with a-Se or AsS_3 / CGL / or by direct generation of carrier using nitrogen-pumped dye laser. For years it was assumed that silicon atoms cannot make long polysilane chains similar to carbon polymers, but oligomeric polysilanes have been known since more than forty years [62]. They received more attention in the studies of West [63]. Later on, permethylpoly-silanes have been used as precursors for silicon carbide fibers [64]. In the early 80's, first soluble polysilanes were synthetized [65,66] and soon photo-conductivity in poly methylphenylsilane /PMPS/ was discovered [67]. This discovery made them the object of studies in view of their application in electro-photographic systems. The same concerns other poly-silanes, e.g. poly(methylcyclohexylsilane), poly(me-thyl-n-propylsilane), poly(methyl-n-octylsilane) and copolymer of dimethyl and methylphenyl silan /1:1 ratio/. Films of these polymers can be cast from toluene, forming clear transparent colourless layers. Charge transport in poly(methylphenylsilane), a typical representative of these polymers, is thermally activated and relatively non-dispersive over a wide range of temperatures [$68, 69$]. The charge carrier mobility μ is of 10^{-4} cm^2/V·s at E= 10^5 V/cm. The μ values are field dependent at T<T_g. The polysilanes are characteristic of hopping transport among disc-rete states. It is sensitive to the nature of side groups and the molecular weight. The time-of-flight experiments carried out in the wide range of tempe-ratures and electrical fields. Current transients were also generated intrinsically using pulses of a dye laser. The discussion of obtained results have shown that μ is temperature dependent according to Arrhenius relation, taking into consideration effe-ctive temperature T_{eff} [70]. Similar behaviour has been found for PVK and other organic glasses [71]. Recent numerical simulation studies provide an alter-native model for hopping transport in organic glas-ses, which assumes that carriers are injected and transported in a disorder – induced gaussian distri-bution of localized states [72]. This model predicts a non-Arrhenius temperature dependence of mobility in the long time limit as well as linear dependence of logμ on electric field. The question regarding the microscopic origin of drift mobility in organic

solids is not resolved but the studies on photoconductivity of this new class of polymers make it possible to obtain more information on charge migration
in glassy organic polymers. It seems interesting to
note that in PMPS, charge transport is not associated
directly with the aromatic side groups since similar
efficient transport has been found in polysilanes
with aliphatic pendent groups with no π electrons.
Therefore transport involves states associated with
the silicone backbone itself. Charge transport in
PMPS is insensitive to the molecular weight of this
polymer, which seems to indicate that transport states are not really extendet over the whole length
of the chain. They remain localized in short segments of the chain. This is a new and very interesting result.

Natural polymers, e.g. cellulose and chitosan, have
been studied mostly in order to evaluate their photoconductivity for crosslinking. More attention
must be focused to modify these polymers in such
a way to increase their photoconductivity and/or
photosensitivity, because of an interesting and promising variety of such polymeric systems.

Photoconductivity in heterogeneous
polymeric systems

It was mentioned before that photoconductive
N-vinylcarbazole copolymers having electron donor
or electron acceptor groups in the same polymer
chain show interesting electrophotographic properties; e.g., with acrylate monomers [24] . The incorporation of such comonomers reduces the stiff brittle
character of PVK in favour to increase flexibility,
yielding better film properties. As expected, these
polymers show lower T_g values but they are less pho-
toresponsive in general, which may be understood on
the basis of a diminished concentration of photoactive groups.

Blending is now a widely used method to obtain a
variety of polymeric materials with modified properties. It was expected that this technique can be
used to improve mechanical properties of PVK without depressing the photoresponse to a high extent.
We have taken up the studies on photoconductivity
of blends consisting of PVK and poly(bis-phenol A

polycarbonate) (PC) . This system is known to be
immiscible on segmental level; thus this blend is
heterogeneous at very broad concentration range. In
this case one should expect that the generation of
charge carriers , which occurs in the PVK phase
/or in the PVK-rich phase/ will not be modified to
a high extent but the different trapping phenomena
can be influenced by the interface /or interphase/
with PC. PC is not photoactive and PVK is the best
known polymeric photoconductor; thus their blend can
be considered as a model system for photoconductivity.

Recently we have investigated carrier trapping
and recombination effects in PVK-PC films using
thermally stimulated current /TSC/ and thermally
stimulated luminescence techniques /TL/ [79] . TSC
method is often applied in studies of charge transport and trapping phenomena in photoconductive materials. These informations are important for understanding of photoconductivity process, especially
in the case of photoconductive mixtures (measurements of light emitted as result of charge carrier
recombination of trapped carriers). TL is independent of transport processes which usually affect
TSC in low-mobility materials. In the case of polymers, both processes are additionally complicated by
molecular relaxations if they become active in the
investigated temperature range. This is due to the
fact that carrier detrapping events are caused by
molecular motions. So called "wet dog" effect has
been shown in several systems − see, e.g.,[74,75] .

Without going into discussion of the experiment
conditions, it is important to note that the TSC and
TL measurement were carried out simultaneously in
a vacuum chamber. TSC and TL spectra with different
drain fields show a strong influence of electric
field. The TL maximum consists of two peaks with
different dependence on the field applied. Low temperature peak decreases with the increase of electric field but its position on temperature scale
is field independent. This result shows that this
peak can be attributed to geminate recombination
according to the prediction of the Onsager model
[76] . The high temperature TL maximum around 200 K
is not detected at low fields but it is enhanced
by electric field and its position shifts towards
lower temperatures. This fact suggests that it is
related to non-geminate, random recombination. Its

position depends on the diffusion of the released
carriers which have avoided the geminate recombination or other recombination centers in PVK. Looking
at the TSC, however, one can not detect a maximum
in this temperature range. It appears at higher temperatures. It leads to a conclusion that carriers
released from the traps at low temperature in the
PVK phase are not removed from these areas but they
recombine or are retrapped. These retrapped carriers
can again contribute to the current when, at higher
temperature, they are free. The current maximum in
TCS experiment is observed when the major part of
those carriers reaches the electrode; thus its position is determined by transport properties of the
sample. This conclusion is in agreement with the
theory and experiment related to the transport peak
[76,77].

The measurements of TL and TSC can be analysed by
carrying out partial heating experiments [78], which
make possible to determine the activation energies.
These studies have revealed a broad spectrum of
activation energies in which several energy levels
can be distinguished. These levels are similar in
TSC and TL at low temperature region. These energy
levels agree well with the activation energy spectrum for the β-molecular relaxation in pure PC, as
determined by thermally depolarization method [79].
Taking into consideration that we have found higher
TL maxima in PVK-PC blend as compared with pure PVK,
we can arrive to an important conclusion. The dominant fraction of trapping centers in the blend under
discussion is situated at the PVK-PC interface.
Positive charge carriers are released from these
traps by molecular motions, then recombine with ionized photogeneration centers in PVK giving rise to
TL or contribute to the current. Trap-controlled
transport results in TSC maximum with much higher
apparent activation energy /c.a. 0.5 eV/. It seems
interesting to note that the intensity of TL is the
highest for 50:50 composition of the components and
it is equal for 20:80 or 80:20 ratio of PVK:PC. It
shows that at the same surface ratio the release
of trapped carriers occurs of the interface. It
shows also that the population of trapping sites is
proportional to the total interface.

This short description of our recent results
shows that the simultaneous studies of TSC and TL
in PVK-PC blend can provide complementary informa-

tion about charge carrier detrapping and carrier
transport which is also trap dependent and witnesses
the importance of the interface in photoconducting
two component polymeric systems.

PVK exhibits, as it was mentioned before, high
photosensitivity within the UV region. By combination with suitable dopants its photoelectric sensitivity can be extended to the visible region. This
modification of photoresponse can also be used in
preparation of PVK blends taking advantage of different solubility of dopants in the components of
the blend.

It is known that pure PVK undergoes photooxidation, which leads to a change of its photoconductivity. In the blend the influence of photooxidation
can be controlled, to a certain extent, because of
the difference of equilibrium solubility of oxygen
in the components of the system.

These remarks show that the preparation of photoconducting blends has a large and not yet explored
prospect.

Conclusions

This review and the discusion of our results
reveals that a large number of polymers has been
studied in regard to their photoconductive properties, and sufficient useful and significant
information have already been accumulated. This
concerns both theoretical and experimental aspects
of many polymeric systems. There are many methods of
 introducing chemical moieties to obtain materials with enhanced photoconductivity and photosensitivity.

Structure-property correlations in photoconductivity and/or photosensitivity have been established only for a limited number of polymeric materials, including pure and doped or sensitized systems.
It seems that possibilities exist for an effort
in this direction. Conclusive information can be
reached on the influence of such factors as glass
transition, tacticity, crystallinity, supermole-
cular structure and other properties, on photoconductivity of polymeric materials.

Further research in this direction is needed to
obtain a variety of such polymers which will be
appropriate for scientific exploration and prac-

tical applications.

It is proper to add that the investigation of photoconductivity has an influence on understanding of electronic conduction in insulating polymers /e.g. the photoinjection hole of surface states, space charge trapping phenomena,etc./.

It is worthwhile to say that the photoconductivity in heterogeneous polymeric systems has not been intensively investigated yet. There are still many problems to be solved in view of the availability of an interesting and promising variety of such polymers for appropriate applications. There are many methods of blend preparation which will result in enough fine dispersion of components to comply with requirements for practical applications.

Evolving from the studies on homogeneous and heterogeneous photoconductivity systems have come to understanding and unexpected advances which in principle have additional technological importance. Here, competitive technologies and materials already exist and thus serve to focus attention on scientific issues where progress and understanding are required for an accurate assessment of further technological applications of photoconducting polymeric materials.

It seems that general lines for further research on photoconductivity of high-molecular-weight compounds, at least in the near future, may be regarded as well established, but new materials which can be obtained may introduce very new viewpoints to the knowledge and application of photoconducting polymers.

Valuable science and technology grow in an unexpected way and narrow planning sometimes will fail to the fruits of imagination in research and applications.

Acknowledgements

It is with pleasure and gratitude the author acknowledges the valuable discussions and collaboration with his coworkers, Drs. Jeszka,J.K.; Tracz.A.; Ulański, J.

This work was subsidized by Polish Academy of Science, Research Project CPBP 01.14.

References

[1] Pennwell, R.C.; Ganguly, B.N.; Smith, T.W.
     Macromol. Rev. 1978, 13, 63.
[2] Stolka, M.; Pai, D.M. Adv. Polym. Sci. 1979,
     29, 1.
[3] Kryszewski, M. "Semiconducting Polymers"; PWN
     Warsaw, 1980
[4] Pearson, J.M.; Stolka, M. "Poly/N-vinylcarba-
     zole/", Polymer Monographs, vol. 6; Gordon
     and Breach Science Publishers, Inc. New York
     1981
[5] Seymour, R.B.,Ed.; "Conductive Polymers",
     Plenum Press: New York, 1981
[6] Seanor, D.A., Ed.;"Electrical Properties of
     Polymers"; Academic Press: New York, 1982
[7] Mort, J.; Pfister, G., Eds.; J.Wiley and Sons:
     New York,1982
[8] Itaya, A.; Okamoto, K.; Kusabayashi, S. Bull.
     Chem. Soc. Jap. 1979, 52, 2218.
[9] Agarwal, S.K.; Hemmadi, S.S.; Pathak, N.L.
     Polym. J. 1979, 20, 867.
[10] Terrell, D.R.; Evers, F. German Pat: 3.032.425.
[11] Schiller, J. Chem. Abstr. 1980,93, 150977h
[12] Nespurek, S.; Cimrowa, V. 31 IUPAC Macromolecu-
     lar Symposium, Abstracts VII p. 199.
[13] Mort, J.;Morgan, M.; Grammatica, J.; Noolandi,J.
     Hong, K.M. Phys. Rev. Lett. 1982, 48, 1411.
[14] Stolzenburg, F.; Ries, B.; Bässler, H. Material
     Science 1987, 13, 259.
[15] Richert, R.; Bassler, H. J.Chem.Phys, 1986, 84
     3567
[16] Schorf, W.; Betz, E.; Port, H.; Wolf, D.C.
     Chem. Phys. Lett. 1986, 123, 306.
[17] Bässler, H.; Schönherr, G.; Abkovitz, M.;
     Pai, D.M. Phys. Rev. B. 1982, 26 3105.
[18] Schein, L.B.; Rosenberg, A.; Rice, S.L. J.Appl.
     Phys. 1986, 60, 4287.
[19] Jones, E.G.; Das, S.K. Brit. Polym. J. 1982,
     15, 52.
[20] Pochan, J.M.; Gibson, H.W. J. Polym. Sci.,
     Polym. Phys. Ed. 1982, 20, 2059.
[21] Pai, D.M.; Yanus, J.F. Photogr. Sci. Eng. 1983
     27, 14
[22] Yokoyama, M. Electrophotography Jap. 1986, 25,
     286
[23] Wojciechowski, P.; Kryszewski, M. Acta Phys.
     Pol. 1979, 56, 89.

[24] Chang, D.M.; Gromelski, S.; Mulvaney, J.E. J. Polym. Sci., Polym. Chem. Ed. 1977, 15, 517.

[25] Biswas, M.; Das, S.K. J. Polym. Sci., Polym. Lett. Ed. 1981, 20, 933.

[26] Block, H.; Bowker, S.M.; Walker, S.M. Polymer 1978, 19, 531.

[27] Mulvaney, J.E.; Gromelski, S.; Rupp, R.J. Polym. Sci., Polym. Chem. Ed. 1981, 19, 305.

[28] Pillai, P.K.C. Chemdreskhar Polymer 1979, 20, 505.

[29] Halm, J.M.; Delorme, J.H. Photogr. Sci. Eng. 1979, 23, 252.

[30] Biswas, M.; Urgu, T. J.Macromol. Sci. Rev., Macromol. Chem. Phys. 1986, C26, 249.

[31] Tanikawa, K.; Enomoto, T.; Hatano, M.; Motegi, K.; Okuno, Z. Macromol. Chem. 1975, 176, 3025.

[32] Kitayama, H.; Yokayama, M.; Mikawa, H. Mol. Cryst. Liq. Cryst. 1981, 69, 257.

[33] Chapoy, L.L.; Sethi, R.K.; Sorensen, P.R.; Rasmusen, K.H. Polym. Photochem. 1981, 1, 131.

[34] Chapoy, L.L.; Spaseska, D.; Ramussen, K.; Du Pre, D.B. Macromolecules 1979, 12, 680.

[35] Kryszewski, M.; Swiderski, T. unpublished results.

[36] Penwell, R.C.; Prest, M.M. J. Polym. 1978, 19, 537.

[37] Oshima, R.; Kumanotani, J. Polym. Prepr. Am. Chem. Soc. Div., Polym. Chem. 1979, 20, 522.

[38] Hatano, M.; Tanikawa, K. Am. Chem. Soc. Div., Org. Coat. Plast. Chem. Preprints. 1979, 40, 288.

[39] Chapoy, L.L.; Biddle, D. ; Halström, J.; Kovacs, K.; Brunfeldt, K.; Quasim, M.A.; Christiansen, T. Macromolecules 1983, 16, 181.

[40] Chapoy, L.L.; Biddle, D. J. Polym. Sci. Polym. Lett. 1983, 21, 621.

[41] Chapoy, L.L.; Munk, D.K. J. de Phys. 1983, 44 /Coll c.3, Supl. 6/697

[42] Mc Gibbon, G.; Rostron, A.J.; Sharples, A. J. Polym. Sci. 1971, A 29, 569.

[43] Tien, H.T. Nature 1970, 227, 1232.

[44] Aizawa, M.; Hirano, M.; Suzaki, S. Electrochim. Acta 1978, 23 11, 85.

[45] Turner, R.S.; Pai, D.M. Macromolecules 1979, 12, 1.

[46] Vermulen, J.A.; Wintle, H.J.; Nicodemo, D.A. J. Polym. Sci. 1971, A1 9, 544.

[47] Guillaud, G.; Maitzot, M. J. Phys. Lett. 1982, 43, L 559.

[48] Czain, J.P.; Yelon, A. J. Appl. Phys. 1983, 51 2106.

[49] Rostogi, A.C.; Chopra, K.L. Thin Solid Films 1973, 18, 187.

[50] Sumita, O.; Fukada, A.; Kuze, E. J. Polym. Sci. Polym. Phys. Ed. 1980, 18, 877.

[51] Sumita, O.; Fukada, A.; Kuze, E. J. Appl. Polym. Sci. 1981, 26, 1659.

[52] Skotheim, T.A. Ed. "Handbook of conducting polymers", Marcel Dekker Inc. New York, 1986.

[53] Yee, K.C.; Chance, R.R. Am. Chem. Soc. Div. Org. Coat. Plast. Chem. Pap. 1978, 38, 611.

[54] Siddiqui, A.S.; Wilson, E.G. J.Phys.C. 1979, 12, 4237.

[55] Kern, R.J. J. Polym. Sci. Polym. Chem. Ed. 1969, 7, 621.

[56] Opferman, J.; Hörhold, H.H. J. Phys. Chem. 1982, 259, 1089.

[57] Hörhold, H.H.; Rabe, D. Acta Polymerica 1979, 30, 86.

[58] Hörhold, H.H.; Helbig, M. 31 IUPAC Macromolecular Symposium, Abstracts VII, 156

[59] Gilland, G.; Rosenberg, N; Maitrot, M. J. Polym. Sci. Polym. Phys. Ed. 1980, 18, 523.

[60] Freeston, D.B.; Nichols, C.H.; Pailtrope, M. Polym. Photochem. 1981, 1, 85.

[61] Kurtz, S.; Hughes, R.C. J. Appl. Phys. 1983, 54, 229

[62] Kipping, F.S. J. Chem. Soc. 1924, 125, 2291.

[63] Brough, L.F.; Matsumura, K. R West Agew. Chem. Int. Ed. Engl. 1979, 18, 955.

[64] Yajima, S.; Hagashi, J.; Omori, M. Chem. Lett. 1975, 931.

[65] Wesson, J.P.; Williams, T.C. J.Polym. Sci. Polym. Chem. Ed. 1980, 18, 959

[66] West, R.; David, L.D.; Djurovich, P.I.; Stearley, K.L.; Srinvasan, K.S.V.; Yu, J. J. Am. Chem. Soc. 1981, 103, 7352.

[67] Kepler, R.G.; Ziegler, J.M.; Harrah, L.A. Bull. Am. Phys. Soc. 1984, 29, 509.

[68] Yuh, H.J.; Abkovitz, M.; Mc Grane, K.; Stolka, M. Bull. Am. Chem. Soc. 1984, 31, 381.

[69] Stolka, M.; Abkovitz, M. J. Polym. Sci. Polym. Chem. Ed. in press.

[70] Gill, W.D. J. Appl. Phys. 1972, 43, 5033.

[71] Lemus Santos, S.S.; J. Hirsh. Phil. Mag. 1986, B 53, 125.

[72] Bässler, H., Phil. Mag. 1984, 50, 347

[73] Głowacki, Z.; Ulański, J.; Jeszka,J.K.; Kryszewski, M. Material Science 1987, 23, 83.

[74] Nakakita, T.; Ito, D. J. Appl. Phys. 1980, 51, 3273.

[75] Kryszewski, M.; Ulański, J.; Jeszka, J.K.; Zieliński, M. 1987, Polymer Bulletin 8, 187.

[76] Plans, J.; Zieliński, M.; Swiderski, T.; Kryszewski, M.; Ulański, J. J. Appl. Phys. 1982, 53, 3103.

[77] Plans, J.; Zieliński, M.; Kryszewski, M. Phys. Rev. 1981, B 23, 6557.

[78] Creswell, R.A.; Perlman, M.M. J. Appl. Phys. 1970, 41, 2365.

[79] Kryszewski, M.; Ulański, J. J. Appl. Polym. Sci. Polymer Symposia 1979, 35, 553.

ELECTRONIC TRANSPORT IN POLYMERIC PHOTORECEPTORS: PVK TO POLYSILYLENES

M. A. Abkowitz and M. Stolka

Xerox Corporation, Webster, NY

ABSTRACT

Organic electrophotographic photoconductors (OPC) are typically polymer based multilayer structures each layer of which is optimized for a specialized function. The charge generation layer (GL) absorbs incident radiation and supports separation of the hole-electron pair under the action of the applied electric field. The polymeric charge transport layer (TL) accepts one sign of charge from the GL and must be capable of completely transiting that charge in a time dictated by machine-cycle requirements with essentially no loss to deep traps. The nature of the electrophotographic process requires that the TL polymer have negligible d.c. conductivity and high dielectric strength. At the same time it is required to exhibit satisfactory (a) film forming characteristics at a specified thickness (b) mechanical and (c) adhesive properties. In this chapter the role of mechanistic understanding in the historical development of viable OPC's is emphasized. Three materials systems are described. Early mechanistic studies of poly(N-vinylcarbazole), PVK, established the pivotal role played by the carbazole sidegroups which furnish the sites among which photoinjected holes undergo thermally assisted hopping. Understanding the latter suggested the concept of molecularly doped polymers (MDP) in which the active small molecule is dispersed in an inert binder and mobility is directly controlled by concentration. There is presently interest in exploring the large class of silicon backbone polymers, polysilylenes. These exhibit efficient hole transport whose qualitative characteristics are remarkably similar to the previously studied materials systems. In this case the indication is that carrier transport proceeds by hopping among electronic states associated not with sidegroups, but rather with the chain backbone. The nature of these states is of considerable current scientific interest.

1. **Technological Background**

1.1   The Electrophotographic Process [1].   The electro-photographic process is initiated by deposition of a uniform surface charge from corona on the xerographic photoreceptor drum or belt in the dark (Figure (1a)). The charge should be retained on the photoreceptor for the duration of a xerographic cycle which means that the material is required to be a relatively good insulator. An electrostatic latent image is then generated on the photoreceptor, either by projecting an image reflected from an original document or by a computer controlled narrow laser beam. The areas exposed to light become conductive and the surface charge in those areas is neutralized (Figure (1b)). Charge is retained on the non-exposed areas. The latent image is then developed by depositing charged toner particles on the photoreceptor (Figure (1c)). If the toner particles carry charge of an opposite sign, they adhere to the unexposed dark areas, if they carry the same sign charge, they are attracted to the discharged areas. The latter method is applied in some laser printers, s.c. "write black" method, or reversal development. The developed image is then transferred to paper (Figure (1d)).

1.2   Materials Requirements. Interest in organic polymeric photoreceptors arose from the need to develop inexpensive, flexible photoreceptors for belt architecture. Materials capable of both efficient generation and transport of free carriers following the absorption of light such as a-Se and its alloys with Te and As, or a-Si:H, etc., are not flexible enough for applications requiring small belts. On the other hand, organic polymers with good film forming properties that are capable of supporting high surface potentials of the order of 500-1000V across $20\mu m$ are usually not efficient charge carrier generators in the visible range of light (400-700 nm) even though some can transport injected charges very efficiently.

The need to satisfy two distinct requirements led to the concept of a layered photoconductor [2] where one layer (usually very thin) absorbs the incident radiation and generates free carriers and the second layer provides the desired mechanical properties, charge retention and has the ability to transport carriers of at least one sign, injected from the generation layer.

The scheme of a layered electrophotographic photoreceptor is shown in Figure (2). The charge generation layer (GL) is usually very thin, just thick enough to absorb most incident light, typically $0.4\text{-}2\mu m$, and the polymeric charge transport layer (TL) is $10\text{-}25\mu m$ thick so that it can support high potentials needed

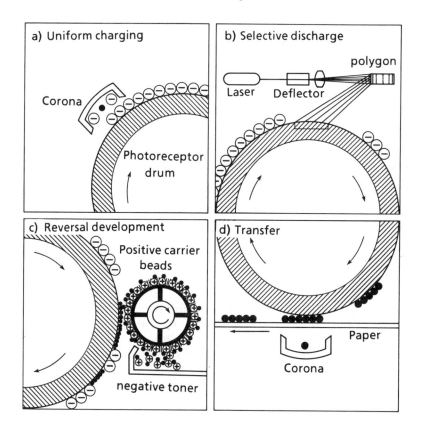

Figure 1. Schematic of the electrophotographic process.

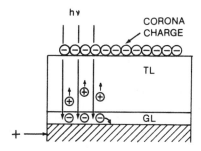

Figure 2. Scheme of layered electrophotographic photoreceptor.

for the development of electrophotographic images. Since the GL is usually sandwiched between the TL and the substrate electrode for protection from the corrosive action of corona and the abrading action of paper and developer, the TL material must in turn be highly transparent to the imaging radiation.

A number of essential requirements must be fulfilled. First, the carrier generation efficiency which is a function of the GL composition and the instantaneous value of the electric field in the generation zone must be high. Second, the photogenerated carriers must be injected into the TL before recombination can occur within the GL. The injection efficiency depends on both GL and TL materials. The combination of both processes determines the overall device sensitivity (supply efficiency). Finally the basic electronic requirements for the TL materials are sufficiently high carrier mobility, μ, and an absence of deep (i.e., release time >> electrophotographic cycle) trapping centers to minimize image degrading space charge buildup.

The importance of μ can be understood as follows: For an increase in the light exposure intensity ΔF, the final decrease in surface potential ΔV is proportional to the number of injected carriers (supply) and the distance they travel within the TL (assuming the GL is negligibly thin and carriers leave it instantly). Carrier displacement or range R in a given field E is controlled by $\mu\tau$, the product of mobility and deep trapping lifetime. Thus,

$$R = \mu\tau E \qquad (1)$$

The mobility is the mean velocity of photoinjected carriers per unit field which may itself be field dependent.

$$\mu = \mu(E) \qquad (2)$$

The time to traverse the TL of thickness L is called the transit time t_{tr}. Clearly

$$t_{tr} = L/\mu E \qquad (3)$$

During xerographic discharge a CV_0 of charge (where C is the film capacitance and V_0 the charge voltage) traverses the bulk inducing time dependent variation in the electric field behind the leading edge of the injected carrier front. Thus as the fastest carriers transit the TL, the electric field behind them is reduced and the slower carriers transit at lower field which in turn makes their mobilities lower. Xerographic discharge, which is a highly space charge perturbed process, is therefore characterized by significant dispersion in the arrival times of photoinjected carriers. In some cases, the transit times of the slowest carriers are ten times or more the transit times of the fast carriers. It is, however, required that they too exit the TL before the photoreceptor reaches the development zone, typically 0.3-1.0s after exposure. In practice, it is desirable that even the low field carrier mobilities significantly exceed 10^{-7}-10^{-6}cm^2/V.s.

For example, a 25μm thick TL (in which there is no deep trapping) with mobility $\mu = 10^{-6}$cm^2/V.s. at $E = 10^4$V/cm in a photoreceptor charged initially to 1000V, still retains a residual voltage of 20V after 0.3s, or 7V after 1s, even when enough incident light is used to photogenerate all carriers needed (i.e., CV_0) to discharge the device. A device with $\mu = 10^{-7}$cm^2/V.s. at the same field will retain as much as ~60V one second after exposure, which is probably unacceptable.

Finite carrier range [3] reflects the influence of deep traps and is controlled by the mobility-lifetime product $\mu\tau$. Using the above numbers, $\mu\tau$ in practical polymeric TL should substantially exceed 10^{-6}cm^2/V.

Practical transport layer materials must also satisfy the requirements of chemical integrity (resistance to corona effluents, etc.), photochemical stability, mechanical integrity and abrasion resistance. The TL polymer must also match the selected generation material, i.e., should accept charges photogenerated in the GL. Modern GL materials are either one component vapor deposited or solvent deposited thin layers of a-Se, a-As$_2$Se$_3$, various azo dyes or solvent-coated dispersion of photogenerating pigments such as trigonal (crystalline) Se, squarylium, or perylene derivatives, in a polymer binder. The overall effective thickness of the GL is 0.3-0.5μm. In laser printers where the latent image is generated by a solid-state laser emitting in the 780-830 nm range (near IR), the photo-generating pigment must be sensitive to near IR radiation. Various specific phthalocyanines, and squarylium pigments, have been developed for that purpose [4]. The transporting polymers must be electronically compatible with the photogenerator of choice.

Many charge transporting polymers have already been identified [5]. The most thoroughly studied charge transporting insulating polymer is poly(N-vinylcarbazole), PVK [6]

The first layered photoconductor with a PVK TL was reported by Regensburger [2]. Many other charge transporting polymers were then synthesized but none was studied as extensively as PVK. Structurally, all charge transporting polymers are derived from monomers containing an ionizable group (mainly oxidizable), with a stable oxidized form where charge is delocalized in an aromatic unit [5]. The most predominant structural moiety in these polymers is an aromatic amine group

in all imaginable substitutions. PVK is an example of a substituted aromatic amine polymer. Polymers with complex aromatic groups such as anthracene or pyrene have also been studied as potential transport materials but they are generally unstable, oxidatively or photolytically [5].

It was initially assumed [7] that the ability of PVK to transport charges is related to the spatially crowded structure of PVK where the carbazole groups, due to their covalent bonding backbone, are forced to interact with each other to the point of significant orbital overlap creating a band structure. However, the observed mobilities were found to be rather low ($\sim 10^{-7}$ - 10^{-6} cm^2/V.s.), with field and temperature dependences which were not characteristic of band transport alone. Moreover, it was later discovered that the same charge transport phenomena could be achieved simply by dissolving in inert polymeric binders functional groups of the above polymers [8,9]. This observation led in turn to the concept of molecularly doped polymers. Thus it became obvious that the carbon backbone in PVK played only the secondary roll of holding carbazole groups together [9,10]. It is now widely recognized that charge transport in amorphous sigma bonded carbon backbone polymers is a hopping process which involves electric-field driven charge exchanges among discrete sites which are dissolved molecules or covalently bonded functional pendant groups.

Most practical polymer based electrophotographic photoreceptor systems now in use are in fact solid solutions of an active species in a binder host polymer [11]. This system concept embodies the notion of full chemical control of the transport process. Thus the concentration of dopant molecules directly controls the drift mobility which is in turn controlled by wavefunction overlap between active sites. The host polymer binder is then specialized for its mechanical and adhesive properties. *Understanding key features of small molecule transport mechanistics provides guidelines for the optimization of injected carrier range. The key point is the understanding of how chemically induced traps arise in such systems -- how molecularly doped materials containing various contaminants in substantial quantities can still efficiently transport charge. (Even ppm concentrations of contaminants are extraordinarily high by conventional semiconductor standards).*

Many practical small molecule doped polymeric systems have been studied. Two notable examples are: N,N'-diphenyl-N,N'-bis(3-methylphenyl)-[1,1'-biphenyl]-4,4'-diamine (TPD) [12], and p-diethylamino-benzaldehyde-diphenyl hydrazone (DEH) [13] both dispersed in polycarbonate. Most investigated small molecule systems have also been complex substituted aromatic amines.

There is continuing interest in designing single component polymer transport materials to completely eliminate solvent extraction, diffusional instability and crystallization of the small molecules. One obvious route that has not been successful to date is the design of yet another aromatic amine containing carbon backbone polymer. An alternative may be to explore the large class of glassy silicon backbone polymers, polysilylenes [14], polyphosphazenes [15] etc.

Polysilylenes have been known for many years but were not considered interesting until the mid 1970's when Yajima [16] and West [17] used them as precursors for Si-C ceramics. In 1980 the first soluble high molecular weight polymers were synthesized [18]. Photoconductivity and charge transport in poly(methylphenylsilylene)/TNF was first observed by Kepler et al [14]. Preliminary charge transport studies showed that hole mobility in poly(methylphenylsilylene) (PMPS) is near 10^{-4} cm^2/V.s., at about 10^5 V/cm, quite high for a polymer [14,19,20].

Polysilylenes are synthesized by Wurtz coupling reaction of dichlorodialkyl or dichloro arylalkylsilanes with molten sodium in an aromatic solvent

$$\begin{array}{ccc} R_1 & & R_1 \\ | & & | \\ n\ \text{Cl-Si-Cl} + 2n\text{Na} \rightarrow 2n\text{NaCl} + \text{-(-Si-)-}_n \\ | & & | \\ R_2 & & R_2 \end{array}$$

Poly(methylphenylsilylene), PMPS, used in this study was synthesized in refluxing toluene. Typically, the PMPS synthesis results in a mixture of three products; a cyclic oligomer (n≈5), a low molecular weight linear polymer (M_n~4-20,000) and a high molecular weight polymer (M_n~300,000 or more).

The molecular weight distribution factor of the high molecular weight fraction is typically 1.5-3.0. When the reaction is carried out according to Zeigler [21] (sodium dispersion steadily added to the solution of monomer in a mixture of aliphatic and aromatic solvents), the narrow distribution low molecular weight fraction is so well separated from the narrow distribution high molecular weight fraction that acetone or hexane precipitation easily isolates the monomodal high molecular weight fraction. It is speculated that two different polymerization mechanisms proceeding in parallel are responsible for the unusual molecular weight distribution of the product.

The high molecular weight PMPS is soluble in common solvents and is a good film former. Qualitatively, the overall handling characteristics resemble those of polystyrene. Films of PMPS were cast from toluene solution by a solvent evaporation technique. Other polysilylenes, poly(methylcyclohexylsilylene), poly(methyl-n-propyl-silylene), poly(methyl-n-octylsilylene) and a copolymer of dimethyl- and methylphenylsilylene (1:1 ratio) were also prepared by Wurtz coupling, according to West et al [22]. Films of these polymers were also cast from toluene. All polymers form clear transparent colorless films.

2. Transport Measurements

Xerographic discharge involves photogeneration and injection of the mobile carrier type (transport is typically unipolar) into transport states, transport of the photoinjected carrier and fractional loss to, and immobilization in deep traps. Measurement techniques fall into two subclasses, either (1) steady state, or (2) transient. Transient measurements themselves are further classified according to whether they are carried out under small or large signal conditions. Xerographic discharge is itself a large signal process in that the charge in transit, a full CV_0's worth, significantly disturbs the field distribution in the transport layer causing it to become time dependent. *The result is to force the transiting charge to become spatially dispersed even when the mobility is itself uniform and time independent for each transiting carrier.* Under small signal conditions, the transiting carrier packet is constrained to contain much less than a CV_0 of charge which then as a result moves in a uniform field. Any complicated time dependence of mobility [23] which becomes experimentally visible under these conditions must, we believe, reflect microscopic and fundamental features of transport in the disordered medium and should be distinguished from the macroscopic dispersion induced by carrier motion under highly space charge perturbed conditions [24].

2.1 What's Wrong with Steady State Measurements? The
operational definition of ohmic conductivity is well known

$$\sigma = ne\mu \qquad (1)$$

where n is the thermal equilibrium carrier density, e the electronic charge and μ the conductivity mobility, which is, the mean velocity of a carrier per unit applied field under steady state conditions. In relatively good glassy polymeric insulators, despite popular but naive arguments to the contrary, it is very difficult to unambiguously determine the mobility directly from current density j vs. electric field E curves. The problem is always the uncertain role of contacts which are required to be neutral in the sense that during measurement the bulk carrier density which exists under zero field conditions, is not perturbed [25]. The latter is rarely the case. The observation of linear j vs. E curves in fact guarantees nothing and this behavior is often observed when currents are contact emission limited and not bulk controlled [26] (i.e., controlled by injection rate from the contact rather than the bulk of the glassy film). Even if ohmic contact can be made, it still remains to deconvolute from the j vs. E curves the respective contributions of n and μ. It is difficult to determine n independently in insulating systems. (Thermopower measurements--Seebeck coefficients commonly carried out to determine n in doped crystalline semiconductors are not practical in good insulators). The only steady state measurements free of ambiguity are carried out in the trap-free space charge limited current (SCLC) regime [25]. These measurements can be carried out when the average transit time t_{tr} of any excess injected carrier across the sample bulk is shorter than the time required for the bulk to locally neutralize the carrier [24]. The transit time t_{tr} has been defined in equation (3) as $t_{tr} = L/\mu E$ where L is the sample thickness, μ the injected carrier drift mobility and E the applied field. The time to neutralize any excess injected carrier is the bulk dielectric relaxation time $\tau_r = (\rho\epsilon)^{1/2}$ where ρ is the bulk resistivity and ϵ the bulk dielectric constant. The condition for the observation of SCLC is given by the expression

$$t_{tr} = L/\mu E < \tau_r = (\rho\epsilon)^{1/2} \qquad (4)$$

The trap free space charge limited current [24] (TFSCLC) is proportional to mobility as follows:

$$L_{TFSCLC} = 9/8 \ \mu\epsilon \ E^2/L \qquad (5)$$

The observation of SCL currents requires that the contacts on the polymer be able to act as an infinite carrier reservoir with respect to bulk demands which is just an alternate operational description of an ohmic contact. *To reiterate; unambiguous determination of mobility directly from steady state electrical measurements requires ohmic contacts and space charge limited conditions. The observation of linear j vs. E curves is usually of no significance.*

2.2 Transient Measurements.

Time resolved injection photocurrents have the particular virtue in the present context that they simulate the conditions under which carriers transit a polymer film during xerographic discharge. However, we limit ourselves here to the case where the integrated number of carriers q in the transit pulse is small enough ($q \ll CV_0$) that transit occurs in a uniform field which significantly simplifies interpretation of data. This is the condition that describes the cannonical small signal time-of-flight (TOF) experiment [27]. The field is such that equation (4) applies. For TOF, the electrical contacts must block dark injection. This condition is relatively easy to satisfy. With the field applied, the photoreceptor is excited with a weak flash of visible light which penetrates to the generation layer. The flash typically supplied from a nitrogen pumped dye laser is of short duration compared to the transit time. A schematic of the apparatus is shown in Figure (3). The series resistor senses current. When the overall circuit RC time constant is short compared to t_{tr}, the voltage

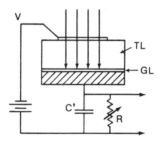

Figure 3. Schematic of TOF apparatus. R is the current sensing resistor. C′ is the total shunt capacitance.

developed on the sensing resistor is proportional to the instantaneous current in the specimen. Ideally, the current drops to zero when the transiting carriers reach the collecting electrode as a coherent sheet. In practice, rounding of the transit pulse is observed because some dispersion in arriving carrier transit times is always induced by thermal diffusion and mutual electrostatic repulsion. Superimposed on these pervasive minimal effects are dispersion phenomena which are more fundamentally related to the degree of disorder in any given system. For example, in a disordered molecular solid it is useful to separate the overall effects of steric disorder into a self consistent site variable random potential on the one hand and randomly varying inter and intramolecular transition matrix elements which directly encompass the systems geometric (i.e. positional) disorder on the other. These respectively contribute to the diagonal and off-diagonal terms in the systems effective Hamiltonian and so are referred to as diagonal (energetic) and off-diagonal (positional or site) disorder. Transport in many disordered systems can be thought of as a chain of discrete events (e.g., hopping or detrapping) where each event is associated with a waiting time on a given site. When the waiting time distribution function is Gaussian, the mean position of a transiting carrier packet initially injected as a thin sheet is proportional to time while the width of the spread increases as only the square root of time. In a current mode time of flight experiment, the time resolved current is constant until the leading edge of the transiting packet reaches the absorbing boundary. The subsequent decay of current in effect maps the distribution of arriving carrier transit times. In units of time normalized to the transit time (t/t_{tr}) the width of this decaying tail scales as $(t_{tr})^{-\frac{1}{2}}$ and is thus actually smaller at lower fields which gives rise to longer transit times (t_{tr}). However, when the waiting time distribution is very much broader, it is more appropriately represented by an algebraic function [23] of a form for instance encountered in phenomenological descriptions of pair recombination luminescence in crystalline semiconductors ($\psi(t) \alpha t^{-(1+\alpha)}$) where the dispersion parameter α varies between zero and one. Under the latter circumstances the maximum of the carrier distribution can remain (depending on α) close to the plane of origin even when the leading edge of the distribution has penetrated to the absorbing boundary [23]. In a TOF experiment, under the latter circumstances the time resolved current will acquire two algebraic branches which intersect at a "statistically defined" t_{tr}.

For $t<t_{tr}$ $\qquad\qquad$ $i(t) \sim t^{-(1-\alpha)}$ $\qquad\qquad\qquad$ (6)

While for $t>t_{tr}$ $\qquad\qquad$ $i(t) \sim t^{-(1+\alpha)}$

A key result is the prediction [23] of a unique scaling of this statistically defined transit time with applied field and sample thickness which leads to an anomalous thickness dependent mobility. "Anomalous" (i.e., compared to Gaussian transport) behavior ceases only in the limit that the dispersion parameter

α approaches one. The transit time t_{tr} obeys the scaling law

$$t_{tr} \sim (L/E)^{1/\alpha} \qquad (7)$$

As α decreases from a maximum of one, the current mode transit pulse acquired in real time becomes increasingly featureless while retaining a clearly defined kink when the log of current is replotted versus the log of time. When the log-log plot is of the current, normalized to its value at the transit time, versus time, normalized to the transit time, then current transients collected at different fields superimpose displaying the distinctive feature of "universality". A key feature of systems in which transport is characterized by an algebraic distribution of waiting times is the correlation between the respective slopes of these universality plots (equation (6)) and the scaling law for the transit time (equation (7)) both of which are determined by the degree of dispersion as manifested in α [23]. It should be emphasized that some confusion in terminology has already arisen regarding the issue of dispersion in time of flight experiments. "Relative dispersiveness" as specifically manifested in the shape of TOF transients in polymeric glasses, for example, is characteristically observed to vary with temperature, sample purity, sample surface condition and even the details of the initiating injection step. It must be emphasized that in some cases, rather dispersive looking TOF transients cannot be clearly identified with anomalous scaling. Experimentally observed dispersion in transit times thus may or may not be associated with the so-called "anomalous features" which are a direct consequence of a specific phenomenological representation of a very broad algebraic distribution of carrier waiting times--broad enough to in fact encompass the actual time of observation.

3. Transport Processes - General

In glassy polymeric media, electronic transport can be phenomenologically thought of as a series of discrete steps characterized by a distribution of waiting times $\psi(t)$ which may or may not extend into the time scale of observation. In the former case, the mobility will itself always appear to be time dependent. In the latter case, mobility can be characterized by an averaged value for most of the transit event though it will exhibit thermalization at early times which may be resolvable under certain experimental conditions [28,29]. There are several more microscopically detailed pictures which can correspond to this phenomenological description of electronic transport.

3.1 Trap Controlled Band Transport. Carrier displacement in the field direction takes place in an effective density of extended states N_{eff} in the band where the microscopic mobility μ_0 is controlled by scattering processes. In disordered solids where mean free paths are of the order of atomic spacing, carrier motion in "extended" states would likely be diffusive, not ballistic, and microscopic mobilities would fall in the range 1-10 cm^2/V.s. at room temperature [30]. The observed drift mobility in a TOF experiment is much smaller because the carrier undergoing motion in the band is frequently interdicted by a trapping event. Trapping, thermally activated release and retrapping control the temperature dependence of the drift mobility which is characteristically activated [31]. For example, in the simplest case of control by N_t nearly monoenergetic traps displaced from the transport level by energy E_t, the drift mobility is written:

$$\mu = \mu_0 \left[1 + N_t/N_{eff} \exp\left(E_t/kT\right)\right]^{-1} \qquad (8)$$

In this situation, there is a single waiting time which is simply the detrapping time and transport is nondispersive. As the trap distribution broadens so does the distribution of trap release times. In the limiting case that the traps are organized into a distribution falling off exponentially from a demarcation level in the band called a mobility edge, transport can become highly dispersive [30]. The featureless transit pulses seen in a-As$_2$Se$_3$ are thought to conform well with this picture of multiple trapping in an exponential band tail and in fact, manifest all the features characterizing dispersive transport [31].

3.2 Hopping in a Manifold of Localized States. If the hopping (i.e., quantum mechanical tunneling) sites are monoenergetic and the carriers do not relax with their surroundings to form a polaron during each wait at a discrete site, then transport is temperature independent and the mobility scales experimentally with average intersite hopping distance. In glassy polymers, a random disorder potential at each site, should cause site energies to reorganize into a Gaussian distribution [32]. The simplest description of the ensuing transport is phonon assisted hopping involving an activation ϵ related to the Gaussian width. The mobility is proportional to $\rho^2 \exp(-2\rho/\rho_0) (\exp-\epsilon/kT)$ where ρ is the average intersite distance and ρ_0 is the localization radius [33]. Recently, the time-resolved behavior of a carrier packet injected into such a Gaussian distribution of localized states has been considered analytically by Movaghar and coworkers [34] and has been extensively simulated in computer experiments by Bässler and his coworkers [32,35]. Novel results are as follows: Relaxation of the excess injected carriers to a temperature dependent level lying below the average energy in the distribution occurs in a relatively few hops. Observation of transport occurs in the long time limit where the drift mobility is predicted to exhibit a non-Arrhenius type of temperature dependence given by

$$\mu(T) = \mu_0 \exp(-T_0/T)^2 \qquad (9)$$

T_0 is the width of the Gaussian distribution expressed in temperature units and μ_0 is the mobility in an analogous system with no disorder (i.e., the Gaussian distribution collapsed to a single energy level). A carrier hopping in the field direction gains energy $eE\rho$, where ρ is the hopping distance, and can therefore gain access to more states in the distribution. The net effect of an electric field is to reduce the effective width of the Gaussian distribution. It can be shown that the combined field and temperature dependence of the drift mobility is therefore contained in the expression

$$\mu(E,T) = \mu_0 \exp(-T_0/T)^2 \exp(E/E_0) \qquad (10)$$

The key point is that the full behavior of the drift mobility can be expressed in terms of the mobility of an ordered microscopic analogue system together with a parameter T_0 which expresses the effect of disorder in terms of energy broadening of the level into a Gaussian distribution. It should be emphasized that this model, because it presumes a time independent mobility, makes no allowance for the observation of anomalous dispersion. (In fact, anomalous dispersion in this picture must be identified with additional traps lying well outside the Gaussian manifold, i.e., trap controlled hopping).

3.4 Trap Controlled Hopping. In this case, the scenario described for trap controlled band mobility applies. However, the microscopic mobility is now associated with carriers hopping in a manifold of localized states. Overall temperature and field dependence reflects the complicated convolution of the temperature and field dependence of both the microscopic mobility and the trap kinetic processes. Clearly, the observed behavior can now range from non-dispersive to anomalously dispersive as before depending on the energy distribution of transport interactive traps.

3.5 Small Polaron Hopping. The microscopic hopping mobility even among identical sites in the absence of disorder now acquires an extra activation associated with the relaxation of the carrier and its polarized surroundings--a self trapping process. As before, all other processes would now be expected to convolute with the extra temperature dependence associated with polaron formation.

4. Some Key Examples

4.1 Poly(N-Vinylcarbazole), PVK. The following are the most relevant characteristics of transport in PVK and PVK complexed with TNF.

4.1a Transit Pulse Shapes. Figure (4) displays the current mode hole transit pulse shapes in PVK and Br substituted PVK from 264K to 490K together with the corresponding Arrhenius plots of mobility [36]. In the case of brominated PVK the transit pulses remain relatively featureless over the entire experimentally accessed range. For PVK, however, the transit pulses display a well developed shoulder above 414K

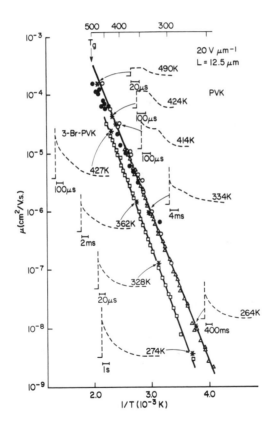

Figure 4. Log μ vs 1/T for hole transport in PVK and 3 Br - PVK together with current mode transients at indicated temperatures. Ref. [36].

characteristic of non-dispersive transport of the photoinjected
carrier sheet. A striking feature of the PVK data is the absence
of any evidence of a change in the temperature dependence of
mobility on passing through a range where transit pulses clearly
undergo a significant increase in dispersion. The relative
increase in dispersiveness at high temperatures suggests that
the chemical disorder introduced by Br substitution in effect
broadens the waiting time distribution function. Gill [37]
demonstrated electron transport in PVK complexed with TNF. In
this system Seki [10] demonstrated that room temperature
electron transits involve hopping motion among uncomplexed
TNF molecules. Figure (5) demonstrates that at room
temperature superposition with respect to applied field is
observed for electron transit pulses at room temperature.
Superposition is a key feature of anomalous dispersion as
described earlier.

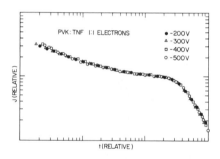

Figure 5. The superposition characteristic for 1:1 TNF:PVK at room
temperature. Ref. [10].

4.1b Strong Field Dependence of Drift Mobility. Hole drift
mobility increases sharply with field exhibiting in PVK a
characteristic $E^{\frac{1}{2}}$ dependence of log μ as originally reported by
Pai [38] whose data are displayed in Figure (6). Drift mobility
values are low falling into the range 10^{-9}-10^{-6}cm^2/V.s. at room
temperature.

4.1c Strong Temperature Dependence of the Drift Mobility.
Figure (7) is an Arrhenius representation of hole drift mobility
data on a 0.2:1 TNF:PVK film [10]. The apparent activation in
this representation is field dependent as illustrated in Figure (8)
for both electrons and holes. The activation in the zero field
limit is close to 0.7eV for both carriers. In Arrhenius

representation the family of log μ vs. $10^3/T$ curves parametric in field tend to converge at a finite $10^3/T$ value corresponding to temperature $T=T_0$. T_0 is 416K which is in the vicinity of Tg. The effect of small molecule additives, which act as plasticizers in PVK, on the transport has been noted by Fujino et al [39]. Their

POLY (N-VINYL CARBAZOLE)

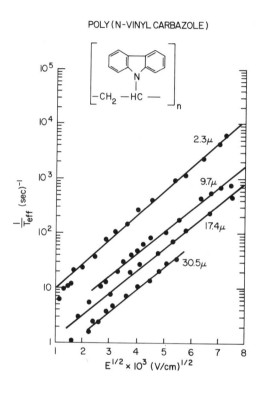

Figure 6. Inverse effective transit time versus square root of the applied field for four film thicknesses. Ref. [38].

data suggests that a strong correlation exists between T_0 and the glass transition temperature of the PVK based polymeric system.

An alternative framework for interpreting the temperature dependence of drift mobility is provided by the model suggested by Bässler and coworkers [35] and Movaghar and his coworkers [34]. To reiterate: it assumes that a carrier initially injected into a Gaussian manifold of localized state of width σ(ev) or equivalent temperature width T_0 establishes thermal equilibrium within a few hops after which the mobility achieves

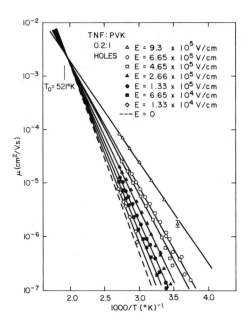

Figure 7. Arrhenius representations of hole drift mobility data on a 0.2:1 TNF:PVK film parametric in field. Ref. [37].

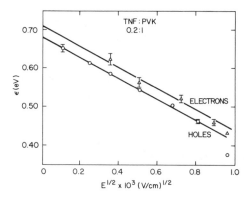

Figure 8. Activation energies of hole and electron drift mobilities vs. field in a 0.2:1 TNF:PVK film. Ref. [37].

its stationary value. The model does not treat the effect of additional states outside the Gaussian manifold and presumes that in the absence of diagonal and off- diagonal disorder the manifold would collapse into a single monoenergetic level among which the mobile carrier would undergo activationless hops with mobility μ_o. A key prediction as stated earlier is an inverse quadratic temperature dependence. In a recent paper Bässler [32] represents Gill's hole transport data [31] in this framework as reproduced in Figure (9). From the slopes of this family of Bässler plots parametric in field one can estimate the width of the Gaussian distribution by extrapolating to the zero field limit. We specifically return to this point later when making overall comparisons with polymethylphenylsilylene.

4.1d Mobility Scales with Average Intersite Hopping Distance.
Gill originally identified an exponential dependence of electron drift mobility on average TNF intersite hopping distance and noted an associated decrease in hole drift mobility [37]. This

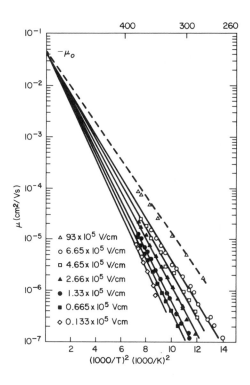

<u>Figure 9</u>. Data of Figure 7 replotted as log μ vs $(10^3/T)^2$. Ref. [32].

seminal observation led to work on a host of molecularly doped systems which are essentially small molecules dissolved in an inert binder matrix. Gill's subsidiary observation that TNF addition also decreased hole mobility precisely because complexed carbazole is removed as a hole transport site provided the first direct evidence for the key role of the discrete carbazole groups in hole transport.

4.2 Molecularly Doped Polymers (MDP). A number of hole transport studies have been carried out on substituted aromatic amines [13,33,40-44] with polycarbonate used as an inert binder. Results on the system N,N'-diphenyl-N,N'-bis(3-methyl phenyl)-[1,1'-biphenyl]-4,4'-diamine (TPD) in bisphenol A polycarbonate (PC) are illustrative [12].   Three important results are:

4.2a Concentration Dependence of TPD in PC. Figure (10) shows the concentration dependence of hole mobility expressed [12] in terms of the average intersite distance between TPD transport active sites, $\rho(\text{Å})$. The localization radius ρ_0 is a

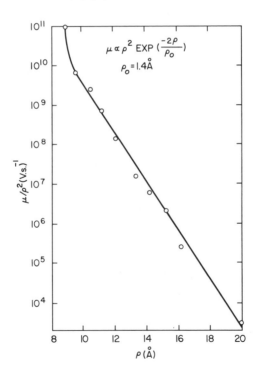

Figure 10. Concentration dependence of hole mobility in TPD/PC. ρ is the intersite distance. Ref. [12].

measure of the attenuation of the molecular wave function associated with the transport state. The value $\rho=1.4\text{Å}$ was similar to values [33] reported for other MDP's. Figure (11) demonstrates that the apparent activation energy determined from Arrhenius plots of mobility are concentration as well as field dependent. For most of the concentration range, μ/ρ^2 varies exponentially with ρ. The average intersite distance varies from 20Å in a 9 wt% sample of TPD/PC to 9Å estimated for amorphous TPD. The average distance between TPD sites was computed assuming

$$\rho = (M/(Ad))^{\frac{1}{3}} \qquad (11)$$

where M is the molecular weight (516), d is the density (1.2 gm cm^{-3}) and A is Avogadro's number. Figure (11) displays a significant change in overall activation energy between the glassy TPD film and a solid solution. For example, a slight change in intermolecular distance from 9 to 9.6Å between a-TPD and a film containing 80 wt% diamine produces an approximate twofold increase in activation energy. Further dilution of TPD in the solid solution, which induces a significant change in intermolecular distance, induces only a gradual change in activation. A corresponding steep increase in mobility between the 80 wt% solid solution and glassy TPD is clearly indicated in Figure (10). The concentration region of maximum change corresponds to the case that average intersite distance approaches the maximum molecular dimensions. At these high concentrations percolative effects may contribute significantly to transport. The large change in mobility and its apparent activation may signal an alteration in transport mechanism.

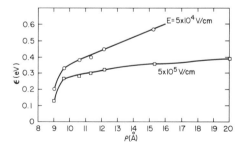

Figure 11. Concentration dependence of activation energy at two fields in TPD/PC. Ref. [12].

4.2b Scaling of Mobility with Thickness and Field. Figure (12) shows that the mobility in a 50/50 TPD/PC film is independent of film thickness in the range 5-90 μm. The absence of anomalous scaling is generally reflected in hole transport behavior over a wide temperature range. The system TPD/PC was particularly free of space charge memory effects visible in TOF and did not develop significant residual potentials (i.e., voltage remaining on photoreceptor after discharge which is due to bulk space charge) even after repeated xerographic photoinjection. The latter indicates strongly that the system is trap free, that states able to act as traps outside of the manifold of states among which carriers hop are totally absent. This behavior was clearly manifested in the shapes of space charge limited dark current injection transients [45]. Figure (13) demonstrates that steady state current equilibrium is rapidly achieved (i.e., within several transit times). In trap containing insulators, molecular crystals for example, the space charge limited current is expected to decay rapidly after the kink which occurs at approximately $0.8t_{tr}$. The behavior illustrated in Figure (13) is in fact quite extraordinary.

Plots parametric in temperature of log μ vs. E for a 60/40 TPD/PC film clearly demonstrate a linear field dependence. This behavior contrasts with the square root behavior of drift mobility on field observed in PVK [38], PVK:TNF [37], N-isopropylcarbazole [46] dispersed in PC but has also been reported in other substituted aromatic amines [42].

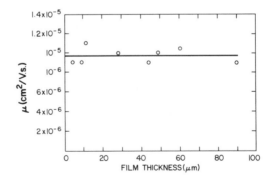

Figure 12. Thickness dependence of hole mobility in 50/50 TPD/PC film. Ref. [12].

While field dependence of drift mobility in organic glasses below their respective Tg's appears to be pervasive, it now appears that for reasons not yet understood the dependence of log μ on field ranges in behavior from $E^{\frac{1}{2}}$ to E in otherwise closely related systems.

4.2c   Trap Controlled Hopping. Physical defects which are associated with intermolecular conformations that deviate in some specific way from the normal statistical distribution of molecular arrangements and extrinsic chemical species can generate additional localized states which lie outside of the distribution of bulk states. The bulk states control the primary transport channel - the "microscopic mobility" while the additional extrinsic states can superimpose the effects of multiple trapping much as bandtails do in inorganic amorphous semiconductors. According to Bässler [32] the distribution of primary transport states in organic glasses controlled by diagonal and off-diagonal disorder (geometric and electrostatic fluctuations from site to site) are typically insufficient to induce anomalous dispersion. Anomalous dispersion, it is then argued, is symptomatic of additional multiple trapping in the system and can thus vary in degree according to sample purity. However, relative energetics are key in determining whether an impurity or structural defect can act as a trap or will be electrically inert and act as an "anti-trap". For example, because the ionization potential of TPD is lower than that of PVK, TPD can introduce

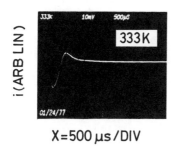

X=500 μs/DIV

Figure 13. Dark current response to step field excitation in 44 wt.% TPD/PC film with Au (ohmic) contacts. T = 333K, L = 26 micron. 500 V applied. Cusp occurs at approx. 0.8 t_{tr} as determined independently using TOF technique. Current equilibrium achieved in several transit times indicates absense of deep trapping. Ref. [45].

hole traps [47] into PVK. As the concentration of TPD is
increased beyond a threshold level it can proceed to provide
the dominant transport channel. The latter is clearly
demonstrated in Figure (14) which is a plot of log μ in a PVK
host as a function of TPD dopant molecules. At concentrations
of TPD below 5×10^{19} molecules/cm^3, TPD traps progressively
limit the transport which proceeds by hole hopping among
carbazole pendant groups. At higher TPD concentrations the
TPD molecules provide first a parallel, then the dominant
transport channel. In the latter case PVK becomes an inert
binder.

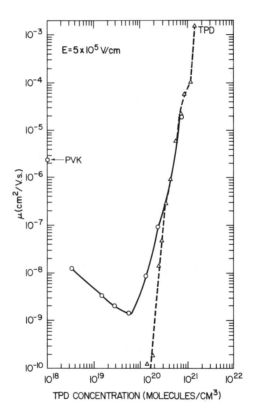

Figure 14. Log μ vs concentration of TPD dopant molecules. E =
5×10^5 V/cm, T = 295K. Solid curve is TPD/PVK, dashed curve is
TPD/PC. Ref. [47].

5. Transport in Polysilylenes

5.1 Experimental. Injected carrier drift mobilities were measured directly over a wide range of electric fields and temperature by the time of flight (TOF) technique using thin layers (0.5μm) of a-Se or As$_2$Se$_3$ as photogenerators. The top electrodes were evaporated semitransparent aluminum layers. Small signal current mode TOF transients were produced by hole injection from a-Se or As$_2$Se$_3$ excited at 450 or 550 nm by attenuated pulses from a nitrogen pumped dye laser. The current transients were also generated intrinsically using highly attenuated 337 nm pulses from the nitrogen laser. The absorption coefficient of PMPS at 337 nm is high, approaching 10^5; consequently most carriers were produced at the top of the film. The samples were examined for evidence of photodegradation, particularly during the uv photoexcitation experiments. No sign of degradation was observed.

5.2 Results and Discussion. Dark conductivity in PMPS and other investigated polysilylenes is negligible. In xerographic experiments where the top surface is corona charged instead of using the semitransparent electrode, the dark decay of the potential was extremely small, typically <<1% per second, as with good dielectric polymers. Non-ohmic (i.e., contact controlled) dc steady state dark injection currents were also infinitesimal.

In the experiments using layered structures [20] and visible excitation (to which PMPS is transparent), transient currents were observed only when the top electrode was biased negative

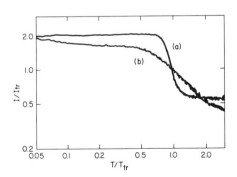

Figure 15. Log of current I normalized to its value at the transit time t_{tr} vs. log time in units of the transit time. (a) 8 micron PMPS film at 292K, (b) same PMPS film at 184K. E = 75V/micron. Ref. [20].

with respect to the substrate. The substrate was composed of a visible photoconductor (CGL) overcoated aluminum ground plane. When the polymer top surface was directly (intrinsically) photoexcited using pulsed 337 nm excitation, current transit pulses were observed only when the top electrode was positively biased. Thus under existing experimental conditions, only hole transient transport could be directly observed. Transit pulses were nondispersive over a wide range of temperature. In Figure (15) the relative increase in dispersion encountered in cooling from 292K to 184K is illustrated. There is no evidence for anomalous thickness dependence at the transit time even at the lowest experimentally accessed temperature. Transit pulses are represented as the log of current normalized to its value at the transit time. In highly dispersive transport media like a-As$_2$Se$_3$ such a plot would as discussed earlier exhibit two algebraic branches corresponding to the power law curves $t^{-1+\alpha}$ when $t<t_{tr}$ and $t^{-1-\alpha}$ when $t>t_{tr}$ with α the dispersion parameter $\alpha \leq 1$. In dispersive media at fixed temperature transit pulses represented in this form show the universality exemplified in Figure (5) for PVK:TNF[10]. Universality is not observed in PMPS [20]. Neither was it observed in the TPD/PC system [12]. Figure (16) shows the field dependence of hole drift mobility in PMPS at various temperatures. Field dependence grows progressively stronger as temperature is lowered. At temperatures approaching the glass transition measured calorimetrically, experimentally

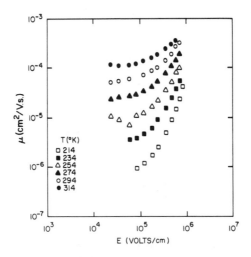

Figure 16. Log μ vs. log E in PMPS at various temperatures.

resolvable field dependence of the drift mobility essentially disappears. These characteristics parallel the behavior described for PVK:TNF and MDP's generally.

The program for analyzing hole transport data in PMPS is to first determine the functional form of the field dependence of hole drift mobility. That information is then used to analyze the temperature dependent behavior of the mobility extrapolated to the zero field limit. In this limit, mobility data is analyzed in two phenomenological frameworks. First we treat the mobility as simply and singly activated following the procedure used by Pai [38], Gill [37] and others to represent transport data in PVK. The resulting activation is field dependent. Next we treat the temperature dependence of mobility in the manner suggested by numerical simulation studies of Bässler and coworkers [35] and analytical calculations of Movaghar and coworkers [34]. This procedure provides a means for establishing the width of the distribution of sites among which the carriers hop. A square root type of field dependence was observed in PMPS [20] at 292K. A family of such plots is displayed in Figure (17). From the latter, the temperature dependence of the slopes β can be determined. These are displayed in Figure (18).

When hole mobilities are represented as a family of Arrhenius plots parametric in field, the apparent activation energies exhibit the characteristic field dependence illustrated in Figure (19) and represented more explicitly in Figure (20). The zero field activation energy for a large number of sample films is found to cluster in the 0.34-0.37 eV range. The latter is substantially less than the corresponding 0.7 eV reported in PVK:TNF and displayed in Figure (8) [37].

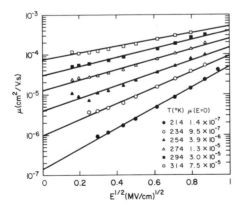

Figure 17. Log μ vs. $E^{1/2}$ at different temperatures in PMPS.

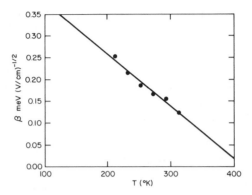

Figure 18. Slope of the curves in Figure 17 vs. temperature.

It is interesting to note that in poly(methylcyclohexyl-silylene) at $E=1\times10^5$ V/cm, ϵ is only 0.15 eV, about half of that in PMPS at the same field, even though the drift mobility at room temperature is the same as in PMPS [48].

Figure (19) shows that the Arrhenius plots of log μ vs $1/T$ parametric in field tend to converge near $T_0=416K$ which is in the vicinity of the glass transition temperature for PMPS. Recall that a similar coincidence of T_0 with Tg was observed in PVK plasticized with a small molecule [39]. Preliminary data appear to also indicate that both the apparent activation energy and

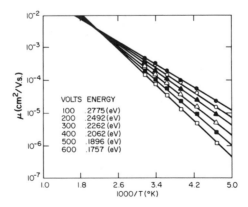

Figure 19. Arrhenius plots of μ for PMPS parametric in field.

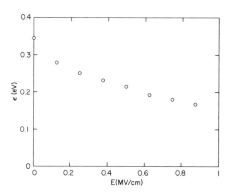

Figure 20. Field dependence of apparent activation energy of hole transport in PMPS.

the drift mobility are field independent at $T>T_O=T_g$. The extrapolated μ_O at T_O is $4\times10^{-3}cm^2/V.s$.

Generally, the field and temperature dependent behavior is as described earlier for PVK and molecularly doped polymers. All of these data are at least phenomenologically consistent with the expression.

$$\mu = \mu_O \exp[-(\epsilon_O-\beta(T)E^{\frac{1}{2}}/kT_{eff}] \qquad (12)$$

where $1/T_{eff} = 1/T-1/T_O$.

The numerical simulation studies [35] provide an alternative model for hopping transport in organic glasses. It is assumed that carriers are injected and transported in a disorder induced Gaussian distribution of localized states of width σ or equivalent temperature $T_O=7400\ \sigma$. According to this model, the temperature dependence of μ which is non-Arrhenius in the long time limit, takes the form given by equation (9).

This model, however, predicts a linear field dependence of $\log \mu$.

In PMPS the dependence is, as already indicated, closer to $E^{0.5}$. Bässler's model [32] predicts that log mobility vs. $1/T^2$ plots should intersect at a common point on the mobility axis corresponding to the value μ_o, the mobility at the limit of no disorder. In Figure (21), the temperature dependence of hole drift mobility at various fields, when replotted in conformity with the predictions of Bässler's model, leads to the value $\mu_o\sim10^{-2}cm^2/V.s$.

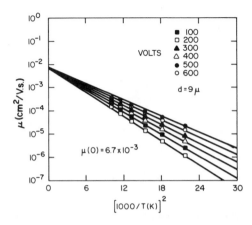

Figure 21. Log μ vs $(10^3/T)^2$ at different fields in PMPS.

From analysis of the field dependent slopes of the curves in Figure (21), the underlying disorder induced Gaussian width σ of the hopping states can be calculated. The resulting field dependent width expressed in terms of energy and equivalent temperature units is displayed in Figure (22) together with the analogous data obtained for PVK/TNF 1:0.2.

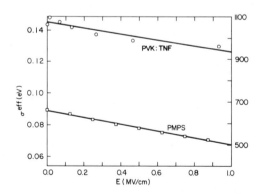

Figure 22. Width of the Gaussian distribution of hopping sites expressed in terms of energy σ and equivalent temperature T_0. PMPS and PVK/TNF (1:0.2) are compared.

The hopping site distribution, which is represented by the value σ extrapolated to zero field, appears in this analysis to be very narrow (i.e., 0.09 eV) compared to PVK/TNF. This result should be compared with the observation of relatively non-dispersive transit pulses at room temperature in PMPS (Figure (15)), and the corresponding dispersive transits in PVK/TNF - (Figure (4)).

5.3 **Transport Mechanism.** It might appear that the similarity of transport behavior in PMPS and other aromatic sidegroups containing polysilylenes like poly(methyl-p-methoxyphenylsilylene) to PVK and to molecularly doped polymers could at least be schematically rationalized on physical grounds. In the case of PVK it is well established that hopping transport is controlled by π electron overlap among the carbazole sidegroups and the carbon backbone therefore plays a minor role [10]. Extensive concentration dependent studies of molecularly doped polymers which have demonstrated that transport involves hopping among active dopant molecules only, bear this out. An analogous argument might be made for the role of phenyl and other aromatic sidegroups on polysilylenes. Hole drift moblity values in the vicinity of room temperature are however surprisingly similar to those in aromatic sidegroups containing polysilylenes and in those like poly(methylcyclohexylsilylene) which contain only aliphatic sidegroup (no π electrons) and in which, on the basis of simple chemical reasoning, no significant wavefunction overlap among sidegroups is to be expected [14,48-50]. Thus the preliminary indication from comparative studies would suggest:

(1) transport in polysilylenes involves hopping among states associated directly with the silicon backbone rather than pendant sidegroups.

(2) Charge transport is observed to be insensitive to the molecular weight of PMPS at least in the range $5 \times 10^3 < M_n < 1 \times 10^6$, that is from about forty repeat units up. It is interesting that recent studies [31] of polarized excitation and fluorescence spectra on poly (di-n-hexylsilylene) also suggest in this case that the chain is separated into a series of localized chromophores. Calculations further suggest all trans segments of approximately 10-15 repeat units separated by gauche links. We suggest that transport states in PMPS are also not fully extended over the polymer chain but remain relatively localized.

6. Summary - Key Points

(1) Technological interest in organic polymeric photoreceptors arises from the need to develop inexpensive flexible media for belt architecture. In conventional inorganic photoreceptors photogeneration and transport occur in the same medium while in most organic polymer-based systems photogeneration and transport occur in compositionally distinct layers each of which needs to satisfy a set of specialized requirements. The present review has focused on transport layer materials and mechanistics of charge carrier motion.

(2) A key requirement on the transport layer is that charge accepted from the excited photogeneration layer pass through a dielectric thickness, which is fixed by system developability criteria, in a time short compared to the interval between exposure and development of the latent image. The latter imposes a requirement on charge carrier drift mobility. In addition, it is required that the fractional loss of transiting charge to deep traps be minimized because trapped space charge in the bulk produces unacceptable image degradation. It should be understood that relatively high mobility for the photoinjected carriers must be achieved simultaneously with high charge retension in the dark (negligible conductivity) and high dielectric strength. It should also be emphasized that the requirement on space charge buildup during repeated xerographic photoinjection is extraordinarily stringent for an organic polymer. In practical terms for typical transport layer thicknesses (10-25 microns), the trap density should not exceed 10^{14} cm^{-3} corresponding to no more than 1 active trap site per 10^7 monomer units. The latter is far below the expected level of chemical contamination associated with typical synthetic and purification procedures. Proper choice of transport media renders unavoidable resident impurities electrically inactive. For example, in molecularly doped polymers which transport holes, the oxidation potential of the transporting species in a given host polymer must be less than that of resident background impurities. A similar requirement applies for the carbazole groups in PVK.

(3) PVK was an early candidate for photoreceptor application. The embodiment deployed in copiers was the complex of PVK with TNF. The mobilities in this system are not sufficient to satisfy contemporary requirements of high speed copiers/duplicators and laser printers. In addition PVK is too brittle for belt architecture. Mechanistic studies on this

system however were of seminal importance in providing design criteria for closely related molecularly doped polymers.

(4) Deployment of molecularly doped polymers which are solid solutions of a transport active species (typically in the range 30-50 wt.%) in an inert host polymer embodies the concept of full compositional control of mobility. Transport molecules are chosen with the aim of rendering accidental contaminants electrically inactive. The concentration of small molecules controls drift mobility. Appropriate choice of a compatible host polymer can then focus on mechanical properties and environmental stability.

(5) Molecularly doped polymers can suffer the limitation of phase separation, crystallization and leaching by contact with the development system, cleaning solvent, etc. The latter continue to motivate the search for single component transport polymers.

(6) Scientifically motivated studies of glassy silicon backbone polymers have revealed their relatively high hole drift mobility at room temperature. Unlike PVK where hole transport clearly involves hopping of photoinjected holes among carbazole sidegroups, transport in glassy polysilylenes appears to involve holes hopping among states which derive from the saturated Si main chain. Though it appears that transport active states are not fully delocalized on the polymer chain, their detailed nature is as yet uncertain. The nature and origin of hole traps likely to exist in these systems, which is of key practical importance, are as yet unknown.

References

[1] For comprehensive review see Schafert, R.M., *Electrophotography*, 1975, (The Focal Press, New York).

[2] Regensburger, P.J., *Photochem. Photobiol.*, 1968, **8**, 429.

[3] Rose, A., *Photoconductivity and Applied Problems*, 1963 (Interscience, New York).

[4] Loutfy, R.O., Hor, A.M., and Rucklidge, A., *J. Imaging Sci.*, 1987, **31**, 31, and references therein.

[5] Stolka, M. and Pai, D.M., *Adv. Polym. Sci.*, 1978, 29, 1.

[6] Pearson, J.M. and Stolka, M., *Poly (N-vinylcarbazole)*, 1981 (Gordon and Breach, New York).

[7] Williams, D.J., *Macromolecules*, 1970, **3**, 602.

[8] Regensburger, P.J., *Can. Pat. No. 932199, iss. Aug. 21*, 1973.

[9] Gill, W.D., *Proc. 5th Int. Conf. on Amorphous and Liquid Semicond., Garmisch-Partenkirchen*, 1974 (Taylor and Francis, London), 901.

[10] Seki, H., *Proc. 5th Int. Conf. on Amorphous and Liquid Semicond., Garmisch-Partenkirchen*, 1974 (Taylor and Francis), p. 1015.

[11] Murayama, T., *Electrophotography (Jap.)*, 1986, **25**, 290.

[12] Stolka, M., Yanus, J.F. and Pai, D.M., *J. Phys. Chem.*, 1984, **88**, 4707.

[13] Schein, L.B., Rosenberg, A., and Rice, S.L., *J. Appl. Phys.*, 1986, **60**, 4287.

[14] Kepler, R.G., Zeigler, J.M. and Harrah, L.A., *Bull. Am. Phys. Soc.*, 1984, **29**, 509.

[15] Fatori, V., Bürge, G., and Geri, A., J. *Imaging Technology* 1986, **12**, 334.

[16] Yajima, S., Okamura, K., and Hayashi, T., *Chemistry Letters*, 1975, 1209.

[17] Mazdiyasni, K.S., West, R., and David, L.D., *J. Ceram. Soc.*, 1978, **61**, 504.

[18] Trujillo, R.E., *J. Organomet. Chem.*, 1980, **198**, C27.

[19] Stolka, M., Yuh, H.-J., McGrane, K., and Pai, D.M., *J. Polymer Sci., Polym. Chem. Ed.*, 1987, **25**, 823.

[20] Abkowitz, M.A., Knier, F.E., Yuh, H.-J., Weagley, R.J., and Stolka, M., *Solid St. Comun.*, 1987, **62**, 547.

[21] Zeigler, J.M., *Polymer Preprints*, 1986, **27**, 1, 109.

[22] Zhang, X.-H. and West, R., J. Polymer Sci., *Polym. Chem. Ed.*, 1984, **22**, 159, ibid., **22**, 225.

[23] Scher, H. and Montroll, F.W., *Phys. Rev. B*, 1975, **12**, 2455.

[24] Lampert, MA. and Mark, P., *Current Injection in Solids*, 1970 (Academic Press, New York).

[25] Abkowitz, M. and Scher, H., *Philos. Mag. B*, 1977, **35**, 1585.

[26] Abkowitz, M., *J. Appl. Phys.*, 1979, 50, 4009.

[27] Dolezalek, F.J., in *Photoconductivity and Related Phenomena*, Mort, J., and Pai, D.M., editors, 1976 (Elsevier Sci. Publ. Co., New York), p. 27.

[28] Muller-Horsche, E., Haarer, D., and Scher, H., *Phys. Rev. B.*, 1987, **35**, 1273.

[29] Orlowski, T.E., and Abkowitz, M., *Solid St. Comun.*, 1986, **59**, 665.

[30] Mott, N.F. and Davis, E.A., *Electronic Processes in Non-Crystallline Materials*, 1971, (Clarendon Press, Oxford).

[31] Enck, R. and Pfister, G., in *Photoconductivity and Related Phenomena*, Mort, J. and Pai, D.M., editors 1976 (Elsevier Sci. Publ. Co., New York), p. 27.

[32] see Bässler, H., *Philos. Mag. B*, 1984, **50**, 347 and references therein.

[33] Pfister, G., *Phys. Rev. B*, 1977, **16**, 3676.

[34] see for example Movaghar, B., Würtz, D., and Pohlmann, B., *Z. Phys. B - Condensed Matter*, 1987, **66**, 523.

[35] Schönherr, G., Bässler, H., and Silver, M., *Philos. Mag. B.*, 1981, **44**, 47.

[36] Pfister, G. and Griffiths, C.H., *Phys. Rev. Lett.*, 1978, **40**, 659.

[37] Gill, W.D., *J. Appl. Phys.*, 1972, **43**, 5033.

[38] Pai, D.M., *J. Chem. Phys.*, 1971, **52**, 2285.

[39] Fujino, M., Kanazawa, Y,., Mikawa, H., Kusabayashi, S., and Yokoyama, M., *Solid St. Commun.*, 1984, **49**, 575.

[40] Bässler, H., Schönherr, G., Abkowitz, M., and Pai, D.M., *Phys. Rev. B.*, 1982, **26**, 3105.

[41] Mort, J., Pfister, G., and Grammatica, S., *Solid St. Commun.*, 1976, **18**, 693.

[42] Pai, D.M., Yanus, J.F., Stolka, M., Renfer, D., and Limburg, W.W., *Philos. Mag. B.*, 1983, **48**, 505.

[43] Borsenberger, P.M., Mey, W., and Chowdry, A., *J. Appl. Phys.* 1978, **49**, 273.

[44] Tsutsumi, M., Yamamoto, M., and Nishijima, Y., *J. Appl. Phys.*, 1986, **59**, 1557.

[45] Abkowitz, M., and Pai, D.M., *Philos. Mag. B.*, 1986, **53**, 193.

[46] Santos Lemus, S.J. and Hirsch, J., *Philos. Mag. B.*, 1986, **53**, 25.

[47] Pai, D.M., Yanus, J.F., and Stolka, M., *J. Phys. Chem.*, 1984, **88**, 4714.

[48] Yuh, H.-J., Abkowitz, M., McGrane, K., and Stolka, M., *Bull. Am. Phys. Soc.* 1986, **31**, 386.

[49] Kepler, R.G., Zeigler, J.,M., Harrah, L.A., and Kurtz, S.R., *Phys. Rev. B*, 1987, **35**, 2818.

[50] Abkowitz, M., Stolka, M., and Yuh, H.-J., *Bull. Am. Phys. Soc*, 1987, **32**, 885.

[51] Klingensmith, K.A., Downing, J.W., Miller, R.D., and Michl, J., *J. Am. Chem. Soc.*, 1986, **108**, 7438.

ELECTROCHEMICAL EVALUATION OF THE

p-Si/CONDUCTING POLYMER INTERFACIAL PROPERTIES

G. Nagasubramanian
S. DiStefano
J. Moacanin

Jet Propulsion Laboratory
CALIFORNIA INSTITUTE OF TECHNOLOGY
Pasadena, CA 91109

A. INTRODUCTION

π conjugated organic conducting polymers are being used for a variety of applications, such as batteries[1] corrosion protection coatings[2,3] While a high bulk conductivity is essential for these applications, a variety of specific criteria have to be met for different applications. For example, for battery applications, structural robustness (integrity) and reversibility are most important. For solar cell applications (specifically for the formation of an ohmic contact) interface resistance (semiconductor/conducting polymer) and the alignment of band edges are important, and additionally optical transparency requirements must be met.

Most conducting polymers such as polypyrrole exhibit strong absorption in the visible region when in the doped or conducting state[4]. However, more recently it has been shown that by a judicious choice of monomer the absorption maxima of the doped polymer can be shifted into the U.V., thus polyisothianapthene does not absorb (to any great extent) in the visible region when doped[5].

We report below the results of our investigation on the contact resistance and interfacial properties of p-Si/conducting polymer interface for solar cell applications. Electrochemical studies indicate that poly (isothianapthene) PITN and (polypyrrole) PP can be p-doped only. By studying the electrochemical behavior of both PITN and polypyrrole in acetonitrile solution on p-silicon electrodes the electronic nature of the polymer/semiconductor junction has been determined.

B. EXPERIMENTAL SECTION

The solvent acetonitrile (MeCN), and supporting electrolyte tetra-n-butylammonium perchlorate (TBAP), were purified as previously described,[6a] while tetra-phenyl phosphonium chloride (ϕ_4PCl) (Aldrich) was used as received. The monomer isothianaphthene (ITN)[6b] was obtained by dehydrating its precursor, dihydroisothianaphthene S-oxide, which subsequently was electrochemically polymerized to obtain poly(isothianaphthene) (PITN)[6b].   All chemicals were stored inside an argon-filled glove box.   Single crystal p-Si working electrodes were prepared as described elsewhere[5].   Si surface was etched by dipping for 5 seconds in HF containing 2 drops of Bromine[2a]. Conducting polymer films were deposited on the working electrode surface utilizing a two-compartment cell of 15 ml capacity, with a large Pt flag counter-electrode separated from the working and reference (Ag quasi-reference, here after denoted AgRE) electrodes by a medium porosity glass frit. In a typical experiment, inside the glove box, 15 mL of MeCN solution containing 30 mM of ITN, or 1M pyrrole and 0.1 M supporting electrolyte was used and the films were grown potentiostatically. Impedance measurements were made in the frequency range 500 Hz to 5000 Hz.   Plots of G_p vs V at three different frequencies for bare and PP coated p-Si are shown in Figures 1 and 2.   Typically plots of conductance G_p, where G_p is the measured conductance, vs. potential (V) were made.   For bare p-Si, the plots show frequency dependant spectral features with a peak potential shifting towards V_{fb} (flat-band potential) as a function of frequency. This result implies a continuous distribution of surface states.   However, poly-pyrrole coated silicon electrodes show an improved behavior as evidenced by the absence of spectral features in the G_p vs V plots. Unfortunately, PITN coated silicon electrodes show no such improvements and exhibit a G_p vs V plot similar to that of bare silicon.

As previously outlined, the contact resistance between the conducting polymer and p-Si should be small so that the voltage drop across the interface is negligible if these materials are to be used as current collectors.   With this in mind, we evaluated the interfacial resistance using ohm's law. The test structure and experimental procedure have previously been described[7].   The data show that while polypyrrole forms a quasi-ohmic contact with p-Si, PITN

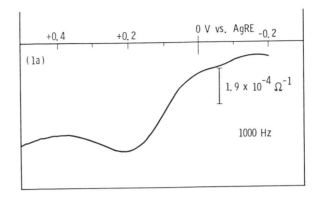

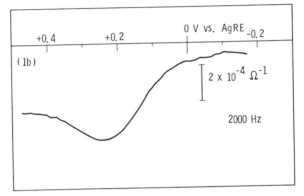

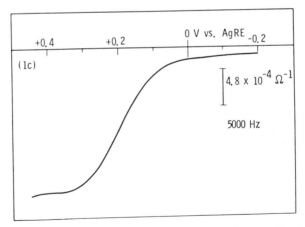

Fig. 1 – G$_p$ (equivalent-parallel conductance) vs. V plots in acetonitrile – 0.1 M TBAP solution, at three different frequencies for naked p-type single crystal silicon electrode. Scan rate 2 mV/sec.

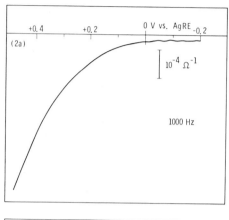

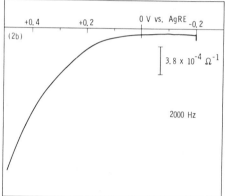

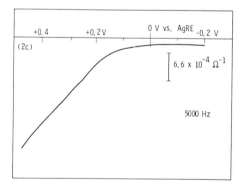

Fig. 2 – G_p vs. V plots in acetonitrile 0.1 M TBAP after depositing poly(pyrrole) onto p–Si, Counterion; ClO_4. Scan rate 2 mV/sec.

exhibits rectifying behavior and the corresponding plots are shown in Figure 3. One of the necessary thermodynamic criteria for the formation of an ohmic contact between a p-type semiconductor and a metal is the position of the flat band potential of the semiconductor with respect to the Fermi level of the metal. In particular, an ohmic contact is possible only when the Fermi level of the metal is more positive than the V_{fb} flat-band of the semiconductor. Flat-band potentials culled from Mott–Schottky plots for p-Si and conducting polymers are shown in Figure 4. Our data show that only in the case of poly-pyrrole the V_{fb} determined from the Mott–Schottky plots satisfies the thermodynamic condition.

C. CONCLUSION

Results of this study show that electrochemically deposited polyprerole on p-Si passivates surface states and forms ohmic contact. But both contact resistance and optical absorbance is rather high.

PITN on the other hand, although intrinsically more conductive than polypyrrole neither passivates surface states nor forms ohmic contact. However, Heeger et al showed that whereas polythiophene is highly colored, PITN a substituted polythiophene(1.a) becomes transparent in the visible region when doped[5]. Recently Grubbs reported the synthesis of polyvynilidinemethylisoindoline(1.b) .

1b   1a

This substituted polypyrrole, polyvinylioenemethylisoindoline, we anticipate would show improved electrical and optical properties. Furthermore we would expect this polymer to be more electropositive than polypyrrole (Fig. 4) and therefore should form ohmic contact. The evaluation of the properties of polymer (1.b) are underway in our laboratory.

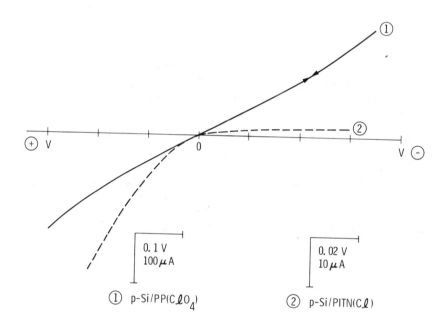

Fig. 3 – I–V traces of p–Si interface for PP and PITN. I–V
Curves were recorded as follows: after depositing PP or PITN
onto p–Si (mounted as electrode; for description see
Experimental) surface, a Hg contact was made to the film
after evaporating the solvent. A copper lead was connected
to both auxiliary and reference of the PAR potentiostat, the
working through a Pt wire dipping in the Hg pool is
connected to polymer films. The potential was scanned at 20
mV/sec and output current was recorded as a function of
potential.

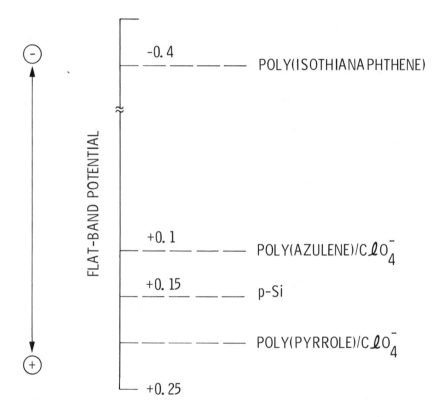

Fig. 4 - Comparison of flat-band potentials and the thermodynamic requirements for ohmic contact with p-Si. (eg. flat-band must be more positive than +0.15V for ohmic contact to be thermodynamically possible.

ACKNOWLEDGMENT

The work described in this paper was carried out at the Jet Propulsion Laboratory, California Institute of Technology and was sponsored by U.S. Department of Energy, through an agreement with the National Aeronautics and Space Administration.

REFERENCES

1. (a) K. Kaneto, M. Maxfield, D. P. Nairms, A. G. MacDiarmid, and A. J. Heeger, J. Chem. Soc., Faraday Trans. 1, 78, 3417 (1982); (b) R. B. Kaner and A. G. MacDiarmid, ibid., 80, 2109 (1984); (c) "Conducting Polymers: Polymeric Organic Metals and Semiconductors." Course offered by the Institute in Materials Science, State University of New York (Nov. 12-14, 1984).

2. (a) F. R. F. Fan, B. L. Wheeler, A. J. Bard, and R. N. Noufi, J. Electro Chem. Soc., 128, 2042 (1981); (b) R. N. Noufi, A. J. Frank, and A. J. Nozik, J. Am. Chem. Soc., 103, 1849 (1981); (c) R. N. Noufi, A. J. Nozik, J. White, and L. F. Warren, J. Electro Chem Soc., 129, 2261 (1982).

3. R. Noufi, D. Tench, and L. F. Warren, J. Electro Chem. Soc., 128, 2596 (1981).

4. E. M. Genies, G. Bidan and A. F. Diaz, J. Electroanal. Chem. 149, 101 (1983).

5. H. Yashima, M. Kobayashi, K.-B. Lee, O. Chung, A. J. Heeger, and F. Wudl, J. Electrochem. Soc., 134, 46 (1978).

6. (a) G. Nagasubramanian, S. DiStefano, and J. Moacanin, J. Electron. Mats. 15, 21 (1985). (b) F. Wudl, M. Kobayashi and A. J. Heeger, J. Org. Chem., 49, 3382 (1984).

7. G. Nagasubramanian, S. DiStefano and J. Moacanin, J. Electrochem Soc., 133, 305 (1986).

8. R. H. Grubbs, Private communications.

INTEGRATED LANGMUIR SENSORS: AN OVERVIEW

C.W. Fu, D.A. Batzel, S.E. Rickert

W.H. Ko, and M.E. Kenney

Depts. of Macromolecular Science, Electrical Engineering
and Chemistry
Case Western Reserve University
Cleveland, Ohio 44106 U.S.A.

Abstract

Interest in the Langmuir-Blodgett ("LB") technique for production of ultra-thin organic crystals (i.e. "Langmuir films") has led to a number of investigations into different types of material that can be deposited in the form of mono-molecular layers. For example, the classic long chain fatty acids, alcohols, polymerizable molecules, aromatic hydrocarbons and dye substances can all be produced in monolayer form. Unfortunately, few of these materials fulfill the requirements of mechanical and thermal stability necessary for commercial, military, or aerospace use. Phthalocyanine compounds, which are well known for such stability appear to be promising in this regard. However, due to the planar structure of the molecule, nobody, to date, has been able to form multilayer crystals (i.e. "foils") from these compounds.

It has long been known that the conductivities of phthalo-cyanine and derivatives are very sensitive to the presence of certain gases, particularly to the oxides of nitrogen. The increased conductivity of such materials has been demonstrated to be confined to the surface of the crystal: hence many phthalocyanine gas detector systems have been based on thin films. However, these films are highly disordered, leading to difficult data interpretation, slow response times, and long recovery times. Techniques being developed at the Polymer Microdevice Laboratory ("PML") allow us to prepare true foils of phthalocyanines. The response time of these foils to NO_2 is on the order of milliseconds, and recovery times of under 5 seconds are observed. Similar data is becoming available for other gases such as O_2, CO_2, and NH .

A general review of solid state gas sensors is presented in this article. The gas sensors are classified into four types: Semiconducting gas sensors, Metal-Oxide-Semiconductor ("MOS") gas sensors, Surface-Acoustic Wave ("SAW") gas sensors, and Calorimetric gas sensors. Gas sensing mechanisms based on the adsorption gases at the solid surface and the consequential effect on the conductivity of the semiconductors, the work function of the MOS structures , the frequency function of the SAW devices and the temperature function of the calorimetric sensors are covered.

A Langmuir-Blodgett film micro gas sensor is given as an example. The micro sensor integrating an interdigitated electrode, heating element and temperature diode on a chip has a size of 3 mm by 2 mm. Some preliminary results are reported.

I. Introduction

In the last two decades, the progress made in microelectronics has had a great impact on every aspect of human life [1,2]. The monolithic circuit has not only been reduced in size but has also increased in functionality which has resulted in reducing cost and improving reliability. These developments have created many applications in industrial automation, robotics, health care and information processing. The barrier in these applications lies not in microelectronics, but rather in the area of interfacing with the non-electronic world. Sensors need to improve in accuracy, resolution, speed, reliability, cost and process compatibility with integrated circuit processes before full advantage of microelectronics can be taken.

The gas sensor is designed to selectively convert a gas concentration into an appropriate electrical signal. They are used to detect noxious or dangerous gases [3-5] in the environment, to monitor the air-fuel ratio in the automobile engine [6-7], to monitor gases in hospitals and as process controls in manufacturing [8,9]. With advancements in engineering and chemistry, a new family of gas sensors based on the use of thin [10] and thick film technologies [11] has emerged. A superior design for sensors is made possible through a variety of available materials and local modification of their composition by using techniques such as thermal diffusion and ion implantation.

II. Classification of Gas Sensors

Solid-state gas sensors depend on the reaction of a gas at a solid surface and the consequential effect of this reaction on a measurable property or condition of the solid. A comprehensive classification of gas sensors based on their measurable properties is illustrated in this section. A review of performance and prospects in the design of sensors is given.

A. Semiconducting Gas Sensors

It has been known that adsorbed gases can have a marked effect on the electrical conductivity of semiconductors. Semiconductors suitable for use as gas sensors must have properties such as conductivity which varies with the concentration of the measured gas. Furthermore, the conductivity must return to its original value in the absence of the gas. There are many ways that ambient gases can change the conductivity of a semiconductor. Irreversible surface reactions such as etching or the growth of oxide films are unacceptable. In contrast, surface reaction involving reduction/oxidation, anion exchange or adsorption is reversible. Many proposed or commercial sensors are based upon a combination of these reactions while the dominant one is still not cleared. Transition metal oxide semiconductors have been widely studied as sensors in the past, but organic semiconductors are gaining more attention.

1. Metal Oxide Gas Sensors. It has long been known that gas adsorption on metal oxides influences the conductivity of the oxides. The metal oxides have defects arising from a stoichiometric excess of the metal or of oxygen. The metal oxide sensor utilizes a reaction between the gas of interest and oxygen on a solid surface. In titanium dioxide, for example, there is excess titanium in the structure which results in donation of electrons to the conduction band. On the other hand, cuprous oxide often has excess oxygen which accepts electrons from the valence band. The empty valence bonds (holes) provide high conductivity by moving freely throughout the materials [12-13].

ZnO exhibits promising surface properties and is applied in the chemical industry as a catalyst. However, SnO_2 is more widely used for gas sensors at present. One reason may be that the operation temperatures for SnO_2 are lower than for ZnO. ZnO and SnO_2 are used as typical examples, as they are n type conductors due to an excess of metal (oxygen vacancies) in their structures. In air these oxides are

covered by chemisorbed oxygen and exhibit decreasing conductivity near their surfaces as a result.

The band gaps for ZnO and SnO_2 are 3.2 eV and 3.5 eV,
respectively. The energy band for each is shown in Figure 1.
Intrinsic donors in these systems are connected to a stoichiometric excess of metal. The concentration of electrons
increases with increasing temperature in that a growing fraction of the electrons normally bound in donors and traps
enters the conduction band. The surface conductivity can be
modified by adsorbing donors (reducing gases) or acceptors
(oxidizing gases). A depletion layer (shown in Figure 1) is
induced by adsorbed acceptors binding with electrons from
donors below the surface. On the other hand, an accumulation
layer is generated by adsorbed donors donating electrons to
the surface layer.

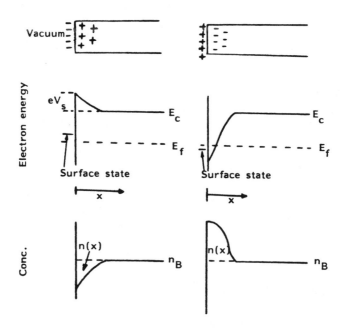

Figure 1. Schematic diagram of space charge layer on the
surface of metal oxide. Energy band diagram. Charge concentration.

In air the surface of sensors is covered by adsorbed
oxygen which binds their conduction electrons. A reducing
gas will first react with the oxygen and return electrons to
the oxide. In the second step, surface donors may be produced by chemisorption, causing further increases in carrier

density. Finally diffusion processes must be considered, where donors from the surface penetrate into the uppermost atomic layer of the oxide.

2. Organic semiconductor gas sensors. There is a need to develop more sophisticated gas sensors in order to improve their sensitivity, selectivity and response. It is well known that conductivity changes in organic semiconductors may be many orders of magnitude when a gas adsorbs on their surfaces. In addition, the organic molecule can be easily modified to improve its gas selectivity. The ability to form uniform well ordered thin layers of such organic semiconductors gives additional improvement in sensitivity and response.

There are several p-type organic semiconductors such as phthalocyanines, polyaromatic compounds, carotene and rhodamine [14] which are well known to respond to several gases. For example, strong electronegative gases like Cl_2 and NO_2 show a strong interaction with p type semiconductors.

There are three types of interactions which gases can have with the interface material. These are physisorption, chemisorption, and coordination chemistry. Physisorption involves only electrostatic van der Waal interactions between the gas and surface material. Chemisorption is the specific interaction between a gas and a surface which involves chemical bond formation. Intermediate interactions can also be distinguished, some of which belong to the area of coordination chemistry or charge transfer complex formation. A coordination compound usually consists of a central metal ion M surrounded by a neutral or charged organic ligand. In the ligand, one or more donor atoms interact with the metal ion. The coordinate bond is stronger than the physisorptive interaction (0-40 kJ/mole) but weaker than a true covalent chemical bond (~ 300 kJ/mole) [14].

As far as sensitivity and selectivity are concerned, chemisorption is preferred. Chemisorption, however, exhibits a slow recovery back to the baseline value in the absence of the testing gases. In physisorption, no bonds are broken, so that the baseline is immediately restored when the test gases are removed. Coordination chemistry provides a reasonable compromise between sensitivity, selectivity and speed. The synthesis of a specific molecule in coordination chemistry offers the possibility to improve selectivity as well as speed.

B. Metal-Oxide-Semiconductor Gas Sensors

Gas sensors based upon Metal Oxide Semiconductor (MOS) structures were first reported by I. Lundstrom in 1975 [15]. Pd/SiO$_2$/Si structure sensors such as Schottky diodes [16], capacitors [17], and Field-Effect Transistors (FET) [18] for detection of various low molecular weight, hydrogen-containing gases have been of great interest.

The principle operation of these devices is believed to be via a change in the work function difference between the metal gate electrode and the semiconductor, which results from chemisorption of a reagent gas onto the gate of the MOS structure. Although the precise nature of the processes involved in the adsorption and desorption of the reagent gas is not clear, it has been suggested that a chemisorbed hydrogen-containing gas undergoes catalytic dissociation on the Pd-gate causing a direct change in the metal work function (a bulk effect). Alternatively, atomic hydrogen may migrate through the metal and form a "dipole layer" at the gate metal/insulator interface, thus altering the effective work function of the metal (a surface effect). The dipole layer model is the more widely accepted theory to date.

A conducting MOS diode is a rectifying contact which has an ultra-thin insulator layer (20-30 Å) inserted between the semiconductor and metal. This insulator layer is thin enough to allow carriers to communicate back and forth between the metal and semiconductor but its presence can modify transport and barrier height. A MOS capacitor has a non-conducting contact having an oxide thickness greater than 100 Å. The flat band voltage and interface state density can be measured by the capacitance method or the conductance method [19]. A MOSFET has a conduction path between the drain and the source region. When the gate-source voltage is above the threshold voltage, a conduction channel is formed. This conduction channel can be modulated by varying the gate voltage. When the devices are exposed to a gas, the adsorption process will change the effective barrier height of the MOS diode, the flat band voltage of MOS capacitor, or the threshold voltage of the MOSFET.

C. Surface-Acoustic-Wave Gas Sensors

Surface acoustic waves (SAW) are elastic waves propagating along the surface of a substrate, having amplitudes which decay exponentially with substrate depth. Surface waves are usually generated and detected in a piezoelectric substrate. STX quartz or YZ lithium niobate are commonly used as sub-

strates. In homogeneous substrates, the phase velocity and
the amplitude of a SAW device are determined by its elastic,
piezoelectric, dielectric and conductive properties along
with its mass.   If one of these properties is altered by the
gas´ interaction with the substrate, the effect of sensing
can be quantitatively measured.

One type of SAW device called the SAW delay-line oscillator is of great interst in gas sensor applications [20].   The
device shown in Figure 2 consists of two interdigitated
electrodes, one acting as an emitter of surface acoustic
waves and the other as a detector of the waves.   Since the
electrodes are located at some distance from each other, the
device operates as a delay line.   The acoustic path of the
delay line is covered with a selective chemical interface.
Interaction of the gas molecules with this chemical interface
causes a velocity change in the acoustic signals, resulting
in a phase shift between the input and output signals of this
delay line.   This phase shift leads to a frequency change in
the oscillator frequency.   The concentration of the gas can
be determined by the frequency shift.

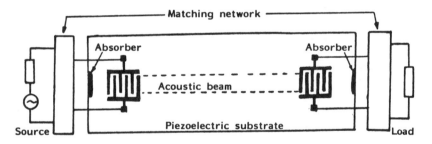

Figure 2.   Surface Acoustic Wave (SAW) delay line with two
interdigitated electrodes on a piezoelectric substrate.

D. Calorimetric Gas Sensors

The calorimetric gas sensor measures gas concentration as
a function of the temperature rise produced by the heat of
reaction on a catalytic surface.   Calorimetric gas sensors
are generally used to monitor flammable gases in air.   For
instance, in the oxidation of one mole of methane to carbon
dioxide and water, 800 kJ of heat are liberated. To sustain
such a reaction at a reasonable temperature, a catalyst is
required.   Since the reaction occurs at the catalyst surface,
it is convenient to supply heat to the catalyst rather than

the reacting gas, and to measure the heat of reaction as a temperature rise of the catalyst. Thus, the catalytic gas sensors consist of a catalyst surface, a temperature sensor, and a heater to maintain the catalyst at the operating temperature. The combination of high heats of oxidation and a ready oxygen supply is ideally suited for this type of device [21].

Two modes of operating this type of gas sensor can be identified. The first mode (the more widely used among commercial gas detectors) is non-isothermal. In this mode, the temperature of the sensing element is allowed to rise as a result of chemical reaction at the catalyst surface. The reaction, rate, and hence the concentration of the flammable gas, is derived from the increase in temperature. In the second mode, the temperature of the sensing element is maintained constant during the reaction and the concentration of the gas is obtained in terms of a difference in electric power dissipation of the element under reacting and nonreacting conditions.

III. A Langmuir-Blodgett Film Micro Gas Sensor

The general consideration in a gas sensor design is optimization of sensitivity, selectivity and reversibility. In addition, cost, reliability and electronic compatibility of the sensors are important. A batch electronic process is preferred because of cost reduction and reliability improvement. Aspects such as fabrication technology, device structure and chemical interface preparation are also important in the design of a gas sensor. Silicon is the choice substrate for gas sensing devices since its properties are well understood. In addition, integrated circuit technology based upon silicon provides electronic compatibility for on-chip signal conditioning.

The goal of device design should be simple fabrication but versatile functionality. A micro sensor designed to operate at several different temperatures or to scan temperature was constructed and consists of an interdigitated electrode, a temperature sensing diode and a heating element on a chip. Because gas sensitivity is a function of temperature, gas selectivity can be improved by the proper choice of temperature for any given gas [22].

The choice of a chemical interface is the most important of all the sensor considerations. The general rule of thumb is to increase the surface to volume ratio, since first order gas responses happen at the surface. The Langmuir-Blodgett film technique is attrative in preparation of the interface because the thickness of the interface can be controlled down to a monolayer [23]. It is expected that the LB film method will improve the sensitivity as well as the response time.

Sensor Design

The chip is shown in Figure 3, and the interdigitated electrode consisted of three gold fingers, each 20 micrometers wide with a gap of 20 micrometers between the fingers and the finger overlap distance was 320 micrometers. The heater is made of a p-type diffusion resistance Si material. It can be used to heat the silicon substrate to 300°C in a

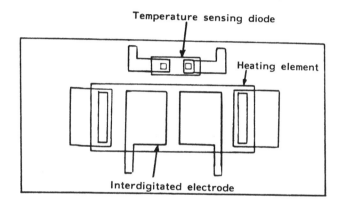

Top view

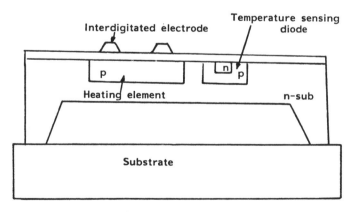

Cross section view

Figure 3. An interdigitated electrode with temperature management devices.

few seconds. A silicon diaphragm is located just beneath the heating area which thermally isolates the substrate from the package and thus speeds up the temperature rising time and reduces the heat required to raise the sensor to a specified temperature. The temperature sensing diode is made of a p-n junction and located in good thermal contact with the sensing film. The diode provides a real time monitor for the temperature variation of the sensor. It can be used for temperatures well above 300°C if needed eventually. The overall chip size is 3 mm by 2 mm.

Phthalocyanines, known to be thermally stable to 400 ° C, are attractive as potential gas sensing materials. The properties of phthalocyanines can be modified by substitution either on the ring or the central atom. The phthalocyanines we tested were deposited by LB film technique. In order to use this method, the material should be soluble in a suitable spreading solvent and possess chemical functional groups conducive to the desired preferential orientation. However, usual phthalocyanines are insoluble in water and common organic solvents. The solubility in organic solvents of phthalocyanines can be improved by ring substitution [24] or axial substitution. In this paper, an unsymmetrical axially substituted phthalocyanine having a hydrophilic group on one side of the ring and a hydrophobic group on the other side showed a good solubility in chloroform. This phthalocyanine was synthesized at Case Western Reserve University. The thin film was deposited by the LB film technique.

B. Experimental Results

The phthalocyanine film was transferred from the gas-water interface to a hydrophobic silicon wafer. The deposition ratio was approximately 1:1. This indicates that the film on the substrate probably has the same packing as on the gas-water subphase. The first material (the monomer, Hex SiOSi-PcOH) had a co-area of 130 Å^2, somewhat smaller than those expected for molecules packed parallel to the water surface (typically about 160 Å^2). It is probable that the rings are somewhat stacked or tilted on the water subphase. The second material (the dimer, $Hex_3SiOSiPcOGePcOH$) had a co-area of 165 Å^2 , indicating that the rings may be deposited parallel to the substrate (perfectly flat on the water surface).

The dimer films were exposed to various gases such as hydrogen, nitrous oxide, carbon monoxide, carbon dioxide, ammonia, sulfur dioxide, chlorine, nitrogen dioxide and air. Preliminary data indicates that gases such as chlorine, nitrogen dioxide and ammonia can be sensed with a good response at room temperature. A nitrogen ambient was used as the reference background. The conductance of the electrode increased with increasing concentration of the gas until saturation was reached. A four order magnitude change in the conductance between absence and saturation with the gas was observed for chlorine (Figure 4) and nitrogen dioxide (Figure 5).

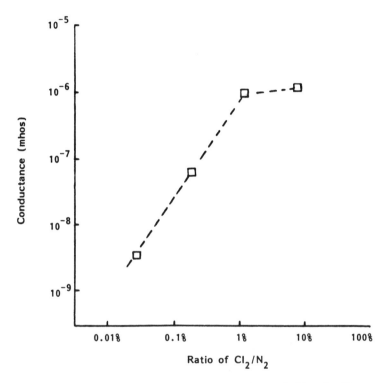

Figure 4. Conductance vs. concentration of Cl_2 in N_2 at room temperature.

A plot of conductance vs. temperature for the dimer in nitrogen and air indicates that the film is sensitive to oxygen (Figure 6).

The change in conductivity of the monomer with concentration of NO_2/N_2 was measured (Figure 7). The conductivity increased from 10^{-7} ohms/cm to 10^{-4} ohms/cm as concentration increased from 10 ppm to 10000 ppm.

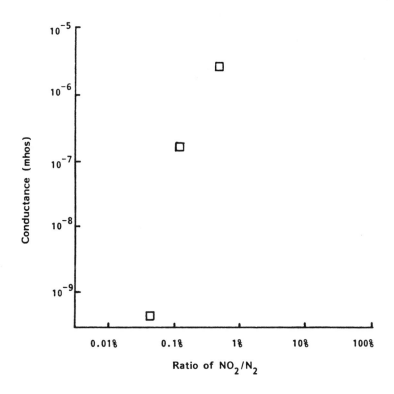

Figure 5.   Conductance vs. concentration of NO_2 in N_2 at room temperature.

The   silicon substrate device was fabricated by microelectronic technology.   The voltage of the diode was found to be
a linear function of temperature (Figure 8).   The heating
element can heat the substrate to $240°C$ in a few seconds
using a 7.5 volt power supply.

IV. Conclusions

In this paper, an overview of solid-state gas sensors was
given.   Properties such as conductivity, work function, frequency and temperature were sensitive to gases when the
sensor structure was changed to chemisistor, MOS structure,
SAW device and calorimeter, respectively. However, the principle gas sensing mechanism is based on adsorption of gas
species onto the chemical interface in all cases.   It is

therefore critical to prepare a proper chemical interface for
all sensor structures.

Phthalocyanine LB films were used as the gas sensing
interface and preliminary results have been reported. A
general purpose microsensor having a size of 3 mm by 2 mm was
fabricated by integrated circuit technology. A uniform and
near defect-free multilayer film was deposited on the sensor
by the LB method. Isotherm studies suggested that the dimer
is flat on the gas-water interface under close pack conditions. It was shown that the films had large sensitivity to
gases such as NO_2 and Cl_2.

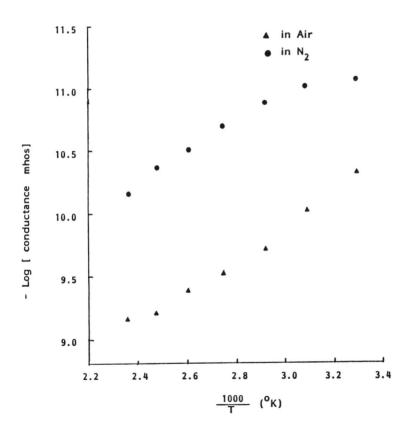

Figure 6. The conductance vs. temperature for dimer in N_2 and
air.

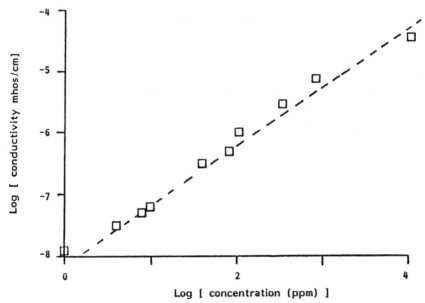

Figure 7. Conductivity vs. concentration for NO_2/N_2 gas at 17° C.

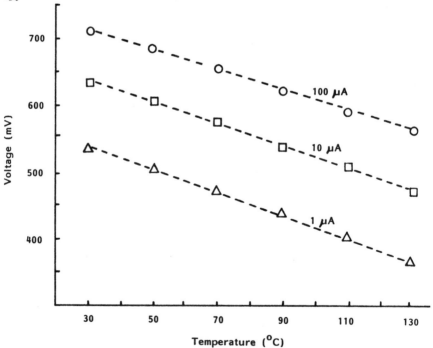

Figure 8.   Diode voltage vs.   temperature at a constant diode current.

References

[1] Wise, K.D. "Integrated Sensors: Interfacing Electronics to a Non-Electronic World", Sensors and Actuators, 1982, 2, 229-237.

[2] Ko, W.H.; Fung, C.D. "VLSI and Intelligent Transducers", Sensors and Actuators, 1982, 2, 239-250.

[3] Bott, B.; Jones, T.A. "A Highly Sensitive NO_2 Sensor Based On Electrical Conductivity Changes in Phthalocyanine Films", Sensors and Actuators, 1984, 5, 43-53.

[4] Beitnes, H.; Schroder, K. "Detections of Trace Concentrations of Gases with Coated Piezoelectric Quartz Crystals", Analytica Chemica Acta, 1984, 158, 57-65.

[5] Ross, J.F.; Obins, I.; Webb, B.C. "The Ammonia Sensitivity of Platinum-Gate MOSFET Devices: Dependence and Gate Electrode Morphology", Sensors and Actuators, 1987, 11, 73-90.

[6] Ohsuga, M.; Ohyama, Y. "A Study on the Oxygen-Biased Wide Range Air-Fuel Ratio Sensor for Rich and Lean Air Fuel Ratio", Sensors and Actuators, 1986, 9, 287-300.

[7] Logothetis, E.M.; Vassell, W.C.; Hetrick, R.E.; Kaiser, W.J. "A High Sensitivity Sensor for the Measurement of Combustible Gas Mixtures", Sensors and Actuators, 1986, 9, 363-372.

[8] Pocock, R.E. "Location Oxygen and/or Combustible Analyzer", Sensors, 1986, 3(2), 35-38.

[9] Kobayashi, T. "Solid-State Sensors and Their Applications in Consumer Electronics and Home Appliances in Japan", Sensors and Actuators, 1986, 9, 235-248.

[10] Morrison, S.R. "Semiconductor Gas Sensors," Sensors and Actuators, 1982, 2, 329-341.

[11] Prudenziati, M.; Morten, B. "Thick-Film Sensors: An Overview", Sensors and Actuators, 1986, 10, 65-82.

[12] Heiland, G. "Homogeneous Semiconducting Gas Sensors", Sensors and Actuators, 1982, 2, 343-361.

[13] Gentry, S.J.; Jones, T.A. "The Role of Catalysis in Solid-State Gas Sensors", Sensors and Actuators, 1986, 10, 141-163.

[14] Nieuwenhuizen, M.S.; Barendz, A.W. "Processes Involved at the Chemical Interface of SAW Chemosensor", Sensors and Actuators, 1987, 11, 45-62.

[15] Lundstrom, I.; Shivaraman, S.; Svensson, C.; Lundkvist, L. "A Hydrogen-Sensitive MOS Field-Effect Transistor", Appl. Phys. Letts., 1975, 26(2), 55-57.

[16] Fonash, S.J.; Huston, H.; Ashok, S. "Conducting MIS Diode Gas Detectors: The $Pd/SiO_x/Si$ Hydrogen Sensor", Sensors and Actuators, 1982, 2, 363-369.

[17] Reihua, W.; Fortunato, G.; D'amico, A. "The Correlation of the Transient Current Response of H_2 Absorption and Desorption with the H_2 Sensitivity in Pd-Gate MOS Capacitors", Sensors and Actuators, 1985, 7, 253-262.

[18] Ross, J.F.; Robins, I.; Webb, B.C. "The Ammonia Sensitivity of Platinum-Gate MOSFET Devices: Dependence on Gate Electrode Morphology", Sensors and Actuators, 1987, 11, 73-90.

[19] Nicollian, E.H.; Goetzberger, A. "The Si-SiO$_2$ Interface-Electrical Properties as Determined by the Metal-Insulator-Silicon Conductance Technique", The Bell System Technical Journal, 1967, 12, 1055-1133.

[20] Venema, A.; Nieuwkoop, E.; Vellekoop, M.J. "Design Aspects of SAW Gas Sensors", Sensors and Actuators.

[21] Gentry, J.G.; Jones, T.A. "The Role of Catalysis in Solid-State Gas Sensors", Sensors and Actuators, 1986, 10, 141-163.

[22] Wu, Q.H. Private communication.

[23] Roberts, G.G., "Langmuir-Blodgett Films", Contemp. Phys., 1984, 25, 2, 109-128.

[24] Barger, W.R.; Wohltjen, H.; Snow, A.W. "Chemiresistor Transducers Coated with Phthalocyanine Derivatives by the Langmuir-Blodgett Technique", 1985, International Conference on Solid-State Sensors and Actuators, 410-413.

STRUCTURAL ASPECTS OF PYROELECTRIC AND PIEZOELECTRIC POLYMERS: POLY(VINYLIDENE FLUORIDE) AND ITS DERIVATIVES

Gwo-Shin Swei, Scott E. Rickert and Jerome B. Lando

Department of Macromolecular Science
Case Western Reserve University
Cleveland, Ohio 44106 U.S.A.

I. Introduction

Poly(Vinylidene Fluoride), a pyroelectric and piezoelectric polymer, exists in at least six crystalline phases. Crystal structures for Phase I (β phase), [1,2] Phase II (α phase), [2-4] Phase III (γ phase) [5,6] and Phase IV (δ phase) [7] have been determined. Phase V (ϵ phase) [8] has been reported to be a nonpolar phase having the same chain conformation as Phase III, while Phase VI (ζ phase) [9] is obtained mixed with Phase II and III when transcrystallized film [10] in Phase II is electrically poled. It is clear from the determined crystal structures that Phases I, III, and IV have highly polar unit cells in the ab plane perpendicular to the chain axis and thus are pyroelectric and piezoelectric. What is not always realized is that the unit cells of the so-called nonpolar Phase II as well as Phases III and IV would have considerable dipole strength along the chain axis if it were not for statistical packing [5,6,8] of up and down chains. Thus one could assume that if a film that would not break down could be made of, for example, Phase II with the chain axis perpendicular to the film, it would be pyroelectric and piezoelectric after poling in a strong electric field.

Recently we have become interested in a derivative of poly(vinylidene fluoride) produced by the dehydrohalogenization of that polymer yielding fluorinated polyacetylene that not only may be pyroelectric and piezoelectric but would be far more stable than polyacetylene and quite possibly could be doped to a high conductivity. The synthesis and characterization of this material is discussed below.

The properties of electrically conductive polyactylene have been studied extensively [11]. However, it is a material of limited practical significance due to its instability and intractability. Attempts at preparing conjugated polyenes which exhibit both processibility and conductivity include the synthesis of graft copolymers [12,13], blends [14], substituted poly(diacetylene) [15], and modification of nonconducting polymers [16,17]. Reports of phase transfer-catalyzed dehydrofluorination of solid poly(vinylidene fluoride) (PVF$_2$) have recently appeared [17,18]. Kise and Ogata have reported the electric conductivity of an iodine-doped film sample to be 1.3 x 10^{-5} s/cm. The dehydrofluorination of PVF$_2$ in solution has been reported by Dias and McCarthy [19]. The heterogeneous phase transfer catalyzed dehydrofluorination of PVF$_2$ surfaces using aqueous NaOH and tetrabutylammonium bromide (TBAH) as the phase transfer catalyst has been studied by Kise [17]. Bonafini et. al. [20], have performed surface dehydrofluorinations of PVF$_2$ films followed by bromination and developed a quantitative gravimetric technique to characterize the reaction depths.

We have performed the dehydrofluorination of PVF$_2$ with aqueous NaOH and TBAH both in solution and in a heterogeneous system, i.e., PVF$_2$ powder and films. The poly(fluoroacetylene) (PFA) was prepared by the heterogeneous dehydrofluorination of poly(vinylidene fluoride) (Equation 1),

$$-\left(C-\underset{\underset{F}{|}}{\overset{\overset{F}{|}}{C}}\right)_n \quad \xrightarrow[\text{PTC}]{\text{OH}^-} \quad -\left(C=\underset{}{\overset{\overset{F}{|}}{C}}\right)_n \qquad (1)$$

and subsequently doped by iodine. In another study, it was also doped by suspension of NOPF$_6$. The doping of polyacetylene with nitrosyl salts dissolved in a mixture of nitromethane and methylene chloride was demonstrated by Gau and coworkers [21]. The nitrosyl salts are attractive dopants because they are strong oxidizing agents and their NO+ cation is reduced during doping to NO gas which readily escapes from the reaction.

In this paper, we report the successful synthesis and characterization of poly(fluoroacetylene) "PFA", and the application of both vapor phase (with iodine) and in suspension (with NOPF$_6$) doping of polyfluroacetylene to render an electrically conductive polymer. Important physical properties of this conductive polyfluoroacetylene are also reported.

II. Experimental

A. Materials. Polyvinylidene fluoride "PVF_2" powder was obtained from Pennwalt Corp. PVF_2 films (3 mil, 0.0762 mm) were provided by Westlake Plastics Co. PVF_2 films were extracted with CH_2Cl_2 for one hour to remove the plasticizer and then dried under vacuum. The films were checked by ultraviolet (UV) absorption spectroscopy to insure complete removal of the plasticizer. PVF_2 powder was purified by precipitation from DMF solutions with methanol.

B. Dehydrofluorination of PVF_2. The dehydrofluorination of PVF_2 was performed both in DMF and dimethyl sulfoxide (DMSO) solutions using as a base either potassium tert-butoxide or 50% aqueous NaOH (which is not miscible in DMF or DMSO) and tetrabutylammonium hydrogen sulfate (TBAH). Dehydrofluorinations were also performed on the PVF_2 surface (powders and films). The solution dehydrofluorinations of PVF_2 with potassium tert-butoxide in DMF were performed according to Dias and McCarthy [19]. The surface dehydrofluorinations of PVF_2 powder were carried out by starting with 25 ml of aqueous NaOH, in concentrations of 3% to 27%, and adding 0.4 grams of TBAH. The solution was stirred for 10 minutes before addition of 1 g of PVF_2 powder. The reaction mixture was stirred for 24 hrs at room temperature. The polymer was filtered and washed with distilled water and ethanol. The dehydrofluorinated powder was then dried under vacuum. The PVF_2 films and single crystals were dehydrofluorinated in a similar manner. The reaction rates for the dehydrofluorinations of the PVF_2 powders and films were much slower without stirring of the reaction mixture. The dehydrofluorination of the single crystals was very slow without stirring, and yielded little dehydrofluorination. After three days less than 5% dehydrofluorination took place at room temperature.

C. Techniques. IR spectra were obtained either on a Perkin-Elmer model 1320 IR or a Digilab FTS-14 FTIR spectrometer, using either KBr pellets with PVF_2 powder or using the polymer films directly. Ultraviolet spectra were taken using a Perkin-Elmer Lambda 9 UV/VIS/NIR spectrometer. The ESCA spectra were recorded on a Varian IEE spectrometer with a Mg K source. The Cls ESCA curves were resolved using a Dupont 310 curve analyzer, using a clean peak due to $-CF_2-$, from purifed PVF_2, to calibrate the individual channels. X-ray diffraction patterns were recorded on a Philips Automated Powder Diffractometer with monochromatized Cu K_α radiation and scanned at 1.2 degrees per minute with a 0.02

degree step. The diffraction patterns were transferred to
system memory for further manipulation and spectra plotting.
Electron micrographs were obtained with a JEOL 35 scanning
electron microscope.

D. Doping. 1.0 g of NOPF$_6$ (Alfa Chem.) was dissolved in
50 ml of nitromethane. Exactly 0.50 g of polyfluoroacetylene
was then added and the suspension was stirred for 2 hrs. The
dopant solution was removed by centrifugation and filtration
using a fine fritted glass filter. The doped polymer was
washed four times with pure nitromethane to remove residual
dopant, and dried in a vacuum oven at room temperature for 2
hrs. The amount of dopant was determined from the weight
increase in the polymers.

Exactly 0.50 g of finely powdered polyfluoroacetylene (or
films) was put into a coarsely fritted glass filter, which
was set beside another coarsely fritted glass filter
containing iodine. Both were placed inside in a vacuum
dessicator equipped with drierite, and vacuum (10^{-3} torr)
was applied. The polymer was weighed every day. After 6
days, the weight became constant. The iodine concentration
in doped poly(fluoroacetylene) was determined only from the
weight increase.

E. Electronic conductivity measurement. A four-point
square probe was connected to a constant current supply
(Keithley 220 Programmable Current Source) and a voltmeter
(Keithley 192 Programmable DMM). The current supply was
monitored by a Digital MINC-11 microcomputer. Conductivities
were calculated using the following equation [22] for the
four point square probe:

$$\sigma = (\ln 2) I/2\pi aV) \qquad (2)$$

This equation is accurate provided that the sample thickness
"a" is much smaller than the square dimension. Electrodag
phase was applied to the probes for ohmic contact with the
sample.

The pressure dependence of conductivity was studied using
a sandwich type apparatus. Finely ground polymer powder was
put into an insulator washer with a 1.3 cm diameter hole.
The powder was pressed on opposite sides with two blocks of
stainless steel which were connected to a constant current
supply and a voltmeter.

F. Temperature dependence. Temperature-dependent conductivity studies were carried out in a large four-probe apparatus immersed in solvent slush or hot oil baths. The apparatus was filled with ca. 1 atm. of dry Nitrogen for efficient
heat transfer, and a thermocouple at the sample allowed
accurate determination of sample temperature.

G. EPR spectra. EPR spectra were recorded using a Varian
El12 EPR spectrometer provided by the National Institute of
General Medical Science (grant 27519). Samples for each
measurement were prepared by weighing about 7.0 mg of the
polymer into capillary tubes. Sixty micrograms of diphenyl
picrylhydrazyl (DPPH) was used as a reference to determine
the g factor as well as the spin density. Six mg of DPPH was
disolved in 10 ml of benzene, and 0.1 ml of the solution was
transferred to an EPR tube. The solution was then evaporated
slowly at room temperature. The temperature dependence of
unpaired spin density was measured from $120°K$ to $400°K$. The
temperatures were measured using a copper-constant thermocouple attached to the sample, with a $0°C$ reference. A
Keithley 117 Microvolt DMM was used to read the thermopoten-
tial. The relaxation times were determined using the standard EPR saturation technique. The magnetic field, H_1, was
obtained by the method of perturbing spheres [23]. A stainless steel sphere was put into the center of the EPR cavity
and the resonance frequencies before and after perturbation
were measured. H_1 in gauss was then calculated from
Equation 3.

$$H_1^2 = Q_0[(\nu^2-\nu_0^2)/2\pi a\nu_0^3]W \qquad (3)$$

where a=1.17 mm, the radius of the sphere, ν_0 and ν are the
initial and perturbed resonance frequencies, 9.042 GHz and
9.056 GHz, respectively, Q_0 is the unloaded cavity (Q_0
=6000), and W is the klystron power in watts. Substituting
these values to Equation 3, we have

$$H_1^2 = 0.204W \qquad (4)$$

III. Results and discussions

A.   Dehydrofluorination of PVF$_2$. We have performed the dehydrofluorination of PVF$_2$ both in solution and in hetero-phase. Solution dehydrofluorinations of PVF$_2$ are very fast. The solution turns black after a few seconds, and the dehy-drofluorinated PVF$_2$ appears to remain in solution, as has been reported by Dias and McCarthy [19].   This is an inter-esting result since it gives a soluble block copolymer of PVF$_2$ with poly(fluoroacetylene).   Since the dried polymer does not redissolve, as correctly stated by Dias and McCarthy [19], in order to characterize the obtained dehydrofluori-nated polymer Percec et al. performed a series of in situ NMR studies [24].   They proposed the possible formation of micellar aggregates of the poly(fluoroacetylene) segments (i.e. the PVF$_2$ segments stay in solution while the poly(fluo-roacetylene) segments are insoluble).   Since the polyenic segment comes out of solution, it is expected that both the length and   the distribution of the effective conjugated polyenic segments along the chain will be controlled by this process.   Therefore, we expect that PVF$_2$ with a high degree of dehydrohalogenation obtained in solution will   present a shorter conjugation length than a surface dehydrohalogenated polymer.   This is because in the first case, the length of the polyenic sequence in a zipper dehydrohalogenation mecha-nism is controlled by the solubility of this segment in the reaction mixture. In the second case, the conjugated segment length is controlled through 1) the accessibility of the base to the surface, 2) the PVF$_2$ microstructural defects (i.e., head-to-head structural units) 3) the morphology of PVF$_2$. UV analysis of polymer films cast from the dehydrohalogenation solution presented a λ max at 340 nm (Figure 1A) indicating an average sequence length of five conjugated double bonds [25].   This represents a lower wavelength than the λ max obtained from heterogeneous dehydrofluorinations, indicating that dehydrofluorinated sequences containing more than five double bonds precipitate out of solution.

Heterogeneous phase transfer catalyzed dehydrofluorination of PVF$_2$ has proven to be the much more controllable reaction. These reactions are easily controlled by varying the concen-tration of the aqueous NaOH (Figure 2).   The brown to black powders obtained were totally insoluble.   The extent of the dehydrofluorination was determined from the weight change of the polymer, assuming the loss of an HF molecule per vinyli-dene fluoride unit reacted [17].

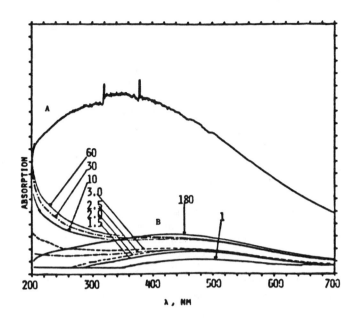

Figure 1.   A) UV/VIS adsorption spectrum of PVF$_2$ dehydrofluorinated in solution, using TBAH, NaOH/DMF, obtained from
film cast directly from reaction solution; B) UV/VIS absorption spectra of dehydrofluorinated PVF$_2$ films.   Reaction
time in minutes is presented on the corresponding absorption
spectrum.

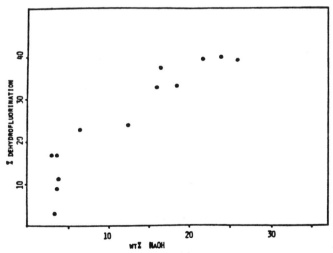

Figure 2.   Plot of % dehydrofluorination vs. wt. % aqueous
NaOH.   All reactions using TBAH at room temperature for 24
hours.

B.   Infrared spectroscopy.   Dehydrofluorinations of the
PVF_2 powder surfaces can reach up to 40% conversion after 24
hours reaction time at room temperature.   IR analysis of the
dehydrohalogenated polymer showed no indication of carbon-
carbon triple bond formation.   Higher degrees of dehydrofluorinations can be achieved with longer reaction times   or
higher temperatures as reported by Kise and Ogata [17].   IR
analysis   of   the   polymers obtained from the   reactions
performed at higher temperatures show a broad but weak
absorption at 2120 cm^{-1}.   This absorption dissappears
after the polymer is placed in an aqueous NaOH solution (4N),
stirred with heat, filtered and washed (Figure 3).   When the
dehydrofluorinations are performed at higher temperatures (90°
C), the TBAH will undergo Hofmann elimination to form tributylamine.   The tributylamine can quaternize with any allylic
fluorines on the partially dehydrofluorinated PVF_2 .   This
quaternization can   occur even in the presence of   a strong
base such as NaOH, if all of the TBAH is degraded.   Without
the presence of a phase transfer catalyst, the hydrophilic OH$^-$
does not compete with the hydrophobic amine in reacting with
PVF_2 , therefore, the polymer behaves as if only in contact
with the amine.   This   quaternary amine can   then   undergo
Hofmann elimination to give a dibutylamine on the   polymer
backbone (Scheme 1).   This amine can then give results similar to those formed in the solution dehydrofluorinations
using DMF, as discussed earlier.

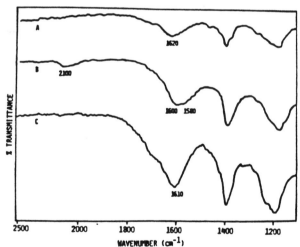

Figure 3.   Infrared spectra of A) dehydrofluorinated PVF_2
(room temperature, 25% aq. NaOH, TBAH); B) dehydrofluorinated
PVF_2 at 90 °C (same conditions as sample A); C) B after
treatment with aq. NaOH at 90 °C.

Scheme 1. Hofmann degradation of TBAH followed by quaternization of the resulting amine with an alluvic fluorine and by Hofmann elimination.

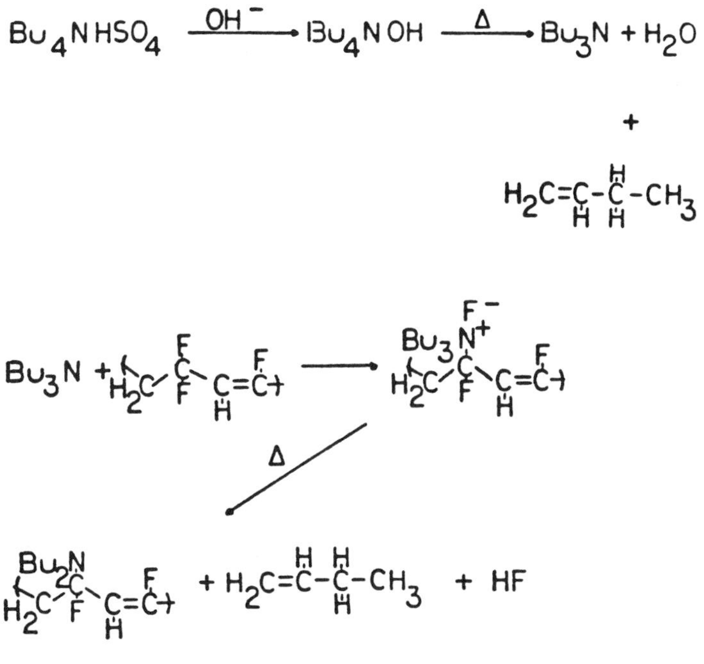

$$Bu_4NHSO_4 \xrightarrow{\ OH^-\ } Bu_4NOH \xrightarrow{\ \Delta\ } Bu_3N + H_2O$$

$$+$$

$$H_2C=\underset{H}{\overset{}{C}}-\underset{H}{\overset{H}{C}}-CH_3$$

C. UV. UV analysis of PVF$_2$ thin films dehydrofluorinated for a period of 5 seconds to 3 hours show only a slight change in the position of the λ max. Only an increase in the intensity of the absorbance is noticeable as shown in Figure 1B. This tells us that although the concentration of the dehydrofluorinated segments is increasing, the length of conjugated carbon-carbon double bond sequences remains about the same. UV analysis of both films and powders (using KBr pellets) which were dehydrofluorinated gives λ max values in the same range, i.e., 470 nm, indicating a sequence length of 12 to 13 conjugated double bonds [25]. Both in this case as well as in previous cases the calculated length of the conjugated polyenic sequence is based on $-(-CH=CH-)_{\overline{n}}$ units. It is expected that for the same λ max. the number of $-(-CH=CF-)_{\overline{n}}$ conjugated units is shorter. 200 MHz H-NMR analysis of the unreacted PVF$_2$ gives a ratio of one tail-to-tail defect for every fourteen head-to-tail units. These two results indicate that the dehydrofluorination probably takes place by a "zipper " type mechanism which might stop at head-to-head defects (Scheme 2).   Since the average distance between head-to-head defects would be fourteen vinylidene fluoride units, this would give an average sequence of fourteen conjugated double bonds after dehydrofluorination.

Scheme 2.   Possible "zipper" type mechanism for the dehydrofluorination of PVF$_2$, stopping at a head-to-head structural unit.

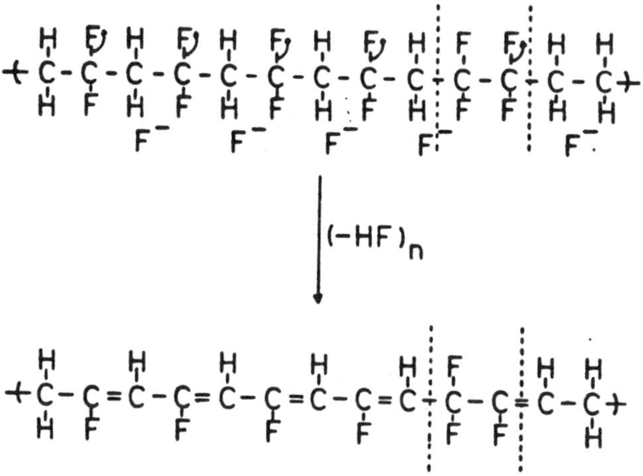

$(-HF)_n$

D.   Film reaction.   When PVF   film was treated with NaOH solution   in   the   presence   of   TBAH   at room   temperature,   the reaction   was   found   to be very slow as shown   in   Figure   4. This   is   probably due to the high crystalline film   surface, which   is hard to penetrate under these conditions   (extruded films   are   known to have a large amount of hedrides   on   the film surface and spherulites in the film).

Figure   5 shows a plot of absorbance (at 410 nm) vs   temperature.   The optimal reaction was found at higher   temperature (70°C),   and the film gradually went from white to brown to black during 15 hrs of reaction.   A schematic representation of the surface-sensitive dehydrofluorination reaction is shown in Equation 5.

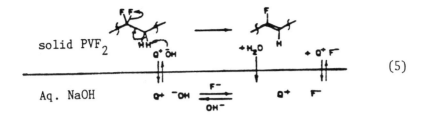

solid PVF$_2$

Aq. NaOH

(5)

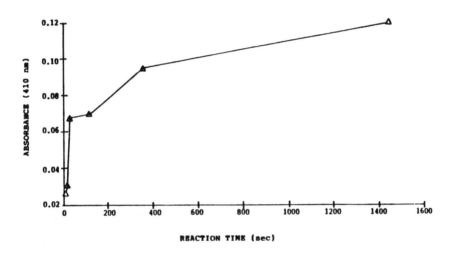

Figure 4.   UV absorption of films vs. reaction time.

PVF$_2$ is known to have low wettability and poor adhesive properties. The critical surface tension of PVDF is reported to be 25 dyne/cm [26].   Applying phase-transfer catalysis in the surface modification of PVF$_2$, the outer few angstroms of the solid organic polymer is regarded as the organic phase, and the hydroxide ion is transported from an aqueous phase in contact with the polymer to this "organic phase" where dehydrofluorination is effected.   In the absence of phase transfer   catalysts,   hydroxide   ion   does   not   induce dehydrofluorination because the basic solution does not "wet" the film.   TBAH ion transports hydroxide ion across this interface PFA and serves as a "wetting agent", facilitating reaction   to   form   PFA.   The eliminated surface is hard to penetrate under these conditions;   hence elimination is most confined to the surface.   However, vigorously stirring the reaction solution, the reaction can gradually get into the film as shown in Figure 5 by the increase of UV absorbance (at 410 nm).

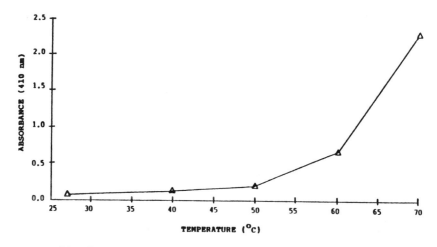

ABSORBANCE (410 nm)

TEMPERATURE (°C)

* Reaction time = 2 hours

Figure 5.   Temperature vs. absorbance at 410 nm.

ESCA surface analysis has further verified the dehydrofluorination reaction on the film surface. Figure 6 shows the Cls ESCA spectrum of PVF_2 film (6A), and dehydrofluorinated PVF_2 film (6B). For PVF_2, the H-C-H peak was found at 286 ev. The F-C-F peak is located at a higher binding energy (291 ev), due to the chemical shift effect (resulting from the highly electronegative fluorine atoms bonded to carbon). After dehydrofluorination, both the orginal F-C-F and H-C-H peaks disappeared. Because of the loss of the F atom, and furthermore the sp^2 character of the =C-F bond, the peak shifts to lower bind energy. As shown in Figure 6B, it merges with =C-H (283.5 ev) peak, and can be seen as a shoulder of the broad peak at 285.5 ev.

We observed a very interesting change of the mechanical properties of PVF_2 thin films as a function of the dehydrofluorination time (Figure 7). Curve A shows the typical ductile stress-strain behavior of a PVF_2 film. Curve B and C show that both fracture strain and fracture stress dropped under reaction, but the modulus increased as the reaction time increased. Finally, after 4 days of dehydrofluorination the films became very brittle as shown in curve D. This

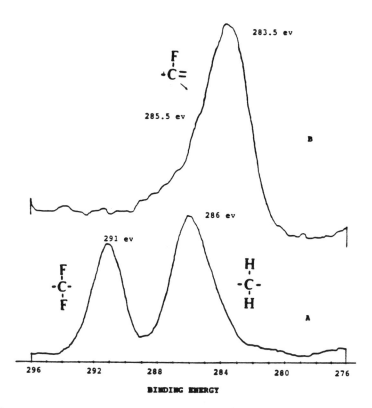

Figure 6.   Carbon 1s ESCA spectra of A) poly(vinylidene
fluoride); B) dehydrofluorinated poly(vinylidene fluoride).

transition phenomenon can be explained in two ways.   1) On
the molecular level, the dehydrofluorination has introduced
rigid double bond segments into the polymer chains.   2) From
the morphological level, the dehydrofluorination must be
occurring on the vulnerable amorphous regions. Thus, the
overall rigidity of the sample has been increased.   The
wide angle x-ray diffraction of film samples (Figure 8) shown
by the sharpening of the PVF_2 diffraction rings in the dehydrofluorinated film confirms that larger, more perfect crystallite remains.   This is further confirmed by the porous
morphology of the film surface (Figure 9).

    E.   Doping of polyfluoroacetylene. The structural resem-
blance to polyacetylene induced us to examine the electrical
conductivity of PFA. Both powder and film samples of PFA
obtained   from heterogeneous reaction have been doped with
iodine and $NOPF_6$.   Kise et. al. [27] has reported that

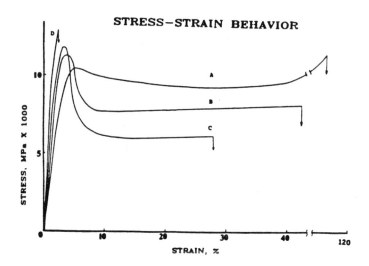

Figure 7.   Stress-strain behavior of poly(vinylidene fluoride and dehydrofluorinated poly(vinylidene fluoride) films.   A) PVF$_2$;   B) 3 hours of dehydrofluorination;   C) 12 hours   of   dehydrofluorination;   D)   4   days   of dehydrofluorination.

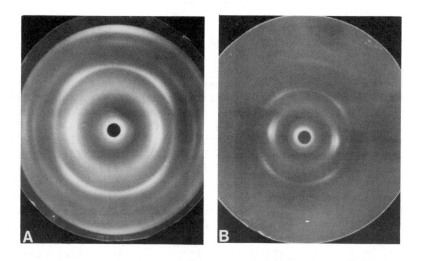

Figure   8.   X-ray diffractions of A) poly(vinylidene fluoride) film; B) dehydrofluorinated PVF$_2$ films (30 hours of reaction).

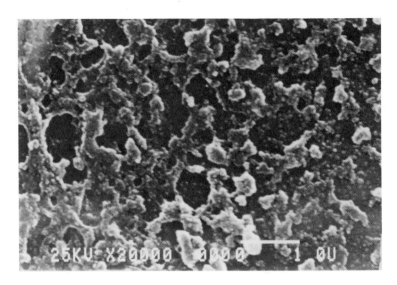

Figure 9. Scanning electron micrograph of the surface morphology on the dehydrofluorinated PVF_2 film surface.

doping of the dehydrochlorinated films of poly(vinylchloride) was often limited to near the surface of the film. On the other hand, powder samples generally provided a large surface area which can easily be accessed by dopants. Due to insolubility the conductivity of the powder was studied in the form of a compressed pellet.

Figure 10 shows the pressure dependence of conductivity for both $NOPF_6$ and iodine doped PFA. The conductivities increased rapidly as pressure increased, and finally level off at high pressure (2.9×10^4 Kg/cm^2). This supports an electronic conduction mechanism, because ionic conductivity decreases as pressure increases [28]. A great deal of effort been devoted to characterization and understanding of electrical transport in conducting polymers [11]. The factor limiting the conductivity is the carrier mobility, not the carrier concentration. The doping process produces a generous supply of potential carriers, but to contribute to conductivity they must be mobile. Experimental determination of mobility in polyacetylene confirms that interchain and interparticle transport are the limiting factors for conduction. The pressure/conductivity behavior of doped PFA can thus be related to the interparticle and interchain charge transport. Initially, the interparticle voids are eliminated by increasing pressure, and the interparticle contact is

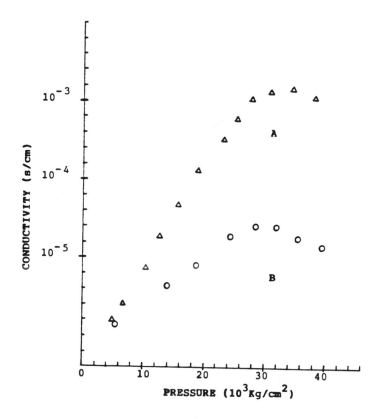

Figure 10. Pressure dependence of conductivity of A) iodine doped PFA; B) NOPF$_6$ doped PFA.

increased. Thus, the contact resistance for the charge carrier is decreased. Under high pressure, the interparticle
contact becomes even tighter, and the spacing between molecular chains is also reduced, resulting in somewhat easier
carrier transport on the interchain level. This may explain
our observation that the conductivities of doped PFA
increased dramatically by two orders of magnitude (from 5 x
10^{-6} to 1 x 10^{-3} s/cm) as the pressure increased from 2500 to
3.5 x 10^3 Kg/cm^2.

The electrical conductivity of PFA as a function of dopant
concentration is represented in Figure 11. The use of NOPF$_6$
yielded lower conductivities, but the doping time of 90
minutes is much shorter than the 6 days required by iodine.
The maximum amount of dopant incorporated into the powder PFA
was 39 mole % for NOPF$_6$ and 23 mole % for iodine doping.

Both curves of Figure 11 displayed changes in slope. However, this change did not correspond to an obvious transition such as the semiconductor–metal transition present in polyacetylene.

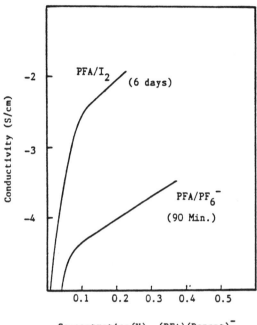

Figure 11. Electrical conductivity of doped PFA as a function of dopant concentration.

The effect of temperature on conductivity is shown in Figure 12. The normal temperature dependence of conductivity is given by the Arrhenius plot of the Equation $\sigma = \sigma \ exp(-\Delta E/KT)$, where K is the Boltzman constant and ΔE is the activation energy, which can be determined from the slope of the line. The activation energy of conduction was calculated 0.271 ev for iodine doped PFA and 0.313 ev for $NOPF_6$ doped PFA. The conductivity for doped PFA is fairly high, although they are still non–metallic. The activation energy is in the semiconductor's range.

The UV/VIS spectra of doped PFA are given in Figure 13. The absorptions are near 209, 257 nm for iodine doped PFA, and 201 , 259 nm for $NOPF_6$ doped PFA. These absorptions are due to the K band absorptions of conjugated double bonds

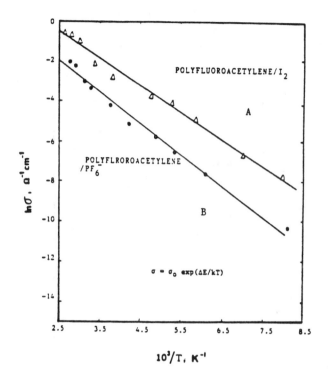

Figure 12.   Temperature dependence of conductivity: A) from iodine doped PFA; B) from NOPF$_6$ doped PFA.

[29].   The structureless spectrum of the doped PFA suggests the presence of free electrons.   The optical band gap, or absorption   threshold of doped PFA (determined from plotting the   absorbance vs wavenumber and extrapolating to zero) was found at 1.4 ev and 1.9 ev respectively for iodine and   NOPF$_6$ doping.   The   fact   that   the NOPF$_6$ doped PFA has a larger band gap than the iodine doped PFA agrees with the results of the   activation energy study determined by the temperature dependence study, where NOPF$_6$ doped PFA also presents a higher activation energy.

Thermal   behavior of PVF$_2$, PFA, and doped PFA by DSC analysis are shown in Figure 14.   Curve A is for PVF$_2$ which has a Tg= -38 ℃ and Tm = 178 ℃, and at room temperature is a flexible   material.   The PFA has a Tg at 193 ° C.   This is related   to the change of the polymer's backbone into   rigid double bond segments.   When heating PFA to higher tempera- ture,   the polymer started to degrade at 275°C.   Rubner et. al.   [30] reported upon doping by nitrosyl salts PPS will be

UV/VIS

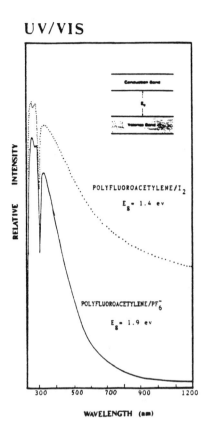

Figure 13.   UV/VIS spectra of A) iodine doped PFA; B) NOPF$_6$ doped PFA.

crosslinked.   We found that NOPF$_6$ doped PFA started to degrade at fairly low temperature (100°C). However, even for the heavily iodine doped PFA, we still can see a Tg at 193°C (Figure 14B).

An IR spectrum of heavily iodine doped PFA is displayed in Figure 15, which shows no iodine addition to the double bonds.   The broadening of the spectrum suggests the presence of free electrons. The C=C–F vibration modes shifted to higher frequency, from 1620 cm^{-1} for PFA to 1650 cm^{-1} for heavily iodine doped PFA. This is attributed to the increase of the force constant for C=C bonds by complexation with the acceptors [31].   The generally broader absorption bands of the spectrum are probably due to plasma electron absorption,

D S C

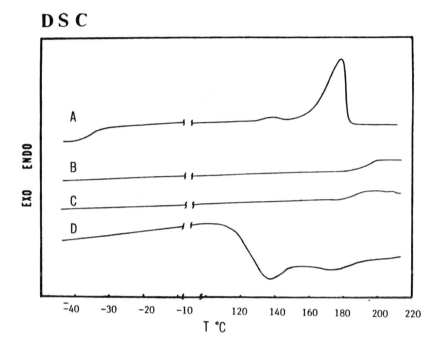

Figure 14.   Thermal   properties   of   A)   poly(vinylidene
fluoride); B) dehydrofluorinated PVF_2 (PFA); C) iodine doped
PFA; D) $NOPF_6$ doped PFA.   (Heating rate: 30° C/min.)

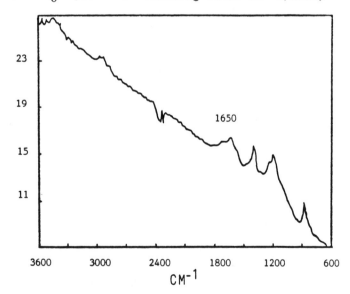

Figure 15.   IR spectrum of iodine doped PFA.

which indicates that significant amounts of free carriers were induced by the heavy doping.

Electron paramagnetic resonance was used to study unpaired spin density and the effect of doping. Lorentzian line shape is found for undoped PFA (Figure 16). The line width increases from 6.1 G to 11.5 G for $NOPF_6$ doped PFA, and 6.1 G to 10.5 G for iodine doped PFA. The line shapes are no longer symmetric after doping, which may imply metallic conductivity as well as paramagnetic impurity contributions.

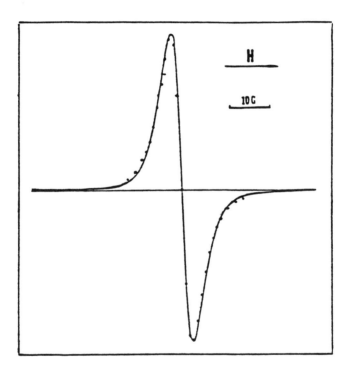

Figure 16. EPR spectrum of polyfluoroacetylene at room temperature, the points represent a Lorentzian derivative curve having the same peak to peak amplitude and line width as the spectrum.

Figure 17 shows the temperature dependence of the line width. For PFA and doped PFA, all line widths decrease as temperature increases, which could be due to either motional or exchange narrowing or both [32].

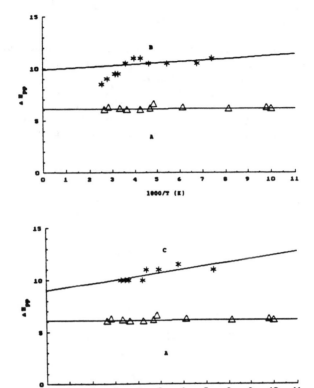

Figure 17.   Temperature dependence of EPR line width: A)
from polyfluoroacetylene;   B) from iodine doped PFA;   C) from
NOPF$_6$ doped PFA.

The temperature dependence of the spin density (Figure 18)
shows that the spin densities of all samples vary linearly as
(a-b)/T.   This can be expected from Equation 6,   if and only
if the

$$n = n_1 + n_2 \exp(-\Delta E_i / KT) \qquad (6)$$

activation energy, ΔE, is much smaller than KT. It can then
be expressed as Equation 7,   where n_1 is trapped spin risen
from a

$$n = (n_1 + n_2) - n_2 \Delta E_i / KT \qquad (7)$$

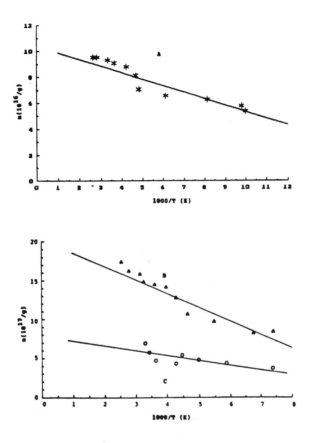

Figure 18. Temperature dependence of EPR spin density: A) from polyfluoroacetylene; B) from iodine doped PFA; C) from NOPF$_6$ doped PFA.

neutral defect in the π system; n_2 thermally excited spin, which is formed by thermo-excitation. Generally the amount of n_1 is much smaller than n_2 .   Therefore, the estimated E_i can be obtained from Equation 7, The values are 0.0042 ev, 0.0073 ev, and 0.0061 ev for PFA, iodine doped PFA and NOPF$_6$ doped PFA respectively.

The values for the g factor, Δ Hpp (line width), n(25°C), n_2 and ΔE are listed in Table 1.   The increase of ΔHpp is probably due to the inhomogeneous broadening induced by doping.   The g value, 2.0028, for PFA is normal for unpaired spin in the conjugated π system of a polymer.   The g factor is shifted by 3.3 x 10^{-3} and 2.5 x 10^{-3} , respectively, for

Table 1. The g factor, line width, spin concentration, and activation energy of doped and undoped PFA

Polymer	g	ΔH_{pp}	$n_{25} \times 10^{17}$ /gm	$n_2 \times 10^{17}$ /gm	ΔE_i
PFA	2.0028	6.1	0.93	1.04	0.0042
PFA/Iodine	2.0061	10.5	14.9	20.25	0.0073
PFA/PF$_6$	2.0053	11.5	6.9	7.82	0.0061

Table 1. The g factor, line width, spin concentration, and activation energy of doped and undoped PFA.

iodine and NOPF$_6$ doped PFA. There is one unpaired spin per every 1.04 x 10^5 units for PFA. and one unpaired spin per every 6.5 x 10^3 and 1.4 x 10^4 units for iodine doped PFA and PF$_6$ doped PFA respectively. Also, we can see the total number of spins has been increased by the doping.

The EPR saturation technique was employed for determining the relaxation times. The spin-lattice relaxation time is determined by the maximum signal amplitude (H_{1m}). The satu-ration value can be used to calculate the spin-lattice relaxation time (T_1), and spin-spin relaxation time (T_2) by the following equations (Equation 8, Equation 9) [33].

$$T_1 = 0.49 \times 10^{-7} \Delta H_{pp}^{\circ} / (g(H_{1m})^2) \quad (s) \qquad (8)$$

$$T_2 = 1.313 \times 10^{-7} / g \Delta H_{pp}^{\circ} \quad (s) \qquad (9)$$

ΔH_{pp} is the smallest line with below saturation. The calculated T_1 and T_2 are listed in Table 2. The decrease of both T_1 and T_2 after doping indicates that there is a strong interaction between the dopant nuclei and the polymer lattices.

Both EPR and IR data suggest that the local electronic environment in doped PFA is that of a metal or semimetal, though the material is a semiconductor. We feel that this may be due to the fact that the low surface energy prevents

Table 2. Relaxation times T and T
                            1   2

Polymer	T_1 (10^{-8} s)	T_2 (10^{-9} s)
PFA	5.97	10.75
PFA/Iodine	2.98	6.23
PFA/PF$_6$	1.60	5.69

Table 2. Relaxation times T_1 and T_2 .

the powder particles to contact tightly. Thus while carrier
mobility may be high within a molecule, the probability of
hopping to another molecule is relatively low. The inter-
granular contact resistence may also be important. Both
effects can be seen in the pressure dependence of
conductivity as discussed previously. The strong electron
withdrawing effect of the fluorine could create carriers in
the main chain, and account for the low activation energy of
the EPR signal.

Figure 19 shows the x-ray diffraction of PFA and iodine
doped PFA (NOPF$_6$ doped PFA is amorphous). Given the number
of observed x-ray reflections, it is not feasible to determine unequivocally the unit cell constants. However, it is
possible to index the observed reflections in an orthorhombic
unit cell with a non-linear least square refinement program.
The a, b, and c(chain axis) axes values and a comparison of
the calculated and observed values of the d spacings, together with the appropriated Miller indices based on this
assumed unit cell is given in Table 3. Both PFA and iodined
PFA have very high crystallinity. The dopant seems to diffuse into the PFA preferentially in the b planes. This can
be seen in Figure 20, which shows the elimination of the
(010) diffractions.

When a dehydrofluorinated PVF$_2$ film was exposed to iodine
vapor a marked increase in conductivity was observed. A
typical time-conductivity curve is shown in Figure 21. Conductivity levels off within 10 hrs. The conductivity depends
on the PFA film samples. For samples from reaction at 70°C,

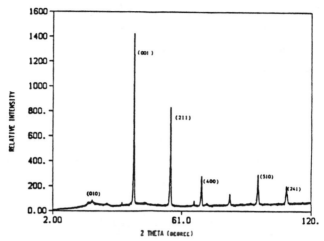

Figure 19.   X-ray diffraction of polyfluoroacetylene and
iodine doped polyfluoroacetylene.

Table 3. X-ray diffractometry data for polyfluoroacetylene

A = 5.363Å   B = 4.461Å   C = 2.322Å   SiGMA = 0.0033

#	D(EXPERI-MENTAL)	INTENSITY	D(CALCU-LATED)	H	K	L	%DELTA
1	4.46	105	4.46	0	1	0	-0.013
2	2.68	65	2.68	2	0	0	-0.052
3	2.32	1482	2.32	0	0	1	-0.079
4	1.64	848	1.63	2	1	1	0.408
5	1.40	84	1.39	3	2	0	0.369
6	1.34	285	1.34	4	0	0	-0.052
7	1.30	52	1.30	2	3	0	-0.025
8	1.16	145	1.16	4	0	1	-0.089
9	1.04	291	1.04	5	1	0	-0.272
10	0.94	217	0.94	2	4	1	-0.133

Table 3. X-ray diffractometry data for polyfluoroacetylene.

for 4 days, the conductivity can be as high as 6.1×10^{-3} s/cm.

IV. Conclusions

We have demonstrated that PFA can be produced by the surface modification of PVF_2, with the heterogeneous phase transfer catalyzed dehydrofluorination having been proven to be much more controllable than solution phase dehydrofluorination.

Surface morphologies and chemical defects of PVF_2 controlled both the reaction rate and sequence length of the conjugated double bonds. Powder reactions reached 40% of conversion at room temperature, but film reactions required higher temperatures. The phase transfer catalysis degraded at $90^\circ C$, therefore the optimal reaction condition was found to be $70^\circ C$.

Iodine doped PFA samples show an electronic conductivity at 5×10^{-3} s/cm. ESR studies suggest strong coupling of the spin system between the dopant nuclei and polymer lattices.

The phase transfer catalyzed surface modification has been shown to be a useful technique in preparing conducting polymer film surfaces. The results indicate a strong potential for applying this technique to a wide variety of polymers and applications (ie. antistatic coatings).

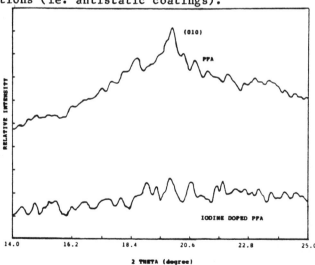

Figure 20. X-ray diffraction of polyfluoroacetylene and iodine doped polyfluoroacetylene (the only difference is the missing of (010) diffraction.

Gwo-Shin Swei et al.

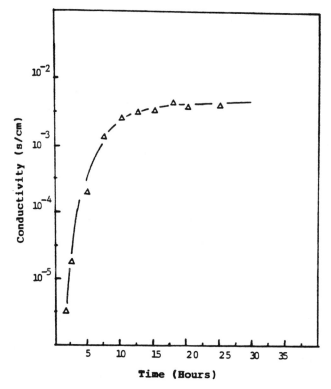

Figure 21.   Conductivity of iodine doped PFA films.

REFERENCES

[1] Lando, J.B.; Olf, H.; Peterlin, A.   J. Polym. Sci., 1966,   Part A-1, 4, 941.
[2] Hasegawa, R; Takahashi, Y.; Chetani, Y.; Tadokoro, H. Polym. J., 1972, 3, 600.
[3] Doll, W.W.; Lando, J.B.   J. Macromol. Sci.-Phys., 1970, B4(2), 309.
[4] Bachmann, M.A.; Lando, J.B.   Macromolecules, 1981, 14,40.
[5] Weinhold, S.; Litt, M.H.; Lando, J.B.   Macromolecules, 1980, 13,1178.
[6] Takahashi, Y.; Tadokoro, H.   Macromolecules, 1980, 13, 1317.
[7] Bachmann, M.; Gordon, W.L.; Weinhold, S.; Lando, J.B.   J. Appl. Phys., 1980, 51, 5095.
[8] Lovinger, A.   Macromolecules, 1987, 15, 40.
[9] Weinhold, S.; Litt, M.H.; Lando, J.B.   Ferroelectrics, 1984, 57, 277.

[10] Weinhold, S.; Litt, M.H.; Lando, J.B. J. Appl. Phys., 1980, 51, 5415.

[11] Fromer, J.E.; Chance, R.R.; "Conducting Polymer", Encyclopedia of Polymer Science and Engineering, Wiley, New York,1985; Vol. 5.

[12] Galvin, M.E.; Wnek, G.E. Polym. Comm., 1982, 23, 795.

[13] Bates, F.S.; Wnek, G.E. Polym. Comm., 1983, 23, 795.

[14] Rubner, M.F.; Tripathy, S.K.; Georger, J.; Cholewa, P. Macromolecules, 1983, 16, 870.

[15] Wenz, G.; Wegner, G. Makromol. Chem. R.C., 1982, 3, 231.

[16] Kise, H. J. Polym. Sci. 1982, Polym. Chem. Ed., 20, 3189.

[17] He, F.; Kise, H. J. Polym. Sci. 1983, Polym. Chem. Ed., 21, 3443.

[18] Dias, A.J.; McCarthy, T.J. Org. Coat. Appl. Polym. Sci., 1983, 49, 547.

[19] Dias, A.J.; McCarthy, T.J. J. Polym. Sci., 1985, Polym. Chem. Ed., 23, 1057.

[20] Bonafina, J.A.; Kias, A.; Guzdar, Z.; McCarthy, T. J. Polym. Sci., 1985, Polym. Lett. Ed., 23, 33.

[21] Gau, S.; Milliken, J.; Prom, A.; MacKiarmid, A.G.; Heeger, A.J. J. C. S. Chem. Comm., 1979, 662.

[22] Valdes, L.B. Preceedings IRE, 1954, 420.

[23] Freed, J.H.; Leniart, D.S.; Hyde, J.S. J. Chem. Phys., 1967, 47, 2762.

[24] Hahn, B.; Percec, V. "Functional Polymers and Sequential Copolymers by Phase Transfer Catalysis", to be published.

[25] Leigenaar, S.; Jurrian, C.; Huevel, M.; Huysmans, W. Makromol. Chem., 1985, 186, 1549.

[26] Brandrup, J.; Immergut, E.H., Eds., Polymer Handbook, Wiley, New York, 1975.

[27] Kise, K.; Sugihara, M.; He, F. J. Appl. Polym. Sci., 1985, 30, 1133.

[28] Blythe, A.R. "Electrical Properties of Polymers", Cambridge, Eng., New York, 1979.

[29] Silverstein, R.M.; Bassler, G.C. "Spectrometric Identification of Organic Compounds, 2nd Ed., 152.

[30] Rubner, M.; Cukor, P.; Jopson, H.; Deits, W. J. of Electronic Materials, 1982, 11.

[31] Kim, O.K. J. Polym. Sci., 1985, Polym. Lett. Ed., 23, 137.

[32] Ruan, J.; Litt, M.H. "Electronic Properties of Poly[8-Methyl, 2,3-6,7-Quinolino](PMQ)", to be published.

[33] Poole, C.P.; Farach, H.A. "Relaxation in Magnetic Resonance", Academic, New York, 1971.

FREQUENCY ANALYSIS OF FERROELECTRICITY:
A TOOL FOR UNDERSTANDING COOPERATIVE PHENOMENA
IN POLYMERS

J. Fuhrmann, U. Jahn and H. Sturm

Institut für Physikalische Chemie, TU Clausthal, FRG

Fachbereich Chemie, Universität Kaiserslautern, FRG

Abstract The frequency region of nonlinear dielectric behaviour can show an extreme limit of nonlinearity which, in the case of ferrroelectrics, characterizes an upper frequency limit at which vibrations of neighbouring charge carriers or orientation of neighbouring dipoles can be brought into ordered phase relations. The tendency to bring charge carriers or dipoles into ordered phase relations prevails up to this frequency limit against other effects. The FFT–Analysis of the generating and the response function in the ferroelectric state shows that a linear phase matching of the fundamental to higher harmonics, i. e. 3rd, 5th, ..., of the response function is necessary criterion for a ferroelectric hysteresis loop. The vanishing linear phase relation implies a changing of the nonlinear behaviour. Coercive field strength and spontaneous polarization are correlated to phase shift and amplitude of the fundamental.

Introduction

In the last years a lot of papers describe the piezo- and ferroelectric behaviour of PVDF [1-5]. The origin of these properties is the macroscopic polarization which can be formed by poling and streching of PVDF. The stretching and the high inhomogeneous local fields causes the changing to β-crystalline structure and the forming of the electret foil. The magnitude of spontaneous polarization is influenced by volume polarization and surface effects, i. e. injection of electrons and the generation of space charge owing to the barrier situation [6,7]. To study the influence and the coupling of surface and volume properties, we made dynamic dielectric experiments. Fourier analysis of the frequency dependent current and PVDF hysteresis loop, known in literature [8,9], allows to discuss cooperative dynamic effects in PVDF.

Experimental

The dielectric experiments with sinusoidal ac fields are performed in a sandwich type electrode configuration given in Fig. 1. The polymers (characterized by IR and DSC) used in this work are PVDF samples provided by Kureha Chemical & Co., biaxially oriented 12 and 25 µm thick, α- and β-modification, about 60 % crystallinity. The average thickness of the amorphous regions is about 3.5 nm and that of the crystallites 7.0 nm [10]. To establish a good contact between the polymer sample and the electrode gold was evaporated on both sides of the polymer. The thickness of the gold layer was some nanometers.

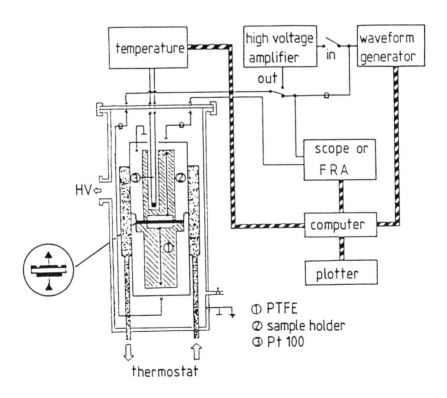

Figure 1. Experimental setup for dielectric measurements.

Results and Discussion

Ferroelectricity can be understood as a special case of dielectric nonlinearity. For reason of symmetry the series expansion of the polarization function with respect to field strength contains only odd powers of E:

$$\underline{P}\,(\underline{E}) = \chi\,\underline{E}\,+\,\zeta\,E^2\,\underline{E}\,+\,\xi\,E^4\,\underline{E}\,+\,\dots \tag{1}$$

If we turn to the dynamic case according to a generating function $E = E_0 \sin(\omega t)$ we get after trigonometric transformation the following response function

$$P(t) = A_\omega \sin(\omega t + \delta)\,+\,A_{3\omega} \sin(3\omega t + 3\delta + \rho_3) \tag{2}$$
$$+\,A_{5\omega} \sin(5\omega t + 5\delta + \rho_5)$$

In the case of PVDF the pure ferroelectric hysteresis loop shows a linear phase correlation, i. e. the phase mismatch ρ_3 and ρ_5 in equation (2) vanishes. In order to describe the damping of the frequency dependant polarization in that case we choose as an appropriate characteristic decay function:

$$e^{-b\,|\sin(\omega t + \delta)|} \tag{3}$$

leading to the following polarization response function:

$$P(t) = A \sin(\omega t + \delta)\,\,e^{-b\,|\sin(\omega t + \delta)|} \tag{4}$$

with the reference function :

$$A \sin(\omega t + \delta) \tag{5}$$

The expansion of this polarization function in terms of higher harmonics shows phase matching according to equation (6):

$$P(t) = A_\omega \sin(\omega t + \delta)\,+\,A_{3\omega} \sin(3\omega t + 3\delta) \tag{6}$$
$$+\,A_{5\omega} \sin(5\omega t + 5\delta)$$

In Fig. 2 a graph of the expansion of equation (4) is given in terms of higher harmonic amplitudes, phase matching and ferroelectric hysteresis loop as well as the generating and the polarization response function.

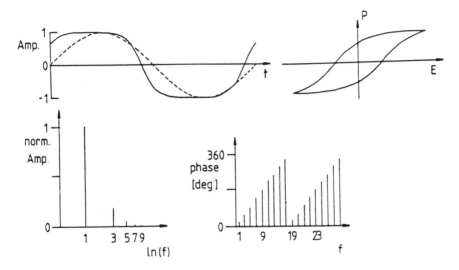

Figure 2. Simulation of generating and polarization response function, hysteresis loop and amplitude and phase correlation.

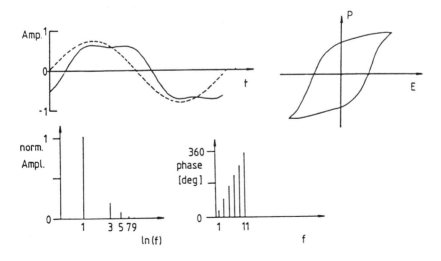

Figure 3. Fourier-analysis of PVDF ferroelectric hysteresis loop at 140 °C, 55 MVm and 0.5 Hz generating frequency [8].

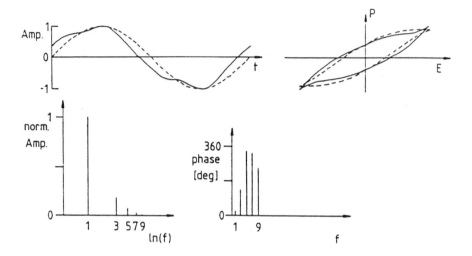

<u>Figure 4</u>. Ferroelectric behaviour of PVDF (Kureha, 12 µm) at 65 °C,
33.3 MVm^{-1} and 0.15 Hz generating frequency. Dotted
hysteresis loop represents a simulation without phase mismatch.

The fourier analysis of a PVDF ferroelectric hysteresis loop, measured by J. C. Hicks, T. E. Jones and M. L. Burgener, shows the linear phase relationship between the fundamental and the higher harmonics (Fig. 3).

Evaluation of the experimental data with the help of the fourier analysis shows that all PVDF-experiments carried out with field strenths below coercive field yield phase mismatching and the frequency dependent polarization has to be fitted by equation (2).

Figure 4 shows the generating and the normalized polarization response as well as the hysteresis loop measured at 12 µm PVDF Kureha sample. The dotted hysteresis loop is calculated starting from the experimental data (full line hysteresis loop) after subtraction of the phase mismatching ρ_3 and ρ_5, i. e. we assume in this case that the phase shift of the fundamental is ferroelectric active according to equation (4). The comparison with the calculated hysteresis loop illustrates the influence of the higher harmonic amplitudes and of the phase mismatching. The procedure of calculating the mismatch renders it possible to plot the mismatch as a function of field strength, temperature and frequency in order to find the corresponding region in which pure ferroelectric behaviour is evident.

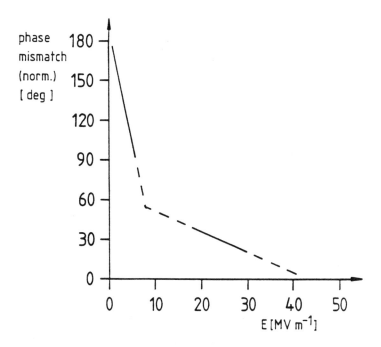

Figure 5. Phase mismatch ρ_3 (normalized to 0 - 180 deg) versus field strength of ac field of PVDF (Kureha, 12 μm) at 65 °C and 10 Hz generating frequency.

———— measured ------- calculated

Another view of the same fact is that in this regions the validity of equation (4) can be proved. The decreasing phase mismatch of the third harmonic with increasing field strength is shown in Fig. 5, the values of which have been measured below coercive field strength at 10 Hz and 65 °C. The relative third harmonic amplitude of poled and unpoled samples belonging to 25 MVm^{-1} field strength is given in Fig. 6 as a function of frequency. It is important to notice that all experiments carried out with unpoled samples in this low field strength region show ac-history dependent results.

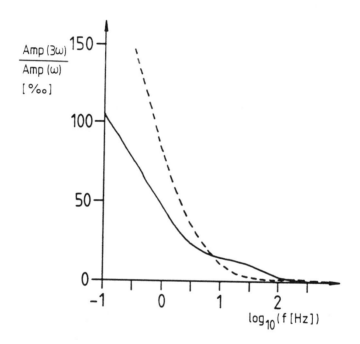

Figure 6. Frequency dependence of the normalized 3rd harmonic amplitude at 25 MVm^{-1} ac field strength of PVDF (Kureha, 12 μm)
at 65 °C
[Prehistory: ---- 4.5 d shorted, —— 4.2, 8.3 and 16.7 MVm^{-1}]

Conclulsion

Dynamic studies of the nonlinear dielectric response allow to estimate the significant data of ferroelectric hysteresis on the basis of fourier analysis. The frequency dependence of the ferroelectric hysteresis discussed in terms of phase shifts and amplitudes of higher harmonics render it possible to find a correlation function which describes cooperative phenomena. Equation (4) is proposed as an appropriate type of function to describe the ferroelectric behaviour of PVDF. Equation (5) is the respective reference function.

References

[1] M. G. Broadhurst and G. T. Davis, Ann. Rep., Elec. Insul. Diel. Phen. **48**, 447 (1979).

[2] A. J. Lovinger, Development in Crystalline Polymers - 1 (Appl. Sci. Publishers) p. 195 (1982).

[3] H. Kawai, Jpn. J. Appl. Phys. **8**, 1975 (1969).

[4] N. Murayama, K. Nakamura, H. Obara and M. Segawa, Ultrasonics **14**, 15 (1976).

[5] Y. Wada and R. Hayakawa, Jpn. J. Appl. Phys. **15**, 2041 (1976).

[6] R. Hofmann and J. Fuhrmann, Coll. Polym. Sci. **259**, 280 (1981).

[7] J. Fuhrmann, R. Hofmann, H.-J. Streibel and U. Jahn, IEEE Trans. on El. Ins. Vol. EI-21 N$\underline{0}$ 3, 529 (1986).

[8] J. C. Hicks, T. E. Jones and M. L. Burgener, Ferroelectric Lett. **44**, 89 (1982).

[9] A. G. Chynoweth, J. Appl. Phys. **27**, 65 (1956).

[10] M. Haardt, Thesis, Physikalisches Institut der Universität Stuttgart, 1982.

NONLINEAR PIEZOELECTRICITY IN POLYVINYLIDENE FLUORIDE

D.Jungbauer,J.H.Wendorff

Deutsches Kunststoff-Institut ,6100 Darmstadt,FRG

Abstract

The piezoelectric response of uniaxially stretched and poled poly(vinylidene fluoride), PVDF,was found to be nonlinear if the strain or stress were applied perpendicular to the stretching direction.the piezoelectric constant was found to decrease in this case with increasing mechanical loading and to change its sign above a critical load level.This level was found to depend on the test frequency.The decrease of the piezoelectrical constant was accompanied by a phase shift amounting to up to 180 deg.PVDF furthermore displays a second harmonic electrical response, if subjected simultaneously to a static and a sinusoidal dynamical stress.Blends of PVDF with poly (methyl methacrylate) ,PMMA,were found to display a similar nonlinear behavior provided that the PMMA concentration is kept below 20 wt%.No nonlinear behavior was observed for higher concentrations of PMMA.So far no definite model exists which is able to account for the main features of the nonlinear piezoelectrical behavior.

Introduction

Polyvinylidene fluoride is known for its large piezoelectric and pyroelectric responses, which originate from the presence of a noncentrosymmetric polar crystal modification in stretched films of PVDF (1-3).The application of strong electrical fields along the film normal or a corona charging gives rise to a macroscopical electrical polarization of such PVDF films. This polarization is subjected to changes if a mechanical force is applied to the film,causing thus a piezoelectric response ,or if the temperature of the material is changed,causing thus a pyroelectric response.

It is a general observation that a linear relation exists between the mechanical force and the variation of the polarization P or of the induced surface charges Q (4):

$$Q \sim d \cdot T$$

where d is the piezoelectric coefficient and T the applied stress.A similar linear relation holds for the pyroelectric behavior.

PVDF differs from other polymers in that its piezoelectric and pyroelectric coefficients are unusually large and of the same order of magnitude as the ones observed for technically important nonpolymeric piezoelectric materials such as ceramics.Examples in case are shown in Table I.

Material	d(pC/N)
Quartz	2.0
$BaTiO_3$	78
PVF	1.0
PVC	0.7
PVDF	23

Table I
Piezoelectric coefficients of some anorganic and organic materials.

It is because of the unusually large piezoelectric constant of PVDF and because of some other favorable properties of polymeric piezoelectric materials relative to the corresponding properties of ceramics,for instance, (see Table II) that the piezoelectric properties of PVDF have been studied in detail.All these studies were concerned,however, with the linear behavior.

Thin flexible films
Self-supporting films
Large areas obtainable
Layered films possible
Good impedance matching vs. water

Table II
Favorable   properties   of   polymeric   piezoelectric
materials

The   present   paper   is   concerned   with   a   new
phenomenon,namely   the   nonlinear   piezoelectric
behavior   of   PVDF   and   of   blends   of   partially
crystalline   PVDF   and   of amorphous   poly   (methyl
methacrylate) PMMA (5-8).   The   well   known   linear
piezoelectric behavior is described in some   detail
in   the   first part of the paper for   the   polymers
introduced   above   whereas   the second part   of   the
paper is concerned with experimental results on the
nonlinear behavior.

The linear piezoelectric behavior of PVDF and its blends with PMMA

A   linear behavior is defined by the fact that a
linear relation exists between the disturbance- the
mechanical   stress   T   or the strain   in   the   case
considered   here- and   the   response   i.e.   the
variation   of   the polarization or of   the   induced
surface   charges   Q   .   This   is   equivalent   to   a
piezoelectric constant which does not depend on the
applied   stress.   Figure 1 represents   experimental
results   on   the   dependence   of   the   piezoelectric
coefficient d_{31} on the stress T.This coefficient is
obtained   by   applying   the   stress T_1 along   the
stretching   direction of the film and by   measuring
the   surface   charges on the   surface of   the   film
with   the normal along the direction 3 (see   Figure
2).

The   application   of a sinusoidal   stress   gives
rise   to   a   sinusoidal variation   of   the   induced
surface   charges,as   expected   for   a   linear
behavior.The   additional   observation   is   that   no

phase shift exists between the stress and the
surface charge variations.So the piezoelectric
response is ideally linear.

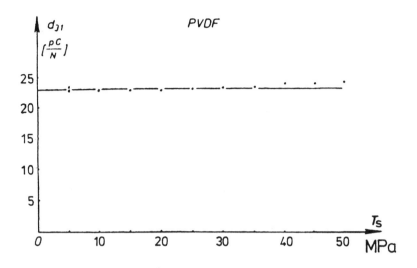

Figure 1
Dependence of the piezolectric coefficient d_{31} of
PVDF on the applied static stress T_s

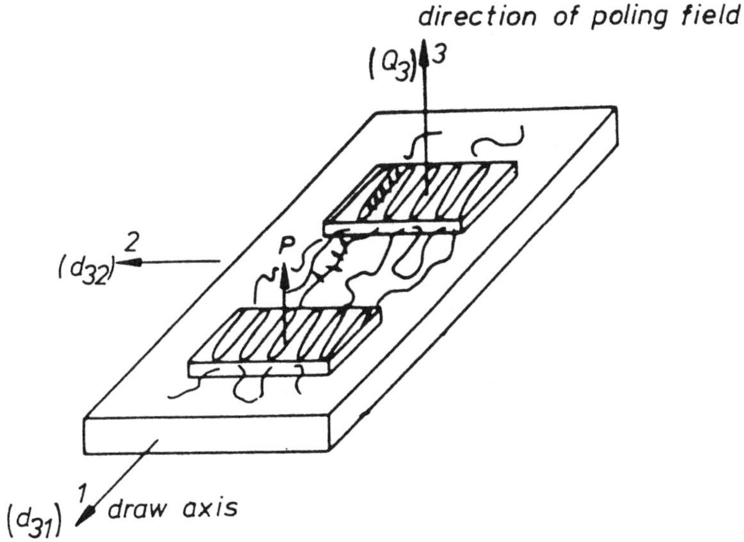

Figure 2
Definition of the axes 1,2,3 for the poled film

A similar behavior is observed for blends of PVDF and PMMA, as is obvious from Figure 3.

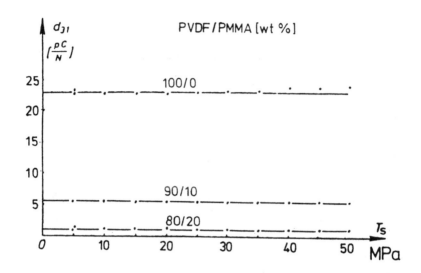

Figure 3
Dependence of the piezoelectric constant d_{31} on the static stress T_s for PVDF and various blends of PVDF and PMMA

It is known from structural studies (9,10) that the PMMA component is introduced into the amorphous regions between the crystalline PVDF lamellae and that a two phase amorphous structure results. A pure amorphous PVDF interphase exists at the surface of the PVDF crystals and a mixed amorphous phase well within the amorphous regions. This structural variation gives rise to a strong decrease of the piezoelectric coefficient d_{31} but does not lead to a deviation from the linear behavior. In addition, again no phase shift is observed between the mechanical loading and the electric response. It has to be pointed out that such ideal linear behavior has also been observed for a large variety of nonpolymeric piezoelectric materials.

The nonlinear piezoelectric behavior of PVDF-and PVDF-PMMA blends

In the following the piezoelectric behavior will be discussed for the case that the mechanical stress is applied along the direction 2, i.e. perpendicular to the stretching direction.The finding is,first of all, that the absolute value of the piezoelectric constant d_{32},which is obtained in this case,is by at least one order of magnitude smaller than the value of the piezoelectric constant d_{31} (8). The surprising finding is,however, that the piezoelectric constant d_{32} is no longer independent of the applied stress T_2.

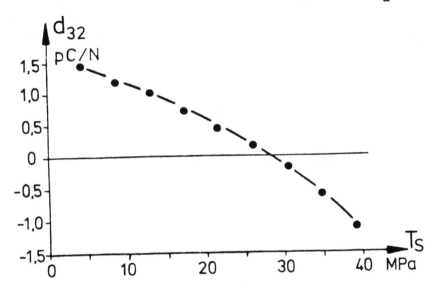

Figure 4
Dependence of the piezoelectric constant d_{32} on the static stress T_s

This is obvious from Figure 4,which shows for the case of pure PVDF that the piezoelectric constant decreases with increasing stress and changes even sign at a critical stress.

The obvious conclusion is that the relation between the induced surface charge Q and the applied stress is no longer linear.

The observed behavior can be represented to a first approximation by:

$$Q \sim d_o \cdot T + \beta \cdot T^2$$

$$\beta = (1/2) \ (\partial d/ \ \partial T)$$

where the second order term β can be derived from the slope of the dependence of the piezoelectric coefficient on the stress (see Figure 4).

The nonlinear piezoelectric behavior should manifest itself not only in a piezoelectric constant which depends on the applied stress but also in the generation of a second harmonic response. Suppose that the applied stress consists of a superposition of a static stress T_s and a sinusoidal stress , according to:

$$T = T_s + T_o \sin \ (2 \pi f \ t)$$

This gives rise to a charge response with a static term and terms with the frequency and twice the frequency f:

$$Q = Q_o + Q(f) + Q(2f)$$

$$Q(2f) \sim \beta T_o^2$$

Figure 5 represents the experimental results, which were obtained in cooperation with Prof.Fukada and coworkers (8) .The amplitude of the dynamic strain was varied in this case and the resulting variations of the stress and the induced charges were determined.The result is that the stress yields only a linear response whereas the electrical charge response is characterized by the occurrence both of a first order response $Q(f)$ and a second order response $Q(2f)$.

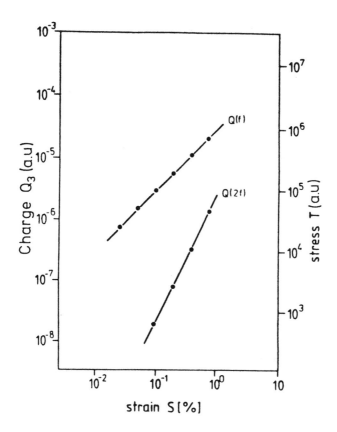

Figure 5
Second harmonic response of PVDF

A further phenomenon which is connected with the
decrease of the piezoelectric constant is the
occurrence of a phase shift between the mechanical
load and the electric response, in contrast to the
case of the linear piezoelectric discussed
above. This is shown in some detail in Figure 6.

The phase shift is found to increase with
increasing stress at constant frequency. The
magnitude of the phase shift may approach 180 deg

D. Jungbauer and J.H. Wendorff

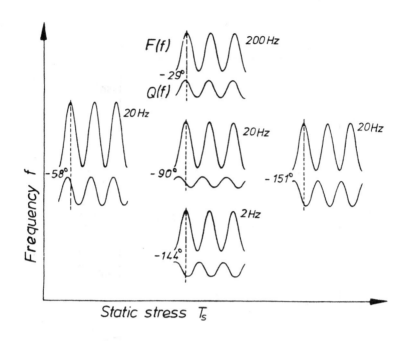

Figure 6
Dependence of the phase shift   between stress and
induced   charges   on   the   stress.Parameter   is   the
frequency used.

at   large   stresses.It is a remarkable   observation
that   the   charge   response seems   to   precede   the
exciting   force,as is obvious from the sign of   the
phase   shift depicted in Figure 5.This aspect   will
be   discussed   below.The phase shift   is   found   to
depend   in addition also on the test frequency f.An
increase of the frequency at constant static stress
results   in   a decrease of the   frequency   shift,as
shown   in Figure 5.Figure 6 shows this   dependence
of the phase shift both on the test frequency f and
on   the   applied   static   stress   in   a   different
representation.The   phase   shift   is   shown   as   a
function   of   the stress and the frequency   is   the
parameter.

In   principal one may speculate that   the   phase
shift   described above results from the   occurrence
of   a   relaxation phenomenon,   related   to   motions
occurring within the polymer such as the α or the ß
relaxation.Such   relaxations   do   not   give

rise,however,to   the occurrence of phase shifts   of
up to 180 deg.

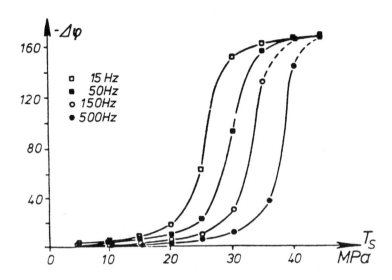

Figure 7
Variation of the phase shift   with the applied
stress.Parameter is the test frequency.

    Such   magnitudes   of   the   phase   shift   are
reminiscent of resonance phenomena.A possible model
is thus that a resonance phenomenon occurs and that
the increase of the stress gives rise to an
increase   of   the   resonance   frequency.   The
contradiction to this assumption is ,however,that
the phase shift goes into the wrong direction,since
the response seems to precede the excitation as
mentioned above.This is of course impossible for
principal reasons.

    A tentative explanation is that a phase shift of
180 deg exists already at low stresses   .The
piezoelectric coefficient would reverse its sign in
this case and the unreasonable phase shift from 0
to -180 deg would be transformed into one from 180
to 0 deg ,so that the excitation would no longer
lag behind the response.The origin of such a
hypothetical phase shift is still unknown to us. So
far we do not have a definite interpretation of the
phenomenon described above.

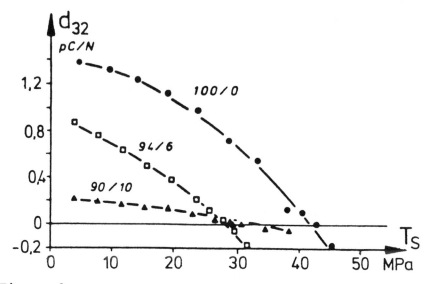

Figure 8
Stress dependence of the piezoelectric   coefficient
d_{32} for   PVDF   and blends of PVDF   and   PMMA   with
different compositions (wt% PVDF/PMMA)

It   is   obvious in any case that   the   amorphous
regions   within the partially crystalline state   of
the   PVDF   play   a major role   in   controlling   the
piezoelectrical   properties .This is apparent   from
the fact that both the linear piezoelectric and the
nonlinear   piezoelectric   behavior   are   strong
functions of the temperature,particularly above the
glass   transition   temperature.In   order   to
investigate the influence of the amorphous phase on
the   nonlinear   response   in   some   more   detail,we
performed   experiments   on   blends   of   PVDF   and
PMMA.Some experimental results are shown in Figures
8 and 9.

The   result   is   that   the   nonlinear   response
becomes   weaker   with increasing   concentration   of
PMMA   and   is   totally   suppressed   once   the
concentration   of PMMA has become about   20 %.This
finding   indicates   the   important   role   of   the
amorphous   phase   in   influencing   piezolectrical
properties.We   are   currently   performing
investigation   on   PVDF   samples   in   which   the
amorphous   regions   have been crosslinked with   the

purpose to further our knowledge on the origin of the nonlinear piezoelectric behavior of PVDF.

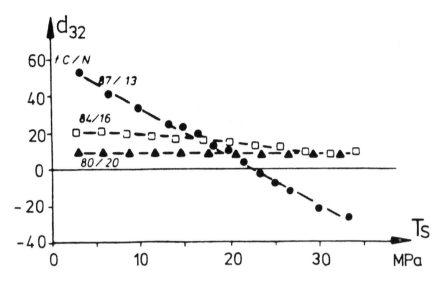

Figure 9
Stress dependence of the piezoelectric coefficient d_{32} for blends of PVDF and PMMA with different compositions (wt % PVDF/PMMA)

References

(1) Kawai,H.J.Appl.Phys.1969,8,975
(2) Broadhurst,M.G.;Davis,G.T.;Mc Kinney,J.E.;
Collins,R.E.J.Appl.Phys.1978,49,4992
(3) Kepler,R.G. Ann.Rev.Phys.Chem.1978,29,497
(4) Sonin,A.S.;Strukow,B.A. In "Einführung in die
Ferroelektrizität";Vieweg Verlag Braunschweig,1979
(5) Hahn,B.R   Thesis,Technical   University
Darmstadt,1983
(6) Hahn,B.R.J.Appl.Phys.1985,57,1294
(7) Neumann,H.E.Master   Thesis,Technical University
Darmstadt,1985
(8) Fukada,E.;Date,M.;Neumann,H.E.;Wendorff,J.H.
submitted to J.Appl.Phys.,1986
(9) Hahn,B.R.;Wendorff,J.H.;Yoon,D.Y.Macromolecules
1985,18,718
(10) Hahn,B.R.;Herrmann-Schönherr,O.;Wendorff,J.H.
Polymer 1987,28,201

THE RANGE OF ALPHA PARTICLES IN SEVERAL POLYMERS

G. Kennedy[*], E. Sacher[+] and M. R. Wertheimer[+]

École Polytechnique de Montréal
C.P. 6079, Succursale "A"
Montréal, Québec H3C 3A7
Canada

[*] Institut de génie énergétique
[+] Groupe des couches minces et département de génie physique

Introduction

The use of ceramic substrates for the mounting of microelectronic chips runs the risk of introducing "soft errors" due to the emission of natural alpha particle radiation from the ceramic. Because these are of relatively high mass and low energy, they are easily stopped by thin sheets of solid material; should this occur within the chips, soft errors may result. However, it is equally probable that they will be stopped by thin sheets of solid material interposed between the substrate and the chip. In terms of pure economics, this interlayer is best applied after chip mounting, dictating its introduction as a liquid which then solidifies, a vapor which reacts to form a solid or a plasma polymer. Another constraint is that this interlayer must withstand the harsh environments experienced during the rest of the fabrication process, as well as the hostile atmospheres encountered during the use lifetime of the fabricated device.

We have determined the minimum thickness necessary to stop alpha particles for the three types of solids refered to earlier, choosing materials whose physical properties suggest they will withstand these constraints. We have chosen the following candidates:

1. polyimide - while insoluble after formation, the polyamic acid precursor is soluble in N-methylpyrrolidone. Once applied and the solvent evaporated, heating converts the polyamic acid to polyimide.

2. poly-p-xylylene - the highly strained p-cyclophane precursor undergoes scission to give xylyl diradicals in the vapor phase which, on condensation, react to give poly-p-xylylene.

3. plasma-polymerized hexamethyldisiloxane (PPHMDSO) - hexamethyldisiloxane vapor undergoes reaction in the plasma to give species which depend on the plasma parameters. The condensation of these species gives the highly crosslinked PPHMDSO.

Experimental

The polyimide was purchased in film form as Du Pont Kapton H. Similarly, the poly-p-xylylene was purchased in film form as Union Carbide Parylene N. The PPHMDSO was deposited in our laboratory as previously described[1]; it was deposited onto Du Pont Kapton H polyimide film at a power setting of 200 W and a substrate temperature of 200°C.

The range was determined with the apparatus in Figure 1,

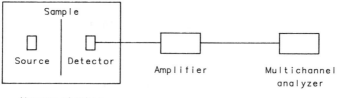

1. Apparatus used for alpha particle ranges in polymers.

using the 5.47 MeV alpha particles from [241]Am. Sheets of several thicknesses were used and the average energies of the transmitted alpha particles were measured. These were then fit to a curve whose slope is known from stopping power calculations[2]; an example is shown in Figure 2 for Parylene N. These ranges, estimated to be accurate to ± 2%, are those which stop 50% of the alpha particles; the range necessary to stop 100% of the particules is difficult to determine but is estimated to be 7% greater.

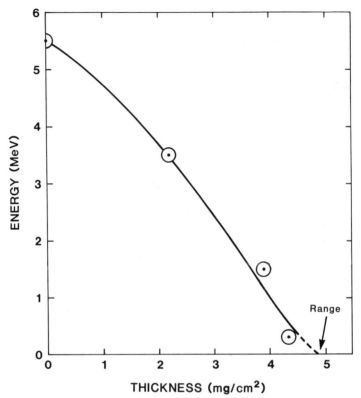

2. A plot of average energies of 5.47 MeV alpha particles as a function of Parylene N film thickness.

Results and Discussion

The ranges for the three materials tested are found in Table I. The ranges in μm were determined by dividing by the

Table I

Ranges for 5.47 MeV Alpha Particles

Material	range found in mg/cm^2	in μm	range calculated in mg/cm^2	in μm	% difference
Parylene N	4.81	43.7	4.16	37.8	-15.6
Kapton H	5.04	35.5	4.66	32.8	-8.2
PPHMDSO	4.16	24.1	5.01	29.0	+20.4

appropriate densities. Calculations were made using the TRIM program[3], a version of which was made available to us by J. F. Ziegler, IBM Corporation. This is a predictive program using LSS range theory[4], as well as Bragg's rule of stopping power additivity[5] (i.e., the stopping power of a molecule is the sum of those of the elements). Note that the TRIM calculation underestimates the ranges (negative difference) for Parylene N and Kapton H, and overestimates that for PPHMDSO.

A possible reason for the underestimations lies in the structures of Parylene N[6] and Kapton H[7]: each is partially ordered, with aromatic rings oriented in the plane of the film and lying stacked perpendicular to the plane. Thus, the possibility exists for the formation of cones of protection, as seen in Figure 3, much as in channeling. A much more

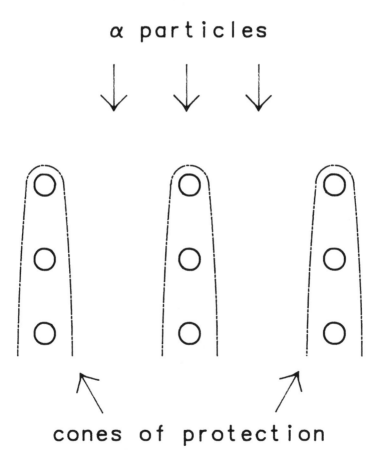

3. Cones of protection formed from molecular ordering in the direction of irradiation.

reasonable explanation[8] comes from the fact that both are aromatic polymers, with extensive electron delocalization, so that Bragg's rule does not hold; in any case, Bragg's rule should be considered an approximation.

The overestimation for PPHMDSO is unexpected. The experimental data were rechecked in an effort to find a possible source (alpha energy, thickness measurement, etc.) for this 20% error; none was found. A TRIM study of the effects of various elements on the predicted range was carried out and this is seen in Table II. Note that C and O cause small range reductions from Si, and even 2 O gives a small decrease over O. It is only when H is added, do we see a large reduction but, even for SiH_4 (= 80 atomic % H), the

Table II

Predicted Ranges for Si-containing Materials

Material	Predicted range (mg/cm^2)
Si	6.27
SiC	5.75
SiO	5.88
SiO_2	5.77
$H_2SiO_3 (=SiO_2+H_2O)$	5.30
SiH_4	4.55
PPHMDSO $(=Si_{21}C_{24}H_{50}O_5)$	5.01

range predicted is substantially higher than that found for PPHMDSO, whose chemical analysis gave $Si_{21}C_{24}H_{50}O_5$, i.e., 50 atomic % H.

It is most probable that the TRIM range overestimation for PPHMDSO is due to low stopping powers used in the calculations. This was previously found in a study carried out on silicons of various densities (amorphous, polycrystalline and single crystal), where it was shown[9] that the stopping power increased with increasing density. PPHMDSO is known to have a higher density than expected[10,11], due to the plasma deposition process; thus, the stopping powers used in the TRIM calculation, based on more "normal" Si-containing compounds, are too low for PPHMDSO. No provisions are easily made for modifying the stopping powers used in the TRIM program, so the determination of the extent of increase necessary to eliminate the overestimation could not be done. It is, however, clear that an increase in stopping power would move the calculation in the correct direction.

Acknowledgments

Thanks are due to the Natural Sciences and Engineering Research Council of Canada and the Fonds FCAR of Quebec for financial support.

References

[1] Sacher, E.; Klemberg-Sapieha, J.E.; Schreiber, H.P.; Wertheimer, M. R. J. Appl. Polym. Sci.: Appl. Polym. Symp., 1984, 38, 163.

[2] Chu, W.-K.; Mayer, J.W.; Nicolet, M.-A. "Backscattering Spectroscopy", Academic Press, New York, 1978, Chapters 3 and 9.

[3] Ziegler, J.F.; Biersack, J.P.; Littmark, U. "The Stopping and Range of Ions in Polymers", Pergamon, New York, Volume 1, 1985, Chapter 8.

[4] Linhard, J.; Scharff, M.; Schiott, H.E. Mat. Fys. Medd. Dan. Vid. Selsk., 1963, 33, No. 14.

[5] Bragg, W.H.; Kleeman, R. Philos. Mag., 1905, 10, 318.

[6] Sacher, E. J. Macromol. Sci., 1987, B25, 415.

[7] Gorham, W.F.; Niegish, W.D. Encyclopedia Polym. Sci. Technol., Wiley, New York, Vol. 15, 1971, p. 98.

[8] Twaites, D.I. Radiat. Res., 1983, 95, 495.

[9] Ziegler, J.F.; Brodsky, M.Y. J. Appl. Phys., 1973, 44, 188.

[10] Wrobel, A.M.; Klemberg-Sapieha, J.E.; Wertheimer, M.R.; Schreiber, H.P. J. Macromol. Sci.-Chem., 1981, 15, 197.

[11] Sacher, E.; Klemberg-Sapieha, J.E.; Schreiber, H.P.; Wertheimer, M.R. J. Appl. Polym. Sci.: Appl. Polym. Symp., 1984, 38, 163.

FURTHER OBSERVATIONS ON TRIBOELECTRIC CHARGING EFFECTS IN KERATIN FIBERS

J. Jachowicz, G. Wis-Surel and M. Garcia

Clairol Inc.
2 Blachley Rd., Stamford, CT 06902, USA

ABSTRACT

Triboelectric charge exchange studies in the rubbing mode of intact and surface-modified keratin fibers have revealed a number of previously unreported effects. The kinetics of triboelectric charging was found to be dependent upon the mechanical and electrostatic history of the fibers. The results are discussed in terms of mechanically-induced permanent polarization of cuticles and the existence of nonequilibrium distribution of electrons injected by previous contacts. The fibers modified by silicon oil films were shown to produce a directional triboelectric effect reversed in comparison with untreated fibers, i.e. charging more negatively in root-to-tip than tip-to-root rubbing. Acid and base treatments, adsorption of cationic polymer and deposition of two layers of different treatments altered the work function of keratin by shifting it to either higher or lower values.

INTRODUCTION

Contact and tribo charging between metal or polymer and insulating surfaces has been the subject of many studies in recent years. The phenomenon has found practical applications in the processes of xerography, electrostatic painting, and electrostatic separations, among others. In other industries, such as in textile manufacturing, it has been a cause of major concern, and considerable effort has been made to study the origin of triboelectric charging and means of eliminating it. The present communication describes triboelectric charging effects in keratin fibers, such as human hair, previously shown to exhibit unusual, directional triboelectric effect (1,2).

EXPERIMENTAL

Triboelectric Charging Measurements

The device shown in Figure 1 was employed. A hair tress
was mounted in clamps in such a way that it formed a smooth
layer 0.03 cm thick (approximately four layers of single
fibers). The fibers within the tress could be positioned
with the cuticle edges pointing either downward or upward.
A rubbing element in the form of a half cylinder was
attached to an adjustable arm that could be rotated by a
variable speed motor. A speed of 70 rotations/minute was
used throughout this work. Static charge was produced by
contact between the rubbing element and the hair fibers and
measured as a function of rubbing time by means of a static
detector probe connected to an electrometer. Two load
cells were used to monitor the forces during each rub in
the upper and lower part of the tress-holding frame. This
allowed us to calculate the frictional coefficient between
the probe and hair. Load cells and electrometer were
interfaced with an IBM PC by means of an analog-to-digital
converter.

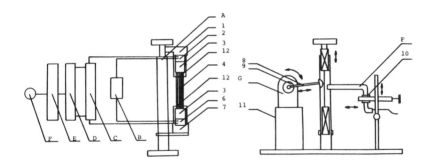

Figure 1: Apparatus for measuring static charge generated
by rubbing: (A) mechanism for changing fiber tension, (B)
power supply, (C) operational amplifiers (Analog Devices,
Model 2B31J), (D) computer and A/D converter (Model DT
2801, Data Translation, Inc.), (E) electrometer (Keithley,
Model 616), (F) static detector probe (Keithley, Model
2503), (G) motor, (1) and (7) load cell holding elements,
(2) and (6) load cells (Sensotronics, Model 60036),

(3) clamps, (4) fiber tress, (8) adjustable arm with
rubbing element, (9) mechanism for adjusting the length of
the arm, (10) mechanism for positioning the static detector
probe, (11) table for motor, (12) mechanism for positioning
the fibers.

Aquisition of data was performed by the use of
Labtechnotebook software (Laboratory Technologies
Corporation) and all subsequent calculations were done with
Lotus 1-2-3 spreadsheet (Lotus Development Corporation).
The entire setup, except for the electrometer and computer,
was housed in a dry box maintained at 25-30% relative
humidity under positive pressure of air passed through
several columns filled with Drierite. The surface charge
density on the fibers was calculated as described before
(2).
   The rubbing element was exchangeable so we could examine
the electrification of hair fibers using a variety of
materials such as teflon, aluminum, gold, stainless steel,
nylon, etc. We also used solution-cast polymer films as
rubbing materials mounted on aluminum half-cylinder. We
found that for 20-30 μm thick films, the support material
has no effect on the process of charge transfer between the
film and the keratin fibers. Charge decay measurements
were performed with the same experimental setup by
following the changes in generated charge density as a
function of time.

FILM PREPARATION

   Films of polycarbonate bisphenol-A (PC), poly(methyl
methacrylate)(PMMA), and chitosan acetate (ChA) were cast
from 5 wt % solutions in CH_2Cl_2–$C_2H_4Cl_2$ (1:1), $CHCl_3$, and
H_2O, respectively. The polymer films were dried for 24
hours in a vacuum oven before use.

Surfactants, Polymers and Emulsions

   The cationic surfactant used in this study was N,N-
dimethyl-N-octadecylbenzenemethanaminium chloride. The
quaternary ammonium polymer employed was poly (1,1-di-
methyl-piperidinium-3,5-diallyl methylene chloride)
(PDMPDAMC).

   Silicon emulsion was Dow-Corning Q2-7224, amine-modified
poly(dimethyl siloxane) emulsified with a nonionic
surfactant system.

Preparation of Fiber Samples for Triboelectric Measurements

Virgin brown hair, purchased from deMeo Brothers, New York, was used throughout this work. It was purified by the procedure described previously (3).

The removal of cuticle cells was performed by a modified Allworden reaction. Hair was exposed to a solution of $NaClO_4$ at pH 3 (adjusted with HCl) for 1 minute, rinsed thoroughly with deionized water, and then treated with alkaline solution of H_2O (pH 12 adjusted with NaOH) for 45 seconds, followed by extensive rinsing and drying at 30% R.H. In order to modify the surface properties of the keratin fibers, hair swatches were placed into large excess of polymer, detergent or emulsion solution of appropriate concentration for 2-120 minutes in case of silicon emulsions and 24 hours in case of detergents or polymers, at room temperature, and stirred occasionally. The fibers were then rinsed under running deionized water and exposed to an excess (2-3 liters) of deionized water for 2-4 hours. The purpose of the prolonged pure water exposure was to obtain fiber samples with irreversibly adsorbed polymers or detergents. Acid or base treatments were performed in HCl/NaOH solutions (liquor/hair ratio 1600) at appropriate pH for 20 hours.

RESULTS

Morphology of Hair Fiber Surface

Hair fibers have two major morphological components, the cuticle and the cortex. The cuticle covers the whole fiber and is composed of layers of overlapping scales each about half a micron thick. A scanning electron micrograph illustrating the surface morphology of a hair fiber is presented in Figure 2a. Each cuticle cell has a substructure and is composed of an external proteinous cell membrane layer (3-10 nm) called epicuticle, a layer of cystine-rich, highly crosslinked protein called exocuticle, and a layer of nonkeratinous protein material called endocuticle. Figure 2b shows a SEM micrograph of fibers with removed cuticle cells. It reveals the smooth outer surface of the cortex consisting of elongated cortical cells, packed tightly together and oriented parallel to the fiber axis. They contain microfibrils, long uniform filaments composed of

highly crystalline material exhibiting a characteristic
α-helical x-ray pattern.

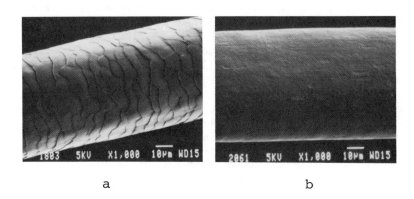

a   b

Figure 2. SEM micrographs of hair fibers: (a) intact fiber
(b) fiber with removed cuticles.

Directional Triboelectric Effect

Our earlier reports (2,3) demonstrated that the process
of electron transfer is affected by the direction of
rubbing (root-to-tip or tip-to-root) and the experimental
data can be explained in terms of simple band model of the
electronic structure of polymers and metals, assuming
certain characteristic values of work function for each
material as suggested by Davis (4). According to this
interpretation, materials with an effective work function
lower than keratin would donate electrons to hair and vice
versa. In order to explain the directional triboelectric
effect, we hypothesized that the electrochemical surface
potential of keratin is modified by a potential appearing
during tangential friction of the cuticle. This additional
potential of supposedly piezoelectric origin (1) may shift
the work function towards higher or lower values depending
on whether the scales are compressed [tip-to-root (t-r)] or
extended [root-to-tip (r-t)] during rubbing. Figure 3
shows the kinetic curves of charge build-up on untreated
keratin

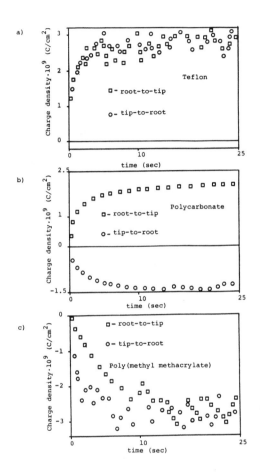

Figure 3. Kinetic curves of charge generation on intact hair fibers by rubbing against various materials.

fibers obtained with teflon, polycarbonate and poly(methyl methacrylate) probes. In agreement with previously reported data, teflon produces positive charges on the fiber surface for both directions of rubbing, polycarbonate charges hair positively and negatively in r-t and t-r modes of rubbing, respectively, and poly(methyl methacrylate) charges hair negatively for both directions of rubbing. Other investigated materials such as polyethylene, polypropylene, polystyrene, aluminum and gold produced positive charges on the fiber surface irrespective of the direction of rubbing while chitosan acetate showed the directional triboelectric effect. Tribocharges generated by rubbing of hair with removed cuticles with gold, aluminum, teflon and PMMA are of the same sign and magnitude as for untreated hair. The first two materials act as electron acceptors, while PMMA behaves as an electron donor. The directional effect disappears and both PC and ChA probes produce tribocharges of the same sign for both modes of rubbing. PC probe acts as an electron acceptor while ChA as an electron donor.

Sequence of Rubbing Effects

It has been suggested that triboelectric charge transfer might be affected by either the mechanical or electrostatic history of an insulator. To explain the ability of nominally identical materials to charge each other when they are rubbed together, Shaw and Hanstock (5) proposed that the triboelectric properties of a surface might be influenced by its state of "strain". In asymmetric rubbing experiments two contacting surfaces might become strained to different extents, so that they become triboelectrically distinct. Lowell and Truscott (6,7) suggested that the charge transfer in asymmetric rubbing of the same materials is associated with an initially non-equilibrium distribution of electrons with some of them in localized, high energy states. These electrons are able to tunnel into lower energy, empty states on the other surface during asymmetric rubbing. Differences in the areas the materials come in contact with each other may then lead to net charge transfer. History-dependent charge transfer effect for keratin fibers was explored by using PC. This probe is the most sensitive to small changes of keratin work functions. Each 30 second charge generation kinetics was repeated three times on the same tress in a sequence: first rubbing-discharge by Polonium element- second rubbing-discharge by Polonium element- third rubbing -slow

discharge and the measurement of decay kinetics. The data
indicate that the kinetics of charge generation depends on
mechanical history of a tress. This is evident especially
in cases when relatively low densities of tribocharges are
generated. The general, qualitative rules observed are the
following:

1. For both positive and negative charging, the rate of
   charge increase as well as the absolute value of
   saturation charge density are always smaller in the
   first rubbing and gradually increase in subsequent
   rubbings to an equilibrium value (Figure 4a).
2. If charge reversal is observed during the first rubbing,
   then this effect usually disappears in the second or
   third kinetic runs although the rate of reaching the
   equilibrium positive or negative saturation values is
   small (Figure 4b).
3. If the first rubbing in the direction tip-to-root
   produces very high density of negative charges (for PC
   probe, maximum charge density is about $3 \cdot 10^9$ C/cm^2) then
   in subsequent root-to-tip rubbing with PC charge
   reversal within a test run is usually observed (Figure
   4c). The initial strokes of the rubbing element,
   generate negative charge on fibers but then its sign is
   reversed to expected positive values if rubbing is
   continued. A similar effect is observed for a reversed
   sequence of rubbing (first root-to-tip and then
   tip-to-root.

Some additional features of the "sequence of rubbing"
effect can be demonstrated by considering the following
sequence of rubbings (each rubbing is followed by a
discharge with a Polonium element:
PC(r-t)-PC(t-r)-PC(t-r)-PC(t-r)-PC(r-t)-PC(r-t)-Al(r-t)-PC(t
-r)-Al(t-r)-PC(r-t)-PC(t-r)-Al(r-t)-PC(r-t) Figure 5). It
is evident from these data that if PC(t-r) follows PC(r-t)
(or vice versa) reversal of the sign of tribocharge is
observed within a test run with the initial generation of
positive (negative for PC(t-r)-PC(r-t) sequence) charges.
A similar pattern is observed for the sequences
Al(r-t)-PC(t-r) and Al(t-r)-PC(r-t) (Figure 5, rubbings 7-8
and 9-10) in which the charge generated by Al(r-t) and
Al(t-r) rubbings is positive. It should also be noted that
the effect of sign of charge reversal is usually much
stronger in the sequence (t-r)-(r-t). If the charge
density in r-t kinetic test run is not very high (lower
than $2.5 \cdot 10^9$ C/cm^2) then a subsequent t-r experiment might
not reveal a sign of charge reversal effect within a test
run (Figure 5, rubbing 11).

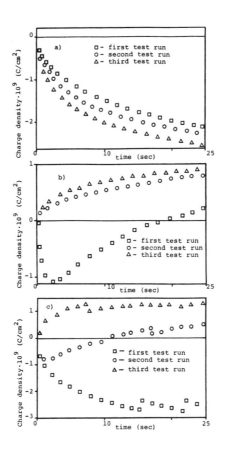

Figure 4. Effect of rubbing sequence on static charge generation by rubbing against PC: (a) three consecutive rubbings in the tip-to-root direction, (b) three consecutive rubbings in the root-to-tip direction, (c) first rubbing in the tip-to-root direction followed by two consecutive root-to-tip rubbings.

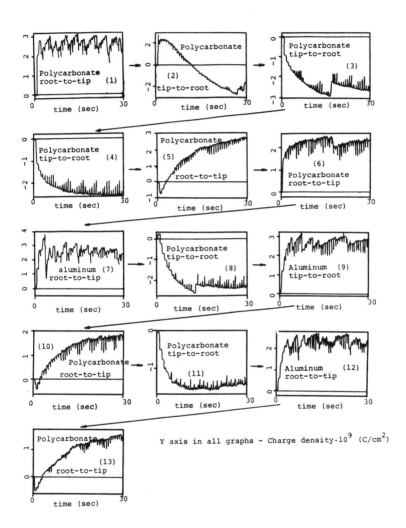

Figure: 5: Effect of rubbing sequence on static charge generation in keratin fibers. Kinetic curves of charging obtained in the sequence of rubbings: PC(r–t), PC(t–r), PC(t–r), PC(t–r), PC(r–t), PC(r–t), Al(r–t), PC(t–r), Al(t–r), PC(r–t), PC(t–r), Al(t–r), PC(r–t).

According to Shaw and Hanstock (5), the explanation of
these effects might be the permanent deformation and
opposite polarization of cuticles induced by stretching or
compressing during rubbing in the r–t or t–r mode,
respectively. For example, PC(t–r) rubbing results in
compression of cuticles and their polarization to the
extent that in subsequent PC(r–t) rubbing negative charges
are generated during initial contacts with the probe.
However, the experiments involving the use of an aluminum
probe seem to contradict such an interpretation and point
to a more complex nature of tribocharge transfer
phenomena. Consider the following sequences of rubbings:
PC(t–r)-Al(t–r)-PC(r–t) and PC(t–r)-Al(r–t)-PC(r–t) (Figure
5, rubbings 8-13). In both cases, initial PC(t–r) rubbing
generated high density of negative charges and subsequent
rubbings with Al probe produced a high density of positive
charges on the fiber surface. After this sequence of two
rubbings, we expected PC(r–t) to show sign reversal only
for the sequence involving Al(t–r) experiment.
Surprisingly, the effect was observed for both sequences.
This seems to suggest that in both cases the negative
charge injected during PC(t–r) test run remains trapped in
the sample, is not removed by subsequent contacts with Al
probe or by discharging with Polonium element, and can
further induce initial negative charging during rubbing in
root-to-tip mode with a PC probe.

Tribocharging Characteristics of HCl and NaOH Treated Fibers

S. P. Hersch et al. (8) studied the generation of
tribocharges on acid and base treated hair fibers. They
concluded that modification of hair using acidic solutions
causes an increase in the generation of negative charges on
fibers (increased electron accepting properties). High
alkaline treatment, on the other hand, gave rise to
generation of positive charges. Different results were
collected by J. A. Mendley (9) and W. E. Morton et al.
(10). They conducted a similar study and found that the
acid-treated keratin shows a higher tendency to be
positively charged, while negative charge accumulation
occurs after the base treatment.
The kinetic curves of charge build-up on keratin fibers
treated in HCl or HaOH solutions at various pH were
obtained with the PC probe. The data indicate that
HCl/NaOH treatments produce very small shifts in work
function of keratin. The hair tress treated with alkaline

solution showed a decreased density of positive charges for
rubbing in the direction from root-to-tip (Figure 6). Even
after three sequential rubbings, the rate of charge
generation is small and saturation level is below $1.5 \cdot 10^9$
C/cm^2. For tip-to-root rubbing, the rate of negative
charge generation is very fast and the maximum charge
density is close to $3.2 \cdot 10^9$ C/cm^2 (Figure 6). In this
case, the discharges with the surrounding atmosphere are
frequent, leading to wide scatter of experimental points.
For hair tresses subjected to HCl treatments, an increase
in electron-donating properties for both r-t and t-r
rubbings was observed. The rates of charge generation are
relatively high for r-t rubbing and low for rubbing in the
opposite direction (saturation density $1.6 \cdot 10^9$ C/cm^2,
Figure 6).

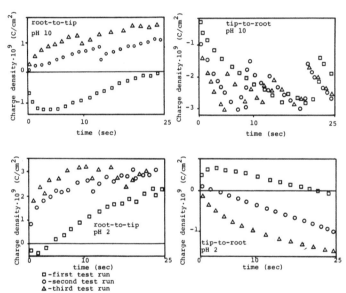

Figure 6. Generation of tribocharges on fibers treated at
PH 2 and 10

The fibers treated at pH 2 and 10 have high rates of
charge decay ($t_{1/2}=2$ and 10 minutes for pH 2 and 10 treated
fibers, respectively). For treatments at intermediate pH
values, the rate of discharge was very small ($t_{1/2}$ of the
order of tens of minutes) and similar to that observed for
untreated fibers. Increased rate of charge dissipation is
caused by adsorption of ions from relatively concentrated
solutions of HCl and NaOH. It should be stressed, however,

that even the fastest rate of charge decay is small in comparison with the time-scale of charging experiment and we do not expect that it might have any effect on the process of charge generation. Thus, the small influence of pH treatment on tribocharging should be related to the changes in work function of the cuticle protein.

Tribocharging of PDMPDAMC Modified Fibers

PDMPDAMC appears to adsorb as a monolayer on hair when the fibers are exposed to 1% solution of polymer for 24 hours (11). The saturation amount of the polymer adsorbed was found to be about 3 mg/g corresponding to a segment surface density of 0.6 segments/A^2 (specific surface area for hair fibers is about 900 cm^2/g). Such a layer of the polymer lowers the effective work function of the fibers making them more electron-donating. The data presented in Figure 7 show a consistent increase in electrochemical potential as detected by PC and PMMA probes. Charging with PC(t-r) was reversed to positive. Other probes, such as Al produce enhanced positive charging of PDMPDAMC-modified fibers.

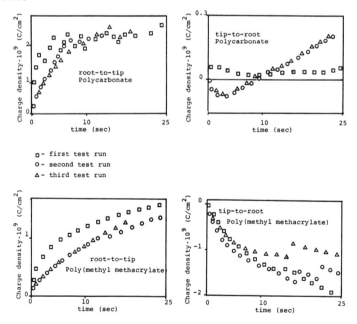

Figure 7. Generation of tribocharges on fibers modified by adsorption of PDMPDAMC.

Reversed Directional Triboelectric Effect

We found that keratin fibers modified by a surface layer of silicon oil exhibit a reversed directional triboelectric effect, i.e. charge more negatively in the root-to-tip than tip-to-root mode of rubbing. Hair samples were exposed to 0.25% silicon emulsion solution for 2 and 120 minutes in order to vary the amount of surface-deposited silicon oil in the range of 2-25 mg/g. The experimental results of tribocharging, exemplified by the data presented in Figure 8, can be summarized as follows:

1. R-t rubbing by PC produced very high density of negative charges on the fiber surface for all samples and all times of treatment. T-r rubbing generates small or intermediate ($0-1.5 \ 10^9$ C/cm^2) density of positive or small density of negative charges, depending on the history of the tress.

2. PMMA acts as a strong electron donor in r-t mode of rubbing and as a weak electron donor or acceptor in t-r mode.

3. The use of teflon or aluminum results in generation of reduced density of positive charges in r-t mode. T-r rubbing, on the other hand, produces a very high density of positive charges.

4. No qualitative differences in triboelectric behavior were detected between hair samples treated for 2 and 120 minutes.

The observed triboelectric characteristics of hair coated with silicon oligomer cannot be explained by a simple shift in modified keratin work function towards higher or lower values. If this was the case, more negative, or more positive charging in both modes of rubbing would be observed. It seemed, thus, necessary to establish the triboelectric properties of adsorbed layers of silicon in the absence of a directional triboelectric effect. Hair fibers with removed cuticles were modified by surface deposition of a layer of silicon oil. Charging with PC (r-t and t-r) and aluminum (r-t and t-r) was positive with some reduction of saturation charge density as compared to untreated decuticled hair. Experiments with PMMA (r-t and t-r) showed positive charging as opposed to high negative charge density generated on untreated hair

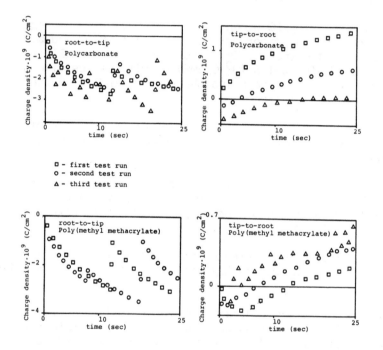

Figure 8. Generation of tribocharges on fibers modified by deposition of silicon oil.

with removed cuticles. Since the directional triboelectric
effect disappears after cuticle removal this result points
to the electron-donating character of silicon treatments.
This is further supported by experiments in which films of
polycarbonate containing 10% silicon oil were used as
contacting probes. They generated negative charges on
intact fibers in both modes of rubbing. For comparison,
N,N-dimethyl-N-octadecylbenzenemethanaminium chloride
treated intact hair and hair with removed cuticles showed a
decreased electrochemical potential and negative r-t and
t-r charging for PC and PMMA or reduced positive charging
for Al probe.

Triboelectric Charging of PDMPDAMC-Silicon Oil Modified Fibers

In order to test the effect of multiple layers of
different treatments on the triboelectric charge generation
on hair, the fibers were pretreated with a cationic polymer
PDMPDAMC before deposition of silicon oil. The procedure
results in the formation of a layer of cationic polymer
(3mg/g) in contact with the fiber surface and an outer
layer of silicon oil. The results of measurements have
shown the additive effect of both layers on the kinetics of
tribocharge generation. R-t and t-r charging for all the
probes was less negative and more positive, respectively,
than for fibers treated with silicone oil. This reflected
the presence of electron-donating layer of PDMPDAMC.

DISCUSSION

The results of triboelectric charging of
surface-modified hair fibers are summarized on the energy
diagram presented in Figure 9. They indicate that the
triboelectric character of the fiber surface can be
controlled by adsorption of polymers, surfactants or oils.
As reported earlier (3), cationic surfactants, composed of
quaternary ammonium head group and a long alkyl chain,
shift the effective work function of the fibers towards
higher values. Adsorption of cationic polymer, PDMPDAMC,
results in lower work function of modified fibers. In both
cases, the effect of the modifying surface layer is
superimposed on the directional triboelectric effect.
Similar dependence, without directional triboelectric
effect, is observed for fibers with removed cuticles. They
can be made more electron-accepting by treatment with
cationic surfactant or electron-donating by modification

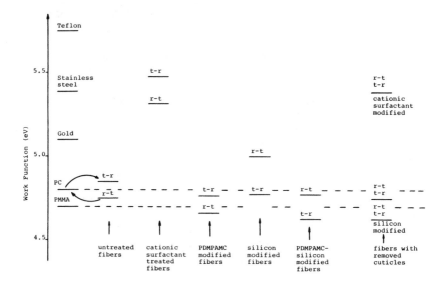

Figure 9. Energy diagram illustrating the effect of surface treatments on the work function of keratin fibers.

with silicon oil. Adsorption of two modifying layers of
PDMPDAMC and silicon oil demonstrate the additive effect of
both components of the treatment. PDMPDAMC shifts the work
function of the fiber towards lower values while silicone
produces reversed triboelectric effect.

All these data are compatible with the idea that in the
space-charge region, extending into the bulk of the polymer
as far as few microns, the equilibration of energy levels
of all contributing layers occurs (12,13). In consequence,
the observed charging patterns reflect the presence of
various species characterized by different work functions.

The influence of silicon oil on the direction of charge
transfer during rubbing of keratin fibers is reminiscent of
the space-charge effects of semiconductive coatings on
tribocharge exchange between metals and polymer films
reported by Pochan et al. (13). Their experiments with
tribocharging of nickel berry coated with poly (vinyl
carbazole) against polycarbonate films showed that the
coating causes charging opposite to what would be expected
on the basis of its electron donating or accepting
capability. It was postulated that this is due to the
transport characteristics of the materials coated onto the
bead and kinetically controlled charging. Short contact
times involved in the charging event between coated bead
and polymer film do not allow to achieve complete charge
equilibration which should produce thermodynamically
controlled results conforming to energy level predictions.
Thus, poly (vinyl carbazole) with the excess of holes
created by contact with nickel acts as an electron acceptor
rather than electron donor in contact with polycarbonate.

Based on our previous arguments, the cuticle is probably
more electron-donating or electron-accepting at the
cuticle-silicon oil junction in root-to-tip or tip-to-root
rubbing, respectively. In the former case, this should
create polarization of the junction with the excess of
electrons in the silicon layer. According to the
suggestion of Pochan et al. (13), polycarbonate probe
should thus acquire negative charges during rubbing against
silicon oil modified keratin which is contrary to what we
observe experimentally.

Another complicating factor, such as surface
contamination, should also be considered. Coagulation of
silicon emulsion on hair results in the formation of a thin
film of oil on the fiber surface raising the possibility of

its transfer to the probe during rubbing. Similar problem
was investigated by Gibson et al. (14). As a result of
transfer of silicon oil from the fiber surface to the
probe, the difference in work function between contacting
surfaces should be reduced, leading to a decrease of charge
generation in consecutive rubbings. As shown in Figure 8
first t-r rubbing of silicon coated fibers by a PC probe
generated positive charge density of $1.4 \cdot 10^9$ C/cm^2. In the
second and third consecutive test runs, the charge density
was reduced to $0.6 \cdot 10^9$ C/cm^2 and $0.02 \cdot 10^9$ C/cm^2,
respectively. This trend could be explained by the
transfer of silicon from the fibers to the probe. In r-t
experiments, however, we could not detect a similar
effect. Generated charge density was very high and
negative. Also, results obtained with PMMA probe do not
suggest the contamination of the probe. In addition to
this, we tested charge generation on silicon coated fibers
by a PC film doped with 10% silicon oil. As in the case of
pure PC films, generated charge was high negative for r-t
rubbing and low positive for t-r rubbing indicating that
incorporation of small amount of silicon oil in the probe
has no effect on the process of triboelectrification.
These results indicate that probe contamination cannot be
the explanation of the observed charging characteristics of
silicon modified fibers.

It is also difficult to account for charging
characteristics of silicon modified fibers by considering
r-t and t-r frictional coefficients. Measurements showed
that both were reduced proportionally with tip-to-root
frictional coefficient remaining higher than root-to-tip as
in the case of untreated fibers. Hence, only corresponding
reductions in friction-induced polarization, but not a
reversal, can be expected.

In summary, we have no good explanation for the observed
reversed directional triboelectric effect. It clearly
reflects the properties of the silicon oil-cuticle cell
junction but a detailed mechanism of the phenomenon remains
unclear in the light of existing interpretations of
triboelectric phenomena.

The explanation of the results of studies on the effect
of sequential rubbing on triboelectric charging is not
straightforward. Permanent polarization of the cuticles,
induced by frictional forces, could be a plausible
explanation of the observed dependencies. This argument is
strengthened by the existence of the

directional triboelectric effect and by the alleged
piezoelectric character of cuticles (1). To the best of
our knowledge, however, unequivocal connection between
piezoelectricity or pyroelectricity and triboelectric
charging has never been demonstrated in the literature.
Another possibility is to invoke a recently published
theory of triboelectrification of identical insulators
(6,7). It assumes the existence of electrons in localized,
high energy states and empty low energy states in polymer
solids. High energy electrons near the interface can make
transitions between identical solids bringing the electron
distribution closer to thermal equilibrium. The model
implies that the solid previously charged to high negative
potential would have tendency to charge positively in
subsequent contacts with the same material. This is,
however, not compatible with our observations that initial
positive or negative charging induces generation of the
charge of the same sign in subsequent contacts.

CONCLUSIONS

Triboelectric charge exchange studies between keratin
fibers and polymer probes have revealed the following
features of the process:

1. A directional triboelectric effect, as detected by
   polymer probes characterized by a work function close to
   that of hair keratin, is caused by the external layer of
   cuticle cells which renders the frictional properties of
   the fibers anisotropic, i.e. different in root-to-tip
   and tip-to-root directions. Triboelectric charging of
   the isotropic surface of the cortex, exposed by the
   removal of the cuticle cells, is not dependent upon the
   direction of rubbing.

2. The kinetics of triboelectric charging is affected by
   the mechanical or electrostatic history of the fibers.
   The mechanical factor can be explained by the
   polarization of cuticle cells resulting in a change of
   the effective work function of the fiber and induced by
   stretching or compressing during rubbing. The
   electrostatic factor is probably a consequence of the
   existence of non-equilibrium distribution of electrons
   injected by previous contacts.

3. The ability of thin silicon oil films to produce a
   reversed directional triboelectric effect. The origin
   of this phenomenon remains unclear.

4. Acid and base treatments impart electron-donating and electron-accepting character to hair fibers, respectively, although the effects are very small. Adsorption of a monolayer of cationic polymer PDMPDAMC, on the other hand, results in a significant decrease of the work function of the fibers. Deposition of a layer of cationic polymer PDMPDAMC and a layer of silicon oil demonstrates the additive effect of both treatments on the process of tribocharging.

REFERENCES

1. A. J. P. Martin, Proc. Phys. Soc. London 53 (2), 186 (1940).
2. J. Jachowicz, G. Wis-Surel and L. J. Wolfram, Text. Res. J., 54 (7), 492 (1984).
3. J. Jachowicz, G. Wis-Surel and M. Garcia, J. Soc. Cosmet. Chem., 36, 189 (1985).
4. D. K. Davis, Brit. J. Appl. Phys. (J. Phys. D.), 21533 (1969).
5. P. E. Shaw and R. F. Hanstock, Proc. R. Soc., 128, 474 (1930).
6. J. Lowell and W. S. Truscott, J. Phys. D:Appl. Phys., 19, 1273 (1986).
7. J. Lowell and W. S. Truscott, J. Phys. D:Appl. Phys., 19, 1281 (1986).
8. S. P. Hersh, P. L. Grady and G. R. Bhat, Pure and Appl. Chem., 53, 2123 (1981).
9. J. A. Mendley, Nature, 171, 1077 (1953).
10. W. E. Morton and J. W. S. Hearle, Physical properties of textile fibers, Butterworth, London, 1962, p. 505.
11. J. Jachowicz, M. Garcia and M. Berthiaume, Coll. Polym. Sci., 263, 847 (1985).
12. T. J. Fabish, M. M. Saltsburg and M. L. Hair, J. Appl. Phys., 47, 930 (1976).
13. J. M. Pochan, H. W. Gibson, F. C. Bailey and D. F. Hinman, J. Electrostat., 8, 183 (1980).
14. H. Gibson, J. Pochan and F. Bailey, Analyt. Chem., 51 (4), 483 (1979).

MOLECULAR AND SUPRAMOLECULAR ORIENTATION IN

CONDUCTING POLYMERS

M. Aldissi

Los Alamos National Laboratory
P.O.Box 1663
Los Alamos, NM 87545, USA

Abstract

Intrinsic anisotropy in electrical and optical properties of conducting polymers constitutes a unique aspect
that derives from π-electron delocalization along the
polymer backbone and from the weak inter-chain interaction. To acquire such an intrinsic property, conducting
polymers have to be oriented macroscopically and microscopically (at the chain level). A review of the various
techniques, including stretch-alignment of the polymer
and of precursor polymers, polymerization in ordered
media, ie, in a liquid crystal solvent, and synthesis of
liquid crystalline conducting polymers will be given.

Introduction

The initial polymeric systems discovered about ten
years ago as conducting materials lacked processability
and easy manipulation and therefore, the field was
materials-limited and the use of processing techniques
known in the case of conventional flexible or rigid chain
polymers was not attempted. It is only in the last few
years that a renewal of efforts and redirection in
research in the area of conducting polymers has occurred.
These efforts focused on (i) technologically important
areas such as processability, orientation and stability;
and (ii) fundamentally crucial aspects to the development
of new systems that consist of defining the various nonlinear excitation states responsible for conduction and a
good understanding of the charge storage mechanisms. A
recent review [1] discusses various aspects including a
materials survey, processability and stability aspects

and electronic states responsible for conduction in the
various doped and undoped materials. Another recent
review [2] discusses some of the potential application
areas. The classic picture of the morphology observed in
most as-grown polymer films has been characterized by
randomly oriented fibrils. This generates many inter-fibril contacts yielding various conduction pathways and
creating barriers for the charge carriers to move freely
in the system. Orientation of conducting polymers is a
very important aspect of the development in this field
and is being investigated actively. It is closely related
to processability and responsible for introducing intrinsic anisotropy in the optical and electrical properties
of conducting polymers. Anisotropy in these systems
derives from the existence of an extended π system on
parallel chains and a weak inter-chain interaction. The
high conductivities that resulted in some cases represent
a strong motivation to pursue efforts in this direction.
Therefore, this paper is devoted to orientation of polymer aggregates and chains, and discusses the various
techniques and materials used for such a purpose.

Stretch-Orientation

Polyacetylene films whose morphology consist of
randomly oriented fibrils are commonly obtained by
Ziegler-Natta catalysis as shown in scheme 1.

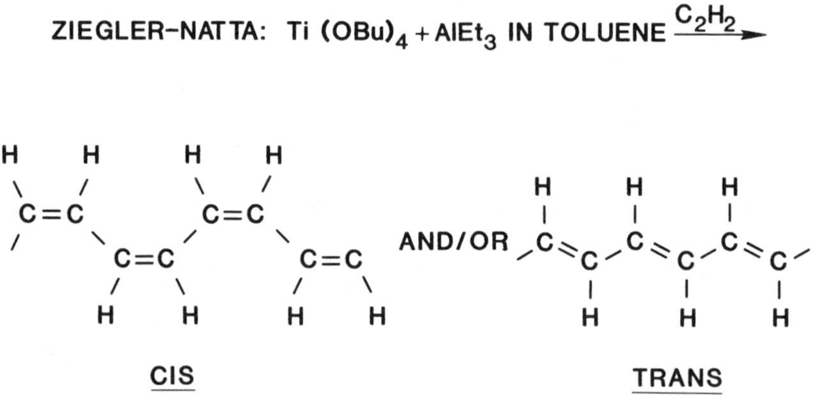

ZIEGLER–NATTA: Ti (OBu)$_4$ + AlEt$_3$ IN TOLUENE $\xrightarrow{C_2H_2}$

Scheme 1

Stretch-orientation was the first attempt to aligning polyacetylene fibrils [3,4]. The technique consisted of an initial drawing of cis-polyacetylene films, synthesized at -78°C. The drawing is then continued at increasing temperatures up to 150°C. A modest draw ratio of 3 could be obtained using this process. The resulting morphology of the film exhibited a somewhat oriented material with partially aligned fibrils along the drawing axis. Polarized IR reflectance studies showed that the reflectance parallel to the orientation direction as a function of frequency behaves similarly to semiconductors while the perpendicular component behaves like an insulator, and therefore, the material exhibits an intrinsic optical anisotropy [5]. The anisotropy is maintained after doping. Electrical anisotropy also resulted from orientation and was translated by an increase in conductivity of the heavily doped samples with AsF_5 along the orientation direction. The $\sigma\|/\sigma\perp$ ratio is higher than 10 and the maximum $\sigma\|$ obtained was approximately 2.8×10^3 S/cm. An improved synthetic technique of highly oriented polyacetylene films developed at BASF (Ludwigshafen, W. Germany) has been reported recently [6]. The method consisted of a thermal treatment of the Ziegler-Natta catalyst and drawing was realized during the polymerization reaction rather than a post-polymerization orientation. Improved optical anisotropy compared to the previous case is obtained, but the most impressive result is that of the electrical anisotropy. The highest Í ever obtained is approximately 1.5×10^5 S/cm which is only 5 times less than that of copper. With Í of 100 S/cm this constitutes the highest electrical anisotropy ever recorded on conducting polymeric systems ($\sigma\|/\sigma\perp \approx 1.5 \times 10^3$). The resulting material is highly dense and consists of almost perfectly aligned fibrils. Because the polymer chains are oriented along the fibril axis in polyacetylene films, the orientation of the chains in BASF polyacetylene must be extremely high. Although, high chain orientation is obtained the crystallinity of the polymer as a whole is almost unchanged [7]. Transparent polyacetylene exhibiting a parallel conductivity of 5000 S/cm was also prepared at BASF using the same catalyst treatment, but the polymer was formed on a thin polyethylene film that was stretched and doped afterwards [6]. The enormous increase in conductivity, as justified by BASF authors is due to reduction of the sp_3 defects to a minimum or to the total absence of such defects by a careful aging of the catalyst.

Precursor Polymers

The Durham route to polyacetylene was the first technique that proceeded via a precursor polymer [8]. This technique offers the advantage of solution processing at the precursor polymer level which then by thermal conversion yields trans-polyacetylene. The reaction sequences are shown in scheme 2.

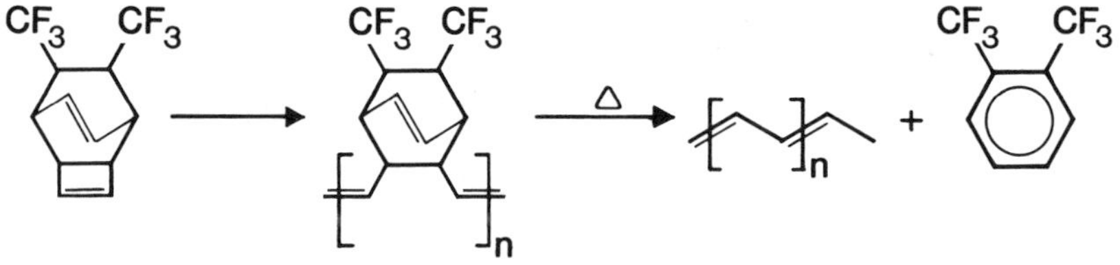

Scheme 2

It has been found that highly oriented dense films of polyacetylene can be obtained by stretching the precursor polymer during the thermal conversion and that the morphology of these films can be controlled by varying the conditions of the transformation step. Draw ratios up to 20 can be achieved [9,10]. This orientation leads to longer straight-chain sequences and motionally narrowed spins rather than immobile spins characterized by a broad esr line in the unstretched polymer [11]. Although, high draw ratios are achieved, conductivities similar to those of the initially stretched Ziegler-Natta polyacetylene (draw ratio $\approx$ 3) are obtained on the doped material. However, the intrinsic optical anisotropy is very high. In fact, the optical properties of the doped oriented Durham-polyacetylene are similar to those of a one-dimensional metal [12,13]. For example, the parallel reflectance is that of a metal and the perpendicular one is that of an insulator.

Poly(phenylenevinylene) was prepared using the precursor technique [14,15]. The technique consists of pyrolysis of the a water-soluble precursor polymer of a

sulfonium salt. The reaction sequences are shown in
scheme 3.

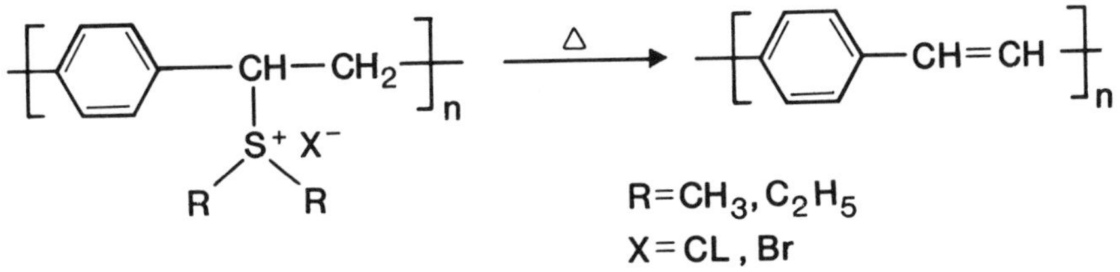

$$R=CH_3, C_2H_5$$
$$X=CL, Br$$

Scheme 3

Draw ratios up to 10 of the precursor polymer during the
thermal conversion can be achieved resulting in a highly
oriented polymer. The highest conductivity achieved on
the H_2SO_4-doped oriented polymer is approximately $4x10^3$
S/cm and the electrical anisotropy is approximately 100
[16]. The optical anisotropy is also high. It is however,
lower than that of stretched Durham-polyacetylene.

Polymerization in Ordered Media

The polymerization of acetylene in an ordered solvent,
ie, a nematic liquid crystal, has become an area of
interest. This method differs from the previous ones in
that the orientation exists already at the molecular
level. For example, a certain degree of order of the
active sites of the Ziegler-Natta catalyst $Ti(OBu)_4$-
$Al(Et)_3$ can be obtained if dissolved in a nematic liquid
crystal that is subjected to a magnetic field as shown in
scheme 4.

The first polymerization attempt [17-19] was made
using N(p-methoxybenzylidine)p-butylaniline (MBBA) which
has a nematic range of 19-30°C in a magnetic field ≥2500
gauss. The polymerization of acetylene using a catalyst
concentration of 0.1 mol/l was carried out under a
magnetic field of 4000 gauss. In these conditions the
liquid crystal molecules are oriented with respect to an
externally defined axis along the catalyst molecules.
This process resulted in a noncrosslinked polymer with a

Catalyst Active Sites, Ti Ⅲ

Oriented Liquid Crystal Molecules

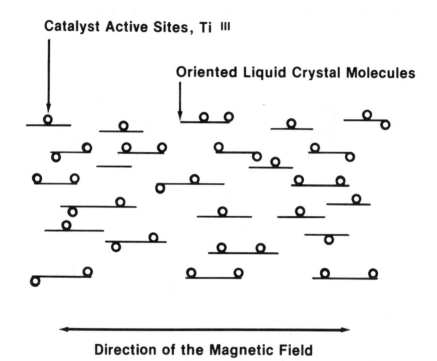

Direction of the Magnetic Field

<u>Scheme 4</u>

minimum of sp_3 defects on the conjugated chains. The
electrical anisotropy was higher than 4 with $\sigma\|$ of
approximately 2×10^3 S/cm for the iodine-doped polymer.
The X-ray diffraction patterns indicated that a preferred
orientation compared to conventional stretched or
unstretched polyacetylene is obtained. The morphology, as
seen by SEM, showed some oriented features and a more
regular structure than in conventional polyacetylene.
These results indicated that orientation at the molecular
level is possible, ie, initiator and propagating species.
This led to considering the alignment of the liquid
crystal itself which is supposed to be perfectly aligned
under the experiment conditions. Adding an impurity to a
liquid crystal is known to disturb its order with or
without a magnetic field, particularly if the volume
fraction of the impurity is $\geq 10\%$. Since the catalyst in
this case acts as an impurity, the alignment of the
medium and therefore of the polymer can not reach its
maximum. Therefore, a smaller amount of catalyst than in

previous experiments, $\leq 10\%$ volume fraction, was used for the polymerization. The higher order of the solution was translated by, at the same magnetic field, by an improved intrinsic anisotropy. In fact, with a $\sigma\|$ of approximately 10^4 S/cm and a $\sigma\|/\sigma\perp \approx 50$ were obtained on the AsF$_5$-doped polymer.

Polyacetylene films synthesized using this technique can be stretched mechanically in a similar manner to that used for conventional polyacetylene. <u>The stretch-orientation (draw ratio $\approx$ 1.5-2) combined with the neat process that involves polymerization in a nematic liquid crystal</u> resulted in an increase of $\sigma\|$ of the AsF$_5$-doped film by approximately 2.5 times, up to 2.5×10^4 S/cm. With $\sigma\perp \approx 10^2$ S/cm, a high electrical anisotropy is obtained ($\sigma\|/\sigma\perp \approx 250$). This result can be explained as follows:

O   A high orientation of the polymer chains is achieved via the oriented liquid crystal molecules which also leads to a high orientation of the fibrils with respect to the externally applied magnetic field.

O   Stretching enhances the orientation of the fibrils by aligning those that were not formed in the direction of the field and increases the density of the films, thus reducing the inter-fibril resistance.

Other experiments followed using 4-(trans-4-n-propylcyclohexyl)-ethoxybenzene or -butoxybenzene as the nematic liquid crystal [20]. Parallel conductivities of 10^4 S/cm and electrical anisotropies ($\sigma\|/\sigma\perp$) of approximately 50 were obtained. In other experiments [21], using the same liquid crystals, the polymerization of acetylene was carried out under a flow of the nematic catalyst solution in the absence of a magnetic field. The polymer obtained in these conditions was oriented but exhibited lower conductivities and anisotropies than those obtained on the films synthesized under the influence of a magnetic field.

By combining the two techniques, flow properties and use of a magnetic field of 47000 gauss with the latter two solvents, Rolland et al. [22] were able to reach conductivities of $2-3\times10^4$ S/cm and a morphology that consists of almost perfectly aligned fibrils. In fact, X-ray diffraction studies showed that the mosaic spread of the fibrils is approximately 20°. Electron diffraction studies on thin samples prepared on a grid showed that

the mosaic spread of the microfibrils that form the
fibrils seen by SEM is 7°. The results of Rolland et al.
indicate that the orientation of the polymer chains is
not perfect and that there is room for improvement. The
optical anisotropy is also very high due to the high
order of the polymer chains.

Conducting Liquid Crystal Polymers

Liquid crystal polymers constitute a very important
class of material for their modulus strength and therefore for their structural applications. This class of
material represents an advantage in that a certain order
exists already in the polymer structure and exhibit
various ordered phases. Structural order is therefore
what makes it different from the polycrystalline or
amorphous conducting polymers that possess a fibrillar
morphology. Combining liquid crystalline properties and
high conductivities is a very exciting approach to making
conducting polymers. 3-Alkylsulfonate and 3-alkylcarboxy-
late derivatives of thiophene or pyrrole can be polymerized electrochemically yielding water-soluble polymers
that are intrinsically conducting due to self-doping by
the polar group or the soap molecule [1,23]. The synthesis reactions of monomers and details of the polymerization are mentioned elsewhere [24]. The general formula
of such compounds is shown in scheme 5.

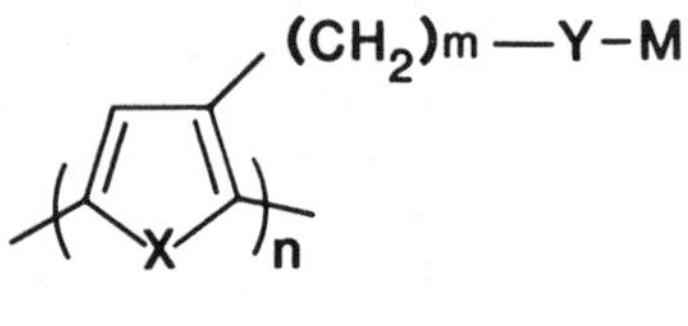

$$(CH_2)m-Y-M \qquad X = S, NH$$

$$Y = SO_3, SO_4, CO_2$$

$$M = H, Li, Na, K, etc.$$

Scheme 5

When $m \geq 10$ a liquid crystalline polymer in the lyotropic
phase can be obtained due to the presence of the polar
group pending from the main backbone which is rigid by
definition.

The possibility of phase separation and ordering in solutions of rigid rodlike polymeric particles has been thoroughly invesigated in theory. For example, Flory [25] demonstrated, by application of the lattice model for polymer solutions, that concentrated solutions of rodlike particles should show phase separation even in the absence of interactions between rods.

The behavior of these polymers is similar to that of fatty acid salts in water. The latter type of molecule was studied by various authors [26-28]. Several equilibrated mesophases exist in the lyotropic solutions that could consist of isotropic micellar solutions and other more ordered phases. These phases exist because of the amphiphilic property of the substituted thiophene repeat unit which has a hydrophilic polar group and a hydrophobic hydrocarbon molecule composed of unsaturated and saturated components. <u>Because the thiophene backbone is of a rigid nature, the mesophases could consist of uniformly elongated cylinders of uniform diameter or micelles arranged in parallel sheets with solvent in between or linear aggregates of spherical micelles</u>. Small angle neutron scattering and electron microscopy studies will be undertaken to study the various phases in the various systems and correlate their morphologies with transport properties. Over-all conductivities similar to those with shorter alkyl groups are obtained. The high conductivity is maintained when m is increased due to the high conjugation of the chain and therefore the highly delocalized character of the electrons. This is the result of the absence of interference of the alkyl group with the chain planarity. Conductivities of approximately 10^{-2} S/cm are obtained on films cast from their aqueous solutions, which can be increased by 3-4 orders of magnitude upon doping with H_2SO_4 or AsF_5. Also, doping of the polymer can be performed in solution in water. Homogeneous films can be obtained by evaporation of the solvent. When the sulfonic acid polymers are oxidized, a loss of a proton occurs and a self-doped polymer is obtained. The sulfonate ion (SO_3^-) acts as the counter anion to the polycation which is the polythiophene backbone.

Conclusions and Outlook

The progress made in the area of conducting polymers within the last 2-3 years with respect to manipulating their morphology and introducing some order in the structure is impressive. With a conductivity almost that of copper and the high electrical and optical aniso- tropies obtained in various cases, conducting polymers are becoming more and more the materials of choice for near-future applications.

Stretch-orientation of the polymer or of precursor polymers provides an orientation of the macroscopic aggregates, ie, fibrils, in which the chains are believed to be oriented along the fibril axis.

Polymerization in ordered media under the influence of a magnetic or electric field constitutes a much more exciting approach to perfection of the chain or fibril alignment since order is introduced already at the initiation reaction level and that order is maintained during polymerization by maintaining the field. Also, an increase in order of the already oriented films is obtained by mechanical stretching. <u>One possibility for the improvement of the over-all orientation involves a thermal treatment history similar to that used at BASF and the use of the flow properties of a nematic liquid crystal under a magnetic or electric field</u> as the poly- merization solvent. Use of low polymerization tempera- tures by choosing the appropriate liquid crystal is also a parameter of importance for high order. Such an approach should result in a perfectly or almost perfectly oriented polymer which is totally defect-free and whose conductivity should exceed that of copper on a volume and weight basis.

Liquid crystalline polymers that are synthesized in their intrinsically conducting form represent a new class of material that combines two technologically important properties which are strength and conduction. Order exists here in the as-synthesized polymer. Due to their solubility in water, the study of the single chain behavior and the inter-chain interaction in functional- ized poly(alkylthiophene)s becomes quite possible and therefore it opens up a number of possibilities for the study of conducting polymers in general.

There are other ways for orienting conducting polymers. For example, epitaxial growth of polyacetylene is possible on single crystals such as biphenyl [29]. However, the orientation is limited to the first layer or few layers of the polymer. Also, fiber spinning of a soluble conducting polymer could be achieved from a blend solution of the conducting polymer and a conventional polymer that is known for its fiber spinning properties. This could be very useful for fiber formation, particularly, when strong, highly crystalline materials are used. This technique would result in conducting polymer composites of a high modulus strength.

The interest in oriented conducting polymers is in part due to the increase in the extent of the uninterrupted π-electron delocalization for which the length of the straight-chain sequences is maximized. This leads to the possibility of studying the intrinsic anisotropic properties of the polymers which can provide detailed information on the microscopic nature of transport mechanisms and the importance of inter-chain and intra-chain interaction for transport.

Acknowledgment

This work is supported by the Center for Materials Science of Los Alamos National Laboratory and the Office of Basic Energy Sciences (DOE).

References

[1]   Aldissi, M. J. Mat. Education, in press.
[2]   Aldissi, M. Proceedings of 1st European Polymer Symposium, 14-18 Sept., 1987 (Lyon, France).
[3]   Shirakawa, H.; Ikeda, S. Synth. Met. 1979/80, 1, 175.
[4]   Aldissi, M. Ph.D. Thesis, July, 1981, Montpellier, France.
[5]   Fincher, Jr., C. R.; Moses, D.; Heeger, A. J.; MacDiarmid, A. G. Synth. Met. 1983, 6, 243.
[6]   Naarmann, H. Synth. Met. 1987, 17, 223; ACS Spring Meeting, April, 1987, Denver, CO, USA.
[7]   Heeger, A. J. Workshop on Low-Dimensional Organic Conductors, June 30 - July 3, 1987, Trieste, Italy.
[8]   Edwards, J. H.; Feast, W. J. Polym. Commun. 1980, 21, 595.
[9]   White, D.; Bott, D. C. Polym. Commun., 1985, 202.
[10] Friend, R. H.; Bradley, D. D. C.; Pereira, C. M.; Townsend, P. D.; Bott, D. C.; Williams, K. P. J. Synth. Met. 1985, 13, 101.
[11] Horton, M. E.; Friend, R. H. Synth. Met. 1987, 17, 395.

-[12] Leising, G. Polym. Bull. 1984, 11, 401; Polym.
       Commun. 1984, 25, 201.
[13]   Leising, G. Workshop on Low-Dimensional Organic
       Conductors, June 30 - July 3, 1987, Trieste, Italy.
[14]   Murase, I.; Ohnishi, T; Noguchi, T.; Hirooka, M.;
       Murakami, S. Mol. Cryst. Liq. Cryst. 1985, 118,
       333.
[15]   Karasz, F. E.; Capistran, J. D.; Gagnon D. R.;
       Lenz, R. W. Macromol. 1985, 17, 1025.
[16]   Murase, I.; Ohnishi, T.; Noguchi, T.; Hirooka, M.
       Synth. Met. 1987, 17, 639.
[17]   Aldissi, M. Symposium on Order in Polymeric
       Materials, 25-26 August, 1983, GTE Laboratories,
       Waltham, MA, USA.
[18]   Aldissi, M. et al. BES/DMS Annual Review, 26-27
       June, 1984, Los Alamos National Laboratory, Los
       Alamos, NM, USA.
[19]   Aldissi, M. J. Polym. Sci., Polym. Lett. Ed. 1985,
       23, 167.
[20]   Akagi, K.; Katayama, S.; Shirakawa, H.; Araya, K.;
       Mukoh A.; Narahara, T. Synth. Met. 1987, 17, 241.
[21]   Araya, K.; Mukoh, A.; Narahara T.; Shirakawa, H.
       Synth. Met. 1986, 14, 119.
[22]   Rolland, M. et al. Proceedings of 1st European
       Polymer Symposium, 14-18 Sept., 1987, Lyon, France.
[23]   Patil, A. O.; Ikenoue, Y.; Wudl F.; Heeger, A. J.
       J. Am. Chem. Soc. 1987, 109, 1858.
[24]   Aldissi, M. U.S. Patent Application, DOE Case #
       S-65,735.
[25]   Flory, P. J. in "Principles of Polymer Chemistry";
       Cornell University Press: Ithaca, New York, 1953.
[26]   Luzzati, V.; Husson, F. J. Cell Biol. 1962, 12,
       207.
[27]   Luzzati, V.; Mustacchi, H.; Skoulios, A.; Husson,
       F. Acta Cryst. 1960, 660, 668.
[28]   McBain, J. W.; Marsden, S. S. Acta Cryst. 1948, 1,
       270.
[29]   Woerner, T.; MacDiarmid, A. G.; Heeger, A. J.
       J. Polym. Sci., Polym. Lett. Ed. 1982, 20, 305.

Part III
Polymers for Advanced Structures

POLYMERS FOR ADVANCED TECHNOLOGIES

H.F. MARK
Polytechnic University
333 Jay Street, Brooklyn, N.Y.

1. Accomplished Facts

The Science of Macromolecules had its origin in the study of important and indispensible natural materials: cotton, wool, silk, leather, keratin, rubber and wood resins. At the end of the 1920's numerous classical theoretical and vigorous exchange of opinions firmly established the concept of linear chains with molecular weights of more than hundred thousands. It immediately encouraged attempts to prepare compounds of similar size and structure and resulted within a few years in the existence of a variety of synthetic polymers, polyvinyls, polydienes, polyesters and polyamides.

Naturally these early synthetics were put to practical uses which were similar to those of their natural models and precursors - fibers, films, rubbers and plastics. Rayon and cellulose acetate, the first commercial "man-made" fibers still used cellulose as their polymeric base but soon polyamides, polyesters and several addition polymers were developed into an enormous variety of "fully" synthetic fibers which dominate today's textile industry and provide for additional improvements and expansions in the foreseeable future. Cellulose nitrate, cellulose acetate and cellophane were the classical precursors of film forming materials for photography, graphic and packaging; they were soon followed by a plurality of polyolefins, vinyls and polyesters which gave these industries new and unexpected dimensions with numerous options for the future. Natural rubber served as an exceedingly successful model for an entire generation of synthetic elastomers which, today, extend the range of rubberiness from 200° down to -70°C and exceed their natural prototype in stability against radiation and chemical agents by orders of magnitude. Finally, wood, rosin and leather were followed up in the domains of their uses by a world of thermoplastic and thermosetting synthetics which pervade our daily life and the trades of home building, transportation, communication and health care.

Whereas, in this manner, synthetic polymers soon achieved a dominating role in a large sector

of useful building materials, they did little to
compete or cooperate with the two large classical
building materials of the past: metals and ceramics.

2.   Trends
a)   Mechanical and Thermal Behavior

The most evident differences between normal organic polymers on one side and metals and ceramics
on the other side are hardness, tensile strength
and softening range (melting point).

Table 1 shows the tensile moduli, tensile
strength and softening ranges of steel, titanium,
aluminum, silica and glass together with those of
typical plastics as they were commercially available in the 1960's and 1970's.   The differences are
very large and make it evident that only new principles and novel technologies could be able to
bridge the gap.   They are now existing in the presence of rigid chain macromolecules and their use
as reinforcing elements in matrices of thermosetting polyaromatic systems.

Table 2 presents the moduli, tensile strengths
and softening ranges of several modern polymeric
fibers which show that their values overlap with
those of Table 1.   Their availability encouraged
curing the last two decades research and development of high performance composites of low specific
gravity which were able, in many instances, to perform favorably in comparison to the much heavier
and much less corrosion resistant metals. Polyaro-
metic thermosetting systems such as polyimides,
polystyrylpyridines and others are reinforced with
aramides or carbon fibers and are manufactured into
a large variety of objects - rods, plates, pipes,
wheels, sheets and bolts - by such new processing
techniques as filament winding (FW), prepregging,
pultrusion and reaction injection molding (RIM).
These new materials were first tailored and tried
for space vehicles, rockets and military aircraft
where they found an adequate and demanding proving
ground for their design and uses.   Presently they
invade on a large scale the construction of airplanes, boats, ships, railroad cars, buses, trucks,
automobiles and home and factory building.   Compared with metals and ceramics the conditions under
which they are produced and processes and also those

under which they may be recycled, require much less energy and time resulting in a much lower insult on the environment. The weight reductions in comparison to the "all metal" precursors - up to 50% in tracked and untracked vehicles and up to 15% in airplanes - permits either a higher payload or a considerable saving in fuel which, in turn, is favorable for the environment.

The introduction of high performance lightweight composites is still in its infancy; design and processing, in many cases, are still tentative and there exists, as yet, little reliable information on such phenomena as aging and fatiguing by mechanical or chemical processes under characteristic environmental conditions. However, the general experience is, at present, highly encouraging; many additional simplifications and refinements are within reach and it is highly probable that these new composites will become a permanent family of building materials with a very wide range of useful applicability.

b) Electrical and Optical Behavior

Traditionally organic substances, including organic polymers, are insulators: they may be non-polar such as gasoline or polyethylene, polar such as sugar or nylon and polarizable such as benzene or polystyrene - but they do not possess any "free" electrons, which could act as carriers for an electric current. But there exist two systems of organic compounds which exhibit considerable electronic conductivity: polyconjugated chain molecules and electron donor-acceptor complexes. The conductivity of intrinsically insulating polymers such as polyacetylene, polyparaphenylene poly-para-phenylene sulfide, polypyrrol and several of their derivatives can be enhanced by about 10 to 15 orders of magnitude right into the metallic or semiconducting range by the addition - doping - of electron donors (alkali metals and organic bases) or acceptors (inorganic and organic acids such as AsF_5, F_3 acid, etc.). These complexes are presetly of great interest because they offer the promise to combine metallic and semiconducting characteristics with the plastic and elastic properties of organic polymers. This could, in principle, lead to sequences of N - P junctions along

the length of very thin and flexible filaments
and to paper thin electrodes of rechangeable storage batteries with a variety of voltages and with
a power to weight efficiency of at least ten times
that of the commercial lead accumulators.

The realistic attainment of these intriguing
prospects is presently pursued by a large number
(probably several hundreds) of scientists and
engineers in several of the leading industrial and
university laboratories. There exist, right now,
two factors which are preventing smooth and rapid
progress in this highly attractive and exciting
field. One is the missing, or at best, incomplete
theoretical understanding of the phenomenon, i.e.
the mobility of current carrying electrons of
"holes" in an organic chemical environment, the
other is the relative instability of the various
"doped" polyconjugated systems. Well organized
and vigorous efforts are now under way to make
progress in both directions but it would be difficult to estimate how soon or how late one may expect that practical applications of these new
phenomena will be forthcoming.

A second, different way to arrive at metallic
conductivity of organic substances is through the
stacking of electron donating and electron accepting compounds. Organic electron donors are mainly
ammonium-sulfonium- and phosphonium bases which
contain condensed aromatic systems such as quino-
linium or thiazonium; they are flat, disc-shaped
molecules with strong polarizability having one
or more polar groups - CO, OH, NH, etc. Electron
acceptors are highly conjugated condensed aromatic
molecules which offer to a donated electron many
energy levels through dislocation and resonance.
Since these donors are also flat and disc-like
systems the complex formations leads to quasi
cylindrical stacks which form small but well developed crystals - usually deeply colored. It was
found that these crystals have an extremely anisotropic conductivity: very high (up to 10^4 reciprocal ohm cm) in the direction of the axis of the
stack and very low (down to 10^{-10} ohm cm^{-1}) perpendicular thereto. Any individual donor molecule

D in such a column is flanked ny two acceptor units
to one of them it is bonded by relatively strong
complexing polar bonds, hydrogen bridges and re-
sonating groups; as a result the "intracomplex"
distances between the two interacting components
are small - usually between 0.23 and 0.26 nanometers. The other acceptor belongs to the next
"sandwich", its distance from the donor D is larger (above 0.28 nm up to 0.35 nm). It has been
found that any factor which diminishes this intercomplex distance - external pressure - or appropriate molecular structure of the two partners of
the complex - increases the conductivity. The
ideal case would apparently be if the intercomplex
distance would be equal to the intracomplex distance; the donor-acceptor column would then possess optimal electron transfer conditions and the
highest conductivity. In fact typical metallic
character and even some type of supraconductivity
have been observed in some of these systems.

Interesting optical properties of certain polymers are known since the days of Polaroid when it
was found that highly oriented linear macromolecules - f.i. polyvinylalcohol - are capable to
give complexes with iodine and organic dyestuffs
which are strongly birefringent and dichroic.
They have found wide applications as polarizers in
microscopes, cameras and sunglasses and are calling
for permanent improvements concerning the degree
of anisotropy and environmental stability.

Another optical use of highly transparent, soft
and flexible polymeric gels - f.i. slightly crosslinked polymethacrylates - has led to the development of very useful contact lenses which are convenient to wear, cannot crack or fracture and represent an important contribution of macromolecules to medicine and health care. Also in this
discipline efforts for further improvements and
simplifications are still going on.

It appears that there will be another use of
synthetic polymers, particularly homo- and copolymers of acrylic and methacrylic systems, in
a new field of technology namely in fiber optics
and light telephony. Fibers or wires of a highly
purified silica composition are already in use
for signal transfer - telephone and TV - over

large distances such as Boston-Washington and Sac-
ramento-San Diego. In these optical cables the
"core" has to be covered with a "cladding" which
reduces the losses of the message carrying light
beam and permits to cover large distances with the
aid of this new method which offers many important
advantages over the classical signal transfer with
electric currents. It appears that for the composition of these claddings synthetic polymers of
high transparency and adequate refractive index
are of interest and it was also found, that for
shorter distances - e.g. inside of a city or or a
large building (bank, school, hospital) - even
the core of the lightcable could be made of an
appropriate synthetic polymer.

There are a few other promising trends but they
are still so much in their infancy that it would
be difficult at this time to forecast their growth
and their ultimate commercial success.

One is the design of radiation resistant thin
films made of poly-para-xylene, polyimides or poly-
phenylhydrazides for the construction of solar
sails and solar space power stations. The other
is the development of films with unusual basic
characteristics for oxygen such as poly-vinyls and
copolymers of ethylene and vinylacetate. They
could be of interest for the tendencies to replace
metals and glass to produce impermeable cans and
other containers for food packaging.

TABLE 1
Mechanical and Thermal Properties of some Metals,
Ceramics and Classical Organic Polymers in psi[*]
and °C

Material	Range of tensile modulus	Range of tensile strength	Temperature of softening or decomposition
Steel	60-100 x 10^6	6 - 7 x 10^4	1553
Titanium	16-20 x 10^6	up to 8 x 10^4	1660
Aluminum	11-12 x 10^6	up to 8 x 10^3	680
Glass	up to 20 x 10^6	up to 12 x 10^3	1200
Bakelite	up to 2 x 10^6		350
Polyamides	2 - 4 x 10^5	up to 12 x 10^3	300
Polycarbonate	3 - 5 x 10^5	up to 11 x 10^3	300

[*] To convert psi approximately into other
scales use

1 kg per cm^2 (atm) = 14.22 psi

1 MPa (Megapascal) = 142.2 psi

TABLE 2
Mechanical and Thermal Properties of some
Polymeric Materials recently developed in psi and
°C

Material	Range of tensile modulus	Range of tensile strength	Temperature of softening or decomposition
Carbon fiber high modulus	80×10^6	2.8×10^5	above 3000
Carbon fiber high strength	40×10^6	up to 4.1×10^5	above 3000
Al_2O_3 fiber	70×10^6	1.1×10^6	above 2300
Aramid 49 high modulus	up to 20×10^6	around 1.0×10^5	around 500
Aramid 29 high strength	up to 8×10^6	around 1.0×10^5	around 500
Superdrawn Polyethylene	4×10^6	up to 0.7×10^5	120

1. Mechanical and Thermal:
      John Delmonte;   "Technology of Carbon Fiber
                        Composites"
                        Van Nostrand, New York   1981

      A.E. Zachariades and
      R. Porter;   "Strength and Stiffness
                       of Polymers"
                       Marcel Dekker,Inc.
                       New York 1983

2. Electrical and Optical:
      Marian Kryszewski;
                       "Semiconducting Polymers"
                       Polish Scientific Publishers
                       Warsaw 1980

      James C.W. Chien;
                       "Polyacetylene"
                       Academic Press, New York
                       1984

      St. E. Miller
      and A.G. Chynoweth;
                       "Optical Fiber Telecommuni-
                       cations: Academic Press
                       New York 1979

3.   Barrier Properties:
      Leonard B. Ryder;
                       Plastics Engineering
                       May, 1985 - pages 41-48

      All comprehensive treatises mentioned above
      contain extensive references to the original
      literature.

POLYMERS FOR ADVANCED STRUCTURES - AN OVERVIEW

David Tanner[1], Vlodek Gabara[2], John R. Schaefgen[3]

E. I. du Pont de Nemours & Co., Inc.
1) Textile Fibers Dept., Nemours Bldg., Wilmington, DE 19898;
2) "Kevlar" Research and Development Laboratory,
Spruance Plant, Richmond, VA 23261
3) Pioneering Research Laboratory, Experimental Station,
Wilmington, DE 19898

Abstract

Significant progress in synthesizing and processing organic polymer molecules has led to an ever growing materials base. These developments, along with theoretical advances, have led to increasing knowledge of structure-property relationships, especially of the role of molecular chain extension and orientation in fibers, of structure in thermally stable polymers, of elasticity in thermoplastic elastomers, and of the mechanism of energy absorption in heterogeneous phase toughened plastics. Therefore, polymers, copolymers and blends with predictable property levels now can be designed and made. The newest step in this ladder of increasing sophistication is the creative integration of this knowledge into designing of systems wherein tailoring of the properties of materials is extended to advanced final structures. This has led to many important applications, eg, in aircraft, aerospace, marine, automotive and ballistics where lightweight, energy efficient structures provide the impetus for radically new designs and performance. This paper will overview advances in polymer science that have enabled development of high performance materials, describe the systems approach to applications, and preview future trends.

1. Introduction

1.1. <u>Human needs and the growth of polymer science.</u> The history of mankind is a tale of a continuing quest to find better ways to meet human needs. In this process new materials constantly replace old as obsolescence takes its toll and the efficiency by which basic needs are met is improved. The progress of this substitution process is driven by advances in science and technology which can be pictured as occurring in a series of S-shaped curves or waves (Fig. 1). In the earlier part of this century, we began to replace naturally occurring materials such as metals, wood, and fibers with synthetic polymeric materials derived from the newly abundant resource - petroleum. This substitution was aided by an explosion of knowledge in polymer science as recorded by H. Morawetz in his recent book [1] which traces the progress of our science to 1960. In his fascinating book, Dr. Morawetz begins with Berzelius's proposal of polymerization reactions, discusses H. Staudinger's uphill battle for acceptance of polymers as high molecular weight molecules, and continues through P. J. Flory's monumental work on all phases of the physics and chemistry of polymers summarized in his classic book "Principles of Polymer Chemistry" [2]. This period also witnessed the growth of a large materials base in synthetic vinyl, diene and condensation type polymers. These were eagerly accepted as replacements for natural materials in predominantly non-structural applications such as fibers for apparel and home furnishings, film for food wrapping, and plastics and thermosets for toys, containers, and castings. Dr. Morawetz aptly ends his book in 1960, which might be considered the end of the steep portion of the first wave of progress during which the foundations of polymer science were laid.

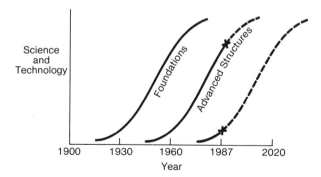

Figure 1. Polymer growth periods.

1.2. <u>Polymers for advanced structures</u>. Dr. H. Mark
recognized that we were positioned to enter a new period in
polymer science when he wrote the following preamble to a
volume of works [3] covering the 1956 IUPAC Symposium in
Israel: "Now that the entire content of the Symposium is
contained in one reasonably handy volume, it can be seen that
it actually represents a complete and critical picture of our
knowledge on macromolecules in solution, and it can be hoped
that starting on this platform new significant progress will
be made in the near future". And indeed it has!

The 1960's represent a turning point in the history of
polymer science. During this period we made the transition
from the foundation phase of polymer science to a period in
which major progress was made in design of polymers for
advanced structures. For the purpose of this paper, we
define "advanced structures" as systems of materials which
meet unusually high performance requirements in, eg,
aircraft, aerospace, automotive, ballistics, and protective
clothing applications.

In the last three decades, breakthroughs have been made
in our understanding of how to design organic polymer
molecules to achieve desired properties and how to process
polymers into preferred morphologies. The result has been a
new generation of polymeric materials providing new levels of
strength, stiffness, environmental stability, and load
bearing capability coupled with elasticity and/or energy
absorption. During this period polymer usage has been
extended from non-structural to structural applications.
Terms like "high performance fibers", "engineering plastics",
and "advanced composites" have become part of our vocabulary.

Polymers for advanced structures have been growing
rapidly in new commercial applications. A clue to what is
going on today and where we are headed is seen in Fig. 2 in
which substitution of structural metals by fiber reinforced
plastics and engineering resins is plotted. The trend
started around 1960, and a 20% replacement already has
occurred.

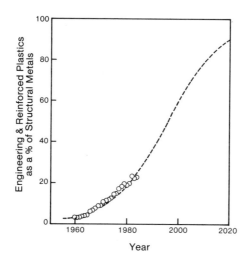

Figure 2. Substitution of Metals by Plastics (volume basis).

1.3. <u>The "systems" approach</u>. Increasing requirements placed on polymers for advanced structures have necessitated a "systems" approach in developing their applications. A simple tetrahedral model provides a framework for thinking about the total system (Fig. 3). This model adds a third

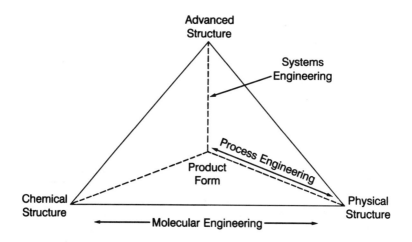

Figure 3. Tetrahedral model.

dimension to the two dimensional model of the past. While integration of three corners of the tetrahedral base - Chemical Structure, Physical Structure, Product Form - requires molecular and process engineering, the creative integration of the three into the fourth corner - Advanced Structures - has necessitated a process referred to as "systems engineering".

This paper will (1) present an overview of advances in organic polymer property development during the past three decades enabling development of high performance materials, (2) describe the "systems" approach to applications, and (3) offer some thoughts on future directions.

2. High Performance Organic Polymers

As polymers achieved significant penetration of markets for natural rubber and fibers, and in non-structural plastics applications, research interest focused on properties of polymeric materials which would stimulate their use as replacements for metals. Resistance to corrosion and ease of fabrication as well as lower energy consumption in their production, are advantages of polymers over metals. To be acceptable as materials for advanced structures, polymers must also be capable of sustaining high loads and stresses over prolonged periods of time and a broad range of temperatures. In many applications good environmental stability, fatigue, and impact resistance also are important.

For this review, we have classified polymers in terms of certain selected properties. We will discuss the development of polymers that provide (1) high tensile strength and stiffness, (2) thermal stability, and (3) high resilience and energy absorption.

2.1. High Tensile Strength and Stiffness

The realization that a nearly perfectly oriented and fully extended polymer chain is essential for high tensile strength and modulus dates back to Dr. Mark's early papers [4]. Theoretical calculations of modulus based on bond force constants [5], or lattice extensions [6], or of strength based on the strength of the weakest bond in the chain led to tensile property estimates which were substantially higher than those subsequently observed experimentally (Table 1). These discrepancies were attributed to polymeric chain folding and entanglement which limit the attainable degree of orientation, and to a finite, often low, molecular weight.

Table 1. Theoretical and observed moduli

Material	Theoretical* (GPa)	Observed**(GPa) (prior to 1960)
Cellulose	60-130	6-13
Polyethylene	260	1-4
Nylon 66	200	8
Poly (ethylene terephthalate)	140	13-19
Poly-*m*-phenylene-isophthalamide	110	12
Poly-*p*-phenylene-terephthalamide	190	-

*Ref, 6
* *Ref, 48

During recent decades, major efforts were directed toward development of approaches to full extension and orientation of high molecular weight chains. Two basic approaches have been followed: (1) development of rigid rod-like polymers which remain essentially fully extended in solution or in the melt, and (2) drawing of semi-rigid and flexible polymers to achieve essentially fully extended structures free of chain folds and entanglements on a molecular basis. Examples of polymers that fulfill these requirements appear in Fig. 4.

2.1.1. <u>Rigid polymers</u>. Discovery of aromatic polyamides (aramids) that form liquid crystalline solutions and high tenacity/high modulus fibers [7] stimulated much research in this area. It clearly demonstrated that theoretical predictions of fiber properties were well founded and approachable. Two primary discoveries were made. The first, in the 1950's, was due to P. W. Morgan, who devised methods for making hitherto intractable high molecular weight polymers of high thermal stability [8]. Fast irreversible reactions at near ambient temperatures were used to synthesize polymers by interfacial and solution methods. The first aramids were made this way as were polyimides and poly-arylates. This first step led to a new materials base.

The second major advance was learning how to organize these new macromolecules into highly oriented extended chain structures to approach the ultimate mechanical properties of the polymer chains. This breakthrough was due to the discovery by S. L. Kwolek of liquid crystalline, nematic solutions of poly-p-benzamide that were processable into fibers with a highly oriented extended chain structure.

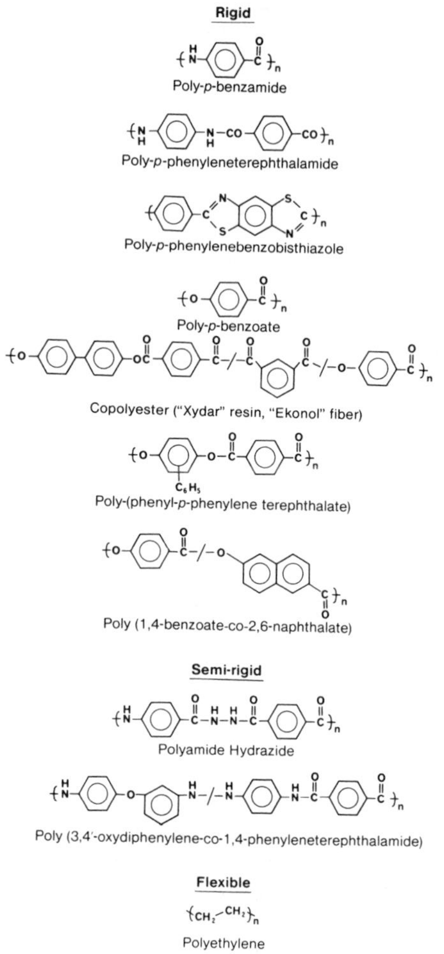

Figure 4. Rigid, semi-rigid and flexible polymers.

These fibers now have moduli 50-75% of the theoretical values [7,9]. This discovery, along with those of her colleagues, especially Morgan and H. Blades [9], led to the first commercial para aramid, introduced by Du Pont in 1971 as "Kevlar" aramid fiber. Following these basic discoveries, a number of polymeric systems forming liquid crystalline solutions and/or melts were uncovered and investigated.

The essential characteristic of polymers forming liquid crystals is their chain rigidity leading to highly extended

chain conformations. For example, poly-p-phenylene-
terephthalamide (PPD-T) exhibits a chain persistence length
of about 40 nm [10] which is equal to about 1/4 of the chain
length of a polymer of molecular weight equal to 20,000. The
conformation of polymer chains in solution depends also on
the solvent and the temperature. Para-aramids in solution
form domains which, under quiescent conditions, are randomly
oriented. During the spinning process the domains orient in
the fiber direction. This occurs mainly through elongational
flow during dry-jet wet spinning.

The crystalline orientation, perfection, and fiber
modulus of the as-spun fiber can be further increased by a
short term heat treatment under tension. For example, the
wide angle X-ray orientation angle of as-spun "Kevlar" fiber
is about 12° and decreases to about 9° after heat treatment.
The fiber modulus increases from about 64 GPa (500 gpd) to
above 150 GPa (1200 gpd).

Application of these principles led to development of
other rigid polymer systems forming lyotropic liquid
crystals. Two of these are cellulose triacetate/trifluoro-
acetic acid, methylene chloride [11], and poly-p-phenylene-
benzobisthiazole (PBZ)/polyphosphoric acid [12]. Cellulose
fibers formed by hydrolysis of heat treated cellulose
triacetate fibers have exhibited a tenacity/modulus ratio of
2.6/60 GPa (20/450 gpd) [13], close to that predicted. PBZ
fiber, dry-jet wet spun from liquid crystalline solution in
polyphosphoric acid and heat treated under tension had a
tenacity as high as 3.9 GPa (28 gpd) [14] and a modulus as
high as 350 GPa (2500 gpd) [15]. This is the highest modulus
observed for an organic polymer fiber.

The excellent properties obtained from aramids led
researchers to synthesize and develop melt processable
thermotropic aromatic polyesters. This has resulted in
commercial resins for self-reinforced moldings, and in
development of strong, high modulus fibers. The initial
efforts in this area involved development by J. Economy and
co-workers [16] of compression moldable poly-p-benzoate. To
improve tractability, more recent work has concentrated on
copolymers containing p-benzoate units. H. F. Kuhfuss and
W. J. Jackson [17] developed liquid crystalline copolymers
with p-benzoate and ethylene terephthalate units. These were
injection moldable. Molded bars had flexural properties
equal to glass reinforced polyester. Dartco Manufacturing
Co. commercialized "Xydar" (see Fig. 4), the first thermotropic copolyester resin for injection-molding. The same
principles of tractability governed development of co-

polyesters for high strength fibers. Several copolymeric
compositions are described in the patent literature [18]. In
addition to p-hydroxybenzoic acid, other monomers with linear
or colinear bonds like hydroquinone, phenylhydroquione,
6-hydroxy-2-naphthoic acid and terephthalic acid have been
used in copolymers. The typical process for fiber formation
involves melt spinning of a polymer of moderately but
sufficiently high molecular weight to exhibit liquid
crystalline properties. The fiber is highly oriented but of
relatively low strength principally because of low molecular
weight. The spinning process is followed by a prolonged heat
treatment of the fiber in a relaxed state in an inert
atmosphere at a temperature approaching the polymer melt or
flow temperature [19]. This treatment leads to an increase
in order, and to a several-fold increase in strength due
principally to a solid phase polymerization. Thus, for poly-
(phenyl-p-phenylene terephthalate), Jackson [20] reported an
increase of molecular weight during such treatment from about
5000 to about 40,000 as tenacity increased from about 0.6 GPa
(5 gpd) to 3.7 GPa (30 gpd). At present, two commercial
fibers have been announced: "Ekonol" by Sumitomo Chemical
Co. and "Vectran" by Celanese-Kuraray [21].

2.1.2. Semi-rigid polymers. Since rigid rod polymers
have limited solubility, a significant amount of research was
concerned with semi-rigid molecules. W. B. Black and J.
Preston [22] concentrated on polymers based on aromatic
polyamide hydrazide structures. These do not form an anisotropic phase under quiescent conditions in amide solvents
because of lower molecular rigidity [23] and were not
commercialized. Teijin, Ltd. recently announced, based on
the work of S. Ozawa [24], an aramid copolymer fiber with a
lower rigidity called "Technora", whose strength is developed
by drawing. The fiber spun from solution has a low
orientation and its final oriented structure and high
properties are achieved by a hot drawing process.

2.1.3. Flexible polymers. Modulus calculations show that
fully extended well oriented flexible polymers can have high
modulus values. At high molecular weights these materials
also should exhibit high tenacity. This potential has been
realized to varying degrees with linear polyethylene by a
number of investigators including H. Blades and J. R. White
[25], A. J. Pennings [26] and, more recently, P. Smith and
P. J. Lemstra [27], whose work on "gel" spinning led to
commercial high tenacity/high modulus fibers. "Gel" spinning
involves extruding a dilute solution (1-5%) of extremely high
molecular weight (about 1 million), polyethylene, preferably
with a narrow molecular weight distribution [28], to produce

on quenching a strong "gel" fiber. The "gel" fiber is dried
and highly drawn, as much as 100X or more at temperatures
near but below the polymer melting point, to produce fibers
with values of tenacity and modulus close to theoretical, eg,
5.8/210 GPa (68/2500 gpd) [29]. The innovative aspect of
this technology is the reduction of chain entanglements below
a critical value by the use of low polymer concentrations.
The technology may be broadly applicable to flexible chain
polymers. High tenacity/high modulus polyethylene fiber is
produced by Allied-Signal Co. as "Spectra" 900 and 1000 fiber
[30], and has been announced by Dutch State Mines-Toyobo Co.
by the trade name "Dyneema", and by Mitsui Petrochemical
Industries as "Tekmilon" fiber [31].

In summary, fibers provide a basis for comparison of
routes to obtain high strength/high modulus materials from
rigid, semi-rigid and flexible chain polymers. The unifying
principle is to avoid formation of molecular entanglements
which severely limit orientation potential. With liquid
crystalline solutions of rigid polymers, rigidity and solvent
prevent entanglement and promote chain extension during
spinning to give directly the final high strength product.
With liquid crystalline melts, the high (neat) polymer concentration fosters sufficient chain entanglement in the high
molecular weight polymer to limit full extension and
attainment of high tenacity; hence, the low molecular weight
polymer with few entanglements is spun and subsequently heat
treated as an oriented solid fiber to increase molecular
weight while maintaining and even improving chain extension
and orientation. Semi-rigid polymers avoid entanglement by
their rigidity and the presence of solvent in moderately
concentrated solutions but cannot attain full tenacity
potential by spinning alone; this is available only through a
hot drawing process. Flexible polymers must rely on very
dilute solutions to avoid entanglements during spinning and
the partially oriented molecular structure must be frozen
into a "gel" state to maintain this alignment. Loss of
solvent collapses the "gel" to a partially aligned state with
a minimum number of entanglements which allow super drawing
to attain an extended chain, well oriented, high tenacity
fiber. Thus, the principal thrust of recent innovative
research in this field has been process engineering.

Progress in the area of high strength and high modulus
materials is shown in Fig. 12B. When properties are
normalized for density, values for the new organic fibers
surpass those of metals and glass. They begin to approach
theoretical levels, more so for modulus (approximately 75%)
than for strength (approximately 25%). One of the challenges

for future research is to achieve strength values as close to
the theoretical as for modulus.

2.2 Thermal Stability

Early research on polymer thermal stability was concerned
with thermal decomposition [32]. This, in turn, was
associated with strength of chemical bonds along the polymer
chain. Bond dissociation energy data (Table 2) favor
thermally stable polymers based on aromatic structures. More
recent development of polymers as structural materials
focused attention on the consideration of change of polymer
physical properties with temperature. Due to this change,
research on thermally resistant polymers moved in two
parallel directions involving synthesis and evaluation of (1)
amorphous materials with a high T_g and (2), crystalline
materials with a high melting point. The factors that raise
T_g, T_m and thermal stability in polymers are qualitatively
the same, namely, chain rigidity due to ring structures
(which is enhanced by para or co-linear chain-extending
bonds), and strong interchain forces conferred by dipolar or
hydrogen bonding interactions.

In addition to thermal stability we need to consider
environmental factors which weaken polymers through chemical
reaction leading to bond scission. The effect of oxidation
and photooxidation is particularly important since most
polymeric products are used in air. Resistance to attack by
solvents and, particularly, by aqueous acids and bases also
is important. Finally, should the polymer be subjected to
high thermal fluxes, the mode of decomposition plays a key
role in many applications. For example, generation of combustible gases and toxic chemicals is usually undesirable
while formation by the decomposed polymer of a coherent char
may be quite desirable as will be discussed later.

The structures and selected properties characterizing
thermal stability of some thermally stable amorphous resins
and crystalline polymers are shown in Fig. 5 [33]. Comparing

Table 2. Bond dissociation energies (kcal/mol)

Bond	Aromatic or Heterocyclic	Aliphatic
C-C	98	83
C-H	102	97
C-O	107	93
C-N	110	82

similar chemical structures, crystalline polymers usually offer higher use temperatures and much better resistance to organic solvents and aqueous solutions, although morphology determined by processing conditions can have an effect. Amorphous materials usually have better dimensional stability, and relatively lower dependence of properties on temperature than partially crystalline materials. Crystalline polymers, because of their higher melting point and better orientability as well as excellent thermal stability generally are more suitable for fiber applications, whereas amorphous resins are generally more suitable for engineering uses in sheet structures and as matrices in composites.

Polymer	Structure	Property	Value*
Poly-p-phenylenebenzobisthiazole	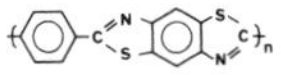	TGA	600°C
Polybenzimidazole	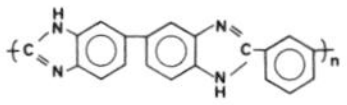	TGA	588
Polyimide	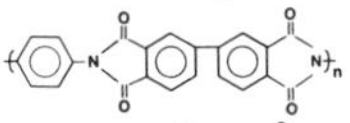	T_g	>500
Polyimide	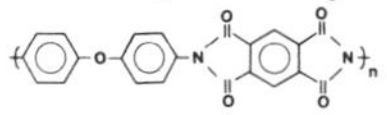	CUT	265
Para-aramid	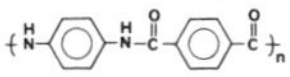	T_g ½ Ten	>375 340
Polyimide (from "Avimid" precursor)	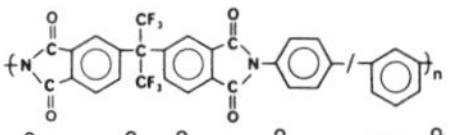	T_g	340-370
Liquid crystalline polyarylate	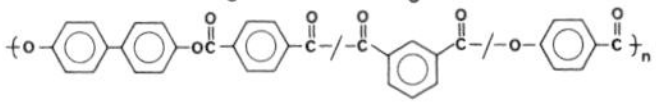	CUT	240
Meta-aramid	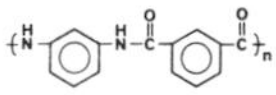	CUT	220
Meta/para-aramid	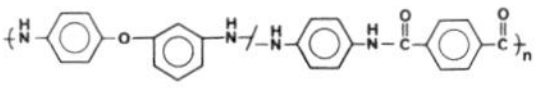	½ Ten	220
Polyetheretherketone	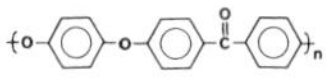	CUT	260
Poly(phenylene sulfide)	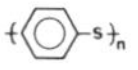	CUT	200
Polyethersulfone	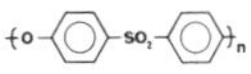	CUT	180
Polyetherimide	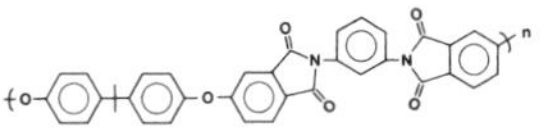	CUT	170
Polysulfone	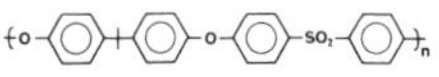	CUT	160

*TGA – Thermogravimetric Analysis, Weight Loss Temperature
CUT – Continuous Use Temperature
½ Ten – Half Tenacity Temperature

Figure 5. Selected polymers with thermal stability.

Retracing development of this field, a major focus has
been to balance good thermal resistance with tractability for
acceptable processing into product forms. Examples of
commercial polymers can be categorized in two groups: (1)
aromatic structures with different spacer groups (eg, amides,
esters, ethers, sulfones) and (2), aromatic heterocyclic
polymers (eg, polyimides, polybenzimidazoles).

2.2.1. <u>Aromatic polymers with different spacer groups</u>.
The first group of thermally stable polymers to attain
commercial importance was aromatic polyamides. The earliest
representative from this class is poly-m-phenyleneisophthalamide, commercialized by Du Pont in 1967 as "Nomex" aramid
fiber. A change from the meta-aramid ("Nomex") to the para
aramid ("Kevlar") leads not only to higher strength and
modulus but also to further improvements in thermal stability,
especially retention of properties at high temperatures.
Teijin's "Technora" aramid fiber is a copolymer containing
meta and para oriented aromatic rings. Replacement of the
amide linkage with an ester linkage leads to a significant
decrease in T_g and T_m in the case of the all meta polymer;
however, for the rigid para structures, the high strength and
modulus are retained over a broad temperature range [16].
High thermal stability allows processing of such high melting
polyesters as "Ekonol" and "Xydar" liquid crystalline
polyester resins.

Excellent mechanical properties of thermally stable
thermoplastic resins combined in many cases with chemical
resistance equal or superior to that of thermoset resins now
permit industry to capitalize on processing advantages inherent in these materials. A series of amorphous, high T_g,
products based on SO_2 as a spacer has been developed by Union
Carbide and ICI. The group includes "polysulfone" and polyethersulfones. The SO_2 linkage leads to a high level of
resonance and thus to good thermal stability while the ether
linkage introduces chain flexibility to improve tractability
and increase toughness. A range of properties is available
by compositional changes. The biphenyl group, for example,
was introduced in some products to decrease the relative
amount of ether linkage and thus further tailor the balance
between thermal properties and tractability. A material
higher on the thermal stability ladder due to its crystallinity is exemplified by polyaryletheretherketone (ICI's
"PEEK"). Sulfur as a spacer between aromatic rings leads to
another crystalline material, poly(phenylene sulfide). Such
a product is sold by Phillips Petroleum Co. as "Ryton" resin.
A stepwise improvement in thermal properties of available
thermoplastic materials is shown in Fig. 6 [34].

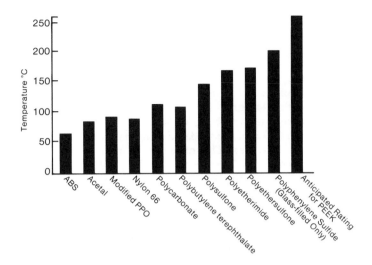

Figure 6. Continuous use temperatures of thermoplastics. Reprinted from Ref. (34). p. 238, by courtesy of Marcel Dekker, Inc.

2.2.2. <u>Aromatic heterocyclic polymers</u>. Aromatic polyimides represent the earliest commercial entry in this group. Preparation of a processable soluble polyamic acid by reaction of a diamine and dianhydride followed by polyimidation in solid form circumvented the problem of intractability of the fully imidized material. Another approach to tractability is to form the partially or completely imidized material from solution in, eg, p-chlorophenol [35]. Finally, commercial products include fully thermoplastic polyimides which can be thermally formed at the expense of some thermal stability, and essentially linear polyimide matrix resins such as are used in composite materials utilizing Du Pont's "Avimid" N or "Avimid" K III polymers. The latter are polymerized as a prepreg and autoclave molded. Because there is minimum cross-linking and T_g is high in these polyimides, laminates have excellent thermal and solvent resistance, and high damage tolerance due to far greater fracture toughness and ultimate strain levels than for epoxy resins. The zero strength temperature of polyimides such as Du Pont's "Kapton" or Ube Industries' "Upilex-R" polyimide film is greater than 800°C. This high thermal stability is coupled with good resistance to organic solvents and aqueous or anhydrous acids. Polyetherimide represents another method of balancing properties and processability. The imide structure is responsible for thermal stability while the flexible ether linkage improves

processability. An amorphous polyimide has been introduced
by the General Electric Company trademarked "Ultem" with a
continuous use temperature exceeding 170°C.

Early work by H. A. Vogel and C. S. Marvel [36] in the
area of polybenzimidazoles opened an important area of
research on high temperature resistant polymers. Polyben-
zimidazole commercialized by Celanese Co. as "Pbi" fiber,
(also paper, and resin) combines thermal and chemical
stability with good textile characteristics. Fiber can be
stabilized by sulfonation to form an insoluble zwitterion
which exhibits lower shrinkage (less than 10%) even at very
high temperatures or after a short exposure to flame.

One of the newest examples from this group is poly(p-
phenylenebenzobisthiazole). Because of its rigid structure
and wholly aromatic character it has a good balance of tensile properties and maintains them at very high temperatures.

Some commercial applications of these thermally resistant
polymers include protective and fire blocking apparel, heat
shields, flue gas filters, electrical insulation, composite
matrices, jet engine fairings and bearing housings, thermoplastic bearings to replace metals, microwave cookware, and
mass transfer tower packings.

2.3. High resiliency and energy absorption

Let's now consider developments in engineering resins that
provide high resilience and energy absorption. In this
category the two types we will discuss are (1) engineering
thermoplastic elastomers and (2) heterogeneous phase
toughened engineering plastics.

2.3.1. Engineering thermoplastic elastomers. In the 1950's
a wide gap existed between the elastic ranges of metals and
plastics on one hand and conventional rubbers on the other.
There was a clear need for a class of polymers that could
combine the melt processability of plastics, the resilience
and elasticity of rubber, and a stiffness and load bearing
capability between plastics and rubber. Random copolymers,
like the N-alkylated nylons, did not meet this need as they
are both a poor plastic and a poor rubber. In the search for
elastomeric fibers, a concept emerged from the early work of
J. C. Shivers and coworkers [37] that a block copolymer with
alternating "hard" and "soft" segments would make a good
elastomeric material. One of the outgrowths of this research
was the discovery of a class of melt processable materials
called engineering thermoplastic elastomers.

Engineering thermoplastic elastomers (ETEs) consist of blocks of "hard" crystallizable or glassy segments alternating with blocks of "soft" amorphous segments with sufficient length and low enough T_g to confer entropy driven elasticity at ambient temperatures. The "hard" and "soft" segments are essentially intimately dispersed separate phases (Fig. 7).

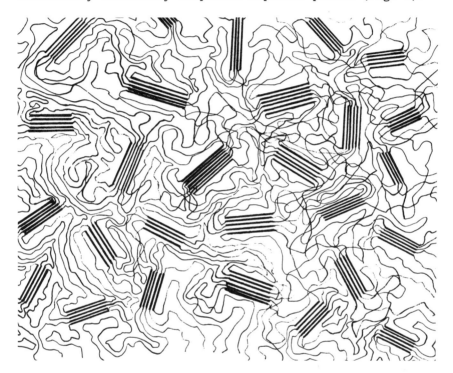

Figure 7. Hard and soft units in thermoplastic elastomers.

Thus each phase preserves to a large extent its own physical properties. The "hard" segments act as polyfunctional cross-links at ambient temperatures to provide a network structure. The continuous phase amorphous "soft" segments with flexible structures provide the elastomeric properties. The elastomers on heating melt and flow above the T_m of the crystalline "hard" segments or T_g of the glassy "hard" segments to provide excellent rapid processability by extrusion or injection molding techniques. At ambient temperatures they show the flexibility, elasticity and good recovery properties of rubbers. The advantage of thermoplastic processability is counter-balanced by a higher hysteresis and permanent set than that of most conventional rubbers because the cross-links become reversible as one

approaches T_g or T_m of the "hard" segment. In principle, any
high melting crystalline or high T_g glassy polymer can serve
as a meltable "hard" segment. Likewise, any flexible
sufficiently long difunctional block with a low T_g could
function as the "soft" segment.

Present day commercial block copolymer ETEs are of two
main types (Fig. 8): (1) polyurethanes, and 2) polyesters.
Polyamide and ester-amide types also are made, but at lower
volume. Polyurethane types have "hard" segments based, for
example, on the condensation of methylene-4,4'-diisocyanate
with 1,4-butanediol in a high molecular weight alternating
copolymer with "soft" segments containing either a macro
polyether glycol or a polyester glycol with molecular weight
above 1000. Polyester types for ETEs are based on the work
of W. K. Witsiepe of Du Pont [38] and M. Sumoto, et al of
Toyobo [39]. These types use poly(butylene terephthalate)
(PBT) "hard" segments because of their rapid rate of, and
high degree of, crystallization, with macro polyether "soft"
segments, eg, $-[O(CH_2)_4]_n-$. In addition to these chemical
composition variables, structure variables such as the length
of the "hard" and "soft" segments in, eg, the PBT and
polyether segments in polyester ETEs, the ratio of PBT to
polyether and the overall molecular weight can influence
elastomeric properties. This often leads to a group of
products with different moduli and degrees of creep and
elasticity. The chemistry and physics of these systems are
well understood, and many composition and property variables
have been studied.

Engineering thermoplastic elastomers are not direct
substitutes for rubber in existing designs. Part redesign is
usually required because the ETEs are generally stiffer and

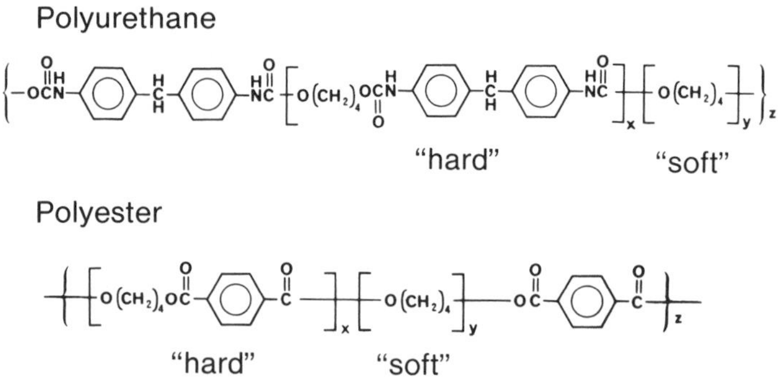

Figure 8. Engineering thermoplastic elastomers.

stronger than rubber. The basis for redesign is to make
parts that are functional and cost effective using light-
weight thin sections to replace rubber.

The ETEs share a number of desirable characteristics with
both rubber and engineering plastics. They are suitable for
use under high loading, are creep resistant, have good
elasticity, elastic recovery, flexibility and flex fatigue,
have a broad service temperature range without significant
property changes and are solvent and chemical resistant. In
addition to these superior mechanical and chemical
properties, they can be processed like thermoplastics into a
variety of useful complex shapes.

Commercial applications for ETEs include: (1) extruded
products such as hoses, films and sheets, wire and cable
coatings, coiled phone cords, (2) cast parts such as tire
inserts, (3) blow molded parts such as constant velocity
joint boots, (4) painted automotive parts such as facia, (5)
industrial shock absorbers and seals, and (6) food contact
and medical applications since potentially toxic vulcanization and compounding residues are absent.

   2.3.2. Heterogeneous phase toughened engineering
plastics. The two-phase toughening of amorphous polystyrene
by the process of polymerizing styrene in a rubber solution
was described about 60 years ago [40]. The real growth of
engineering plastics, however, took place when B. N. Epstein
and co-workers discovered that such relatively tough
engineering resins as nylon [41] and acetal [42] could be
further toughened to achieve notched Izod impact values which
made them viable candidates for metal replacement (Fig. 9).
This combination of resilience, low density, corrosion
resistance, and now toughness approaching that of metals,
together with processing advantages, provided the drive for
such replacement applications.

Let us consider nylon and acetal as representative
members of this group of materials. Nylon is a tough
engineering resin with a high elongation which does not break
in unnotched Izod impact testing. Toughness values decrease
markedly, however, if notches are present, which indicates
great resistance to crack initiation but only modest
resistance to crack propagation [43]. Such resistance is of
importance because seldom can scratches be avoided in molding
or in use. The two-phase toughened system consists of a
nylon continuous phase and a low modulus dispersed phase.
Just as in the case of engineering thermoplastic elastomers,
the properties of the continuous and dispersed phases are

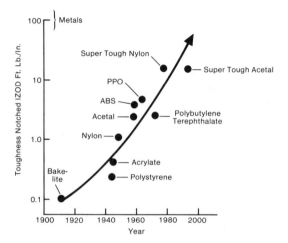

Figure 9. Property improvements in toughened plastics.

only slightly affected, whereas in a homogeneous system the
properties would in most cases be averaged. Thus, the nylon
modulus is reduced by only about 60%, whereas, the notched
Izod toughness is increased some 20 times over that of unmodified nylon. This toughening is a direct result of
reducing the tendency for crack propagation. Both semi-
crystalline and amorphous nylon can be toughened, but the
flexural modulus of the amorphous resin is far less sensitive
to moisture content. The tradeoff of toughness for stiffness
in acetal resin is shown in Fig. 10 [44].

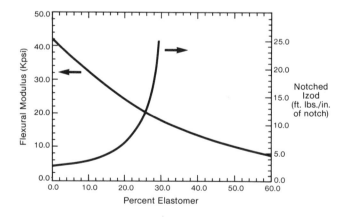

Figure 10. Modulus-toughness trade-off in super tough acetal.
       Courtesy of E. A. Flexman, Jr. of Du Pont.

Studies by S. Wu [43] of specimen bars subjected to analysis after breaking in the notched Izod impact test have helped to clarify the machanism of heterogeneous phase toughening. The two broken parts of a bar of untoughened nylon show brittle fracture, whereas, the toughened bar before final failure shows involvement of a large volume of stress-whitened crazed material around the crack, indicating that a yielding mechanism, mostly in the matrix material, is responsible for the increased toughness of the toughened nylon. Whitening is due to microcracks filled with voids and fibrils formed by yielding of the matrix. Deformation and heating of the crazed matrix material absorbs the impact energy. About 25% of the energy is dissipated by matrix crazing and about 75% as heat by matrix yielding.

The major variable in two phase toughening of engineering plastics appears to be the interparticle distance which is related to particle size and phase ratio. The general condition for toughening is that the interparticle distance must be smaller than a critical value [43].

In addition to toughness measured by the notched Izod impact test, cyclic fatigue short of fracture failure such as might be experienced by an automobile, or a boat hull due to impact from rocks, door openings or falling objects, is important. The importance of impact and fatigue resistance is reflected in the freedom from dents and dings in the surface of an automobile during assembly, and, later, in use by the consumer. The development of exceptional toughness in plastics with little loss in thermal stability, strength, modulus, resilience, corrosion resistance and other desirable properties, especially high specific properties compared to steel is accelerating replacement of automobile exterior parts with heterogenous phase toughened engineering resins.

Commercial applications of heterogeneous phase toughened engineering resins include exterior automotive parts such as vertical panels and bumper components, sporting goods, bicycle wheels, helmets, tool handles, fire extinguisher valves, and snap fasteners.

In conclusion, we see that engineering thermoplastic elastomers and heterogeneous phase toughened engineering plastics have gone a long way toward filling the gap between metals and plastics, on one hand, and rubber, on the other (Fig. 11).

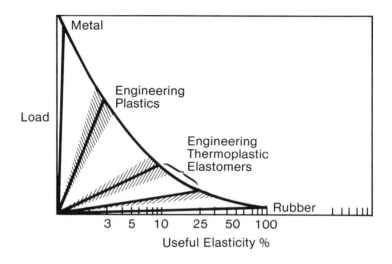

Figure 11. Elastic range of metals, plastics and elastomers.
Adapted from figure of T. W. Sheridan of Du Pont.

3. Applications In Advanced Structures

In the previous sections of this paper we reviewed some
developments of organic polymers and relationships between
structure and properties which allowed design of materials
with high strength and stiffness, high temperature
resistance, and a range of elasticity and toughness. In the
tetrahedral model (Fig. 3), this was represented by a two-
dimensional iteration of chemical structure, physical
structure, and product form. In this section we will discuss
the third dimension, which is the creative integration of
these technologies into advanced structures. The process for
doing this, referred to as "systems engineering", requires
the combined talents of polymer and materials scientists and
engineers as well as many other disciplines. To arrive at
the final advanced structure usually involves much
interaction between each corner of this tetrahedron. It also
often involves multiple polymer structures and product forms
organized into an integrated total system.

To illustrate the "systems" approach to applications for
the polymer properties described in this paper we have
selected four examples: (1) aircraft composites, (2) fire
blocking protective apparel, (3) constant velocity joint
boots in automobiles, and (4) exterior automobile vertical
panels.

3.1. <u>Advanced composites for aircraft</u>. The rapid
improvement in the tensile properties of synthetic fibers
during the last several decades, ie, a 100-fold increase in
stiffness and 3-fold increase in strength gives the systems
engineer in 1987 a much wider selection of materials compared
to 1957 as shown in Fig. 12. Toughness of resin systems also
has improved several orders of magnitude in the same period
and now spans a wide useful temperature range for both
thermoplastics and thermosets. As a result of improvement in
both fiber and resin properties, parts made of composite
materials can now be designed with strength and stiffness
equal or superior to, and toughness approaching that of,
parts made from steel and aluminum.

We will illustrate in an example the use of systems
engineering to optimize the properties and performance of
fibers and matrices working together in structural components
for aircraft.

Structural parts for aircraft require design not only to
attain the full spectrum of required mechanical properties,
but also for damage tolerance, that is, the ability to
tolerate abuse and to survive a catastrophic impact. Carbon
fiber in resin matrix composites confers required high
stiffness and compressive strength. However, because of
their rigid coplanar ring structure, carbon fibers are un-
yielding and fail by brittle fracture; hence, they are unable

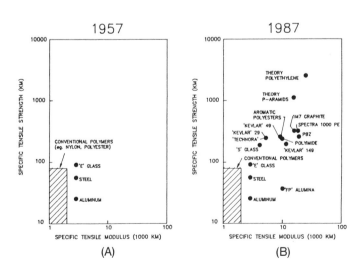

Figure 12. Availability of high performance fibers for
systems engineering.

to survive a catastrophic impact. On the other hand, para-aramid fibers in resin matrix composites have structural features that lead to a ductile compressive failure mode of the type that provides good damage tolerance. This is because at a compressive strain of about 0.5%, we believe buckling of para-aramid molecules occurs due to molecular rotation of amide carbon to nitrogen bonds to accommodate configurational changes which result in yielding to the imposed stress without bond cleavage [45, 46]. The result is an accordion-like collapse on impact. Metal-like ductility of para-aramid compared to brittle carbon fiber structures is illustrated by the flexural stress-strain behavior of unidirectional epoxy matrix composites and of aluminum shown in Fig. 13.

Crashworthy composite systems are being explored by use of a crushed tube test to compare materials. This test involves crushing of filament wound tubes with a drop-weight impact tester. There is a marked difference between the failure modes of tubes wound with "Kevlar" aramid and those wound with carbon fiber. The tube reinforced with "Kevlar" fiber fails by a progressive buckling mode, similar to that observed for an aluminum tube, leaving a structure which is damaged but still intact and able to sustain a load. The carbon fiber-wound tube shatters. However, because the carbon-fiber-wound tube reaches and sustains a higher load

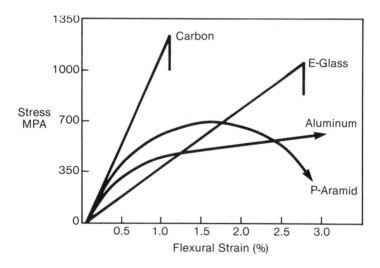

Figure 13. Flexural stress/strain behavior for unidirectional epoxy matrix composites. (Du Pont data)

just prior to failure, this test shows quantitatively that it
absorbs more total energy in the crushing process than does
the tube wound with "Kevlar" fibers.

Thus, hybrid technology that combines para-aramid and
carbon fiber as a wound structure offers a better balance of
properties. High energy absorption, within 7% of that of the
all-carbon tubes, and structural integrity after crushing
approaching that of the aramid-wound tube, is achieved.

Hybrid composites of "Kevlar" and carbon fiber are being
used in commercial aircraft such as the Boeing 767, in a
variety of helicopters, and in commuter aircraft.

3.2. Fireblocking protective apparel. One of the early
incentives for aramid research was to develop polymers that
would meet the need for high temperature stability in
numerous industrial applications. The first polymer
developed was poly-m-phenyleneisophthalamide which gave
fibers with good thermal stability to short term exposure at
temperatures as high as 500°C., and to long term exposure at
temperatures up to 220°C. An unexpected property of the
meta-aramid fiber emerged which suggested a unique contribution in applications such as thermal protective clothing.
When meta-aramid is heated rapidly, as in a flash fire
exposure, trapped moisture and degradation gases expand the
softened polymer to form a carbonaceous insulating protective
foam which is up to ten times thicker than the original
fabric. This property of intumescence combined with high
temperature stability, and self-extinguishing characteristics
when an igniting flame is removed, suggested that this fiber
could be the basis for an outstanding protective apparel
garment. However, further evaluation in actual garments
under simulated flash fire conditions indicated that to
properly protect the wearer and also satisfy other customer
needs, we would need to do extensive "systems engineering".

To achieve full scale realistic evaluation of candidate
materials, an instrumented manikin named "Thermoman" (Fig.
14) was built to test the garment "system" in a simulated
flash fire [47]. Thermoman was equipped with 122 sensors to
measure heat transfer and to predict by computer calculation
the amount of tissue burn resulting from the exposure. In
early testing a serious problem that indicated that our
"system" must be significantly modified was observed. When
the garment on Thermoman was exposed to a flash fire, the
fabric broke open under the restraints of the manikin because
of fiber shrinkage that occurs when the fabric is heated
above T_g of the fiber. Hence, the wearer would be exposed

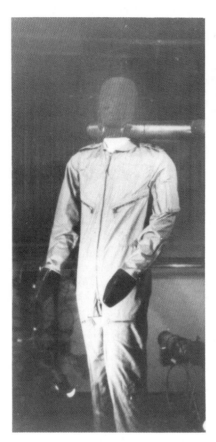

Figure 14. Thermoman outfitted for protection (left) and
under fire (right). Courtesy of W. P. Behnke of Du Pont.

to the direct flames causing him more serious damage than had
been predicted from properties alone.

   Based on the Thermoman results, research was initiated to
modify the system in a way that would reinforce the meta-
aramid. The outcome is a system based on a synergism that
was discovered for intimate staple blends of the meta-aramid
fiber, "Nomex", with the para-aramid fiber, "Kevlar". The
para-aramid, a fully extended rigid polymer which does not
deorient on heating, undergoes a negligible amount of
shrinkage and maintains useable strength up to its decomposition temperature. Hence, the para-aramid fiber provides
the integrity of a strong structural framework while the

meta-aramid provides the insulating foam thermal barrier
matrix which transmits forces between the reinforcing fibers.
Even with a concentration of the para-aramid as low as 5%,
the strength of the fabric is high enough to maintain its
integrity without breaking open, and shrinkage is reduced to
a sufficiently low level so that a protective insulating air
layer protects the maniken when it is exposed to a flash
fire.

Beyond achieving acceptable performance under flash fire
conditions, iterations were required within the framework of
the polymer chemistry, the fiber physics, and the textile
engineering to provide other characteristics important to the
wearer such as comfort and coloration. Today, this
technology is commercial in many applications including
flight and industrial coveralls, fire fighters' turnout
coats, and welders' and race car drivers' apparel.

Recently, flame protection technology has been applied to
protect passengers during an aircraft crash which involves a
fire. The seat cushioning material, when exposed to flames,
decomposes and generates large quantities of smoke and toxic
gases. To meet a set of stringent United States Federal
Aviation Administration requirements, several companies have
engineered "systems" to encapsulate the cushion and provide a
fireblocking layer. The thermally stable fibers participating in this application include aramids, polybenzimidazoles,
carbon and carbonized polyacrylonitrile.

3.3 <u>Constant velocity joint boots in automobiles</u>. One of
the hardest working rubber-like parts in a car is the
constant velocity joint (CVJ) boot [49]. This boot protects
the assembly that transmits power to the front drive wheels
while allowing the freedom of movement required for steering
and suspension motion. It serves as a leak-proof seal to
keep water, dirt and debris out, and the lubricant for the
bearing in. Conventional thermoset rubber boots are
susceptible to weathering, tearing, cracking and flex
fatigue. As one of the most severely stressed elastomer
elements in a car, it had been maintained for years that a
ETE part, which offers superior weathering and fatigue
properties, could not fulfill this exacting function. The
key to utility was engineering design in a "systems"
approach.

As shown in Fig. 15, the rubber version of the boot uses
only a few rounded convolutions to achieve the necessary
flexibility. Substituting a copolyester elastomer in the

Figure 15. Constant velocity joint boot of "Neoprene" (left) and "Hytrel" (right).

original design would make the part too rigid because of the higher modulus of engineering thermoplastic elastomers (ETEs). It would be overstrained and yield on bending. Redesigning in a "systems" approach was required. The redesigned boot has carefully defined hinge points, is molded longer in length than its rubber counterpart, is then compressed 20% on installation. Pre-compressing the boot allows it to bend freely at reduced strain. The new design produces striking results. The new boot is half the thickness and half the weight, and offers six times the flex life of the rubber boot. Also, it has better low and high temperature properties, good weatherability, superior solvent resistance, and is produced by blow molding in one twelfth the cycle time of a thermoset rubber. Engineers capitalized on the fact that the copolyester ETEs have twice the tensile strength and four times the tear strength of rubber. They designed within the elastic limit of the material, used hinge points and thin sections to guide flexibility into the part where needed, and pre-compressed the boot to minimize strain during flexing.

Boots of polyester elastomer are durable, withstand long-term flexing, keep grease in front wheel axle joints and remain functional in temperatures ranging from -40°C to 100°C. Beginning in 1985 major automotive manufacturers have adopted "Hytrel" polyester elastomer for CVJ boots. Note that the automaker used the redesigned part to fit a critical "systems" need.

3.4. Exterior automobile panels. In the automobile market, plastics are beginning to replace exterior metal parts at an increasing rate. Metals have been used for a long time but have reached limitations in their capability for property and cost improvements. They have inherent weaknesses in corrosion propensity, high density, lack of resilience, and high cost due to investment in, eg, stamping machines, and expensive finishing operations. The principal driving force for the automaker is cost in a highly competitive market. Even though the plastic raw materials may cost more, economies due to lower cost processing and higher specific properties more than compensate. An injection molding machine is a lower investment than a stamping machine; retooling costs can be 80% lower for styling changes or face-lifting; resin is easier to store than sheet metal; waste can be recycled; separate steel parts can be integrated into one larger molding; and inventory costs can be reduced. Assembly costs are lower. Compared to steel, plastic exterior panels are about 40% lower weight, allowing increased fuel efficiency. Besides these cost advantages, plastics offer more styling capability. From the consumer standpoint, plastic bodies are said to offer extended life, improved styling, improved fuel economy, reduced maintenance and repair costs, and higher resale value.

Plastics replacement of metals in the automobile market requires a large investment in money, time and personnel, and close cooperation between the plastics company and automobile manufacturer. Cooperation with the automaker begins in the design stage some 3-5 years before production. Parts have to be designed to fit the need and to minimize the cost, performance ratio. Computer aided design is widely used.

Now let us zero in on a single car and a single large plastic exterior panel in that car. The General Motors' Pontiac Fiero plastic exterior car was an instant success. Planned production for the first model year had to be increased 65%. The quarter panel for this car is made of a Du Pont toughened amorphous nylon, "Bexloy" AP C-712 automotive engineering resin. This resin has the advantages

cited above vs. steel and molding cycles are faster than
those of thermosets.

"Systems" design, as in the case of the "Hytrel" boot,
played a prominent role in marketing this part. An
essentially amorphous nylon matrix was formulated to confer
excellent dimensional stability over a wide temperature range
and to minimize changes of modulus with those of relative
humidity, and to give a homogeneous material to avoid
warpage. The higher thermal and moisture induced expansion
of nylon compared to steel had to be designed around to
maintain small gaps between exterior panels for appearance.
This was done by fastening panels at gap points and allowing
either free expansion at other sides or by allowing for
telescoping into neighboring parts. Fastening by use of
flexible nylon rivets allowed this floating type expansion.
Such fastening also effectively inhibited rattle of panels.
Since the panel had a Class A surface (requires no pre-
finishing) directly as removed from the mold, it could be
prepainted (involves lower investment for smaller paint
ovens) and assembled by use of predrilled mounting pads on
the prebuilt chassis. This minimizes damage to surface, and
rejects are few. Because the resin has high resilience over
a wide temperature range and excellent impact fatigue
properties, the panel is resistant to repeated impacts short
of failure and remains dent and ding free in use.

The future for plastics in the automobile industry is
bright. There were over 40 known cooperative plastics parts
programs between carmakers and plastic manufacturers in the
U.S. alone in 1986. The "systems" approach to design and
application of advanced polymer structures appears assured!

To summarize the applications section of this paper, an
understanding of the requirements of a specific market
enables a multi-discipline corps of "systems" engineers to
make use of the principles underlying the tetrahedral model
to design optimum advanced structures which best serve
customer needs.

4. Future Directions

As noted in the introduction, the progress of polymer
science appears to be occurring in waves (Fig. 1). We have
passed the steep portion of the "Foundations" period and are
still advancing along the steep part of the "S" curve of
"Polymers for Advanced Structures". We believe that a third

major wave is beginning that will involve an interdisciplinary future in the world of material science.

Important advances are still occurring in design of polymers for advanced structures. An indication of future
directions is given in the papers that follow this overview.
These include further advances in polymers that provide
higher strength and stiffness, thermal and environmental
stability, and high energy absorption and resiliency. New
directions are unfolding in the exciting fields of conducting
and piezoelectric polymers, molecular composites, and interpenetrating networks. Growth will continue to occur in new
applications due to new polymer compositions, enhanced properties, innovative processes and systems engineering.

In new compositions, the recent synthesis of two hitherto
intractable polymers, poly-p-phenylene and poly-p-phenylene-
terephthalaldimine in film form illustrate the relentless
progress in this area [50, 51]. Biological cloning
procedures and template polymerization at near ambient
temperatures should give rise to entirely new structures with
new levels of properties.

The discovery of a novel means of polymerization called
Group Transfer Polymerization [52] should provide us with an
unprecedented degree of control in engineering polymers at
the molecular level and will likely lead to many unforseen
applications. The essence of the reaction is that a silyl
group transfers from the initiator to the incoming monomer
and reproduces the essential structure of the initiator at
the end of the growing chain. This is a living polymerization and continues until all the monomer is consumed.
Because it is a living polymerization, one can add various
monomers sequentially, and in a controlled way produce a
variety of block copolymers at near ambient temperatures.

Major strides will be made in achieving fiber strength
and stiffness closer to theoretical levels based on advances
in understanding the failure modes of bulk materials. Extension of this work to composites with ever decreasing size of
reinforcing elements down to molecular levels, ie, molecular
composites, should yield even higher properties by circum-
venting some molecular weight limitations, especially for
condensation polymers.

The ever present trade-off between properties and ease of
product formation and processability will continue to fuel
research. The separate reprocessing step in fiber formation
could be eliminated if molecular growth could coincide with

fiber formation. It is conceivable that whiskers, studied so far only for inorganic materials, represent the first practical step in this direction.

Further sophistication of our understanding of the energy absorption by polymers during fracture should lead to further improvements of the toughness of materials by appropriate tailoring of polymeric blends. The range of options in this area includes not only homogeneous and heterogenous blends but also a broad range of interpenetrating polymer networks.

As "systems" engineering of fiber reinforced resins continues to advance in the aircraft industry, and as manu-facturing techniques become more cost effective, this technology will be extended at an increasing pace to automotive parts. New tough thermoplastic engineering resins such as the polyarylates are being designed to better compete with metal, glass, and other plastics in many automotive, electrical and other consumer areas. New advances in short fiber forms will occur and they will gain increasing importance in reinforcing plastics, cement and elastomers.

It is difficult to predict what lies ahead. However, we believe that a new "S" curve is beginning to take shape as multiple disciplines converge in the world of material science (Fig. 16). This convergence will incorporate a continually growing materials base and a deeper understanding on a molecular basis of structure-property relationships. It also will capitalize on the expanding capability of computers

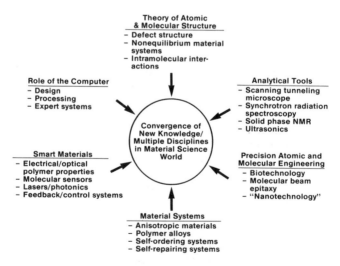

Figure 16. The interdisciplinary future.

to design molecules, predict properties and optimize
structure and manufacturing processes. The development of
new analytical tools and methods, and the integration of new
sensing devices directly into materials will allow development of "smart" systems which will respond to a changing
environment. One can envision that appropriate integration
of components into a system will allow for dynamic changes in
properties of the structure in response to changes of the
environment in which it performs.

We are on the verge of a technological revolution that
will drive materials substitution into the 21st century...a
revolution in the way technologies will converge and
materials systems will be put together. The challenge that
we face is to explore and "artistically" combine new
technologies to find better ways to meet human needs. This
challenge will clearly exceed the limits of a single field,
a single university, a single corporation and even that of
a single country.

Acknowledgements

The authors wish to express their appreciation for
contributions to this paper from other employees of these
departments of E. I. du Pont de Nemours & Co., Inc. These
include W. P. Behnke, J. A. Fitzgerald, M. C. Geoghegan,
R. W. Hess, and A. K. Dhingra of Textile Fibers; B. N.
Epstein, E. A. Flexman, Jr. G. K. Hoeschele, and T. W.
Sheridan of Polymer Products ; K. W. Nelson of Automotive
Products; and S. Wu of Central Research and Development.

References

[1]   Morawetz, H. "Polymers: The Origin and Growth of a
      Science"; John Wiley: New York, 1985.
[2]   Flory, P.J. "Principles of Polymer Chemistry"; Cornell
      University: Ithaca, N.Y., 1953.
[3]   Mark, H. J. Polym. Sci. 1957, 23, 1.
[4]   Mark, H. "Physik und Chemie Der Cellulose"; Julius
      Springer: Berlin, 1932, pp. 61-62. Trans. Faraday Soc.
      1936, 32, 143; Report of the 63rd General Discussion
      held in Cambridge, England, 1935.
[5]   Meyer, K.H.; Lotmar, W. Helv. Chim. Acta 1936, 19, 68.
      Treloar, L.R.G. Polymer 1960, 1, 95, 179, 290. Lyons,
      J. Appl. Phys. 1958, 29, 1429. Holliday, L; White, J.W.
      Pure Appl. Chem. 1971, 26, 545. Fielding-Russel, G.S.
      Text. Res. J. 1971, 41, 861.

[6]   Sakurada, I.; Nakushina, Y.; Ito, T. J. Polym. Sci.
      1962, 57, 651.   Sakurada, I.; Ito, T.; Nakamae, K.
      Makromol. Chem. 1964, 75, 1; Bull. Inst. Chem. Res.,
      Kyoto Univ., 1964, 42, 77. Sakurada, I.; Kaji, K. J.
      Polym. Sci. 1970, C31, 57.   Dulmage, W.J.; Contois, L.E.
      J. Polym. Sci. 1958, 28, 275.
[7]   Kwolek, S.L. U.S. Patents 3 600 350, 1971; 3 671 542,
      1972; French Patent 1 526 745, 1968. Kwolek, S.L.;
      Morgan, P.W.; Schaefgen, J.R.; Gulrich, L.W.
      Macromolecules 1977, 10(6), 1390.
[8]   Morgan, P.W. "Condensation Polymers: By Interfacial and
      Solution Methods"; Interscience: New York, 1965.
[9]   Morgan, P.W. Macromolecules 1977, 10, 138.   Blades, H.
      U.S. Patents 3 869 429, 1975; 3 869 430, 1975.
[10] Erman, B.; Flory, P.J.; Hummel, J.P. Macromolecules
      1980, 13, 484. Arpin, M.; Strazielle, C. Makromol.
      Chem. 1976, 177, 581. Arpin, M.; Strazielle, C.; Weill,
      G.; Benoit, H. Polymer 1977, 18, 262, 591.   Metzger
      Cotts, P.; Berry, G.C. J. Polym. Sci., Polym. Phys.
      Ed. 1983, 21, 125.
[11] Panar, M.; Wilcox, O.B. Belgian Patent 656 359, 1976;
      German Patent Application 2 705 382, 1977; French Patent
      2 340 344, 1977; Chem. Abstr. 1977, 87, 153310. Patel,
      D.L.; Gilbert, R.D. J. Polym. Sci., Polym. Phys. Ed.
      1981, 19, 1231, 1449.
[12] Wolfe, J.F.; Loo, B.H.; Arnold, F.E. Macromolecules,
      1981, 14, 915.
[13] O'Brien, J.P. U.S. Patents 4 464 323, 1984; 4 501 886,
      1985. O'Brien, J.P. Private communication.
[14] Adams, W.W. et al "Ultimate Properties of Rigid-Rod
      Polymer Fibers". Presented in poster session at
      symposium "Approaches to Property Limits in Polymers",
      (Am. Chem. Soc., Div. of Polym. Chem.)
      Scanticon-Princeton, N.J. 11-13 August, 1986.
[15] Allen, S.R.; Filippov, A.G.; Farris, R.J.; Thomas, E.L.
      In "The Strength and Stiffness of Polymers";
      Zachariades, A.E.; Porter, R.S., Eds.; Marcel Dekker:
      New York, 1983; Chapter 9, pp. 357, 371.
[16] Economy, J.; Novak, B.E.; Cottis, S.G. Polym. Prepr.
      (Am, Chem. Soc., Div. Polym. Chem.) 1970, 11(1), 332.
      Economy, J.; Volksen, W. In reference [15] Chapter 7,
      p.293.
[17] Jackson, W.J.,Jr.; Kuhfuss, H.F. J. Polym.
      Sci. 1976,14, 2043. See also Sawyer, L.C. J. Polym.
      Chem., Polym. Lett. Ed. 1984, 22, 347.

[18] Calundann, G.W. U.S. Patent 4 184 996, 1980.
Schaefgen, J.R. U.S. Patent 4 118 372, 1978. Payet,
C.R. U.S. Patent 4 159 365, 1979. Irwin, R.S.;
Logullo, F.M., Sr. U.S. Patent 4 500 699, 1985; Ueno,
K.; Sugimoto, H.; Hayatsu, K. U.S. Patent 4 503 005,
1985.
[19] Luise, R.R. U.S. Patent 4 183 895, 1980.
[20] Jackson, W.J.,Jr. Br. Polym. J. 1980, 12, 154.
[21] Sumitomo Chemical Co. announcement, 24 Oct. 1985. Yoon,
H.N.; Jaffe, M. "Influence of Molecular Interactions on
the Mechanical Properties of Thermotropic Copolyesters".
Presented at symposium of reference [14].
[22] "High Modulus Wholly Aromatic Fibers"; Black, W.B.;
Preston, J. Eds.; Marcel Dekker: New York, 1973.
[23] Bianchi, E.; Ciferri, A.; Tealdi, A.; Krigbaum, W.R.
J. Polym. Sci., Polym. Phys. Ed. 1979, 17, 2091.
Bianchi, E.; Ciferri, A. ibid. 1981, 19, 863.
[24] Ozawa, S. Polym. J.(Tokyo) 1987, 19, 119.
[25] Blades, H.; White, J.R. U.S. Patent 3 081 519, 1963.
[26] Pennings, A.J.; Meihuizen, K.E. J. Polym. Sci., Polym.
Phys. Ed. 1979, 3, 117. Pennings, A.J. Proc. Fiber
Producer Conference 1986, 1-1, Greenville, SC.
[27] Smith, P.; Lemstra, P.J. U.S. Patents 4 344 908, 1982;
4 422 993, 1983; 4 436 689, 1984; 4 430 383, 1984.
Smith, P.; Lemstra, P.J.; Booij, H.C. J. Polym. Sci.,
Polym. Phys. Ed. 1981, 19, 877.
[28] Smith, P. Private communication.
[29] Fiber World 1985, July, p. 12. Japanese Patent
84-216 913, 1984.
[30] Weedon, G.C.; Tam, T.Y. Proc. Fiber Producer Conference
1986, 5A-12, Greenville, SC. "Spectra" trade
literature, Allied-Signal Co.
[31] DSM brochure "New High Performance Polyethylene" 1984,
Feb. Manuf. Chem. 1984, 55(3), 23. Japan Times 1984,
Feb. 25. Jpn. Chem. Week 1984, Dec. 27.
[32] Grassie, N. "The Chemistry of High Polymer Degradation
Processes"; Butterworths: London, 1955. Frazer, A.H.
"High Temperature Resistant Polymers"; Wiley-
Interscience: New York, 1968.
[33] The numerical data used in preparing this table were
taken from "High Performance Polymers: Their Origin and
Development"; Seymour, R.B.; Kirshenbaum, G.S. Eds.;
Elsevier: New York, 1986, pp. 180,201; from "Engineering
Thermoplastics"; ref. [34] pp. 190,207; from ref. [12];
from Ramirez, J.E.; Dwiggins, C.F. Tappi J. 1985,
68(12), 12; from Wilfong, R.E.; Zimmerman, J. J. Appl.
Polym. Sci. 1973, 17, 2039; from ref. [24]; from Mod.
Plast. Encyclopedia, 1985-1986; and from company trade
literature.

[34] Rigby, R.B. In "Engineering Thermoplastics"; Margolis, James M. Ed.; Marcel Dekker, Inc.: New York, 1985, p. 238.

[35] Kaneda, T.; Katsura, T.; Nakagawa, K.; Makino, H.; Horio, M. J. Appl. Polym. Sci. 1986, 32, 3133, 3151.

[36] Vogel, H.A.; Marvel, C.S. J. Polym. Sci. 1961, 50, 511; ibid., Part A, 1963, 1, 1953.

[37] Charch, W.H.; Shivers, J.C. Text. Res. J. 1959, 29, 536; Frazer, A.H.; Shivers, J.C. U.S. Patent 2 929 803, 1960; Shivers, J.C. U.S. Patent 3 023 192, 1962.

[38] Witsiepe, W.K. U.S. Patents 3 651 014, 1972; 3 763 109, 1973; 3 755 146, 1973.

[39] Sumoto, M.; Furusawa, H.; Takeuche, T. Japanese Patent 1 005 108, 1980.

[40] Ostromislensky, I. U.S. Patent 1 613 673, 1927.

[41] Epstein, B.N. U.S. Patent 4 174 358, 1979. Epstein, B.N.; Pagilagan, R.U. U.S. Patent 4 410 661, 1983. Latham, R.A. U.S. Patent 4 536 541, 1985.

[42] Flexman, E.A.,Jr. Annu. Tech. Conf. Soc. Plast. Eng. 1984, 42, 558.

[43] Wu, S. J. Polym. Sci., Polym. Phys. Ed. 1983, 21, 699; Polymer 1985, 26, 1855.

[44] Flexman, E.A., Jr. Private communication.

[45] Tanner, D.; Fitzgerald, J.A.; Knoff, W.F.; Pigliacampi, J.J.; Riewald, P.G. Proc. Int. Symp. Fiber Sci. Tech. Hakone, Japan, August, 1985.

[46] Tanner, D.; Dhingra, A.K., Pigliacampi, J.J. J. Met. 1986, 38(3), 21-25.

[47] Behnke, W.P. Fire Mater. 1984, 8(2), 57.

[48] Morton, W.E.; Hearle, J.W.E.S. "Physical Properties of Textile Fibers"; Butterworths: London, 1962.

[49] Sheridan, T.W. Des. Eng. 1986, Nov., p.81. Bowtell, M. Elastomerics 1986, 118(11), 38. "Hytrel" trade literature.

[50] Ballard, D.G.H.; Courtis, A.; Shirley, I.M.; Taylor, S. submitted for publication in Macromolecules.

[51] Engel, A.K.; Yoden, T.; Sanui, K.; Ogata, N. J. Am. Chem. Soc. 1985, 107, 8308.

[52] Sogah, D.Y.; Hertler, W.R.; Webster, O.W.; Cohen, G.M. Macromolecules, in press.

Part IV
Liquid Crystalline Polymers

LIQUID CRYSTALLINE PARA AROMATIC POLYAMIDES

Stephanie L. Kwolek, Wesley Memeger and James E. Van Trump

E. I. du Pont de Nemours & Co., Inc.
Textile Fibers Department
Pioneering Research Laboratory
Wilmington, Delaware   19898
U.S.A.

Abstract

The early history of the development of high
tenacity, high modulus fibers from liquid crystalline
solutions of para aromatic polyamides is reviewed.
More recent work with emphasis on cost reduction in
aramid synthesis and polymer processing is described.
Structure-property relationships of Kevlar* aramid
fiber are further defined with particular emphasis on
the recently commercialized ultra-high modulus "Kevlar"
149 with its higher crystallinity and unique super-
crystalline structure. Lastly, some industrial applications of "Kevlar" aramid fibers are discussed.

1. Introduction

Since the discovery of high tenacity, high modulus fibers
from anisotropic solutions of para aromatic polyamides
(aramids) in the mid 1960's, there has been a very great
increase in activity in the field of liquid crystalline
polymers as evidenced by the greater than 100-fold growth in
annual publications in this field since 1970. It is difficult to find another family of compounds that has created as
much interest and excitement as liquid crystalline polymers.
Their performance characteristics that include high tensile
strength, high modulus, low creep, low density, high thermal
stability and chemical and flame resistance have opened up
broad new application areas beyond conventional textile
uses.

* Du Pont trademark

This paper will review some of the highlights of the early work on liquid crystalline solutions of para aramids and the resulting high tenacity, high modulus fibers, including "Kevlar" aramid fiber. In addition, it will cover some work related to our continuing effort to find equivalent or better aramid fibers at lower cost. These include (1) vapor phase polycondensation (no solvent) to prepare poly(1,4-phenylene terephthalamide) and (2) melt spinning of shear anisotropic N,N'-dialkylated aliphatic/aromatic polyamides, e.g., poly(ethylene-N,N'-diethyl-4,4'-bibenzamide). An improved "Kevlar" aramid fiber includes the recently commercialized "Kevlar" 149 with its higher modulus, greater crystallinity, reduced creep and moisture regain, and unique super-crystalline structure. Finally, some new industrial applications of "Kevlar" aramid fibers and the process employed to match properties with market needs are discussed.

2. Background

The discovery in 1965 of high tenacity, high modulus fibers from liquid crystalline solutions of synthetic para aromatic polyamides led to the commercial production of "Kevlar" aramid fiber in 1971 [1]. These developments were possible because of technological breakthroughs in polymerization, fiber spinning and fiber heat-treating processes in the Textile Fibers Department, Pioneering Research Laboratory of the Du Pont Company.

Research initiated in about 1950 on low-temperature processes for the preparation of condensation polymers made possible the synthesis of polymers that were unmeltable or thermally unstable, or both [2]. In particular, low temperature solution polycondensation that employs N-alkylated amides alone or with salts as solvent media, was especially useful for the preparation of aromatic polyamides [3]. These weakly basic organic liquids, such as N,N-dimethylacetamide, N,N,N',N'-tetramethylurea, N-methyl-2-pyrrolidone, diethylcyanamide and hexamethylphosphoramide (usually with added LiCl, $CaCl_2$, etc.) served the dual role of solvent and acid acceptor in these condensations involving acid chlorides and diamines. Although these amides are much weaker bases than primary aromatic diamines, the use of an excess will bring the polymerization to completion without loss of diamine as an inactive hydrochloride salt.

When more flexible aromatic polyamides yielded to these new polymerization techniques and solvents, high performance fibers from stiff previously intractable polymers, such as poly(1,4-benzamide) and poly(1,4-phenylene terephthalamide) became our goal. These extended-chain polyamides are characterized by a high persistence length (Table 1) in which

Table 1. Persistence Lengths of Some Polyamides

Polymer[a]	Calc., (Å)[b]	Found, (Å)[c]
PBA	410	400-600
PPD-T	422	150->450
6-6	~10[d]	

a. PBA   - poly(1,4-benzamide)
   PPD-T - poly(1,4-phenylene terephthalamide)
   6-6   - poly(hexamethylene adipamide)

b. Data from Reference 42

c. Data from References 43-46

d. Private communication from the late P. J. Flory

the chain-forming bonds from the aromatic rings are predominantly para oriented and the amide bonds exist preferentially in the trans configuration with considerable double bond character. The latter is conducive to extended-chain conformations in polymers because of the high barrier to free rotation about the carbonyl-nitrogen bond. In this case the rotational activation energy of the carbon-nitrogen bond is approximately 22 kilocalories per mole as compared to 2-4 kilocalories per mole for a simple carbon-nitrogen single bond. Preparations of these polyamides and related copolyamides by low temperature polycondensation are described in a number of patents and publications [4-8,16,17].

In a search for suitable spinning solvents for these essentially non-melting aramids, we found that poly(1,4-benzamide) unexpectedly formed a liquid crystalline solution in tetramethylurea containing 6.5% LiCl. This solution, which consists of randomly oriented domains or arrays of essentially parallel arranged rod-like polymer chains and combined solvent, exhibited the now very familiar stir opalescence, optical birefringence, and phase diagrams which show a critical concentration for separation into anisotropic and isotropic phases. The corresponding AA-BB system, poly(1,4-phenylene terephthalamide) formed lyotropic liquid crystalline solutions at lower concentrations in solvents such as HMPA/NMP/CaCl$_2$. For higher solubility, strongly interacting acids, such as H$_2$SO$_4$ were required [6]. The more concentrated solutions thus obtained promote better alignment of molecules with fewer defects between lamellae and, hence, as-spun fibers with higher tenacities [9]. Structures and properties of these lyotropic aramid solutions are described by Panar and Beste [10] and Baird [11].

Because of a decrease in viscosity as an extended-chain aromatic polyamide solution is transformed from an isotropic to an anisotropic state, spinning solutions (dopes) of higher molecular weight polymer with high solids content and a useful range of fluidity can be prepared. The liquid crystalline domains in these spinning dopes are mainly aligned by extensional flow after extrusion through a spinneret. The high degree of alignment of the polymer domains achieved directly in spinning and low to modest attenuation of the threadline below the spinneret contribute in large measure to the high strength and high modulus of as-spun fibers. In addition, the slower relaxation time of these liquid crystalline solutions allows this orientation to be retained during the coagulation step. This high as-spun fiber orientation is directly related to high fiber tenacity and modulus and differentiates these liquid crystalline polymers from conventional as-spun fibers.

Initially, anisotropic polyamide solutions were dry- and wet-spun [4, 12, 13]; later, air-gap wet spinning was employed for more concentrated sulfuric acid solutions of poly(1,4-phenylene terephthalamide) [14]. The spinning of such polyamide solutions and the related fiber structure and properties are reviewed in detail by Jaffe and Jones [15].

Properties of extended-chain aromatic polyamides extruded from liquid crystalline solutions were improved further by a very brief heat-treatment (seconds) at temperatures above 350°C [14,24]. This heat-treatment which involves some tension but no drawing in the conventional sense appears to be mainly a further crystallization process to remove defects and results in improved perfection of the lateral order of polymer chains.

The heat-treatment response of aramid fibers is dependent on the structure and molecular orientation of the as-spun fiber and the method of spinning. In dry- and wet- spun

fibers, there is generally an increase in both tenacity and
modulus together with increases in crystallinity, crystalline perfection and orientation. In well oriented air-gap
wet spun fibers of poly(1,4-phenylene terephthalamide) there
is virtually no change in tenacity, but there is an increase
in modulus, crystalline perfection and orientation. Copolyamides which are air-gap spun generally exhibit fiber
property improvements more like those obtained with wet-spun
homopolymer fibers.

2.1. "Technora" aramid. A more recent commercial development is "Technora" aramid fiber by Teijin, Ltd. Unlike
the stiff-chain "Kevlar" aramid that is spun from a liquid
crystalline solution directly to oriented high tenacity high
modulus fibers, "Technora" aramid is spun from an isotropic
solution and drawn 10X at 490°C to fibers with tensile properties similar to "Kevlar" 29 (T/E/Mi = 25 gpd/4.4%/570 gpd)
[30]. This semi-stiff-chain copolymer from terephthalic
acid (50 mole %), p-phenylenediamine (25 mole %) and 3,4'-
oxydianiline (25 mole %) is soluble in N-methyl-2-
pyrrolidone/CaCl$_2$ at about 10% concentration and thus does
not require strong acids for spinning. Its homogeneous morphology leads to very good hydrolytic stability while its
greater molecular flexibility results in limited modulus.

3. Molecular Structure of Extended-Chain Aromatic
   Polyamides

   Aromatic polyamides with the potential to form liquid
crystalline solutions and high strength, high modulus fibers, contain a variety of stiff structures, but the common
feature is that the components exhibit essentially coaxial
or parallel extension of the chain-forming bonds. Polymers
can contain aromatic, cycloaliphatic and heterocyclic rings,
and the rings may have substituents, such as chloro, fluoro,
methyl, phenyl and other selected groups. In addition,
other chain-extending units, such as trans vinylene, azo and
azoxy may be used. Within certain limits, structures with a
bend in the backbone or even flexible methylene groups may
be introduced.

   The important feature is that these polyamides form
liquid crystalline solutions, and that these, in turn, are
spinnable directly into well oriented, high strength and
high modulus fibers. Examples of such polymers are found in
Tables 2 and 3. To date, no polyamide that contains only
liquid crystalline side chains has been spun directly into

Table 2. Fibers From Lyotropic Polyamides

Polyamide From	Poly ηinh,[a] dl/g	Fiber
p-Aminobenzoic Acid	1.4	AS HT 525°C
2-Chloro-p-phenylenediamine Terephthalic Acid	3.7	AS HT 445°C
p-Phenylenediamine Terephthalic Acid	6.6	AS HT 350°C
p-Phenylenediamine Terephthalic Acid	"	AS HT 350°C
p-Phenylenediamine 4,4'-Azodibenzoic Acid	2.9	AS HT 500°C
p-Phenylenediamine 2,5-Pyridinedicarboxylic Acid	5.8	AS
p-Phenylenediamine 2,6-Naphthalenedicarboxylic Acid	3.6[b]	AS
2-Methyl-p-phenylenediamine Terephthalic Acid	3.1[b]	AS
3,8-Diaminophenanthridinone Terephthalic Acid		AS HT 450°C

a - Determined in sulfuric acid at 30°C at 0.5 g/100 ml conc.

b - Fiber ηinh

c - gpd (0.8826) (density, g/cc) (0.1) = GPa

oriented high tenacity, high modulus fibers. It is concluded that high level chain extension of the backbone
itself is required for obtaining high mechanical properties.

T, gpd^C	E, %	Mi, gpd^C	Den.	OA°	Ref.
8.3	3.1	509	3.1	18	4
16.9	1.9	1040	2.7	10	"
11.5	5	411	725	31	13
14.4	1.4	1024	674		"
25.1	3.2	779	191	11.2	14
23.9	2.1	1130	177	7.9	"
31.0	4.8	710	1.9	11.2	14
32.0	3.7	920	1.8	7.9	"
4.3	5.6	202	3.1	29	16
7.8	2.2	377	2.6	15	"
18.0	5.8	471	3.5	22	17
17.0	4.9	470	2.0		23
16.0	2.5	780	2.1		23
12.5	3.3	426	3.0		19
25.5	2.3	1090	2.5		"

4. Synthesis of Extended-Chain Aromatic Polyamides

Low temperature solution polycondensation still remains the process of choice for the production of high molecular weight para aromatic polyamides on a commercial scale.

Table 3. Fibers From Lyotropic Copolyamides

Polyamide From	Mole %	Poly ninh,[a] dl/g
p-Aminobenzoic Acid	14	6.2
p-Phenylenediamine	43	
Terephthalic Acid	43	
p-Phenylenediamine	50	6.5
4,4'-Bibenzoic Acid	27.5	
Terephthalic Acid	22.5	
p-Phenylenediamine	50	4.61
Fumaric Acid	10	
Terephthalic Acid	40	
p-Phenylenediamine	50	4.3[b]
t-1,4-Cyclohexanedicarboxylic Acid	12.5	
Terephthalic Acid	37.5	
p-Phenylenediamine	50	2.7[b]
Sebacic Acid	2.5	
Terephthalic Acid	47.5	
p-Phenylenediamine	42.5	3.09
Piperazine	7.5	
Terephthalic Acid	50	
p-Phenylenediamine	40	6.06
3,7-Diaminodibenzothiophene-5,5'-dioxide	10	
Terephthalic Acid	50	
p-Phenylenediamine	40	6.18
3,8-Diaminophenanthridinone	10	
Terephthalic Acid	50	
p-Phenylenediamine	32.5	6.12
5-Amino-2-(p-aminophenyl) benzoxazole	17.5	
Terephthalic Acid	50	

a - Determined in H_2SO_4 at 30°C at 0.5 g/100 ml conc.
b - Fiber ninh
c - gpd (0.8826) (density, g/cc) (0.1) = GPa

Other procedures based on the use of acid, rather than acid chloride, have been developed by Yamazaki, et al. [25] for poly(1,4-benzamide) (ninh 1.7); by Higashi et al. [26,27] for poly(1,4-phenylene terephthalamide) (ninh 4.5), and by Preston et al. [28] for block copolyamides (ninh 4.75).

Fiber	T, gpd^c	E, %	Mi, gpd^c	Den.	OA°	Ref.
AS	9.2	10.1	268	3.4	28	5
HT 610°C	20.1	2.0	1069	2.8	10	"
AS	9.7	8.7	256	3.4	40	5
HT 600°C	22.5	3.2	668	2.8	18	"
AS	24.3	5.72	676		17	18
AS	22.0	4.5	540	1.4		23
AS	16.0	4.1	460	4.0		23
AS	21.1	5.1	504	4.0		20
AS	26.1	4.89	525	3.7		21
HT 450°C	37.5	3.9	971	3.4		"
AS	25.5	4.4	648	3.4		21
HT 500°C	34.2	3.2	990	3.1		"
HT 550°C	36.0	3.2	1053	3.1		"
AS	20.0	8.0	700	2.6		22
HT 420°C	29.0	3.0	1050	2.3		"

These phosphorylation reactions involve the use of molar equivalents of triphenyl phosphite and a mixed solvent of pyridine and N-methyl-2-pyrrolidone with LiCl and CaCl$_2$, (alone or in combination) at about 80° to 120°C.

4.1. Vapor Phase Polycondensation (VPP)

Significant cost reductions would be possible if high
molecular weight para aramids could be prepared without the
use of a solvent. Vapor-phase polycondensation of poly(1,4-
phenylene terephthalamide) by Shin, et al. [29] resulted
from such a goal. In general, the process involves vaporizing the polyamide-forming intermediates (diamine and acid
chloride), diluting with an inert gas that serves as a carrier, feeding to and reacting the mixed intermediates in a
reaction zone heated in the range of about 150° to about
500°C for a very short period of time (about 0.01 second to
about 5 seconds or longer) in equipment ranging from simple
pipeline reactors to fluidized bed reactors. Yields of high
molecular weight polymers (ηinh >3.0) have varied from about
15% for the former type of reactor to nearly 100% for the
latter. Any unreacted intermediates, by-products, such as
hydrogen chloride, and polymer carried from the reactor zone
in the hot effluent stream are separated downline after
cooling. The filtered stream is scrubbed to remove hydrogen
chloride and the inert gas can be recycled.

The success of the VPP process is influenced by certain
factors. A substrate reactor containing a fluidized or moving bed of inorganic (glass beads, sand, sodium chloride) or
organic (polymer) particles gives a higher yield of high
molecular weight aramid than simpler equipment. Although
the molar ratio of polymer reactants is not critical, an
equimolar balance is preferred in actual practice. The
total concentration of monomers in the reaction zone can
range from about 2 to 50 mole % with the remainder being an
inert gas, preferably nitrogen, but for a substrate reactor
a maximum of about 10 mole % is preferred. Higher concentrations of monomers make reaction temperature control more
difficult. The total residence time of monomer vapors and
inert gas in the reaction zone is maintained at about 0.01
to about 5 seconds or longer. However, with temperatures
greater than about 325°C, the gas residence time should be
less than about 1.5 seconds and the polymer residence time
in the reactor should be preferably less than 30 minutes in
order to avoid an excessive amount of branching in the polymer. The preferred temperature in the reaction zone is between about 250°C for the lower limit in order to maintain a
favorable rate of polymerization and about 400°C for the
upper limit in order to avoid excessive formation of branched polymer. For polymer of fiber quality, the preferred
process consists of a monomer concentration of between 3 and
7 mole % at a reaction temperature between 250 and 380°C in
a substrate reactor containing a bed of particles.

With vapor-phase polycondensation it was possible to prepare poly(1,4-phenylene terephthalamide) that was readily spinnable to fibers with tenacities of about 20 gpd as long as polymer inherent viscosity was about 4.0 dl/g. Polymer with higher inherent viscosities contained varying amounts of branching, which increased with intermediate concentration, reaction temperature and reaction time. Such branching affected spinning performance. Table 4 shows the effect

Table 4. Poly(1,4-Phenylene Terephthalamide) By Vapor-Phase Polycondensation Effect of Reaction Temperature and Time on Polymer and Fiber Properties

Reaction[a] Temp., °C	Polymer Residence Time	Polymer η_{inh}	$\eta_B\eta_{BO}$[b]	Yarn[c] T/E/Mi/Den.	Spin[d] Performance
340	15 min.	3.6	1.8	20.2/2.3/895/800	Good
350	"	4.0	2.1	19.6/3.6/520/700	Best
365	"	4.5	2.3	16.5/3.4/465/600	Good
375	"	4.6	2.8	10.2/2.8/430/740	Fair
390	5 min.	4.5		12.8/2.9/450/730	Fair
390	10 min.	5.4		-----	Poor[e]

a - Polycondensation of p-phenylenediamine and terephthaloyl chloride with nitrogen as a diluent in a short pipeline ("cold wall") reactor. Polymer was deposited on a nickel sleeve. Monomer concentration 7%.

b - η_B/η_{BO} = branching ratio, where η_B is the viscosity at 60°C of a 2-4.5 weight percent solution of a sample polymer in 100% sulfuric acid and η_{BO} is the viscosity of a control polymer solution. The control polymer of the same inherent viscosity as the sample polymer is prepared in a solvent at low temperatures that exhibits essentially no branching. Viscosity measurements are made with a Brookfield Viscometer (Model HBT).

c - T and Mi are in gpd.
d - Air-gap wet spinning from 100% sulfuric acid solutions.
e - Required 115°C for jetting; fibers degraded.

of reaction temperature and time on polymer and fiber properties for polymer prepared in a "cold-wall" pipeline reactor. A comparison of poly(1,4-phenylene terephthalamide) fiber obtained from polymer prepared by vapor-phase polycondensation and by low-temperature solution polycondensation is presented in Table 5. The data show that at an inherent viscosity of about 3.7 the two polymers are equivalent.

Spin dopes (19-20% polymer concentration in 100% sulfuric acid) of poly(1,4-phenylene terephthalamide) prepared by VPP

Table 5. A Comparison of Fiber from PPD-T[a] Prepared by Vapor-Phase Polycondensation and Low-Temperature Solution Polycondensation

As-Spun	VPP Polymer[b]	Low-Temperature Solution Polymer[c]
Yarn ηinh (dl/g)	3.7	3.8
T/E/Mi/Den (gpd)	19.6/3.5%/530/700	19.2/3.9%/445/830
X-Ray[d]		
CI	38	39
OA	20°	21°
20°2θ CS	39Å	38Å
23°2θ CS	41Å	40Å

a - PPD-T = Poly(1,4-phenylene terephthalamide)

b - Polymer prepared in a short pipeline ("cold wall") reactor at 350°C with a nickle sleeve. Polymer ηinh = 4.0 dl/g.

c - Polymer prepared in an amide solvent at about 5-10°C.

d - CI = crystallinity index. OA = orientation angle. CS = crystal size.

have higher viscosities than dopes prepared from polymer of similar inherent viscosity prepared by low-temperature solution polycondensation, indicating the presence of branching in the former. This difference in solution viscosity is expressed as a branching ratio (η_B/η_{Bo}), which is defined in Table 4. It is desirable that this ratio be less than 2.0. Figure 1 is a plot of inherent viscosity vs. branching ratio for poly(1,4-phenylene terephthalamide) prepared by vapor-phase polycondensation under a variety of conditions. Although no direct evidence for amidine formation in the vapor-phase polymerization of p-phenylenediamine and terephthaloyl chloride was observed, model compounds similarly prepared exhibit such structures; hence it is assumed that amidine branches are present in VPP polyamides.

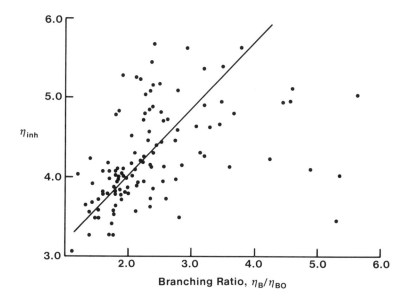

Figure 1. Inherent viscosity vs. branching for poly(1,4-phenylene terephthalamide) prepared by vapor-phase polycondensation (pipe-line cold wall reactor with removable sleeve).

Poly(1,4-phenylene terephthalamide) prepared by the VPP process has been used in the preparation of fiber, films and molded objects, however, branching is a limiting factor for some of these applications.

4.2. Melt polymerization and/or melt spinning. Another approach to cost reduction is through melt preparation and/or melt spinning of polymers. A variety of para-linked linear polyesters and polyazomethines have been found to form optically anisotropic melts from which oriented fibers with high moduli can be melt spun. Analogous melt spinnable polyamides are hard to find because hydrogen bonding produces polymers with melting points too high for melt spinning; however, N-alkyl or N-aryl substitution in polyamides can lower melting points sufficiently to allow melt spinning. In a search for such polyamides, a group of shear anisotropic N-alkyl substituted aliphatic-aromatic polyamides was identified. See Table 6 for their structure and melting points. Of these, poly(ethylene-N,N'-diethyl-4,4'-bibenzamide) (N,N'-DEt2-BB) was further evaluated because a melt of this polyamide exhibited a high degree of order

Stephanie L. Kwolek et al.

Table 6. N-Substituted Aliphatic-Aromatic Polyamides
Forming Shear Anisotropic Melts

$$\left(\!\!-N-R^1-\overset{\overset{\displaystyle R}{|}}{N}--\overset{\overset{\displaystyle O}{\|}}{C}-Ar-\overset{\overset{\displaystyle O}{\|}}{C}\!\!-\right)\ a$$

R	R^1	Ar^b	ηinh^c	m.p., °C
C$_2$H$_5$	-CH$_2$-CH$_2$-	T	0.94	175^d
C$_2$H$_5$	"	BB	1.24	258^e
C$_2$H$_5$	"	"	2.07	264^f
n-C$_3$H$_7$	"	"	0.98	247^f
n-C$_4$H$_9$	"	"	0.53	229^e
n-C$_4$H$_9$	"	"	0.66	240^f
C$_2$H$_5$	-CH$_2$CH $\overset{\|}{\underset{\displaystyle CHCH_2-}{}}$	"	0.53	170^d
C$_2$H$_5$	-CH$_2$C≡CCH$_2$-	"	0.70	175^d
C$_2$H$_5$	-CH$_2$-CH$_2$-	2,6-N	0.69	140^e

a - Polymers prepared from diamine and acid chloride by
    interfacial polymerization in water/methylene
    chloride with Na$_2$CO$_3$ as an acid acceptor.

b - T = terephthalic acid, BB = 4,4'-bibenzoic acid,
    2,6-N = 2,6-naphthalenedicarboxylic acid.

c - ηinh (dl/g) determined at 30°C in conc. sulfuric
    acid at 0.5 g/100 ml conc.

d - Hot stage polarizing microscope.
e - Differential thermal analysis (DTA).
f - Differential scanning calorimetry (DSC).

under small stresses. To illustrate this point, a fiber
with an orientation angle of about 10° was hand-pulled from
a melt at 300°C. The degree of order or orientation, however, was highly sensitive to melt temperature. Figure 2
shows a photomicrograph of a sheared melt of N,N'-DEt2-BB at
265°C between crossed polars. A high degree of birefringence is evident and this persists for an extended period of
time. These characteristics, also present in lyotropic
polyamides, showed promise of utility here.

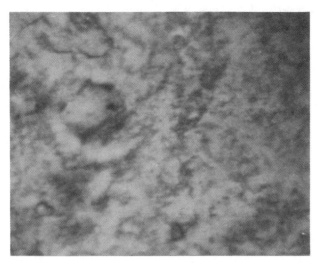

Figure 2. Photomicrograph of N,N'-DEt2-BB melt at 265°C
between cover slips on the hot stage of a polarizing microscope.

These N-alkylated aliphatic-aromatic polyamides were prepared by low-temperature interfacial polycondensation since
this method gave the highest inherent viscosities, and melt
polymerization was unsatisfactory. N,N'-DEt2-BB from N,N'-
diethylethylenediamine and 4,4'-bibenzoyl chloride was obtained in inherent viscosities as high as 2.5 dl/g (H_2SO_4).
Polymer at this molecular weight level exhibited shear
anisotropy in the melt from about 300 to about 360°C.
Polymer with lower inherent viscosity exhibited different
thermal behavior as seen in Table 7.

N,N'-DEt2-BB (ηinh 1.1) was melt extruded to highly
oriented (O.A. 20°) as-spun fibers with T/E/Mi as high as
5.0/20%/85 gpd. Contrary to expectations, higher molecular
weight polymers (ηinh 2.4) did not yield fibers with tensile
properties superior to those obtained for lower molecular
weight polymer as seen in Table 8. This rather low level of
tensile properties can be attributed to a number of factors:
presence of gas bubbles in the fiber, polymer degradation
which increased with spin temperature, and heat-treatment
limited to temperatures below 250°C where fiber softens and
deorients.

In order to reduce polymer melt viscosity to an extrud-
able range, the polymers had to be melt spun above 300°C and

Table 7.
Melts of Poly(Ethylene-N,N'-Diethyl-4,4'-Bibenzamide)[a]

Polymer ηinh	Shear Anisotropic Melt	Isotropic Melt
0.90	297-313°C (fluid)	315°C
2.07	299-310°C (viscous) 311-330°C (fluid)	335°C
2.35	302-325°C (viscous) 330-358°C (less viscous)	360°C

a - Melts were observed on a hot stage of a polarizing microscope.

Table 8. Fiber Properties of N,N'-DEt2-BB

Polymer η inh	Fiber ηinh	Sprt. Temp.	Fiber	Fiber Properties* T	E, %	Mi	Den.	Cryst.	OA
1.24	0.85	313°C	AS	4.0	18	82	---	Low	24°
			HT - 185°C[a]	5.0	7.6	144	---	Med.	16°
1.28	1.07	322	AS						
			HT - 200°C[b]	7.8	9.8	179	---	Med.	12°
1.54	1.13	340	AS						
			HT - 200°C[c]	3.7	8.5	165	58	Med.	13°
2.35	1.06	344	AS	0.6	2.1	29	252	Low	30°
			HT - 195°C[d]	2.8	7.6	127	38	---	---

* T and Mi are in gpd.
Heat-Treatment: a - 1.1X over a 9 cm hot disc plus 0.5 hr taut in oven.
b - 1.1X in tube oven at 20 fpm.
c - 7.7X over a 4 cm hot bar.
d - 5.8X over a 4 cm hot bar.
Codes: AS - As-spun
Poly - Polymer
HT - Heat Treatment
Sprt - Spinneret

as high as 340°C, which is considerably above the DTA melting point of about 260°C. In general, DSC and TGA analyses
(heating rate 20°C/min.) showed good polymer stability to at
least 425°C in nitrogen and to at least 360°C in air. However, isothermal heating of N,N'-DEt2-BB polymer at 300°C in
nitrogen presented a different picture for thermal stability. A thin film of polymer (ηinh 2.35), after soaking in
dilute ammonium hydroxide, water and methanol, was studied
by infra-red between NaCl plates at 300°C in nitrogen over a
period of 2 hours. Up to 300°C the spectra of N,N'-DEt2-BB
were unchanged; however, within less than 10 minutes of exposure at 300°C, the amide carbonyl at 1635 cm^{-1} began to
show a loss in intensity and two other absorption bands began to appear -- one at 1715 cm^{-1} and the second at
2220 cm^{-1}. With further heating the amide carbonyl band at
1635 cm^{-1} disappeared altogether, while the nitrile band at
2220 cm^{-1} increased in intensity. The absorption band at
1715 cm^{-1} also increased in intensity and was joined by a
weaker band at 1770 cm^{-1}. These two absorption bands have
been assigned to imide carbonyl. Another band at about
1390 cm^{-1} also appeared with heating and has been assigned
to an imide carbon-nitrogen bond.

N-alkyl substitution of the aliphatic diamine increases
two to three-fold the possibility for β-hydrogen elimination. A possible scheme (I) for this type of reaction is:

↓ 300°C

↓ 300°C   Secondary Amide

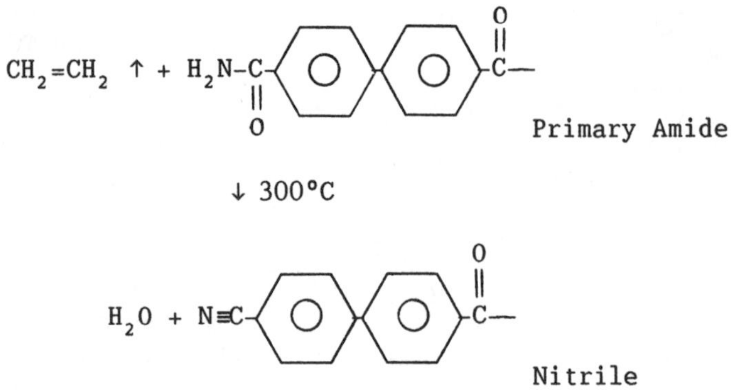

Primary Amide

↓ 300°C

Nitrile

IR spectra of N,N'-DEt2-BB heated at 300°C in nitrogen showed the presence of nitrile but no primary or secondary amide. When off-gases generated during this heating were monitored using a 20 meter IR gas cell, there was spectral evidence of ethylene and carbon dioxide, and nitrile remaining in the cell. If the temperature was increased to 350°C, there was spectral evidence of ethane, carbon dioxide, ethylene and nitrile. Similarly, Bailey and Bird [47] have obtained acetamide and a mixture of two olefins from the pyrolysis of the tert-alkylamide, N-(1,1,3,3-tetramethylbutylacetamide) but at a higher temperature, 510°C.

The presence of imide carbonyl and imide carbon-nitrogen bond in the spectra suggest branching. A possible reaction scheme (II) which would involve the secondary amide formed in scheme (I) is the following:

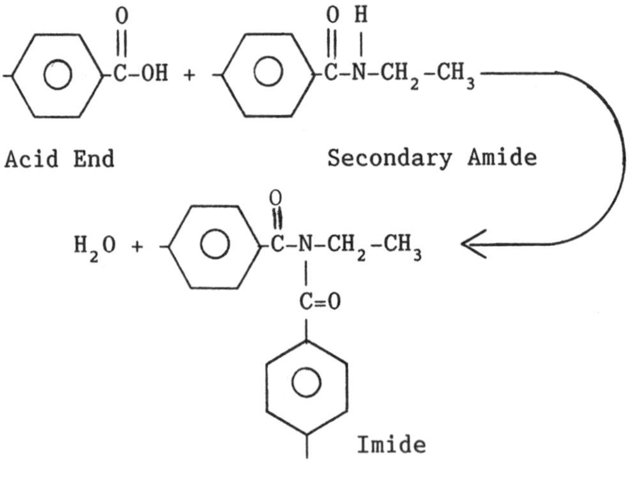

Acid End　　　　　　　　　Secondary Amide

Imide

Weight loss was determined by TGA under isothermal conditions at 300°C in both nitrogen and air. In nitrogen N,N'-DEt2-BB polymer lost 0.8% weight in 15 minutes and 1.30% in 2 hours; in air the polymer lost 1.05% weight in 15 minutes and 6.93% in 2 hours (weight loss based on sample weight at 100°C).

Although poly(ethylene-N,N'-diethyl-4,4'-bibenzamide) fiber exhibits good hydrolytic stability toward base, it loses tensile properties at elevated temperatures. At 150°C the fiber maintained about 38% of tenacity and 26% of modulus based on initial tenacity/modulus values of 4.8/152 gpd.

5. Para Aramid Supramolecular Structure

The currently accepted supramolecular structure of "Kevlar" aramid was first clearly proposed publicly by Dobb, et al. [31,32]. R.G. Scott and P. Antal had introduced similar ideas in the early development of "Kevlar" at Du Pont. Dobb described a structure consisting of radially oriented hydrogen-bonded crystalline sheets pleated about every 500 nm, these pleats being readily observable by optical microscopy. See Figure 3. Fracture of intact filaments, however, revealed a fibrillar structure with elements of about 600 nm and no particular indication of radial or pleated structure.

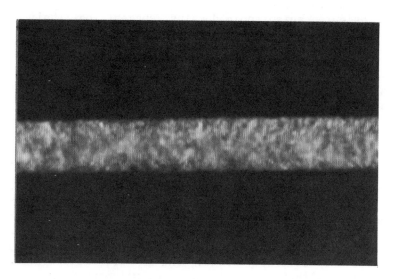

Figure 3. "Kevlar" 29 fiber pleat structure. Optical micrograph at 1500X, crossed polars.

In our work "Kevlar" aramid is generally considered to have a radially symmetric pleated structure and simultaneously a microfibrillar structure. A variety of air-gap wet-spun aramid fibers have been found to have this radially symmetrical hydrogen-bonded sheet structure and the expected positive lateral crystallite orientation and lateral birefringence while possessing a variety of microfibrillar laydown patterns. Commercial "Kevlar" 29 has a tangentially symmetrical microfibrillar laydown with pleating at the expected ~500 nm wavelength (Figures 3 and 8). Other "Kevlar" aramid fibers may have a swirled variant of this structure or even a variation in pleating (spacing, angle) depending on processing conditions and polymer composition. The radial orientation of the hydrogen-bonded crystalline sheets (200 crystal planes) is a defining characteristic of air-gap wet-spun PPD-T regardless of the symmetry of the microfibrillar laydown. It is probably a direct consequence of supramolecular structures in solution being tangentially compressed in the air-gap and during desolvation of the swollen filaments.

Microfibrils have long been postulated as the fundamental units of "Kevlar" structure, but their dimensions have been a topic of some controversy. Panar, et al. found the diameters in broken fibrillated fiber ends to be around 600 nm [33]. However, consideration of the initial nematic dope structure and the fact that microfibrils are generally considered to be the coagulated and solidified residue of the axially oriented anisotropic domains, suggest that their fundamental diameter is in a much lower range. In our study, routinely as-spun "Kevlar" aramid fibers with microfibril cross-sections in the 30-100 nm range have been observed in tension broken (brittle mode) fiber-ends which were mounted, gold coated and examined with a scanning electron microscope. These fibers had been degraded previously by boiling in 20% aqueous NaOH for 1-5 days according to a method described by Horio, et al. [34], that allows for molecular weight degradation without dissolving any significant portion of the polymer. These microfibril samples are laid down in a tangential arrangement analogous to the rings in a tree. Longitudinal splits in them revealed the characteristic pleats with spacing in the expected ~500 nm range. Takayanagi, et al. [48] have found PPD-T microfibrils with diameters of 15-30 nm dispersed in a fractured surface of a composite formed from blends of poly(p-phenylene terephthalamide) and aliphatic polyamides. The latter were prepared by extruding sulfuric acid solutions of polymers into a large amount of water. Sawyer [49] has assigned a value of

50 nm (width) to microfibrils of thermotropic copolyesters,
aramids and ordered polymers in a Liquid Crystal Polymer
Structural Model.

The variety of data on microfibril size suggests that the
latter is dependent on spinning or other processing conditions. The ultimate microfibril diameter may be the width
of a crystallite, particularly if a microfibril is allowed
to split lengthwise between its crystallites.

The other major feature of intermediate aramid structure,
the optically visible pleats, as shown in Figure 3, are a
direct consequence of the anisotropic dope structure and the
spinning process. During coagulation, the tension associated with, and necessary for fiber formation, apparently
causes the formation of a skin-core structure which lowers
fiber strength and toughness and introduces orientational
nonuniformities into the coagulated fiber. One can visualize that on initial contact with the bath, the skin coagu-
lates and is strongly oriented while the core relaxes
rapidly. Since the skin is highly oriented, growth in filament length during coagulation and drying is relatively
restricted. The interior of the filament, basically a
network of microfibrils with off-axis orientation, would
elongate if it could, but since it is restrained by the
skin, folds or crumples instead. This produces pleats with
an angle of about 7° (5-10° range), which pattern is maintained during drying [33,50]. The pleats which involve
folding of elements of significant structural size should
produce voiding in the region of highest strain (bending)
resulting in a region susceptible to hydrolytic attack [51].
It has been observed by scanning electron microscopy that
hydrolysis is favored at the pleat bends.

6. "Kevlar" 149 Aramid With Ultra-High Modulus

6.1. Properties

A new ultra-high modulus form of "Kevlar" aramid, designated "Kevlar" 149, differs significantly from previous
"Kevlar" products in its higher crystallinity and unique
super-crystalline structure [35]. The yarn has tensile
tenacity/modulus of 2.3-2.6/162-175 GPa (18-20/1240-1340
gpd), compared to commercial "Kevlar" 49 with a tenacity/
modulus of about 2.8/121 GPa (22/945 gpd) (depending on type
and denier). Yarn properties for these products are summarized in Table 9.

Table 9. Yarn Properties

	"Kevlar" 149*		"Kevlar" 49*		"Kevlar" 29*	
	GPa	gpd	GPa	gpd	GPa	gpd
Tenacity	2.4	19.0	2.8	22.0	2.8	22.0
Elongation (%)		1.3		2.5		3.5
Modulus	165	1300	125	970		650
Density (g/cc)		1.48		1.45	83	1.45
Moisture Regain** (%)		1		4		7

* "Kevlar" 29, "Kevlar" 49 and "Kevlar" 149 are Du Pont Registered Trademarks.

** 70°F, 65% RH

"Kevlar" 149 yarn can be produced with varying modulus levels up to about 40% higher than the equivalent "Kevlar" 49 aramid, but the higher modulus is obtained with some loss in tenacity. Figure 4 shows the stress-strain curves for a fully developed "Kevlar" 149 yarn and for commercial aramids. "Kevlar" 149 is unusual in that there is no maximum

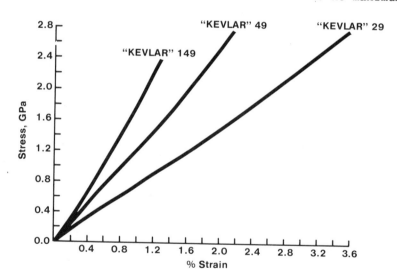

Figure 4. "Kevlar" fiber stress-strain curves. Rate of strain 50%/min.

initial modulus as there is for most organic fibers. Instead there is a steady increase in modulus with strain
level as shown in Figure 5. For a yarn of this type, with a

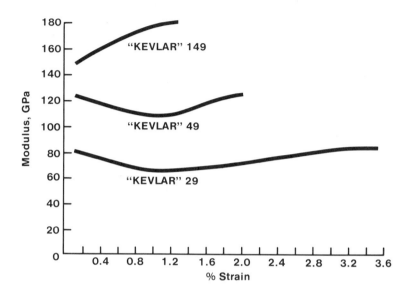

Figure 5. "Kevlar" fiber modulus-strain curves. Rate of
strain 50%/min.

constantly increasing slope of the stress-strain curve, the
point of slope measurement for modulus must be clearly defined. Since the applied stress is more easily defined than
absolute strain, and to simplify comparison with our existing data base, modulus has been defined as the slope of the
stress-strain curve between 2000 and 2450 grams stress for
400 denier yarns. The moduli for other deniers were taken
at approximately equivalent specific stress levels. Under
these conditions, "Kevlar" 149 samples with tested modulus
of >1400 gpd (183 GPa) in yarn and 30 Mpsi (207 GPa) in
strand have been made [39].

This high density fiber has a moisture regain that is
approximately one-third to one-fourth that of "Kevlar" 49.
This lowered moisture regain should be particularly useful
in composites subjected to cyclic extremes of temperature
and/or relative humidity and in applications where low dielectric constant is critical. The greater hydrolytic
stability of "Kevlar" 149 over "Kevlar" 49 in both acidic

and basic environments appears to be primarily the result of higher crystallinity and lower moisture pickup, which factors should reduce the accessibility of the hydrolyzable amide bond. The general lack of a highly regular pleat structure with its hydrolytically sensitive pleat points or bends in "Kevlar" 149 also is an important factor in improved stability. The increase in crystallinity and overall molecular interaction have also improved fiber creep resistance, which has been measured at 0.012% per decade with a coefficient of variation of 10% compared with "Kevlar" 49 aramid yarns which range around 0.02% per decade (50% of ultimate break load, 22°C). This property is of critical importance in any permanently tensioned system, especially in applications where reset is difficult or impossible.

Strength conversion from yarn to composite form is excellent (>130%). While yarn strength is about 10% lower than "Kevlar" 49, strand strength is equivalent. Strand testing in low temperature cure epoxy resin shows a modulus of up to 30 Mpsi (207 GPa), sufficient to provide near perfect modulus and strain matching with most high strength carbon fibers -- a feature very desirable in hybrid composite end uses.

6.2. Structure of "Kevlar" 149. The equatorial wide angle X-ray spectrum shows two major peaks at 20°2θ and 23°2θ, presumably corresponding to the Northolt type structure [36] characteristic of PPD-T and a number of PPD-T copolymer fibers. However, the peak definition indicates a substantially higher level of crystallinity and crystal size than for "Kevlar" 49 as seen in Table 10. Also, the unit

Table 10. Crystalline Parameters

	"Kevlar" 49	"Kevlar" 149
O.A.	12°	12°
C.I.	60	75
20°2θ A.C.S.	65Å	120Å
22°2θ A.C.S.	55Å	80Å
Unit Cell:		
a	7.73Å	7.88Å
b	5.18Å	5.18Å
c	12.82Å	12.94Å
Calc. D (g/cc)	1.54	1.50

cell dimensions for "Kevlar" 49 and "Kevlar" 149 differ
significantly, the "Kevlar" 149 being somewhat larger in
overall volume giving a theoretical density closer to that
observed in "Kevlar" 49. The paracrystalline distortion
parameter g_{II}, based on meridional X-ray reflections, and
the crystalline orientation angle both correlate with modulus up to about 130 GPa, in general agreement with Barton
[37]; beyond 130 GPa no statistically significant correlation was observed. Also in general agreement with Barton,
there is no correlation between axial crystallite length and
modulus.

Longitudinal supercrystalline features of the fiber are
revealed by optical and scanning electron microscopy
(Figures 6 and 7). Little of the highly regular pleat
structure [33] characteristic of the incumbent "Kevlar"
fibers is shown, although larger, irregularly spaced density
and/or orientational differences appear.

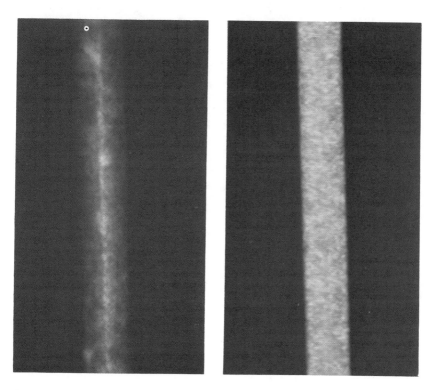

Figure 6. Longitudinal pleat structure of "Kevlar" 149
(left) and "Kevlar" 29 (right) fibers. Optical micrographs
at 1200X, crossed polars.

Figure 7. Longitudinal pleat structure of "Kevlar" 149 fiber
(upper) and "Kevlar" fiber heat-treated at 400°C, 10 seconds
(lower). SEM at 10,000X.

For a study of transverse fiber structural features,
caustic hydrolysis by the method of Horio, et al. [34] was
used to degrade the polymer. Fiber ends obtained by brittle

fracture of the hydrolyzed fiber reveal differences in supercrystalline structure of "Kevlar" 149 and "Kevlar" 29 and 49. In Figure 8, degraded "Kevlar" 29 fiber fractures cleanly across the pleat face, presumably due to faster hydrolysis of the relatively voided, lower crystallinity region at the pleat. The fracture plane steps up or down in integral pleat steps as resistance is encountered. The face appears to be composed of tightly packed discrete particles,

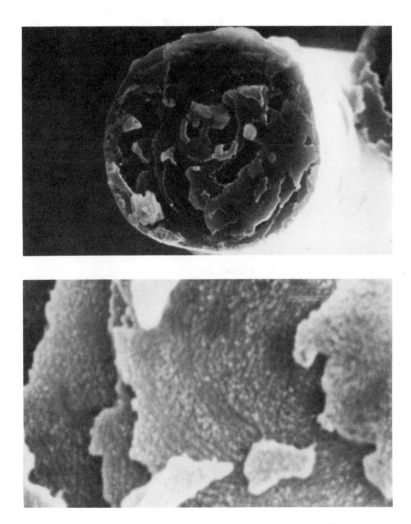

Figure 8. Optical micrographs of "Kevlar" 29 fiber ends at 5000 (upper) and 20,000X (lower) magnifications.

possibly the broken ends of the degraded microfibrillar
fiber elements which comprise the fundamental units of the
supercrystalline network. These microfibrils appear to be
organized in a tangential pattern; this means the well-
established radial order of the crystallites themselves is
contained within individual microfibrils (32,38). The
"Kevlar" 49 fiber (Figure 9) has an intermediate structure

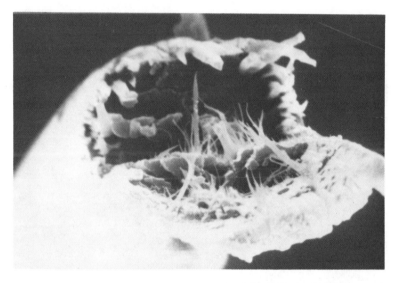

Figure 9. Optical micrographs of "Kevlar" 49 fiber ends at
5000X magnification.

with this pleated structure confined to the fiber core while
the skin becomes a more resistant structure without clean
fracture faces. "Kevlar" 149 itself (Figure 10) appears to
have a substantially different, fine-grained radial structure largely without definable microfibrils or pleats. Also
there is no definable difference between skin and core.

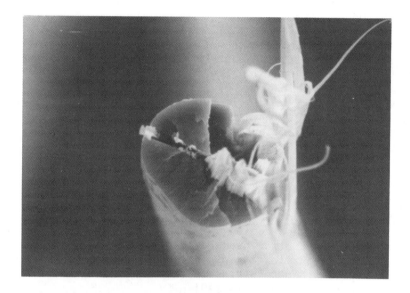

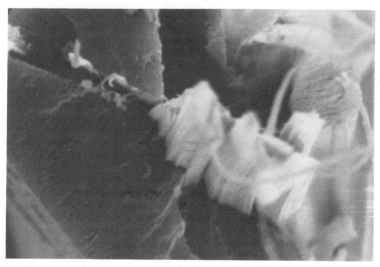

Figure 10. Optical micrographs of "Kevlar" 149 fiber ends at
3000 (upper) and 8000X (lower) magnifications.

7. Applications

Industrial applications of "Kevlar" aramid fiber are the
result of coupling of various attributes of para aramid
fibers with market needs. This process, however, can be
quite complex because it requires a thorough knowledge of
the structure/property relationship of a polymer and/or
fiber, and the use of this knowledge together with selective
design and engineering to adapt these fibers to a particular
application. The latter may have a variety of stringent
requirements. This process which is also called "systems
engineering" requires an early partnership with the potential end-user for on-going feedback. Because aramids have
pioneered the way in a number of industrial areas, it also
has been necessary to develop new processing technologies
suitable to these stiff, low-elongation organic fibers as
contrasted to conventional fibers and metals. In addition,
it has been necessary to overcome industrial prejudices in
favor of the old ways.

"Systems engineering", as described in a previous paper
at this meeting by Dr. Tanner, has been employed successfully in a number of "Kevlar" aramid fiber applications;
namely, reinforcement of belts in radial tires, replacement
of asbestos in friction products, soft body armor, ropes and
cables for use on floating off-shore oil drilling platforms,
and in hybrid fiber reinforced advanced composites for aircraft structural components [40].

For a concise illustration of "systems" technology, the
application of "Kevlar" aramid fiber in the belts of radial
passenger tires is a good example. These fibers in radial
tire belts undergo considerable strength loss from compressive fatigue, especially at belt edges during cornering of
the tire. Two approaches to acceptable fatigue performance
were taken at Du Pont. First, the cord was designed with a
twist level selected to achieve the best balance of tensile
properties and fatigue resistance. This selection was influenced by the knowledge that compressive fatigue resistance is proportional to twist level and that cord tensile
strength and modulus decrease with increased twist.
Secondly, the belt was redesigned by selective folding of
the bottom belt around the edge of the top belt to increase
fiber volume, and, thereby, decrease compressive stresses on
individual filaments in the critical belt edge region normally experiencing greater than 0.5% compressive strain,
i.e., above the compressive yield strain. These design
changes produced a tire with satisfactory performance for

cost effective tire life at a one-to-five replacement ratio
vs. steel cord [40].

More recent developments in "Kevlar" aramid technology
make use of "Kevlar" 149 fiber because of its higher modulus, greater crystallinity, reduced creep, lower moisture
regain and lower dielectric constant vs. current commercial
grades of "Kevlar" aramid. The higher modulus coupled with
high specific strength and damage tolerance inherent in organic fibers should indicate utility in hybrid composites
with carbon fibers, in sporting goods and in fiber optical
cable reinforcement. Its low creep should be useful in
conveyer belt reinforcement, as tension members in large
fabric structures, and in fiber wound metal pressure ves-
sels. The fibers increased hydrolytic stability should be
advantageous in specialty filter applications and other
degradative environments [35].

Some additional new "Kevlar" yarn products include a high
strength yarn with a 20% increase in tenacity ("Kevlar" HT);
light weight "Kevlar" 49 fabric for printed wiring boards;
shorter (1/32-1/4 inch) "Kevlar" fiber forms for reinforcing
molded plastic parts, and low density spun-laced "Kevlar" or
"Nomex" aramid blend structural sheet products for plastic
reinforcement, friction products, thermal and electrical
insulation and fire blocking layers [41].

These are only a fraction of available new "Kevlar"
aramid applications which continue to grow as para-aramid
technology grows. With a continuing interest in light-
weight, energy efficient structures and products, and in
polymer molecular design and super properties, the future
looks bright for para aramid fibers.

8. Acknowledgments

The authors express their appreciation for the work
and/or helpful discussions of P. W. Morgan, J. R. Schaefgen,
T. I. Bair, J. A. Fitzgerald, D. Tanner, P. S. Antal,
H. Shin, R. P. Cellura, F. P. Boettcher, M. R. Samuels,
I. K. Miller, D. E. Plorde, H. Kilkson, R. Barton, Jr.,
R. F. Van Kavelaar, B. D. Chase, C. E. Day and E. Matthews
of E. I. du Pont de Nemours & Company, Inc.

9. References

[1]   Kwolek, S. L. Am. Inst. Chem., The Chemist,
      (Washington, DC), 1980 57(11), 9.
[2]   Morgan, P. W. "Condensation Polymers: By Interfacial
      and Solution Methods"; Interscience, New York, 1965.
[3]   Kwolek, S. L.; Morgan, P. W.; Sorenson, W. R., U.S.
      Patent 3 063 966, 1962.
[4]   Kwolek, S. L. U.S. Patent 3 600 350, 1971; Fr. Patent
      1 526 745, 1968; Chem. Abstr. 1969 71, 4426.
[5]   Kwolek, S. L. U.S. Patent 3 819 587, 1974.
[6]   Bair, T. I.; Morgan, P. W. U.S. Patent 3 817 941,
      1974.
[7]   Kwolek, S. L.; Morgan, P. W.; Schaefgen, J. R.;
      Gulrich, L. W. Macromolecules 1977 10, 1390.
[8]   Bair, T. I.; Morgan, P. W.; Killian, F. L.
      Macromolecules 1977 10, 1396.
[9]   Schaefgen, J. R. In "The Strength and Stiffness of
      Polymers"; Zachariades, A. E.; Porter, R. S., Eds.;
      Marcel Dekker: New York, 1983, Chapter 8, p 327.
[10] Panar, M.; Beste, L. F. Macromolecules 1977, 10, 1401.
[11] Baird, D. G. J. Rheol. 1980, 24(4), 465.
[12] Kwolek, S. L. U.S. Patent 3 671 542, 1972; Reissue
      U.S. Patent 30 352, 1980.
[13] Bair, T. I.; Morgan, P. W. U.S. Patent 3 673 143,
      1972.
[14] Blades, H. U.S. Patent 3 869 430, 1975.
[15] Jaffee, M.; Jones, R. S. In "Handbook of Fiber Science
      and Technology: High Technology Fibers"; Lewin, M.;
      Preson, J., Eds.; Marcel Dekker: New York, 1985;
      Chapter 9, p 349.
[16] Morgan, P. W. U.S. Patent 3 804 791, 1974.
[17] Gulrich, L. W.; Morgan, P. W. U.S. Patent 3 836 498,
      1974.
[18] Morgan, P. W.; Pletcher, T. C. U.S. Patent 3 869 419,
      1975.
[19] Kaneda, T.; Ishikawa, S.; Daimon, H.; Katsura, T.;
      Ueda, M.; Oda, K.; Horio, M. Makromol. Chem. 1982,
      183, 417.
[20] Konomi, T.; Yukimatsu, K.; Katsuo, K.; Yamaguchi, M.
      U.S. Patent 4 011 203, 1977.
[21] Kaneda, T.; Ishikawa, S.; Daimon, H.; Katsura, T.;
      Maeda, T.; Hondo, T. U.S. Patent 4 178 431, 1979.
[22] Nakagawa, Y.; Noma, T; Mera, H. U.S. Patent 4 018 735,
      1977.
[23] Blades, H. U.S. Patent 3 767 756, 1973.
[24] Kwolek, S. L. U.S. Patent 3 888 965, 1975.
[25] Yamazaki, N.; Matsumoto, M; Higashi, F. J. Polym.
      Sci., Polym. Chem. Ed. 1975, 13, 1373.

[26] Higashi, F.; Aoki, Y; Taguchi, T. Makromol. Chem., Rapid Commun. 1981, 2, 329.

[27] Higashi, F.; Ogata, S.-I.; Aoki, Y. J. Polym. Sci., Polym. Chem. Ed. 1982, 20, 2081.

[28] Preston, J.; Krigbaum, W. R.; Kotek, R. J. Polym. Sci., Polym. Chem. Ed. 1982, 20, 3241.

[29] Shin, H. U.S. Patent 4 009 153, 1977.

[30] Ozawa, S.; Nakagawa, Y.; Matsudak K.; Nishihara, T; Yunoki, H. U.S. Patent 4 075 172, 1978. Ozawa, S. Polym. J. 1987, 19(1), 119.

[31] Dobb, M. G.; Johnson, D. J.; Saville, B. P. J. Polym. Sci., Polym. Phys. Ed. 1977, 15, 2201.

[32] Dobb, M. G.; Johnson, D. J.; Saville, B. P. Phil. Trans. R. Soc. Lond. 1979, A294, 483.

[33] Panar, M.; Avakian, M.; Blume, R. C.; Gardner, K. H.; Gierke, T. D.; Yang, H. H. J. Polym. Sci., Polym. Phys. Ed. 1983, 21, 1955.

[34] Horio, M.; Kaneda, T.; Ishikawa, S.; Shimamura, K. Sen-I Gakkaishi 1984, 40(8), T-285-289.

[35] Van Trump, J. E.; Lahijani, J. In "Advanced Materials Technology '87", Vol. 32; Carson, R.; Burg, M.; Kjoller, K. J.; Riel, F. J., Eds.; SAMPE: Covina, CA., 1987, p 917.

[36] Northolt, M. G. Eur. Polym. J. 1974, 10, 799.

[37] Barton, Jr., R. J. Macromol. Sci.-Phys. 1985-1986, B24(1-4), 119.

[38] Dobb, M. G. In "Handbook of Composites", Vol 1; Watt, W.; Perov, B. V., Eds.; Elsevier Science: New York, 1985, Chapter 17, p 673.

[39] Du Pont Kevlar\ 49 Data Manual 1977, p II-B.

[40] Tanner, D.; Fitzgerald, J. A.; Knoff, W. F.; Riewald, P. G. In "Handbook of Fiber Science and Technology: Vol. III, Part B", Lewin, M.; Preston, J., Eds.; Marcel Dekker: New York. In press.

[41] Pigliacampi, J. J.; Zahr, G.E., In "Fifth International Conference on Composite Materials. ICCM-V." [5th: 1985: San Diego, CA); Harrigan, Jr., W. C.; Strife, J.; Dhingra, A. K., Eds.; Metallurgical Soc.: Warrendale, PA, 1985, pp 1545-1556.

[42] Erman, B.; Flory, P. J.; Hummel, J. P. Macromolecules 1980, 13, 484.

[43] Arpin, M.; Strazielle, C. Polymer 1977, 18, 591.

[44] Arpin, M.; Strazielle, C. Makromol. Chem. 1976, 177, 581.

[45] Arpin, M.; Strazielle, C.; Weill, G.; Benoit, H. Polymer 1977, 18, 262.

[46] Metzger Cotts, P.; Berry, G. C. J. Polym. Sci., Polym. Phys. Ed. 1983, 21, 1255.

[47] Bailey, W. J.; Bird, C. N. J. Org. Chem. 1958, 23, 996.

[48] Takayanagi, M.; Ogata, T.; Morikawa, M.; Kai, T. J. Makromol. Sci.-Phys. 1980, B17(4), 591.

[49] Sawyer, L. C. Int. Fiber J. 1987, 2(2), 27.

[50] Roche, E. J., Du Pont Co., Private Communication.

[51] Antal, P. S., Du Pont Co., Private Communication.

STRUCTURE PROPERTY RELATIONSHIPS OF THERMOTROPIC COPOLYESTERS

Michael Jaffe
Hyun-Nam Yoon

Hoechst Celanese Corporation
Celanese Research Company
86 Morris Avenue
Summit, NJ   07901

Introduction

Investigation of the process – structure – property relationships of highly oriented polymers has long been an important technical focus of the Celanese Research Company. Special emphasis has been placed on understanding the behavior of the thermotropic copolyesters, many of which were invented at the Research Company and which form the technical base for the VECTRA(TM) product line. This area has been reviewed by us in the past(see, for example, the work of Calundann(1,2), Jaffe(3), and Wissbrun(4)). The purpose of this paper is to update the status of this research, focusing on the structural origins of mechanical properties.

Background

The importance of mesogenic polymers as high performance materials has been evident since the introduction of aramid fiber by the duPont Company in the late 1960's. Polymer liquid crystals are fully analagous to small molecule liquid crystals, with nematic, smectic and cholesteric textures known. Figure 1 illustrates the molecular order associated with these mesogenic textures. As shown in Figure 2, mesogenicity in polymers can be achieved through a variety of molecular architectures, ie rod-like mainchains, mesogenic side chains off a conventional polymeric backbone and ordered or random block copolymers. The basis of most industrial interest in liquid crystal polymers(LCP) lies in the excellent balance of mechanical properties and processibility inherent in the mainchain nematic types of which VECTRA(TM) and KEVLAR(TM) are well known examples. Figure 3 contrasts the process – structure – property relationships of an LCP with the behavior of a conventional polymer during simple uniaxial flow. Typically, the modulus of an oriented fiber produced from a conventional polymer is about 10 GPa, a fiber spun from a nematic polymer melt or solution at least an order of magnitude higher.

Homopolymer all aromatic polyesters synthesized from para ester monomers, ie terephthalic acid, hydroquinone, p-hydroxybenzoic acid are intractible because they decompose prior to melting and are soluble in no known solvents. Processibility of these polymers is achieved by destroying the backbone perfection through copolymerization with monomers sufficiently straight to preserve mesogenic order. Figure 4 illustrates the chemical approaches which form the patent positions of a number of industrial laboratories working in the LCP area. Figure 5 shows the melting behavior of two thermotropic copolyester compositions as a function of monomer concentration. All of the the thermotropic copolyesters show an alpha transition (glass transition) in the region of 100 to 150 C.

The solid state structure of LCPs has been extensively studied by X-Ray diffraction and optical and electron microscopy. For reviews, see the work of Blackwell et al. (5,6), Windle and Donald(7,8), Sawyer and Jaffe(9). Typical wide angle diffraction patterns are shown in Figure 6. The meridional scatterings of these patterns has been interpretted to reflect the monomer sequences and composition of the chains(very sensitive to order—see work of Blackwell). Overall the X-Ray structure of the LCP's is between two and three dimensional. There have been a number of suggestions as to the nature of the order in the solid state and this question will be dealt with in some detail later.

The morphology of LCP fibers and moldings has been studied extensively by Sawyer and Jaffe(9) and Baer et al.(10,11). The morphology is best described as a hierarchial collection of fibrillar structures spanning more than three orders of magnitude in diameter, as illustrated in the model of the fiber structure shown in Figure 7. Similar models for lyotropic aramid fibers have been published by Panar et al. (12). It has further been shown by Sawyer and Jaffe(9) that the more apparently complex structures noted in moldings and thick extrusions consists locally of the same hierarchy observed in fibers complicated by the flow fields and thermal gradients experienced by the polymer during processing.

Room Temperature Mechanical Properties

Figure 8 shows a plot of the specific modulus of a variety

of fibers as a function of specific strength at room temperature. On this basis, polymers are clearly the structural
materials of choice. In the limit of very high axial orientation always observed in LCP fibers, the tensile modulus
is a function of the inherent chain modulus, the shear
modulus and the chain packing density. Figure 9 shows the
dependence of the tensile modulus of a typical LCP on the
molecular chain orientation and shear modulus as predicted
by the Aggregate Model developed by Ian Ward(13). This
model treats the material as a structure composed of anisotropic elastic units, with all units possessing the elastic
properties of perfectly oriented material and all unit
interaction a function of the shear modulus. All LCPs investigated show tensile modulus levels consistent with the
prediction of the aggregate model. No evidence has been
found that suggests differences in crystallinity play any
significant role in determining the level of tensile
modulus at temperatures well below melting or the alpha
transition (glass transition) temperature of mesogenic
fibers. Table I summarizes the maximum (theoretical limit
is to strong) tensile modulus and modulus values(14) of a
number of high performance fibers. The shear modulus data
are suprisingly similar, given the range of chemistries and
crystallinities represented. The conclusion is that in the
limit of very high orientation, tensile modulus is a function of chain architecture and average chain to chain
interactions, and is not much influenced by changes in
morphology or crystallinity.

The tensile strength of LCP fibers was treated by Yoon and
Jaffe using the lag shear model(15). This model, most often
applied to fiber reinforced composites assumes the fiber is
an assembley of rod-like polymer chains, that the strength
of a chain is much higher than the interchain shear
strength and that all stress transfer is through interchain
shear. The critical parameters of the model are the molecular weight distribution of the polymer and, as in the case
of modulus, interchain interaction. The results shown
plotted in Figure 10 indicate the model accurately reflects
the observed behavior without the introduction of morphological or cystallinity based parameters.

Mechanical Properties as a Function of Temperature

As shown in Figure 5 the melting(mesogenic) transition
temperature associated with a given thermotropic copolyester is clearly a function of the composition of the molecular chain. Figure 11 shows the transition behavior of a
typical phenyl-naphyl LCP as measured by dynamic mechanical

techniques. Yoon(16) and Ward (17) have shown conclusively
that the beta and gamma transition temperatures are associated with the amount of naphyl and phenyl content of the
chain and that the location and magnitude of these transitions are little affected by changes in "crystallinity".
The alpha transition, however, is a cooperative transition
similar to a glass transition and the magnitude of the
modulus loss experienced in heating through this transition
is a function of the thermal history of the sample as evidenced in the stress strain curves shown in Figure 12.
Annealing clearly increases the modulus retention of LCP
fibers. Improvements in fiber strength associated with
annealing have been shown by Yoon(15) to be attributable to
the increases in molecular weight and little affected by
the changes in structural perfection which simultaneously
occur. These changes are evidenced by the sharpened wide
angle X-Ray diffraction pattern shown in Figure 13. Figure
14 shows the typical DSC response of an LCP fiber to
annealing time close to the as-spun melting temperature.
Note the sharpening and movement to higher temperatures of
the melting peak. On the basis of the dynamic mechanical
and DSC results it may be concluded that the temperature
range over which the high modulus and strength characteristics of a given LCP are preserved is very much a function
of the structural perfection of the basic low temperature
LCP ordered unit. Hierarchial details, at least for highly
uniaxial samples tested in axial tension, do not appear to
play a major role.

Solid State Structure

These results suggest that to improve the use temperature range of an LCP the degree of structural perfection in
the solid state need to be increased. Three general models
for the ordering of LCPs in the solid state have been proposed. An ordered glass held together by small homopolymer
units was proposed by Blundell(18). Blackwell(19) proposes
a paracrystalline model in which the entire sample is
treated as a perturbed crystal of the dominant homopolymer
component. Windle(20) has proposed a novel model based on
sequence matching, where similar monomer combinations find
themselves in the solid state and form a series of small
"crystals" of varying compositions. The morphological studies of Sawyer and Jaffe(9) indicate that the basic fibrillar structural unit of the LCP shows variations in perfection along its length, consistent with published results on
mesogenic aramid fibers(12). Diagramatic representations of
these models are shown in Figure 15.

A given thermotropic copolyester can be viewed as a blend of copolymer chains of varying monomer composition (as dictated by the polymerization statistics) which average, over the ensemble, to the original monomer stoichiometry. The nature of composition distribution can be further engineered by blending LCP compositions comprised of similar monomer units. Windle has shown that the length of the primary match length varies with monomer composition in a fashion reminiscent of the melting behavior as a function of composition. Recently, DeMeuse(21) has extended these concepts to LCP/LCP blends. The correlations obtained are shown in Figure 16 a and b. The smooth curves represent the behavior of the thermotropic copolyester system based on HNA and HBA where the points represent 50/50 blends of the 75/25 polymer with the 30/70 composition and the blend of each of these with the 58/42 composition. Figure 17 shows the DSC melting of the 50/50 blend of the 75/25 and the 30/70 polymers contrasted with the melting of the individual components. The consistency of the blend results with the Windle concept implies that this concept may best model the LCP solid state structure.

Conclusions

It has been shown that the low(far from melting or alpha transition) temperature tensile modulus and strength of thermotropic copolyesters is essentially a function of molecular parameters alone and that within similar classes of polymers behavior is dominated by the interchain shear modulus. The temperature range over which the low temperature tensile properties are preserved, however, is a strong function of the degree of order in the basic structural unit (crystal) of the material in addition to the expected dependencies on molecular details of the chains. Thermotropic copolyesters can be modeled as a blend representing a distribution of compositions and it was shown that through the engineering of this distribution by physically blending thermotropic copolyester the location of the melting transition can be affected in a systematic fashion, as suggested by the sequence matching ideas of Alan Windle. Utilizing these concepts, it is expected that thermotropic copolyesters designed on the molecular level to possess a given set of thermal-mechanical characteristics should be possible in the near future.

Acknowledgements

The author would like to thank his Celanese colleagues for sharing their insights and their comments. Special thanks to Drs. G. Calundann, L. Sawyer, K. Wissbrun and M. DeMeuse. The helpful suggestions of Professor Ian M. Ward of Leeds University are also acknowledged with thanks.

References

1. G. W. Calundann and M. Jaffe, Proceedings of the Robert A. Welch Conferences on Chemical Research, 26(1982) 247

2. G. W. Calundann, Amer. Chem. Soc., Polym. Preprints., 27(1986) 493

3. M. Jaffe, High Modulus Polymers, in Encyclopedia of Polymer Science and Engineering, John Wiley and Sons, 7(1987), 699

4. K. F. Wissbrun, Brit. Polym. J.(1980)163

5. J. Blackwell and G. Gutierrez, Polymer, 23(1982) 171

6. R. A. Chivers et al., in Polymer Liquid Crystals, A. Blumstein, ed., Plenum Press, New York, 1985

7. A. M. Donald and A. H. Windle, Coll. Polym. Sci., 1983, 261, 793

8. A. M. Donald and A. H. Windle, J. Matl. Sci.,1984, 19, 2085

9. L. C. Sawyer and M. Jaffe, J. Matl. Sci.,21(1986) 1897

10. E. Baer, A. Hiltner, T. Weng, L. C. Sawyer and M. Jaffe, Polym. Mater. Sci. Eng.(1985)52,88

11. T. Weng, A. Hiltner and E. Baer, J. Mater. Sci.(1986)21, 744

12. M. Panar et al., J. Polym. Sci., Polym. Phys ed., 21 (1983) 1955

13. I. M. Ward, in Mechanical Properties of Polymers, John and Sons, New York(1971)

14. S. J. DeTereas, Air Force Materials Laboratory Report # AFWAL-TR-85-4013 (1985)

15. H. N. Yoon and M. Jaffe, to be published

16. H. N. Yoon, to be published

17. I. M. Ward, private communication

18. D. J. Blundell, Polymer,23(1982) 359

19. J. Blackwell, A. Biswas and R. C. Bonart, Macromolecules (1985)18, 2126

20. A. H. Windle, to be published

21. M. DeMeuse, to be published

TABLE I

MAXIMUM E MODULI AND REPORTED SHEAR MODULUS VALUES
FOR A NUMBER OF HIGH MODULUS FIBERS

MATERIAL	MODULUS(GPa)*	SHEAR MODULUS(GPa)**
TYPICAL THERMOTROPIC COPOLYESTER	200	0.45
POLY(p-PHENYLENETEREPHTHALAMIDE)	200	1.5
POLY(p-PHENYLENEBENZOBISTHIAZOLE)	>300	1.2
POLYETHYLENE	240	0.09

* From reference 3
** From reference 14

FIGURE 1. LIQUID CRYSTAL TEXTURES

Nematic

- Molecules Show Parallel One-Dimensional Order
- Turbid Liquid
- Low Viscosity

Smectic

- Molecules Align Parallel And Stratified; Two-Dimensional Order
- Turbid Liquid
- Highly Viscous

Cholesteric

- Shown By Optically Active Molecules Only
- Nematic Layers Arranged In Helical Structure
- Iridescent Liquid With Optical Rotatory Power
- Highly Viscous

FIGURE 2. MOLECULAR ARCHITECTURE OF LIQUID CRYSTAL POLYMERS

- Rod-Like

  Aromatic Polyamides, Esters, Azomethines, Benzbisoxazoles

- Helical

  Polypeptides, Nucleotides, Cellulosics

- Side Chain Mesogenic ("Comb" Polymers)

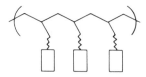

- Block Copolymers (Alternating Rigid/Flexible Units)

FIGURE 3. PROCESS–STRUCTURE–PROPERTY COMPARISON OF FIBER FORMATION WITH A COPOLYESTER AND A CONVENTIONAL POLYMER

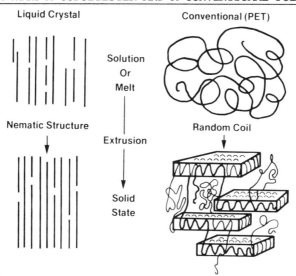

Liquid Crystal   Conventional (PET)

Nematic Structure   Random Coil

Solution Or Melt

Extrusion

Solid State

Extended Chain Structure
- High Chain Continuity
- High Mechanical Properties

Lamellar Structure
- Low Chain Continuity
- Low Mechanical Properties

FIGURE 4. CHEMICAL APPROACHES TO MELT PROCESSIBLE THERMOTROPIC COPOLYESTER

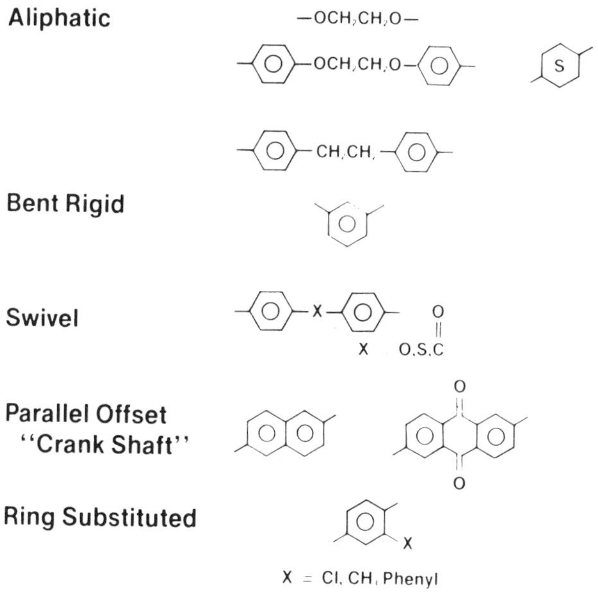

Aliphatic

$-OCH_2CH_2O-$

Bent Rigid

Swivel

X = O,S,C

Parallel Offset "Crank Shaft"

Ring Substituted

X = Cl, CH, Phenyl

FIGURE 5. TYPICAL MELTING BEHAVIOR OF THERMOTROPIC COPOLY-
ESTERS AS A FUNCTION OF COMPOSTION

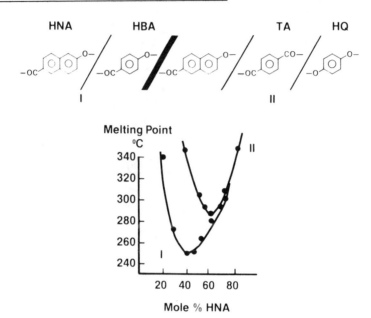

FIGURE 6. TYPICAL THERMOTROPIC COPOLYESTER WIDE ANGLE X–RAY
DIFFRACTION PATTERNS

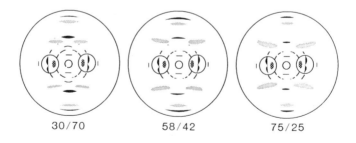

FIGURE 7. HIERARCHIAL   MODEL   OF   THERMOTROPIC   COPOLYESTER
FIBERS

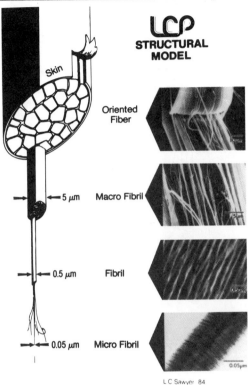

LC Sawyer 84

FIGURE 8. SPECIFIC MODULUS VERSUS SPECIFIC STRENGTH FOR A
VARIETY OF FIBERS

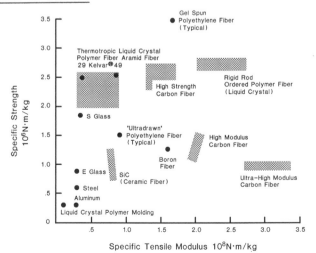

FIGURE 9. AGGREGATE MODEL — TENSILE MODULUS AS A FUNCTION OF MOLECULAR ORIENTATION AND SHEAR MODULUS

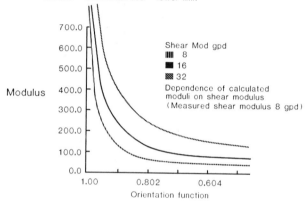

Modulus - Aggregate Model

- Structure composed of anisotropic elastic units
- Unit has elastic properties of perfectly oriented material
- Unit interactions f (shear modulus)
- VOIGT — uniform strain — upper limit
- REUSS — uniform stress — lower limit

Shear Mod gpd
▥ 8
■ 16
▧ 32

Dependence of calculated moduli on shear modulus
(Measured shear modulus 8 gpd)

FIGURE 10. TENSILE STRENGTH OF A THERMOTROPIC COPOLYESTER FIBER AS A FUNCTION OF THE POLYMER MOLECULAR WEIGHT DISTRIBUTION

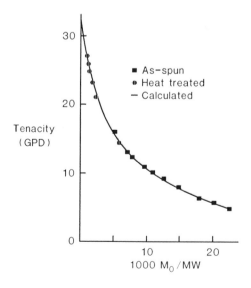

■ As—spun
⬤ Heat treated
— Calculated

FIGURE 11. DYNAMIC   MECHANICAL   SPECTRUM   OF   A   TYPICAL
THERMOTROPIC COPOLYESTER FIBER

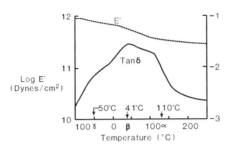

Summary

γ – Transition (~ –70°C)
  – Reorientational motion of phenylene groups
  – Ea = 11 Kcal/mole

β – Transition (~ 60°C)
  – Reorientational motion of naphthalene groups
  – Ea = 28 Kcal/mole

α – Transition (100° ~ 150°C)
  – Rotational motion of several monomer units
  – Ea = ~ 110 Kcal/mole

FIGURE 12. STRESS–STRAIN CURVES OF A TYPICAL THERMOTROPIC
COPOLYESTER FIBER AS A FUNCTION OF THERMAL HISTORY

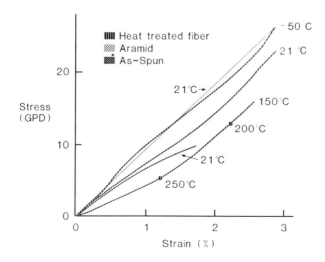

FIGURE 13. WIDE ANGLE X–RAY DIFFRACTION PATTERNS OF AS–SPUN AND ANNEALED THERMOTROPIC COPOLYESTER FIBERS

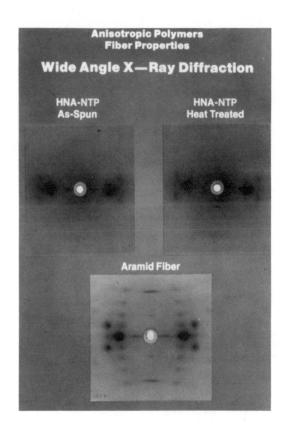

FIGURE 14. CHANGES IN THE DSC MELTING BEHAVIOR OF A THERMOTROPIC COPOLYESTER AS A FUNCTION OF ANNEALING TIME

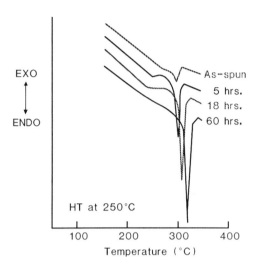

FIGURE 15. PROPOSED MODELS FOR THE BASIC STRUCTURAL UNITS OF THERMOTROPIC COPOLYESTERS

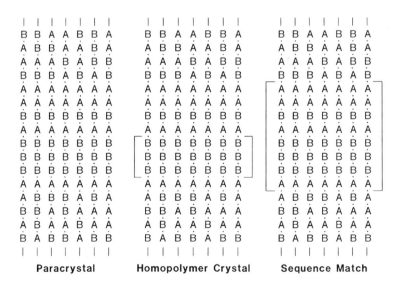

Paracrystal   Homopolymer Crystal   Sequence Match

FIGURE 16. CORRELATIONS BASED ON THE "PRIMARY SEQUENCE LENGTH" OF HBA/HNA THERMOTROPIC COPOLYESTERS A. COMPOSITION B. MELTING POINT – AFTER WINDLE(20)

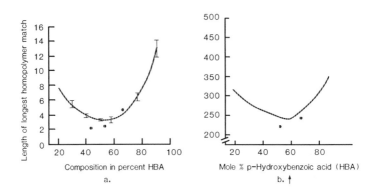

FIGURE 17. DSC MELTING BEHAVIOR OF A 50/50 BLEND OF HBA/HNA 75/25 AND 30/70. THE MELTING OF THE INDIVIDUAL POLYMERS IS INCLUDED FOR REFERENCE

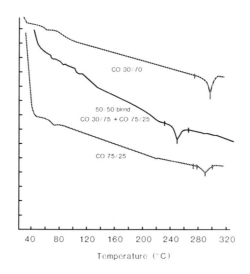

LIQUID CRYSTAL POLYMERS. X. LIQUID CRYSTAL POLYESTER FIBERS

W. J. Jackson, Jr.

Research Laboratories, Eastman Chemicals Division,
Eastman Kodak Company, Kingsport, Tennessee 37662

Abstract

When poly(ethylene terephthalate) (PET) is
modified with p-hydroxybenzoic acid (PHB), the
tensile properties of melt-spun fibers increase
as (1) the PHB content is increased to 30 mol %,
(2) the fiber I.V. is increased, and (3) the spin
draw ratio is increased; as-spun tenacities up to
10 g/den. can be attained. Higher PHB contents
give liquid crystalline polyesters but lower
tenacities (3 to 4 g/den.) along with higher
elastic moduli (up to 330 g/den.). Heat-treatment
of PET/80 mol % PHB fibers to increase their
molecular weight does increase the tenacity, but
only to 10 g/den. Similarly, as-spun tenacities
up to 10 g/den. can be attained with another
class of liquid crystalline polyesters also
containing aliphatic spacers, copolyesters based
on trans-4,4'-stilbenedicarboxylic acid and
2,6-naphthalenedicarboxylic acid and either
ethylene gylcol or 1,4-butanediol; heat-treatment
does not give appreciably higher tenacities. In
contrast, wholly aromatic liquid crystalline
polyesters containing no flexible spacers do not
have this apparent 10 g/den. tenacity limitation,
and tenacities of the as-spun fibers can be
increased to over 20 g/den. by heat-treatment,
which increases the molecular weight. Possible
reasons for these differences in fiber
performance and the effects of fiber processing
conditions on tensile properties are discussed.

Introduction

The first thermotropic liquid crystalline polyesters
(LCP's) to be characterized, injection molded, and spun into
fibers were those which we made by the modification of

poly(ethylene terephthalate) (PET) with p-hydroxybenzoic
acid (PHB) [1-3]. These copolyesters were prepared by the
reaction of PET with various amounts of p-acetoxybenzoic
acid. Muramatsu and Krigbaum [4-6], Sugiyama and coworkers
[7], and Ciferri and coworkers [8,9] have studied the
effects of the spinning conditions on the fiber properties
of our 40 mol % PET/60 mol % PHB copolyester (PET/60 PHB),
and Ciferri and coworkers [9] have also spun the 70/30
PET/PHB copolyester (PET/30 PHB). In our studies, which we
have not reported and which cover wider composition and I.V.
ranges than were available to others, the PET/PHB ratio,
polymer inherent viscosity (I.V.), and spinning conditions
had large effects on the fiber tensile properties. We
observed similar effects in other classes of LCP's containing
flexible spacers from aliphatic glycols, and we obtained
quite different fiber results with wholly aromatic LCP's
containing no aliphatic spacer component.

The objective of this report is to compare the effects of
composition and fiber processing conditions on the tensile
properties of liquid crystal polyester fibers prepared from
copolyesters containing flexible spacers and wholly aromatic
LCP's containing no flexible spacers.

Experimental

PET/PHB copolyesters were prepared by the acidolysis
reaction of p-acetoxybenzoic acid with PET [3]. Copolyesters
of trans-4,4'-stilbenedicarboxylic acid (SDA) with
2,6-naphthalenedicarboxylic acid were prepared from the
dimethyl esters and ethylene glycol [10] or 1,4-butanediol
[11]. Wholly aromatic polyesters were prepared from
aromatic dicarboxylic acids, hydroquinone dipropionate, and
p-acetoxybenzoic acid [12,13].

Inherent viscosities of the PET/PHB copolyesters were
measured at 25°C in a 60:40 wt % mixture of phenol/
tetrachloroethane at a polymer concentration of 0.5 g/100 mL.
I.V's of the other polyesters were determined in a better
solvent, a 25:40:35 wt % mixture of phenol/p-chlorophenol/
tetrachloroethane at a polymer concentration of 0.1 g/100 mL.

The polyesters were melt spun with an Instron melt
rheometer equipped with a capillary. The tensile properties
of the as-spun fibers were determined by averaging ten
breaks of 1-in. gage-length single filaments. Longer gage
lengths or plied yarn gave appreciably lower tenacity and

modulus values. The elastic modulus was the slope of the
stress-strain curve.

The sonic velocity of the fibers was determined with a
Dynamic Modulus Tester, Model PPM-5, from H. M. Morgan Co.
Inc., Norwood, Mass.

Results and Discussion

PET/PHB fibers. When we melt spun liquid crystalline
compositions containing 35 to 80 mol % PHB with various
I.V.'s and spinning conditions, as-spun tenacities up to
only about 4 g/den. and elastic moduli up to 330 g/den. were
obtained. (No higher properties were given in the references
cited in the Introduction.) We obtained appreciably higher
tenacities with PET modified with 25 to 30 mol % PHB.
Table 1 shows the effect of PHB content and I.V. on the
properties of as-spun fibers containing up to 30 mol % PHB.
The fibers were all spun with an Instron capillary rheometer
at the same polymer extrusion rate (plunger speed 0.2 cm/min
in the 3/8-in. diameter barrel, capillary length-to-diameter
ratio 3:1) and taken up at 2,000 ft/min. The spin draw
ratio, calculated as the ratio of the area of the capillary
orifice (0.014-in. diameter) to the area of the fiber cross
section (diameter about 16 microns) was about 500.

Table 1. Properties of PET/PHB Fibers

$$\left(-\underset{O}{\overset{}{C}}-\hspace{-0.5em}\underset{}{\bigcirc}\hspace{-0.5em}-\underset{O}{\overset{}{C}}-OCH_2CH_2O-\right)_x \left(-\underset{O}{\overset{}{C}}-\hspace{-0.5em}\underset{}{\bigcirc}\hspace{-0.5em}-O-\right)_y$$

PHB, Mol %[a]	0		20	25		30	
Spinning Temperature,°C	295	295	275	280	280	275	295
Polymer I.V.	0.60	0.78	0.72	0.74	0.88	0.68	1.06
Fiber I.V.	0.55	0.66	0.67	0.71	0.79	0.57	0.81
Den./Fil.	2.2	3.3	2.2	2.6	2.7	2.6	2.8
Tenacity, G/Den.	1.7	2.2	3.9	7.1	7.5	6.4	9.8
Elongation, %	390	230	32	16	12	12	9
Elastic Modulus, G/Den.	20	17	52	76	109	117	132
Birefringence	0.007	-	-	0.21	-	0.25	-

[a]The PHB component is randomly present in the copolyester, not
blocked as shown in the formula [16].

Appreciably lower spin draw ratios gave lower tensile
properties and higher ratios were of little benefit. The
melt temperatures listed are those which gave the maximum
tenacities, based on an average of ten single filament
samples. At a given PHB content the fibers with the higher
I.V.'s had the higher tenacities and elastic moduli, and at
similar fiber I.V.'s the higher PHB levels gave the higher
tenacities and moduli. Both the higher I.V.'s and higher
PHB levels impart higher relaxation times [14] and,
consequently, the orientation imparted on spinning was
better retained as the polymer relaxation times increased.
This increase in molecular orientation is shown by the large
increase in the birefringence. (Oriented PET for textiles
has a birefringence of about 0.18.) The shrinkage of PET/30
PHB in boiling water is 0%, as is the shrinkage in air at
150°C. (As-spun PET shrinks about 65 to 70% in boiling
water.)

Fiber I.V.'s above about 0.8 in these PET/PHB compositions
could not be attained because the higher melt viscosity,
higher I.V. polymers required higher spinning temperatures,
which caused increased thermal degradation. The 0.81 I.V.
PET/30 PHB fiber was obtained by spinning at 295°C a polymer
having an I.V. of 1.06, so obviously considerable thermal
degradation occurred since the loss in I.V on spinning was
0.25. A 0.78 I.V. fiber (PET/28 PHB not in table) having an
average tenacity of 9.5 g/den. was obtained from a 0.88 I.V.
polymer which could be spun 10°C cooler at 285°C, so the
I.V. loss was 0.10, rather than 0.25 for the higher I.V.
polymer, and the tenacity was almost as high. Thus,
although we could prepare higher I.V. polymers, we could not
obtain higher fiber I.V.'s in order to attain higher
tenacities because of melt viscosity and thermal stability
limitations.

PHB levels above 30 mol % also did not give higher
tenacities (including spinning with higher L/D capillary
ratios than 3:1). Actually, higher as-spun tenacities were
attained with the non-liquid crystalline PET/25 PHB and
PET/30 PHB compositions than the highly liquid crystalline
PET/60 PHB. (About 35 mol % PHB is required for the
copolyesters to become liquid crystalline [3]).

One probable reason for the lower as-spun tenacities of
the LCP compositions is the two-phase nature of the polymer
melt: both a liquid crystalline anisotropic phase and a
non-liquid crystalline isotropic phase are present [15,16].
Another reason is the higher molecular weight of PET/25 and
30 PHB than PET/60 PHB. At a given I.V., the molecular

weights are about 60% higher than in PET/60 PHB, and also higher I.V. polymers of PET/25 PHB and PET/30 PHB can be prepared (by solid-state polymerization) than can be attained with PET/60 PHB (which cannot be solid-state polymerized because of lack of sufficient crystallinity, needed to prevent fusion of the particles at the required polymerization temperature).

When PET/80 PHB was spun, tenacities of about 3 g/den. and elastic moduli of about 330 g/den. were obtained, and heat-treatment of the fiber in an inert atmosphere to increase the molecular weight did increase the tenacity, but only to about 10 g/den. PET/60 PHB fibers cannot be heat-treated to increase the polymer molecular weight because the fibers stick together due to the low degree of crystallinity, as noted above. As will be discussed later, tenacities over 20 g/den. can be attained by the heat-treatment of fibers of wholly aromatic liquid crystalline polyesters. The apparent reason for the tenacity limitation in PET/80 PHB is the presence of the flexible aliphatic glycol component. But a greater increase in tenacity on heat-treatment (to 10 g/den.) was obtained than in the glycol polyesters discussed in the next section, perhaps because only a small amount of a glycol component (20 mol %) is present in PET/80 PHB.

Fibers of SDA and aliphatic glycols. The PET/PHB polyesters were prepared by the reaction of preformed PET with the acetate of p-hydroxybenzoic acid. We could not start with ethylene glycol as a monomer. If an aliphatic glycol is to be used in a conventional polyester melt polymerization, as we desired, to obtain a liquid crystalline polyester the glycol must be reacted with a rod-like dicarboxylic acid of sufficient rigidity and length (aspect ratio) for liquid crystallinity to be imparted in the final polymer. Terephthalic acid is not adequate for this purpose, but trans-4,4'-stilbenedicarboxylic acid (SDA) is effective [17,18].

Because the polyester of SDA and ethylene glycol melts at 418°C, the polyester must be modified to reduce the melting point below 300°C so that the polymer will be sufficiently stable to be melt spun without thermal decomposition. We found that if terephthalic acid is the modifier, over 60 mol % (acids total 100 mol %) is required to reduce the melting point below 300°C and, consequently, the copolyester is not liquid crystalline. 2,6-Naphthalenedicarboxylic acid (N) could be used as an effective modifier, however, and the copolyesters containing at least 30 mol % SDA did exhibit

thermotropic liquid crystallinity. Since the maximum amount
of SDA which can be used is about 40 mol % before the
melting points are above 300°C, in effect the copolyesters
are poly(ethylene 2,6-naphthalenedicarboxylate) (PEN)
modified with SDA. Figure 1 shows the melting points (peak
values of DSC endotherms) obtained from various preparations
of each composition. There was no correlation in any
composition between melting point and I.V. (I.V.'s were 0.5
to 0.9, determined in a mixture of phenol/p-chlorophenol/
tetrachloroethane because some of the polymers were not
soluble in a mixture of phenol and tetrachloroethane).

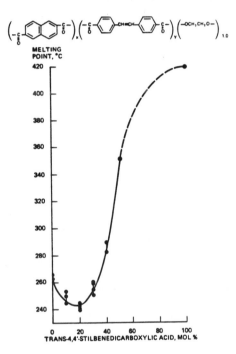

Figure 1. Melting points of PEN/SDA copolyesters.

PEN/SDA fibers were melt spun in the same manner as the
PET/PHB fibers discussed earlier but with capillary L/D
ratios generally of 20 or more (instead of 3) and higher
spin draw ratios (about 650 to 1000) because of the higher
take-up speeds. The longer capillaries permitted
application of a higher temperature to melt any high-melting
particles which might be present while minimizing thermal
decomposition because of the very short time that the

polymer was exposed to the higher temperature as it rapidly passed through the capillary (Table 2). The spinning temperatures listed in the table are those which gave the highest tenacities (average of 10 single filaments).

Even PEN alone gave as-spun fibers having a tenacity of 8 g/den. PEN apparently is close to being liquid crystalline, because modification with only 30 mol % SDA or 25 mol % PHB is required to make the polyester liquid crystalline. The copolyesters containing 30 and 40 mol % SDA do give liquid crystalline melts, but it was necessary to spin them at temperatures above the transition temperatures (Ti) to the isotropic state (Ti is about 262°C for PEN/30 SDA and 330°C for PEN/40 SDA). Since the capillary temperature was high (355°C) for the latter, relaxation of the fiber probably occurred, which would limit the tenacity.

Table 2. Tensile Properties of PEN/SDA Fibers

SDA, Mol %	0	10	20	30	40
Spinning Temperature, °C[a]	290/300	290/325	290/335	300/325	300/355
Capillary L/D Ratio	10	20	20	20	50
Take-up Speed, Ft/Min	3000	3000	2600	2600	4000
Polymer I.V.	0.82	0.83	0.83	0.84	0.86
Fiber I.V.	0.66	0.63	0.69	0.70	0.67
Den./Fil.	1.5	1.6	1.9	1.9	1.3
Tenacity, G/Den.	8.1	9.7	9.6	8.4	6.7
Elongation, %	11	8	8	6	5
Elastic Modulus, G/Den.	152	218	224	228	198

[a]Barrel temperature/capillary temperature.

As with the PET/PHB fibers, higher tenacities (up to 10 g/den.) were obtained with PEN/SDA fibers which did not exhibit liquid crystallinity (Table 2). Also we observed that liquid crystalline polyesters of SDA, N, and 1,4-butanediol spun similarly had somewhat lower tenacities than non-liquid crystalline polyesters containing less SDA and spun at similar take-up rates [19]. And as with the PET/PHB fibers, the maximum tenacities obtained were about 10 g/den. Also we did not obtain higher tenacities with SDA polyesters of other glycols [18].

Since heat-treatment of aromatic LCP fibers in an inert
atmosphere increases the fiber molecular weight and tensile
properties [20], PEN/SDA fiber compositions were clamped in
a stainless steel frame, placed in a glass tube, and heated
at constant length in an aluminum block while argon flowed
into the tube. As an illustration of one of these
experiments, a PEN/30 SDA fiber which had been spun at
3000 ft/min was heated in the tube for 1 hr at 150°C and
then 1 hr at 225°C. The I.V. increased from 0.72 to 1.48,
but since the fibers did not completely dissolve in the
p-chlorophenol/phenol/tetrachloroethane I.V. solvent, the
actual I.V. probably was higher. (Longer fiber heating
times gave lower I.V.'s because of the presence of larger
amounts of insoluble polymer, perhaps because of the
formation of high melting SDA/ethylene glycol blocks.) Even
though the I.V. was increased over 100%, the tenacity
decreased from 8.0 g/den to 6.2 g/den (Table 3).

Table 3. Effect of Heat-Treatment on Properties
of PEN/30 SDA Fibers

	As-Spun	Heat-Treated
Fiber I.V.	0.72	>1.48
Den./Fil.	1.7	1.7
Tenacity, G/Den.	8.0	6.2
Elongation, %	7.3	6.1
Elastic Modulus, G/Den.	213	205
Orientation Angle, Degrees	18.9	13.6
Sonic Velocity, Km/Sec	5.2	4.7

To determine if this decrease in tenacity is due to
disorientation of the fiber by the heat-treatment process,
the X-ray orientation angle and the sonic velocity were
determined before and after heat-treatment. The orientation
angle, determined as by Kwolek and Luise [21], is the width
in degrees of one half of the arc of an X-ray diffraction
pattern from equatorial reflections that are in alignment
with the fiber axis. The smaller orientation angle obtained
after the heat-treatment indicated an increase in orientation
of the crystalline portion of the fibers (Table 3). (The
orientation also increased but to a lesser degree
(orientation angle 17.6°C) when the fibers were heated
unclamped.) Thus there was no loss in crystalline
orientation during the heat-treatment process.

Since we thought that perhaps the orientation of the amorphous component of the fibers might have decreased during the heat-treatment process, we measured the change in sonic velocity, which increases with total overall orientation, crystalline and amorphous [22]. The sonic velocity decreased slightly (Table 3), indicating some decrease in overall orientation. (For comparison, our PET partially oriented yarn had a sonic velocity of 3.6 km/sec.)

A 20% increase in tenacity (5.5 to 6.6 g/den.) did occur on heat-treatment of a 80 SDA/20 N/1,4-butanediol liquid crystal polyester fiber for 1 hr at 180°C and 4 hr at 205°C. The I.V. increased 60% from 1.24 to 2.00 with incomplete solution of the fiber in the I.V. solvent, the orientation angle decreased from 29.2° to 23.7°, and there was very little change in the sonic velocity (about 4.8 km/sec). In general, with SDA copolyesters of ethylene glycol, 1,2-propanediol, and 1,4-butanediol little change in tensile properties occurred during the heat-treatment processes, orientation angles decreased, and sonic velocities were somewhat lower.

Loss in orientation on heat-treatment, as indicated by sonic velocity measurements, adversely affects tensile properties. Also, the presence of the flexible glycol component might permit such entanglement of the polymer chains of the as-spun fibers that an increase in molecular weight gives only a limited increase in tenacity. It would be expected that the greater the extended chain character of a polymer the more an increase in molecular weight should increase the tenacity. Fibers of wholly aromatic LCP's which we have spun have appreciably lower X-ray orientation angles (e.g., 8°) than fibers of SDA LCP's containing flexible spacers, indicating less entanglement of chains and appreciably higher fiber orientation, and this may be the key to the difference in performance of the fibers when heat-treated.

Support for this hypothesis is the observation by Muramatsu and Krigbaum [24] that the tensile properties of fibers of a wholly aromatic LCP (based on p-hydroxybenzoic acid and 2-hydroxy-6-naphthoic acid) were improved by heat-treatment only if the as-spun fibers were highly oriented. Fibers having lower orientation because of lower spin draw ratios did not show increased tensile properties on heat-treatment, even though they did have higher I.V.'s.

Fibers of wholly aromatic LCP's. The effects of
composition on the fiber properties of wholly aromatic LCP's
have been discussed in several of our earlier papers
[12,14,23], but very little was said about the effects of
processing conditions on fiber properties. Therefore, this
area is now briefly discussed. Our papers cited above also
give references to Eastman Kodak, DuPont, and Celanese
patents which describe LCP fibers. A DuPont patent
(Schaefgen) of particular significance, in addition to the
heat-treatment process patent of Luise [20] cited earlier,
discloses fibers of wholly aromatic LCP's [25]. Calundann
and Jaffe [26] at Celanese have spun and studied aromatic
LCP's containing naphthalene rings, and Muramatsu and
Krigbaum [24], cited in the preceding section, have studied
the effects of fiber processing conditions on the fiber
properties of one of these Celanese compositions. Demartino
[27] has modified aromatic LCP's to lower their spinning
temperatures. In addition, Kwolek and Luise [21] have
studied the fiber processing of chlorohydroquinone/
1,4-cyclohexanedicarboxylic acid LCP's. Such polyesters are
not wholly aromatic, of course, but the trans-cyclohexane
isomer does permit liquid crystal formation in the
polyesters, and the heat-treated fibers can attain
tenacities similiar to those of wholly aromatic LCP fibers.

We were aware that liquid crystalline polyesters have
appreciably lower melt viscosities than similar polyesters
which are not liquid crystalline [3,14]. Consequently, we
were interested in determining the effect of polymer melt
viscosity and molecular weight on the melt spinnability and
fiber properties of as-spun wholly aromatic liquid
crystalline polyesters. Our results can be illustrated with
the composition shown in Figure 2, 15N5I(HQ)80(PHB), DSC mp
318°C. The small amount of the isophthalic acid (I)
component (5 mol %, based on all of the carboxylic acids)
facilitated the preparation of the polyester. Neither the
molecular weight nor the inherent viscosity could be
measured because of insolubility of the polymer in our
solvents, so the melt flow was determined according to ASTM
D1238, using a 0.04-in. capillary and 2160-g load. The melt
flow values are the number of grams of polymer extruded at
350°C in 10 minutes. The lower melt flow values, of course,
indicate the higher molecular weights, and Figure 2 shows
the effect of melt flow and shear rate on the melt viscosity
at 340°C of three polymer samples. The values obtained at
the lower shear rates were determined with a Rheometrics
mechanical spectrometer in the eccentric-rotating-disk mode

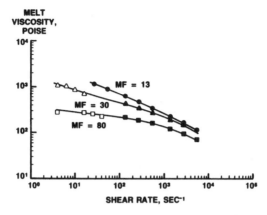

Figure 2. Effect of melt flow rate of
15N5I(HQ)80(PHB) on melt viscosity.

(open symbols), and the viscosities obtained at the higher
shear rates were determined with an Instron capillary
rheometer. (PET used for spinning fibers has a melt
viscosity of about 1000 poise at 300°C and a shear rate of
100 to 1000 sec.$^{-1}$)

The same three polymers were melt spun at 350°C (higher
melt flow sample) and 360°C (other two samples) with an
Instron melt rheometer fitted with capillaries of constant
diameter (0.35 mm) and different lengths (volumetric flow
rate 0.142 cm^3/min). Figure 3 shows the maximum take-up
speed versus the capillary length-to-diameter ratio. The
higher take-up speeds were achieved with the higher L/D
ratios and the lower molecular weight (higher melt flow)
compositions. In our experience, this observation applies
to LCP's in general.

In contrast, the capillary L/D ratio did not affect the
maximum take-up speed of PET (I.V. 0.64) spun with the same
capillaries at 290°C (about 1200 m/min vs. maximum speeds of
about 600 to 1100 m/min for the three liquid crystal
polyesters with different melt flows). The L/D ratio also
did not affect the take-up speed of the non-liquid
crystalline polyesters containing flexible segments from
aliphatic glycols, such as in those polyesters discussed
earlier.

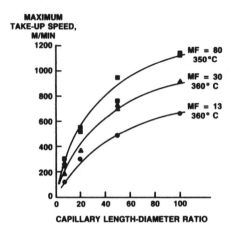

Figure 3. Effect of capillary l/d ratio on 15N5I(HQ)80(PHB) fiber take–up speed.

Figure 4 shows the effect of polymer molecular weight (indicated by melt flow measurements) and take–up speed on the fiber modulus and tenacity of the same three compositions spun at 350°C (higher melt flow sample) and 360°C. Above a take–up rate of 200 m/min there was only a

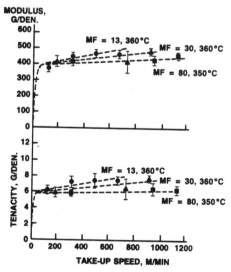

Figure 4. Effect of take–up speed on properties of 15N5I(HQ)80(PHB) fibers.

small increase in properties. Also the higher molecular
weight (lower melt flow) composition gave only a small
increase in tenacity and modulus compared to these properties
from the lower molecular weight composition. For all fibers
the volumetric flow rate was constant (0.142 cm^3/min), so
one benefit of the higher take-up rates is the production of
lower denier fibers (about 2.5 den./filament when the
take-up rate is 640 m/min and 1.5 den./filament at
1100 m/min). Similar results were obtained with other
wholly aromatic LCP's.

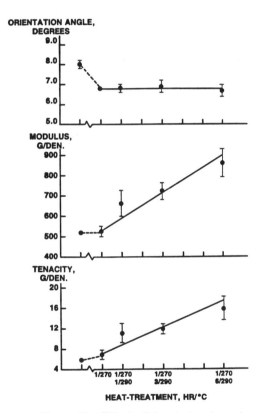

*Figure 5. Effect of heat-treatment on
properties of 15N5I(HQ)80(PHB) fibers.*

We were interested in determining the importance of
structural changes in the fibers when the tensile properties
are increased by fiber heat-treatments. Figure 5 shows the
effect of heat-treatment conditions on the X-ray orientation

angle, modulus, and tenacity of 3 to 5-denier filaments of the same composition, 15N5I(HQ)80(PHB), melt spun at 345°C and taken up at 2500 ft/min. The as-spun fibers had a tenacity of 5.9 g/den. and a tensile modulus of 520 g/den. The low orientation angle (8.0°), determined by X-ray diffraction measurements, indicated that the fibers were highly oriented. For the heat-treatment experiments the fibers were wound unconstrained around a stainless steel frame and placed in a 3-l. resin reaction flask with an argon atmosphere. After a minimal heat-treatment of 1 hr at 270°C the orientation angle decreased from 8.0° to 6.8°, indicating a slight increase in molecular orientation. Very little change occurred in the tensile properties, however. With subsequent heat-treatments (each sample 1 hr at 270°C plus 1 to 6 hr at 290°) the tenacity increased to about 16 g/den., the modulus increased to about 850 g/den., and the orientation angle remained constant. Also the fiber density did not increase. Structural changes, therefore, appeared to play little role in the increase in tensile properties of the fibers, and the increase in molecular weight appears to be responsible for the enhancement in properties. (A higher heat treatment temperature gives higher tenacities, e.g., 20 g/den. after 3 hr at 300°C.)

We were interested in determining the correlation between the fiber properties and the fiber I.V. or molecular weight after heat-treatment of the fibers. Since the fibers were isoluble in our I.V. solvents, we used melt viscosity as a measure of the molecular weight. An aromatic liquid crystal copolyester [13], 25N25T(HQ)50(PHB), DSC mp 327°C, was melt spun at 346°C to give 5 denier/filament fibers with a tenacity of 6.6 g/den. The fibers were heat-treated in bundles in an argon atmosphere at 290°C for various times, and the tensile properties were measured. The fibers were then chopped and compression molded into sample disks for use in determining the melt viscosity of the remelted fibers with a Rheometrics mechanical spectrometer in the eccentric-rotating-disk mode. Figure 6 is a log-log plot of the tenacities versus the corresponding melt viscosities. It is apparent that the increased tenacities obtained on increasing the fiber heat-treatment times are due to the increased molecular weights, which caused the increase in melt viscosities. Similar results were obtained for fiber modulus. It is noteworthy that when the tenacity is 20 g/den., the melt viscosity is over 10^5 poise, which is over an order of magnitude too high for melt spinning (even considering that the higher shear rate during spinning gives a lower melt viscosity).

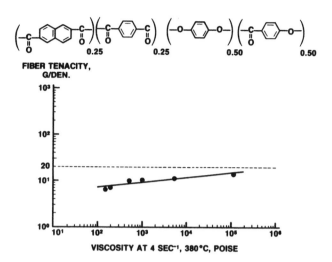

Figure 6. Correlation of fiber tenacity and melt viscosity of 25N25T(HQ)50(PHB).

A similar plot was made with fibers of a polyester prepared from terephthalic acid and phenylhydroquinone diacetate [23,28]. Since this polymer was soluble in our phenol/p-chlorophenol/tetrachlorethane I.V. solvent, it was possible to correlate I.V.'s and tenacities (e.g., I.V. of 1.9 for 4.8 g/den. fibers and I.V. of 12.0 for 20 g/den. fibers). Since the polymer also was soluble in an isorefractive solvent (1:1 by weight o-dichlorobenzene and p-chlorophenol), it was possible to correlate tenacities and molecular weights (determined by low-angle laser light scattering): about 6 g/den. for 30,000 molecular weight and 20 g/den. for 300,000 molecular weight.

Figure 7 shows the effect of the heat-treatment temperature on the tenacity of a 1.5-denier filament of the same polymer heated for 30 minutes in a nitrogen atmosphere. Increased tenacities were also obtained on heat-treatment in air, but the values were lower than those of fibers heated in nitrogen, presumably because of oxidative crosslinking. (Films of this polymer heated in air at 300°C for 1 hr become insoluble in solvents, an indication of crosslinking.)

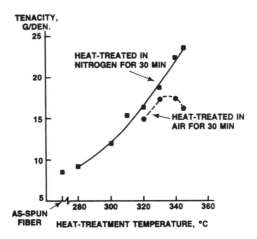

Figure 7. Effect of heat—treatment temperature on tenacity of T(ϕHQ) fibers.

Conclusions

Liquid crystalline polyesters have low melt viscosities and long relaxation times [3,12,14,23]. The low melt viscosities facilitate melt spinning, and the long relaxation times permit the extended chain orientation which can be achieved on melt spinning to be maintained until the polymer solidifies. Because of the extended chain orientation, as-spun fibers have higher tenacities and moduli and lower elongations than these properties in PET. By modification of PET or certain other glycol polyesters with a para-oriented aromatic component which will reduce the chain flexibility, as-spun fiber tenacities up to about 10 g/den. can be achieved even without liquid crystal formation. Because of the presence of the flexible aliphatic component, however, appreciably higher tenacities cannot be obtained by heat-treatment of the fibers in an inert atmosphere, even though the molecular weight is increased. The tenacities of wholly aromatic liquid crystalline polyester fibers, on the other hand, can be increased to values above 20 g/den by heat-treatment, which increases their molecular weights. The reason for the tenacity limitation on heat-treatment of the LCP's containing flexible spacers may be primarily due to their higher degree of chain entanglement, because the crystalline orientation in these fibers is appreciably lower than in fibers of wholly aromatic LCP's which do increase in tensile properties on heat-treatement.

Acknowledgments

I would like to acknowledge numerous contributions of
H. F. Kuhfuss, J. C. Morris, and J. R. Bradley, who
prepared, characterized, and spun the polyesters,
P. D. Griswold, who studied a number of the spinning
parameters, and S. L. Hess, who measured the sonic
velocities.

References

[1] Kuhfuss, H. F.; Jackson, W. J., Jr. (to Eastman Kodak
Co.) U.S. Patent 3,778,410 1973.

[2] Kuhfuss, H. F.; Jackson, W. J., Jr. (to Eastman Kodak
Co.) U.S. Patent 3,804,805 1974.

[3] Jackson, W. J., Jr.; Kuhfuss, H. F., J. Polym. Sci.
Polym. Chem. Ed., Ed. 1976 14, 2043.

[4] Muramatsu, H.; Krigbaum, W. R. J. Polym. Sci. Polym.
Phys. Ed. 1986, 24, 1695.

[5] Muramatsu, H.; Krigbaum, W. R. J. Polym. Sci. Polym.
Phys. Ed. 1987, 25, 803.

[6] Muramatsu, H.; Krigbaum, W. R. J. Polym. Sci. Polym.
Phys. Ed. 1987, (in press).

[7] Sugiyama, H.; Lewis, D. N.; White, J. L.; Fellers, J.
F. J. Appl. Polym Sci. 1985, 30, 2329.

[8] Tealdi, A.; Ciferri, A.; Conio, G. Polymer Comm. 1987,
28, 22.

[9] Acierno, D.; LaMantia, F. P.; Polizzotti, G.; Ciferri,
A.; Valenti, B. Macromolecules 1982, 15, 1455.

[10] Morris, J. C.; Jackson, W. J., Jr. (to Eastman Kodak
Co.) U.S. Patent 4,414,382 1983.

[11] Morris, J. C.; Jackson, W. J., Jr. (to Eastman Kodak
Co.) U.S. Patent 4,459,402 1984.

[12] Jackson, W. J., Jr. Macromolecules 1983, 16, 1027.

[13] Jackson, W. J., Jr.; Morris, J. C. (to Eastman Kodak
Co.) U.S. Patent 4,169,933 1979.

[14] Jackson, W. J., Jr. Br. Polym. J. 1980, 12, 154.

[15] McFarlane, F. E.; Nicely, V. A.; Davis, T. G.
"Contemporary Topics in Polymer Science"; Pierce, E.
M.; Schaefgen, J. R., Eds.; Plenum: New York, 1977,
Vol. 2, p. 109.

[16] Nicely, V. A.; Dougherty, J. T.; Renfro, L. W.
Macromolecules 1987, 20, 573.

[17] Meurisse,P.; Noël, C.; Monnerie, L.; Fayolle, B. Brit.
Polym. J., 1981, 13, 55.

[18] Jackson, W. J. Jr.; Morris, J. C. J. Appl. Polym. Sci.
Appl. Polym. Symp., 1985, 41, 307.

[19] Jackson, W. J. Jr.; Morris, J. C. J. Polym. Sci.
     Polym. Chem. Ed. 1987 (in press).
[20] Luise, R. R. (to DuPont) U.S. Patent 4,183,895 1980.
[21] Kwolek, S. L.; Luise, R. R. Macromolecules 1986,
     19,1789.
[22] Moseley, W. W., Jr. J. Appl. Polym. Sci. 1960, 3(9),
     266.
[23] Jackson, W. J., Jr. "Contemporary Topics in Polymer
     Science", Vandenberg, E. J., Ed., Plenum: New York,
     1984, Vol. 5, p.177.
[24] Muramatsu, H.; Krigbaum, W. R. Macromolecules 1986,
     19, 2850.
[25] Schaefgen, J. R. (to DuPont) U.S. Patent 4,118,372 1978.
[26] Calundann, G. W.; Jaffe, M. Proc. Robert A. Welch
     Found., Conf. Chem. Res. 1982, 26, 247.
[27] Demartino, R. N. J. Appl. Poly. Sci. 1983, 28, 1805.
[28] Payet, C. R. (to DuPont) U.S. Patent 4,159,365 1979.

RELATIONSHIPS BETWEEN POLYMER STRUCTURE AND LIQUID CRYSTALLINE PHASE BEHAVIOR FOR THERMOTROPIC POLYESTERS

R. W. Lenz, A. Furukawa[a] and C.N. Wu[b]

University of Massachusts
Amherst, MA 01003 USA

This paper is concerned with the effect of both main chain and pendant flexible spaces on the liquid crystalline, LC, properties of polyesters. Six different series of polymers were prepared in this study, including both rigid rod polyesters and polyesters containing flexible units to lower their transition temperatures.

Introduction

A principal focus of this laboratory, in our investigations on the structure-property relationships of thermotropic polyesters, is on the effect of flexible spacers on the LC properties of such polymers. We have previously reported that polyesters with ethyleneoxy oligomer spacers in their backbones have significantly different LC phase behavior than those with polymethylene spacers, and this report describes the properties of additional members of the earlier series as well as the behavior of other closely related series. In addition, the effects of pendant ethyleneoxy vs. n-alkyl substituent groups on LC properties was investigated.

a. Present address: Mitsubishi Paper Mills, Ltd., Kyoto, Japan
b. Present address: Department of Chemistry, Southeastern Massachusetts University

In general, but not including rigid rod
polymers, the polyesters with polymethylene
spacers, either in the backbone or present as
substituents, formed only one type of LC phase in
the temperature range above melting, T_m, and below
isotropization, T_i, and this phase was either
nematic or smectic depending on the spacer length.
In contrast, the equivalent polymers with
polyethyleneoxy spacers generally formed
successively both smectic and nematic phases above
their melting points. In all cases, the length of
the spacer had an important effect on controlling
both the melting temperature and the temperatures
for transitions of the liquid crystalline phase or
phases formed.

For rigid rod polymers (that is, those without a
main chain flexible spacer) the type of LC phase
formed on melting could not be identified by the
usual methods, but all such polymers formed a
higher temperature nematic phase, which was stable
up to at least 350°C.

Flexible Spacers In The Main Chain

In an earlier report we compared the LC
properties of two series of main chain polyesters
of the following general structure, in which R is a
flexible spacer consisting of either polymethylene
or polyethyleneoxy units [1]:

Series I: $R = (CH_2)_n$

Series II: $R = (CH_2CH_2O)_n$

The results of the studies an these polymers are
compiled in Table 1.

For the polymers in Table 1 containing
polymethylene units, Series I, those with spacers
having up to 7 methylene groups formed alternately
either a smectic or a nematic phase on melting,
while those with 8 through 12 methylene groups
formed a smectic phase [1]. The isotropization
temperatures of these polymers showed a regular
trend, with an odd-even effect, for the decrease in
T_i with the length of the flexible units up to n=9
in R, but then T_i increased for the 9, 10 and 12

polymers, as seen in the data in Table 1. Both the T_m and T_i transition temperatures for the polymers with an even value of n were generally higher than those with n odd, as has been observed in several other series of liquid-crystal polyesters [2]-[4], but since the effect of increasing the spacer length was greater on T_m than on T_i the odd members had a wider temperature range for liquid crystallinity, ΔT.

The liquid-crystal behavior of the polymers of Series II in Table 1, those with the polyethyleneoxy flexible units, also showed a very strong dependence on the length of the polyethyleneoxy spacer, n [5]. In this series, however, the polymers with shorter spacer units

Table 1. Effects of Structure and Length of Flexible Spacer Unit on the Liquid-Crystal Properties of Main-Chain Polyesters of Series I and II.

n	T_m, °C	T_i,[a] °C	ΔT,[b] °C
Series I: 2	340	(N)365I	25
3	240	(S)314I	75
4	285	(N)345I	60
5	175	(S)267I	92
6	227	(N)290I	63
7	176	(S)253I	77
8	197	(S)220I	55
9	174	(S)233I	59
10	220	(S)267I	47
12	212	(S)245I	33
Series II: 1	342	(N)365I	
2	185	(S)222(N)288I	
3	180	(S)203(N)257I	
4	121	(S)211(N)245I	
8.7	102	(N)242I	
13.2	91	[c]	

a. N, nematic; S, smectic; I, isotropic.
b. $\Delta T = T_i - T_m$. c. Not liquid crystal.

formed smectic as well as nematic phases, while the
polymer with the longest spacer in the series, in
which n is an oligomer with an average value of
about 9, formed only a nematic phase. When the
length of the polyethyleneoxy spacer was further
increased to n=13.2 (which was obtained from the
next glycol monomer available at the time) the
polymer did not form a liquid-crystal phase.

These observations emphasize the fact that both
the thermal stability and the nature of the
mesophase strongly depend on a combination of both
the structure and the length of the flexible spacer
unit. Perhaps the ability of the relatively short
diethyleneoxy spacer (n-2) to cause the formation
of a smectic phase indicates that the presence of
an oxygen atom in the spacer may have exerted a
specific polar effect, which strengthened the
lateral intermolecular attraction between adjacent
polymer chains, thereby helping the formation of
smectic layers of the mesogens.

Unfortunately the changes in the ability to form
either the smectic phases, or even to undergo an
enantiotropic transition with increasing spacer
length, occurred at intermediate values where
polymers were not available at that time, so it was
difficult to draw specific conclusions on these
effects. Recently, however, we prepared the
intermediate member of this series with n=5, 6 and
7. The LC properties for these polymers are
collected together in Table 2.

Table 2. Properties of Polymers in Series II
    with Penta-, Hexa- and Heptaethyleneoxy
                   Spacers.

n	$[\eta]^a$ dl/g	T_m °C	T_i^b °C
5	0.19	135	(S)194
6	0.18	117	(S)178(N)203
7	0.23	119	(S)173

a. Intrinsic viscosity in p-chlorophenol.
b. See footnotes in Table 1.

The new polymers of Series II with n=5 and 7, in contrast to the lower members, formed only the smectic phase between melting and isotropization, but the polymer with n=6 formed both smectic and nematic phases as previously observed for those with n=2-4. Nevertheless, the odd-even behaviors of T_m and T_i were maintained in these newer members of Series II as shown by the data in Table 2.

The identification of the types of LC phases formed by the polymers in Table 2 are, so far, only based on visual observations of the textures of samples placed on the hot stage of a polarizing microscope, so the results are still tentative until confirmation is obtained by WAXD. It is possible, however, that the polymers with n=5 and 7 have monotropic nematic phases relative to their smectic phases.

Flexible Spacers as Pendant Groups

Long-chain polymethylene or polyethyleneoxy spacer groups can also be placed in the polymer as lateral pendant substituents on the mesogenic group. For the former, we have observed that similar results are obtained when the polymethylene group is so attached in terms of the affect of both spacer length and odd-even structures on the thermal properties and the type of liquid-crystal phase formed. In those investigations two different series of polymers of this type were prepared and characterized for these effects, as follows [6],[7]:

Series III: -O-⬡-OC-⬡-C-
              (CH$_2$)$_n$CH$_3$

Series IV: -OC-⬡-CO-⬡-OC-⬡-CO(CH$_2$)$_{10}$-.
             (CH$_2$)$_n$CH$_3$

The first series was based on the rigid-rod polymer poly(hydroquinone terephthalate), which has a melting point well above 600°C. Substantial decreases in T_m values were only achieved when

fairly long pendant alkyl substitutents were used,
so polymers containing n-alkyl groups ranging from
pentyl to dodecyl (n=5-11 in Series III) were
prepared for this purpose, with the results shown
in Table 3 [5],[6]. On melting, the polymers
formed a type of liquid-crystal phase which we have
not yet fully characterized, but which may be a
biaxial nematic. This phase was converted into a
nematic phase at a higher temperature as indicated
in Table 3, in all cases.

Because the polymer samples in Table 3 varied
quite widely in molecular weight, as indicated by
their solution viscosities, it was not possible to
make exact comparisons of the effects of
substitutent length on their liquid-crystalline
properties. However, as the data in Table 3 show,
there was only a surprisingly small variation in
both melting point, T_m, and in the temperature for
the transition between the initial LC phase formed
and the nematic phase; this temperature is
designated as $T_{P/N}$ in Table 3. None of the
polymers in Series III showed an isotropization
temperature below 350°C by either DSC or microscopy
studies.

Table 3. Physical Properties of
Series III Polymers.

n	η_{inh} [a] dl/g	T_m, °C	$T_{P/N}$ °C
5	1.88	300	345
6	0.48	257	302
7	0.47	257	307
8	0.32	237	291
9	1.38	297	323
11	0.37	228	292
12	0.25	217	277

a. Inherent viscosity in
   p-chlorophenol at 0.2 gCm^{-1}
   and 45°C.

For the Series IV polymers containing a flexible decamethylene spacer and an aromatic triad ester mesogenic group there was an unusually large effect of the size of the alkyl group on both T_m and T_i for the methyl, ethyl and propyl groups, but little change in these properties occurred for larger groups [7]. That is, on replacing a hydrogen atom in the central hydroquinone unit with a methyl group in the Series IV polymer, very large decreases occurred in both T_m (from 231 to 154°C) and T_i (from 267 to 190°C). Similar decreases were found with the ethyl group (T_m=71°C and T_i=127°C) but the larger n-alkyl groups caused only minor changes in these properties. It appears, therefore, that the additional increase in the length of the alkyl group beyond four carbon atoms did not produce any additional steric effect that could interfere with the molecular packing in the solid state, and indeed a more or less constant melting point was observed for the polymers with the longer alkyl groups, possibly attributable to crystallization of the side-chain alkyl groups themselves rather than the polymer main chains.

The clearing temperatures of the polymers in this series decreased steadily with increasing length of the substituent, although the contribution of each additional methylene unit to the depression of this transition temperature became much smaller for the butyl, pentyl and hexyl substituents. These reduced effects may be the result of the gradually decreasing contribution of each additional methylene unit to the molecular diameter of the mesogenic units.

Finally, in Series IV, when the alkyl group was lengthened to more than six carbon atoms no thermotropic behavior was observed, possibly because either (1) the clearing temperatures of these polymers may have been depressed so much that they were lower than the melting point (monotropic behavior) or (2) the polymers were incapable of forming a liquid-crystalline mesophase. All of the polymers in this series were nematic, but unlike those in Table 1 with backbone spacers, the polymers with an even number of carbon atoms had a wider temperature range of thermotropic behavior than those with an odd number.

Because of the large differences in the LC properties between the polymers with the main chain polymethylene units, I, compared to those with polyethyleneoxy units, II, and because of the ability of the latter to greatly reduce the melting transition of their polymers and to induce the formation of a smectic phase, we prepared polyesters closely equivalent to both Series III and Series IV which had ethyleneoxy pendant groups; these are designated Series V and VI, as shown below:

Series V: $-O-\langle\bigcirc\rangle-OC\!\!\overset{O}{\underset{}{\parallel}}\!\!-\langle\bigcirc\rangle-\overset{O}{\underset{}{\overset{\parallel}{C}}}-$ n=1 or 2
            $O(CH_2CH_2O)_nR$ R=CH_3 or C_2H_5

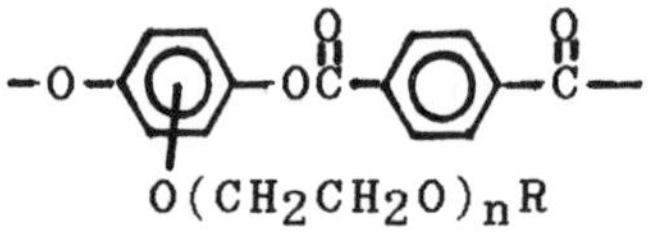

Series VI: $-O\langle\bigcirc\rangle-\overset{O}{\underset{}{\overset{\parallel}{C}}}O-\langle\bigcirc\rangle-O\overset{O}{\underset{}{\overset{\parallel}{C}}}-\langle\bigcirc\rangle-O(CH_2)_{10}-$
            $O(CH_2CH_2O)_nR$
                                        n=0, 1 or 2
                                        R=CH_3 or C_2H_5

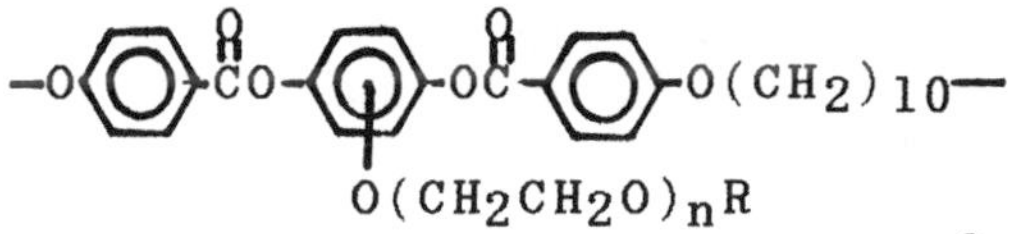

The thermal transition data for these two new series are collected together in Table 4. None of the polymers in Series V showed an isotropization temperature below 350°C, and the identity of the intermediate LC phase formed between T_m and $T_{P/N}$ is again unknown. As with the Series III polymers, this phase could conceivably be a biaxial nematic.

For Series VI, the first two polymers with n=0 formed only a nematic phase between T_m and T_i, but the other four polymers in that Series again showed intermediate endotherms, which are designated as $T_{P/N}$ in Table 4. Above the temperature of this endotherm, all four polymers formed nematic phases as indicated from their textures by visual observation of samples on the hot stage of a polarizing microscope. As before, however, the type of LC phase present between T_m and $T_{P/N}$ in this Series is unknown.

Table 4. Physical Properties of the Polyesters
of Series V and VI.[a]

	n	R	η_{inh}, dl/g	T_m, °C	$T_{P/N}$, °C	T_i, °C
Series V:	1	CH_3	0.21	180	224	–
	1	C_2H_5	0.20	160	233	–
	2	C_2H_5	0.59	145	257	–

a. See footnotes in Tables 1 and 2.

	n	R	η_{inh}, dl/g	T_m, °C	$T_{P/N}$, °C	T_i, °C
Series VI:	0	CH_3	0.23	90	–	222
	0	C_2H_5	0.23	80	–	180
	1	CH_3	0.25	79	137	167
	1	C_2H_5	0.26	74	125	154
	2	CH_3	0.20	70	117	149
	2	C_2H_5	0.25	72	108	137

Acknowledgement

The authors are grateful to the National Science
Foundation, under NSF Grant DMR-8317949 for the
support of these investigations and to the
Mitsubishi Paper Mills Ltd. for the support of A.
Furukawa.

References

[1] C.K. Ober, J-I. Jin and R.W. Lenz, Makromol.
Chem., Rapid Commun., 1983, 4, 49.
[2] R.W. Lenz and J-I. Jin, in "Liquid Crystals and
Ordered Fluids", A. Griffin and J. Johnson,
Eds., Plenum Press, New York, 1984; p. 347.
[3] A.C. Griffin and S.J. Havens, J. Polym. Sci.,
Polym. Phys. Ed., 1981, 19, 951.

[4] L. Strzelecki and D. van Luyen, Eur. Polym. J.,
    1980, 16, 299.
[5] G. Galli, E. Chiellini, C.K. Ober and R.W.
    Lenz, Makromol. Chem., 1982, 183, 2693.
[6] J. Majnusz, J.M. Catala and R.W. Lenz, Eur.
    Polym. J., 1983, 19, 1043.
[7] Q-F. Zhou and R.W. Lenz, J. Polym. Sci., Polym.
    Chem. Ed., 21, 3313, 1983.

REVERSIBLE OPTICAL INFORMATION STORAGE IN LIQUID CRYSTALLINE POLYMERS

Manfred Eich and Joachim H.Wendorff

Deutsches Kunststoff-Institut,6100 Darmstadt,FRG

Abstract

Flexible chain polymers which carry rigid anisometric groups in the side chain have a tendency to display liquid crystalline phases,such as smectic of nematic phases. Polymeric liquid crystals offer, in comparison to low molar mass liquid crystals ,the advantage that the liquid crystalline state can be frozen-in in the glassy solid state.

By appropriate means monodomain films can be obtained from such polymers,which display strongly anisotropic electrical ,magnetic and optical properties.By means of external electrical and optical fields it is possible to vary the local structure and thus macroscopical properties such as optical properties of such monodomain films in a reversible manner.This effect can be exploited for the manufacturing of optical components such as lenses as well as for optical information storage , using holographic techniques, for instance.

Introduction

The development of media whose optical properties can be varied locally in a reversible way is currently an area of intense research activities.Variations of the absorption,the reflectance,the refractive index ,the birefringence or of combinations of these are explored.It is the intention to use these media in optical applications such as optical information storage , optical information processing or in optical display technology (1-3) .The manufacturing of wave guides or the manufacturing of optical components , such as lenses , would also benefit from such materials (4).

The usefulness of materials for such applications depends on parameters such as the energy required to induce local variations of the optical properties, the spatial frequency bandwidth (spatial resolution) which is attainable, the efficiency of the induced variation of the optical properties (1-3) and finally , for certain applications ,the reversibility as well as the long term stability of the induced variations.It seems that polymer materials may play an increasingly important role in optical applications.

Polymer materials have traditionally been used in applications requiring particular mechanical and, to a lesser degree, particular thermal or electrical properties. The use of polymeric materials for optical applications offers many advantages in comparison to the use of inorganic materials, among them the low weight of optical components ,the good mechanical properties of polymers and the ease of manufacturing technical parts even with complex geometry.

Liquid crystalline polymers offer particularly favorable optical properties which result directly from a combination of properties inherent to liquid crystalline phases and properties inherent to polymer materials (5-11). Liquid crystalline phases exhibit, for instance, the interesting feature that their optical properties can be varied locally or macroscopically by means of small external disturbances resulting from electric or magnetic fields, from mechanical forces or from surface forces (5-8). This has been exploited sucessfully in optical displays and similar applications (9-11). These induced variations will relax in low molar mass systems, once the external disturbance is removed. It can, however, be frozen in within a liquid crystalline glassy state in most polymeric liquid crystals due to the strong tendency of many polymers to form a solid glassy rather than a solid crystalline state (9-11).

We have recently shown (12-14) that polymeric liquid crystals may be used as media for optical information storage as well as for manufacturing optical components.It is known , for instance, that the specific effect of optical components , such as a lens or a prism, consists in modifying the optical path of the beam passing through them. This is , in principal, achieved by modifying the length of the physical path , i.e. by geometric means , while the refractive index is kept constant. However, one may induce an appropriate variation of the optical path by keeping the physical path constant and varying the refractive index spatially in the required fashion. Thin films with parallel surfaces will consequently display the optical properties of a lens or a prism. The requirement is ,of course, that the variations which can be induced for the refractive index are suffiently large.

It is obvious that materials are needed for such applications which are transparent, in which sufficiently strong variations of the refractive index can be induced by appropriate means and in which such variations of the optical properties can be stored permanently. Photodielectric polymers have been considered for such applications (1). Refractive index variations result in this particular case from density variations which, in turn, are controlled by variations of the degree of polymerization. The reason is that the polymerization leads to a volume contraction, amounting to several %, on the average.The induced changes are obviously irreversible ones.

Liquid crystalline materials differ from such materials in that the induced variations of the optical properties may relax and in that variations not only of the refractive index but also of the reflectance, the absorption or the birefringence may be induced , as will become apparent below. In the following, the optical properties of liquid crystals will be described.

Optical properties of liquid crystalline phases
===

Structural and dynamical properties of liquid
crystalline side chain polymers have been the
subject of several review articles (9-11,15).
These were concerned with the influence of the
nature of the mesogenic groups, of the spacer
length and of the nature of the main chain on the
kind of mesophases which are formed,on their range
of thermal stability and on their dynamical
properties. It has become apparent that the
equilibrium structures of low molar mass and
polymeric liquid crystals agree in most respects
rather closely while the dynamical properties
differ in their characteristic time scale. The
latter quantity has been found to be larger by a
factor of 10^3 to 10^5 for polymer systems in
comparison to that of corresponding low molar mass
systems

It is not necessary, for the purpose of
understanding the optical properties both of low
molar mass and polymeric liquid crystals, to
consider details of their chemical structure and of
the physical structures of the various mesophases
displayed by them. Only the orientational order has
to be considered. This is done for a domain in
which the molecules are more or less orientationaly
ordered parallel to a prefered direction ,
characterized by the socalled director $\vec{n}$, a unit
vector pointing along the prefered direction.
(Figure 1). This holds both for low molecular
weight as well as for side chain polymers.

The perfection of the order can be described in
the most simplest way by the scalar order parameter
S, defined by:

$$S= \quad (3 \cos^2\theta-1)/2 \quad = \int_0^{\pi/2} ((3 \cos^2\theta-1)/2)f(\theta)\sin\theta \; d\theta$$

where θ is the angle between the director and the
long axes of the rod-like molecules and where $f(\theta)$
is the orientational distribution function. The
order parameter S assumes a value of 1 for a
perfect long range orientational order and zero for
a random distribution characteristic of the
isotropic melt.

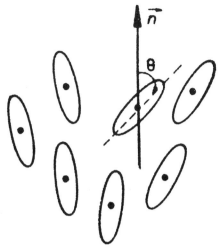

Figure 1
Orientational order in a nematic monodomain

By appropriate methods, such as the application of an electric or a magnetic field (5-11), one is able to induce macroscopical orientational monodomains, in which the director is constant throughout the macroscopical sample or varies continously in a specific way.

The nematic and the smectic A phase are known to be optically uniaxial and positive while the cholesteric phase , where the director varies in space in a helocoidal way, is optically uniaxial and negative.The optical properties of a uniformly oriented liquid crystalline material displaying a nematic or smectic A phase ,particularly the birefringence :

$$\Delta n = n_e - n_o$$

where n_e and n_o are the extraordinary and the ordinary refractive index respectively , depend on the order parameter S and the molecular polarizibilities parallel ($\alpha_\parallel$) and perpendicular ($\alpha_\perp$) to the molecular long axes:

$$\alpha_e = \bar{\alpha} + (2/3)S \Delta\alpha$$
$$\alpha_o = \bar{\alpha} - (1/3)S \Delta\alpha$$
$$\bar{\alpha} = (1/3) (\alpha_e + 2 \alpha_o) \ , \Delta\alpha = \alpha_\parallel - \alpha_\perp$$

The polarizibilities parallel and perpendicular to

the director are given by α_e and α_o, in this case.The orientational order characteristic of the nematic phase or the smectic A phase results in a refractive index which depends strongly on the direction along which the light beam travels and on the polarization direction of this beam relative to the coordinate system defined by the axis of the anisotropic domain.

The birefringence of liquid crystalline phases is known to be a strong function of the temperature through the temperature dependence of the order parameter.This is shown in Figure 2 for the particular case of a side chain polymer displaying a smectic A, a nematic and at elevated temperatures an isotropic phase.

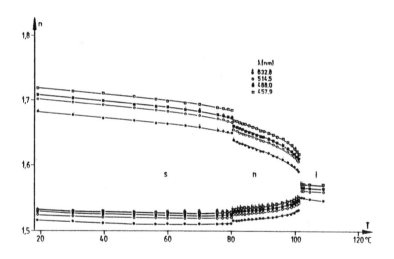

Figure 2
Dependence of the refractive indices n_e and n_o of a side chain polymeric liquid crystal on the temperature .

It is obvious that the birefringence decreases stepwise at the smectic to nematic transition and at the nematic to isotropic transition due to a jump of the order parameter. There is , of course, no birefringence in the isotropic phase. Strong variations both of the birefringence as well as of the absolute value of the refractive indices of the ordinary and the extraordinary beams occur in a limited temperature range close to the nematic-isotropic transition.

In the following, we will be concerned with means of varying the optical properties of liquid crystalline materials by external forces acting on them. It is apparent from the previous discussions that the optical properties can,in principal, be varied by influencing the absolute magnitude of the order parameter by appropriate means or by varying its direction relative to the beam direction and to the direction of polarization.By using cholesteric phases or by adding appropriate dyes one is able to modify not only the refractive index or the birefringence in this way but also the reflectance and the absorption (5-11).

Variations of Optical Properties-Induced by Low Frequency External Fields

A rather strong variation of the magnitude of the order parameter at constant temperature can be induced by external electrical and magnetic fields already in the isotropic phase (16-19) of systems displaying a low temperature nematic or cholesteric phase. This is known as the quadratic electro-optical (Kerr) or magneto-optical (Cotton Mouton) effect . The external fields result in a partial reorientation of the molecular axis, causing thus a birefringence which increases linearly as a function of the square of the applied field (16):

$$\Delta n = B \lambda E^2$$

Experimental results are shown in Figure 3 for a set of different temperatures for a particular side chain polymer, displaying a nematic phase.The observation is that the induced birefringence increases linearly with the square of the electric field and that the Kerr constant B increases strongly with decreasing temperature as the nematic phase transition is approached.This is obvious from Figure 4. The induced birefringence seems to diverge in the neighborhood of the transition into the liquid crystalline phase.

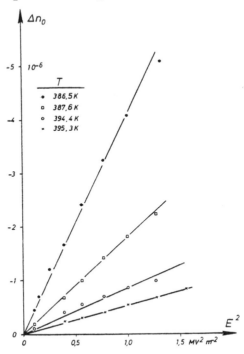

Figure 3
Dependence of the electrically induced
birefringence on the square of the applied electric
field for a nematic side chain polymer. Parameter
is the temperature.

This socalled pretransitional effect has been
treated in terms of the phenomenological Landau
theory of phase transitions by de Gennes . The free
energy density F(T) of a mesogenic system can be
expressed, according to this model, in the vicinity
of the isotropic nematic phase transition as a
function of the order parameter S as follows
(8,16,20):

$$F(S,T) = F_o(T) + A(T)S^2/2 - B(T)S^3/3 + C(T)S^4/4 + ..$$

where the quantities A,B,C etc are expansion
coefficients which depend on the temperature and
pressure as well as on the symmetry of the system
considered.

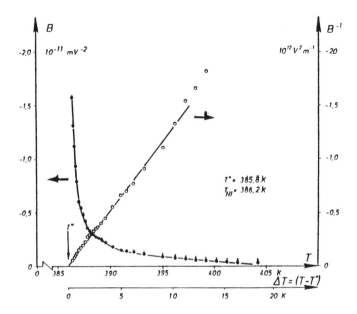

Figure 4
Dependence of the Kerr constant B on the
temperature for a nematic liquid crystalline side
chain polymer

The coefficient of the term quadratic in S,
namely A(T) is taken in this model to be :

$$A(T) = a(T - T^*)$$

T is a hypothetical second order phase transition
temperature.The cubic term B has been included
since the free energy for S > 0 is unequal to the
free energy for S < 0. The prediction is that the
order parameter assumes a nonzero value in the
isortopic state due to the interaction with the
electric field .The induced order parameter S_E
amounts to:

$$S_E = \varepsilon_O \, \Delta\varepsilon \, E^2 / \, (3 \, a_O (T - T^*))$$

where ε_O is the influence constant and $\Delta\varepsilon$ the
anisotropy of the dielectric constant of the
ideally oriented nematic state.So the induced
changes of S are fairly small but diverge as a
characteristic temperature T^* is approached, in
agreement with experimental results. The reason is

that order parameter fluctuations occur in the isotropic phase and diverge at a second order phase transition T^*. So it is obvious that strong changes of the optical properties can already be induced in the isotropic phase of substances able to display nematic or cholesteric phases. This effect is ,however,in general not suitable for most optical applications.The reason is that the Kerr relaxation response times of side chain polymers are large and diverge as the temperature   is approached (17-19).Response times going up to minutes have been observed for side chain polymers.

A much stronger and possibly faster response to external fields may be envisioned to occur within liquid crystalline phases, due to collective effects.It turns out, however, that a direct variation of the magnitude of the order parameter by weak external fields is no longer possible below the isotropic-nematic transition temperature in the nematic phase.The reason is that the molecules become so strongly correlated in the nematic phase that only correlated responses consisting in a reorientation of the director rather than in a variation of the order parameter can take place (8). Such reorientations of the director induced by external fields as well as the variations of the optical properties resulting from them have been treated extensively on the basis of a continuum approach,taking the particular elastic properties of liquid crystalline phases into account (5-11,21).

It is known that the state of lowest free energy corresponds to the one in which the director is constant throughout the macroscopical sample,as it is,for instance,the case in the homeotropic or homogeneous state where the director is either parallel to the layer normal or within the layer parallel to a given direction.Particular modes of orientational deformations are possible within the liquid crystalline state, the allowed deformations depending on the local symmetry of the phase. Figure 5 defines the splay,bent and twist deformation, characteristic of the nematic phase. The corresponding elastic constants are given by:

K_{11} : splay K_{22} : twist K_{33} : bend

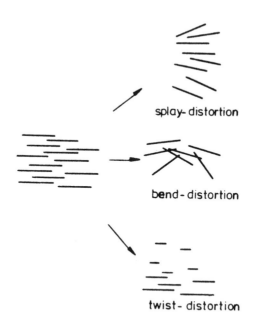

splay-distortion

bend-distortion

twist-distortion

Figure 5
Basic elastic deformations in liquid crystalline
low molar mass and polymeric systems.

These constants are of the order of 10^{-6} to 10^{-7}
dynes both for low molecular weight liquid crystals
as well as for side chain liquid crystals within
the nematic phase.Totally different values are
expected for semiflexible and rigid main chain
polymers (9-11). which are , however, not
considered here. These elastic constants control
the extent to which variations of the director
distribution can be induced by external fields as a
function of the relative orientation of the
director and of the applied field.This is shown in
Figure 6 for the particular case of a magnetic
field acting on a homeotropic monodomain.

The reorientation was monitored in this case by
measuring the capacitance C of the cell. The
characteristic elastic constants ,in this case the
bent constant K_{33} ,are found to control the extent
of director orientation as well as the threshold

above which the reorientation-the Fredericksz
transition -sets in:

$$H_c = (\pi/d)(K_{33}/\Delta\chi)^{1/2}$$

where $\Delta\chi$ is the anisotropy of the diamagnetic
susceptibility and d the thickness of the film.

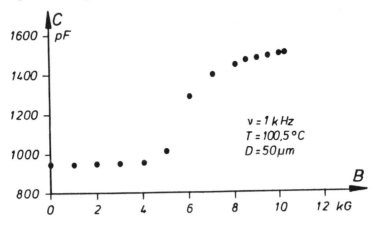

Figure 6
Director reorientations induced by magnetic fields
in a side chain liquid crystalline polymer.The
capacitance C of the cell was used to monitor the
reorientation.

The phase shift induced for the light beam
passing though the liquid crystalline films is
determined by the phase shift between the ordinary
and the extraordinary beams in this case.The
characteristic response times t_r and t_d of the film
in the presence (rise) and in the absence (passive
decay) of the external field are found to be
controlled by the elastic constants introduced
above, the film thickness d ,the magnitude of the
applied voltage U relative to the threshold voltage
U_c and a reorientational viscosity (22-24). For a
planar cell one obtains:

$$t_r^{-1} = t_d^{-1}((U/U_c)^2 - 1)$$

$$t_d^{-1} = (\pi/d)^2 K_{11}/\eta$$

The rheological properties of liquid crystals
are known to be rather complex. More than one

viscosity , among them a reorientational viscosity, are needed to characterize the flow behavior (5-11,25).The corresponding viscosity parameters are found to be of the order of 10 to 100 Centipoise for low molecular weight nematics ,giving rise to a relaxation time of about 1 to 10 s in the presence of the optical field and of 10 to 100 s after the optical field was switched off.The corresponding viscosity parameter of liquid crystalline side chain polymers has been found to be at least three or four orders of magnitude larger in comparison to the values found for low molecular weight liquid crystals (22-24). The relaxation times are thus expected to be of the order of minutes or less,depending on the boundary conditions, in agreement with the experimental findings.

Up to now the case was considered that low frequency electric or magnetic fields were used to induce variation in optical properties.It should be possible to induce such variations also by high frequency fields such as the optical field.The intention is thus to vary the optical properties of the material by the light passing through it or by a second light source.It is obvious that nonlinear optical properties come into play in this case.

Variations of Optical Properties Induced by Light

Light is known to induce ,in general, a polarization which increases linearly as the optical field E increases

$$p = \varepsilon_o \; X \; E$$

where ε_o is the influence constant and X the susceptibility.The refractive index n,which is related to the susceptibility by:

$$n = (1 + X)^{1/2}$$

will depend on the frequency but not on the magnitude of the optical field or on the intensity of the light.This holds,however,only for small optical fields.

One leaves the regime of linear optics and

enters the regime of nonlinear optics when the magnitude of the optical field is increased appreciably (26,27). The polarization starts to depend in a nonlinear way on the optical field:

$$p = \varepsilon_0 (X_1 E + X_2 E E + X_3 E E E \ldots)$$

where X_2 is the second order susceptibility, X_3 the third order susceptibility etc.The refractive index is consequently a function of the intensity of the light at constant frequency.

Nonlinear optical phenomena have met with ever increasing interest particularly over the last decade (1-3,28-31). A large number of industrial and university laboratories are currently investigating the physical mechanism giving rise to optical nonlinearities. In addition,it is the aim to develop new materials with improved nonlinear properties and to explore possible applications for optical information processing integrated optics etc. The physical processes giving rise to nonlinear optical properties may be divided roughly into those involving intramolecular and into those involving intermolecular effects.

The intramolecular effects result from the nonlinear deformation of the electron clouds of the molecules as induced by strong optical fields (26). It is characteristic of this intramolecular nonlinear effect that it is very fast,being an electronic effect.The disadvantage is that the optical nonlinearity is rather weak.So one may need intensities of the order of GW/cm^2 to induce nonlinear optical effects.It should be mentioned that the symmetry controls which of the higher order susceptibilities is nonzero and which is zero.X_2 is zero ,for instance, in systems with an center of inversion.

Photorefractive materials such as $Ba\,Ti\,O_3$ or $Bi_2Ge\,O_{20}$ are known for their large optically induced refractive index variations,originating from a redistribution of charges trapped at various sites in the crystal (1).Intermolecular effects are thus characteristic of these materials.Intermolecular effects connected with structural variations induced by the optical

field,such as ,for instance, variations of the
density or of the molecular orientation, have been
studied extensively for the liquid crystalline
state ,in particular for the nematic phase (28-31).
Such phases have been found to display nonlinear
optical properties which may be larger by a factor
of 10^6 or even 10^8 in comparison to those involving
electronic effects.Nonlinear effects are observable
already at relatively low magnitudes of the optical
field,of the order of $1o^0$ to 10^2 W/cm^2.

It is obvious that this kind of nonlinearity has
to be related to the peculiar molecular arrangement
in the liquid crystalline phases ,to their
dynamical properties and to their collective
response to external electrical or optical fields.
These features will be discussed in detail ,in the
following.It is,at once , apparent that the time
scale of the response will differ strongly from
that of the electronic effects and it will become
apparent later that this influences possible areas
of applications.

Nonlinear Optical Responses of Liquid Crystals

A very simple way of inducing order parameter
variations consists in inducing temperature changes
by the light beam.This will occur for intense light
beams even for optically transparent samples. This
effect can be enhanced by adding appropriate dyes
as additives or as comonomer units in the case of
liquid crystalline polymers.The particular
distribution of the optical field within the laser
beam impinging on the liquid crystalline material
will control the spatial variation of the
refractive index.The refractive index within the
area crossed by the beam will increase relative to
that of the surrounding medium for the case of the
ordinary ray and will decrease for the
extraordinary beams,as is evident from Figure 2.

This variation will be of the same order of
magnitude ,namely about 10^{-3} ,as in the case of a
normal isotropic fluids ,at temperatures much below
the nematic isotropic transition temperature and at
least one order of magnitude larger at temperatures
close to the transition temperature.This is

apparent from Figure 2 .Thermal lens effects will occur in thin films due to this effect ,giving rise to self focusing in the case of the ordinary beam, characterized by:

$$dn_o/dT > 0$$

and defocussing for the case of the extraordinary beam, characterized by:

$$dn_e/dT < 0$$

The focusing and defocusing effect will become apparent only in the near field case whereas diffraction effects originating from self phase modulations occur in the the far field (32,33).

The self phase modulution gives rise to the occurrence of diffraction rings the number of which increases with increasing intensity of the light. The characteristic rise and decay times of the nonlinear effects discussed here are controlled by the thermal conductivity:

$$\tau_T = d^2/(\pi^2 D)$$

where D is the heat diffusion constant and d the film thickness.

Relaxation times of the order of 0.1 s were observed for low molecular weight and polymeric liquid crystals (32,33).They are of about the same order of magnitude since the thermal conductivity is roughly of the same order of magnitude for low and high molecular weight systems.

Nonlinear optical effect may arise in liquid crystalline films even at constant temperature.Optical fields may act in a similar way as static electrical fields causing a reorientational motion of the director.One straightforward result is that the optical field necessary to induce nonlinear reorientational effects will increase with decreasing thickness of the film (28-31).

The threshhold value can also be influenced by the presence of a static external field . The

director will partially be reoriented along the
direction defined by the polarization of the
light.The induced average birefringence was found
to be of the order of 10^{-2} to 10^{-1} .

The reorientation will again,give rise to self
focussing and self phase modulation effects.Other
effects will be treated below.The intensity
required to induce reorientational motions is of
the order of 100 W/cm^2 and thus still rather
large.In the following an intermolecular effect
will be considered which results from electronic
properties of the particular molecules used but
which gives rise to variations both of the order
parameter and of the director distribution,as we
have found recently (12-14).

Photo-optic Effects in Polymeric Liquid Crystals

Some molecules of appropriate chemical structure
are able to display conformational changes such as
a trans-cis conformational change due to the
interaction with light.Azobenzene derivatives are
such molecules (6,12-14) . The trans-cis
transition may be induced by light with a given
wavelength such as for instance 353 nm and the
reverse transition may be induced thermally-which
is a slow process,or by light with a different
wavelength of about 440 nm. These changes are
obvious from the variations in the absorption
spectra,as is evident from Figure 7. The reaction
rates were found to be controlled by the molecular
mobility of the matrix to a great extent, in
agreement with results obtained on nonmesogenic
polymers (34,35).

The important fact is that such conformational
transitions lead to strong variations in the
geometry of such molecules .They are rod-like and
thus mesogenic - in the trans-conformation and more
spherically shaped and thus nonmesogenic in the cis
conformation.This gives rise to two different
effects.The nematic state is diluted by
nonmesogenic units,which causes a depression of the
nematic-isotropic transition temperature,and thus a
decrease of the order parameter at a given
temperature within the nematic phase .

We have reported recently that the trans-cis isomerisation gives ,in addition ,rise to an induced birefringence (12-14).The reason is that the transition moment is oriented perpendicular to that of the polarization of the light so that the changes of the molecular geometry happen along preferred directions. The induced birefringence arises aparently from a collective response of the neighborhood of the molecule to the trans cis isomerization.This effect contrasts thus to the case of the direct reorientation of the director by the optical field, discussed above.The induced birefringence was found to be of the same order of magnitude in both cases,whereas much smaller intensities of the order of $10^{-3} W/cm^2$ are required.

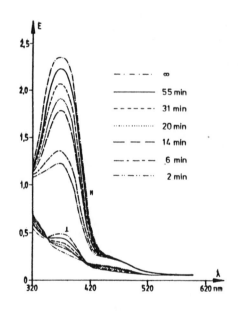

Figure 7
UV dichroism of the azobenzene group within a liquid crystalline side chain polymer directly after irradiation and for various annealing times.

One important difference exists between polymeric and nonpolymeric systems with respect to the last mechanism.The induced variations of the local optical properties will relax in low molar mass liquid crystals due to the thermally induced cis-trans-transition .So the final state correspond to the case that all molecules assume again their trans configuration and the local variations of the optical properties are erased.

The induced variations of the optical properties can, on the other hand freeze- in in the glass liquid crystalline state in the case of side chain polymers.In the following some possible applications of the nonlinear optical response will be described.

Applications of Liquid Crystalline Polymers

Possible applications are in the area of integrated optics and reversible optical information storage.Liquid crystalline side chain polymers (PLCs) have in the past been used to store optical data by a local optical heating of a preoriented PLC film into the isotropic phase (36,37).The resulting macroscopically unoriented light scattering spots were frozen-in by subsequent cooling .This technique is refered to as thermo-recording.The disadvantage is that high intensities of the order of $10^2 W/cm^2$ are required and that an amplitude object rather than a phase object results.A phase object has the advantage of a much better efficiency,as will be discussed below.

We have been concerned with investigations on such applications for liquid crystalline side chain polymers using the optically induced trans-cis isomerization.Preoriented PLC films were used in the storage process as optical storage media.Storage was always performed at room temperature, employing linearly polarized green light of a wavelength of λ = 514.5 nm, which is absorbed by the azobenzene moieties of the PLC. The writing intensity was 100 mW/cm^2 and the exposure time amounted to of about 10 seconds.Red light with a wave length λ = 632.8 nm was used for a nondestructive read-out of the stored information.

To erase the stored information the PLC was heated up to temperatures above the clearing temperature T_{NI} and subsequently oriented. Total

erasure of the stored information was always
achieved by this procedure and no degradation
effects could be observed.

In principal, two different kinds of storage
experiments were performed. A simple mask exposure
was used in order to investigate the possibility of
recording optical information of binary content.An
etched chromium test pattern was sandwiched, for
this purpose, to the storage cell and it was
illuminated with a linearly polarized plane wave. A
replica of the test pattern was obtained as a phase
object. Optically induced variations of the local
birefringence allowed to recognize the stored
information in the polarizing microscope,as shown
in Figure 8. The induced birefringence amounted to
$\Delta n = 7 \cdot 10^{-3}$.

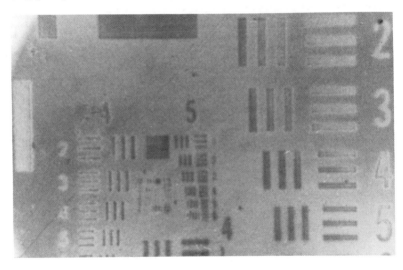

Figure 8
Reproduction of a mask,recorded reversibly within a
polymeric liquid crystalline monodomain film

The smallest features of the test pattern (5
µm) could not be transmitted to the PLC-layer due
to diffraction effects, caused by the optical path
length (glass substrate thickness of the PLC
storage cell). Higher resolutions should be
achievable by reducing the mask layer distance,
preferably by sandwiching the mask directly onto
the PLC layer.

The optically induced uniaxial order was proved by rotating the storage cell with respect to the polarizer.The stored information was found to become invisible in those cases, in which the polarization was parallel or perpendicular relative to that of the writing beam.A maximum contrast was observed for a read-out polarization of 45 degrees relative to the writing beam polarization This is in agreement with the assumption discussed above that the trans-cis transition causes a reorientation of the local axis in a direction perpendicular to the polarization of the writing beam.

Figure 9
Interference pattern obtained for the superposition of planar waves ,as recorded in a polymeric liquid crystalline monodomain film.

In addition, holographic experiments were performed.They consisted in recording interference pattern of intersecting plane waves and spherical waves.The reason is that every possible field distribution can be achieved by an appropriate

superposition of suchsimple waves.Two linearly
polarized plane waves were used to generate the
sinusoidal interference pattern with periodicities
down to 1 um.Figure 9 shows an example of a
grating obtained by double exposure with different
intersecting angles.   The read-out of the
holographic content can be achieved by illuminating
the stored gratings with a He Ne laser beam.Figure
10 displays the multiple diffraction pattern of a
symmetric 10   µm crossgrating.

A superposition of spherical waves and plane
waves was used to generate Fresnel zone type
intensity distributions.Fresnel zone phase plates,
exhibiting excellent focusing properties, were
obtained after the storage process . Figure 11
shows a micrograph of the center of a stored
Fresnel lens,   as obtained by polarizing
microscopical observations.The illumination of this
lens with a He Ne laser gave rise to a diverging
spherical wave and a converging real spherical wave
showing a sharp focus.

Figure 10
Scattering pattern obtained for the cross grating
displayed in Figure 11

A Fresnel zone plate corresponds to a simple
optical device . Depending on whether one uses the

first or minus first diffraction order the zone
plate one obtains a virtual and a real image and
the plate acts either as a convex focusing lens or
as a concave defocusing lens (14).The zone plate
can be used to enlarge or reduce the size of one of
the digits of an LED display out of the total
number of digits of the optical display unit.

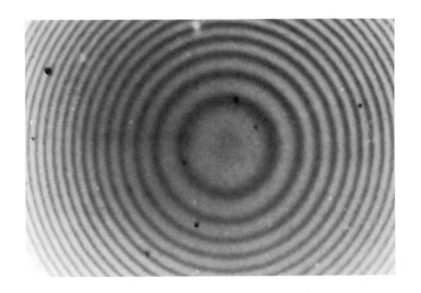

Figure 11
Center zone of a Fresnel zone plate as recorded in
a polymeric liquid crystalline monodomain

Since the basis of the application of liquid
crystalline films described above relies on a
recording and reconstruction process of elementary
waves it is tempting to find out whether the
intensity pattern of much more complex
superpositions of waves may be recorded in
polymeric liquid crystalline films.Such a case
occurs,for instance,if one illuminates an object
with the coherent light of a laser and superimposes
the reflected light and a reference wave
originating from the same laser well within its
coherence length (1-3). The recording of such a
pattern ,which corresponds to a hologram, leads to
an optical light modulator which is capable of

reconstructing the wave front characteristic of the original object,when illuminated with the reference beam.This is actually the case ,as was shown in a recent publication (14).

The results obtained so far indicate that side chain liquid crystalline polymers posess unusual optical properties which may favorably be exploited in a variety of different optical applications.It is obvious from the fact that even details such as the sharp edges of the objects recorded holographically are apparent in the reconstruction pattern that the spatial frequency bandwith of the polymeric material is large. Grating experiments which were performed (12,13) separately showed a spatial resolution of more than 3000 lines/mm which corresponds to a resolution of 0.3_2 μm or to a storage capability of 1 Gbit /cm^2.The maximum diffraction efficiency amounted to 50 % in thick phase gratings. The intensity required to store the pattern amounts to about 1mW/cm^2. Long time experiments extending over many weeks have shown that the pattern are stable at room temperature.

The polymeric liquid crystalline films offer in contrast to conventional silver halide emulsions the advantage that the registration process is reversible . The induced variations of the optical properties can be completely erased within seconds by heating the film up to temperatures above the respective glass transition temperature.The polymer films offers the additional advantage that no developing steps are required and that a high diffraction efficiency can be achieved due to the formation of phase gratings.

References
(1)   Hariharan,P. In"Optical Holography"; CambridgeStudiesInModern Optics,Cambridge University Press : Cambridge ,1984
(2)   Khanarian, G.,Ed. In "Molecular and Polymeric Optoelectronic   Materials:Fundamentals and Applications";Proc.SPIE 1986,682

(3) Collier,R.J.;Burckhardt,C.B.;Lin,L.H. In "Optical Holography"; Academic Press:New York,1971

(4) Adams,M.J. In "An Introduction to Optical Waveguides"; J.Wiley:Chichester,New York,1981

(5) Kelker,H.;Hatz,R. In "Handbook of Liquid Crystals";Verlag Chemie:Weinheim,1980

(6) De Jeu,W.H. In"Physical Properties of Liquid Crystalline Materials";Gordon and Breach:New York ,1979

(7) Haas,W.E. Mol.Cryst.Liq.Cryst.1983,94,1.

(8) De Gennes,P.G. In "The Physics of Liquid Crystals";Clarendon Press:Oxford,1975

(9) Ciferri,A.;Krigbaum,W.R.;Meyer,R.B. Ed. In "Polymer Liquid Crystals"; Material Sci.Ser.;Academic Press:New York,1982

(10) Blumstein,A. ,Ed. In "Polymeric Liquid Crystals";Plenum Press:New York, 1985

(11) Gordon,M.;Plate,N.A.,Ed. In "Liquid Crystalline Polymers I,II,II" ,Adv.Polym.Sci.59-61 ;Springer Verlag:Heidelberg 1984

(12)Eich,M.;Wendorff,J.H.;Reck,B.;Ringsdorf,H. Makromol.Chem. Rapid Commun.1987,8,59.

(13)Eich,M.;Wendorff,J.H.;Reck,B.;Ringsdorf,H. Proc.SPIE 1986,682,93.

(14)Eich,M.;Wendorff,J.H. Makromol.Chem.Rapid Commun.1987,in press

(15)Lipatov,Yu.S.;Tsukruk,V.V.;Shilov,V.V. Rev.Makromol.Chem. Phys.1984,C24,173

(16) Dunmur,D.A. In "Molecular Electro-Optics",Krause,S. Ed.;Plenum Press: New York ,1981

(17) Eich,M.;Ullrich,K.H.;Wendorff,J.H. Polymer 1985,25,1271

(18) Ullrich,K.H.;Wendorff,J.H. Mol.Cryst. Liq.Cryst. 1985,313,361

(19) Eich,M.;Ullrich,K.H.;Wendorff,J.H. Progr.Colloid Polym.Sci. 1984,69,94

(20) De Gennes,P. Mol.Cryst.Liq.Cryst.1971,12,193

(21) Frank,F.C. Disc.Farad.Soc.1958,25,19

(22) Meier,G. Ber.Bunsenges.Phys.Cehm. 1974,9,905

(23)Finkelmann,H.;Kiechle,U.;Rehage,G. Mol.Cryst. Liq. Cryst.1983,94,343

(24)Haase,W.;Pranoto,H. Progr.Colloid Polym.Sci.1984,89,1229

(25) Leslie,F.M. In"Adv.Liquid Crystals", Brown,G.H.Ed.;Academic Press: New York,1979

(26) Williams,D.J. Angew.Chem.1984,96,637

(27) Garito,A.F.;Wong,K.Y. Polymer J.1987,19,51

(28) Shen,Y.R. Phil.Trans.Soc.Lond. 1984,A 313,327

(29) Durbin,S.D.; Arakelian,S.M.; Cheung,M.M.;
Shen,Y.R. J.Physique 1983,44,C2-161

(30)Tabiryan,N.V.;Zel'dovich,B.Ya. Mol.Cryst.Liq.
Cryst.   1980,62,237

(31) Barnik,M.I.; Blinov,L.M.; Dorozhkin,A.M.;
Shtykov,N.M.   Mol.Cryst.Liq.Cryst.1983,98,1

(32) Shibaev,V.P.;Kostromin,S.G.; Plate,S.A.;
Ivanov,S.A.; Vetrov,V.Yu.; Yakolev,I.A. In
"Polymeric Liquid Crystals",A.Blumstein Ed.,Plenum
Press:New York,1985,345

(33)Eich,M.;Wendorff,J.H.;Reck,B.;Ringsdorf,H.
Makromol.Chem.Rapid.Commun.1986,

(34) Eisenbach,C.D. Makromol.Chem.1978,179,2489

(35) Eisenbach,C.D.Makromol.Chem.1979,180,565

(36) Coles,J.H.;Simon,R. Polymer 1985,26,1801

(37) Shibaev,V.P.; Kostromin,S.G.; Plate,N.A.;
Ivanov,S.A.;   Vetrov,V.Y.;   Yakolev,I.A.
Polym.Commun.1983,24,364

PHOTO- AND THERMOCHROMIC LIQUID CRYSTALS

Ivan Cabrera, Valeri Krongauz, and Helmut Ringsdorf[+]

Department of Structural Chemistry,
The Weizmann Institute of Science,
Rehovot 76100, Israel.
[+]Institut fur Organische Chemie,
Universitat Mainz (FRG).

1. Introduction

Many of the physical properties of liquid crystals, such as birefringence, viscosity, optical activity, and thermal conductivity are sensitive to relatively weak external stimuli. Magnetic fields, electric fields, heat energy, and acoustical energy can all be used to induce optical effects [1]. Almost no attention, however, has been paid to the possible modification of such properties by reversible photochemical reactions, mainly because of the synthetic problems to "engineer" suitable photoactive liquid crystals.

Spiropyrans are an interesting class of molecules for the preparation of light-sensitive liquid crystals because the reversible photo- and thermochromic spiropyran to merocyanine conversion [2]

SPIROPYRAN   MEROCYANINE

leads to drastic changes in such molecular properties as polarity, tendency to aggregate, molecular geometry, etc. One could expect, therefore, that the combination of spiropyran and mesogenic groups in a single molecular structure should lead not only to photo- and thermochromic mesophases but that many of the physical properties could be controlled by heat and light.

Our efforts in this direction include the synthesis of both monomeric and polymeric materials containing spiropyran and mesogenic groups.

In section two we show that in the low molar mass compounds the combination of photo/thermochromic and mesogenic moieties in one molecule led to impairment of one of the properties. Which one of the properties was lost depended markedly on the position and type of "bridge" by which the mesogenic group was connected to the spiropyran molecule. The T-shaped molecules of the general formula

show good photo- and thermochromic behavior but do not exhibit mesomorphic properties. However, the melt of one of them, that with R=CN-, reveals a very unusual and intriguing photocontraction effect [3]. This melt, produced by the irradiation and heating of amorphous films of the material contracts markedly under u.v. irradiation.

The rod-shaped "hybrid" molecules, on the other hand,

are thermochromic and give a very peculiar metastable mesophase (we named them quasi-liquid crystals (QLCs)), which proved to be a nematic-like mesophase with a structure that is strongly affected by the presence of the highly dipolar merocyanine groups [4].

QLCs generate second harmonic on irradiation with a laser beam [5]. The studies on second harmonic generation indicated that in contrast to conventional nematic liquid crystals, where polar ordering has never been observed, in quasi-liquid crystals a polar ordering of molecules exist in the absence of external fields. By applying an electrostatic field, the polar ordering can be modified in a manner predictable by a mean field theory.

In section three we report the synthesis of photochromic liquid crystals by preparing polyacrylate [6] and polysiloxane [7] copolymers containing spiropyran and mesogenic side units. We also show that structural changes of the mesomorphic systems can be induced by the reversible spiropyran merocyanine dye conversion. At temperatures at which the segmental mobility is limited, u.v. irradiation of the copolymers results in the formation of isolated merocyanine molecules. At higher temperatures the aggregation of the dye moieties takes place, leading to a network formation and a new rheo-optical effect [8], observed above the clearing point. The reverse photoconversion of merocyanines to spiropyrans occurs on irradiation with visible light.

Finally, in the last section we summarize our results and emphasize that since these materials combine photosensitivity with the physical properties of liquid crystal polymers, they open many possibilities for applications in optical technology.

2. Monomeric Spiropyrans with Mesogenic Groups

The low molecular weight spiropyrans containing mesogenic groups are thermo- and photochromic in solution. Their crystals have sharp melting points and give green or greenish-blue isotropic melts. Thin amorphous films of the compounds can be obtained by casting from solution. These films are metastable and remain in the amorphous state from hours to days. We ascribe the formation of the amorphous films to the presence of merocyanine molecules, which apparently act as impurities, retarding the crystallization of the films. As it is discussed below the properties of the amorphous films depend markedly on the position of the mesogenic substituent.

2.1 Photocontraction of liquid spiropyran-merocyanine films. Thin amorphous films of T-shaped spiropyrans with mesogenic side groups do not exhibit mesophase properties: on heating they give isotropic melts of blue color like the melt from the crystals. However, if an amorphous film of spiropyran with R=CN- is irradiated during slow heating from room temperature to about 60-70°C, it acquires an uniform, stable, cherry-red color.

The photocontraction effect [3] occurs when this red fluid film is irradiated with intense u.v. light at temperatures in the range 85-150°C. The net volume of the films is reduced by 10-20% depending on the film thickness, light intensity and conditions of preparation. The thicker, the film the less pronounced is the effect.

On subsequent standing in the dark the films expand again. The contraction and expansion lag after, respectively, the start and shut-off of the illumination. These lags in change of the film volume on illumination by chopped light at temperatures of 120-130 °C produce an effect which is reminiscent of the pulsation of unicellular organisms. However, after several such cycles the rate of contraction of a given film slows down, and eventually no volume changes are detected in finite times of observation.

Spectroscopic and X-ray examinations of the material allowed us to suggest that the photocontraction is due to the stacking of the merocyanine molecules, formed on illumination, into H-aggregates [3]. In essence the photocontraction arises from replacement of the Van der Waals distances between bulky spiropyran molecules by the distances between merocyanines in closely packed molecular stacks. The stacks are not stable at high temperature, and the back merocyanine-spiropyran reaction proceeds in the dark. Apparently the mesogenic groups provide an arrangement in which head-to-head positioning of the spiropyran moieties facilitates formation of the merocyanine stacks.

2.2 The quasi-liquid crystal structure. The metastable amorphous films of the rod-shaped compounds give a birefringent texture upon heating. Appearance of such texture, characteristic of a mesophase, coincides with a sharp increase in the merocyanine concentration, resulting in a color change of the films from yellow to green, due to the thermochromic spiropyran - merocyanine conversion. Apparently, the thermochromic conversion induces the phase transition to the mesophase. Further increase in temperature leads to the disappearance of the texture but it appears again upon cooling. The temperature range of the birefringent texture is rather wide (for example, spiropyran with R=CH$_3$O- has a texture between 50 – 130 ^{o}C) and lies much below the melting point of the spiropyran crystals ($\approx$ 200 ^{o}C). The mesophase is not monotropic.

The quasi-liquid crystalline films crystallize with time. The films can be aligned in an electrostatic field of more than 0.5 kV/mm. The alignment of the films stabilizes the quasi-liquid crystalline state and spontaneous crystallization no longer occurs. Under supercooled conditions at room temperature the mesophase with homogeneous alignment is preserved practically indefinitely.

Different methods usually employed for investigation of conventional liquid crystals were adapted for the studies of QLCs films. We examined polarization absorption and fluorescence spectra in the visible and u.v. regions of pure QLCs and with different additives, and polarization ESR spectra of a paramagnetic probe in the aligned films. FTIR spectroscopy was used for investigation of the polarization of different groups in the spiropyran molecules aligned in the films. Differential scanning calorimeter measurements and microscopic studies were used for establishing the miscibility and phase diagrams of QLCs with conventional liquid crystals of different nature.

The analysis of all these data allowed us to suggest that [4]: the QLC phase represents an intrinsic two-component mesophase, with properties that are substantially determined by strong interactions between the molecules of these components. The benzopyran groups of spiropyran molecules, interacting with each other, create a site with some features of a smectic phase and at the same time distort the parallel arrangement of the molecules due to the non-planar attachment to the mesogenic rest of the molecules (Fig. 1). The merocyanine molecules, on the other hand, improve the directional order of the spiropyran molecules in their environment and destroy smectic-like sites, promoting the nematic properties of the mesophase. Obviously the nematic- and smectic-like sites are in a dynamic equilibrium which relates to the equilibrium between merocyanine and spiropyran molecules. The two component nature of the mesophase is also a major factor in stabilization of this metastable state which occurs much below the crystal melting point.

Remained unanswered however, was the question whether the QLC state, stabilized by the electrostatic field, is centrosymmetric, as the conventional nematic state, or non-centrosymmetric with a ferroelectric-type arrangement. Optical second harmonic generation [9] is an ideal tool for such study because, within the electric-dipole approximation, the second-order nonlinear process is forbidden in a centrosymmmetric medium.

2.3 Non-linear optical properties of QLCs. A second harmonic generation (SHG) study of QLC films [5] indicated that QLCs aligned in an electrostatic field acquired a non-centrosymmetric structure with a polar axis pointing along the director (i.e. the average molecular orientation) defined by the

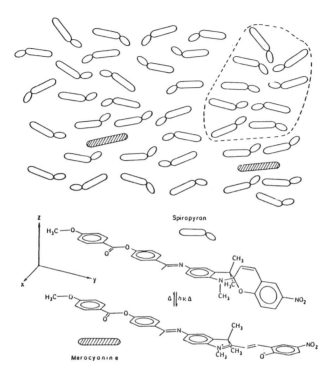

Figure 1. Simplified structural model of the QLC oriented mesophase. The area enclosed by the dotted line shows a possible layer arrangement.

direction of the field. This symmetry should arise from a polar alignment of the molecular dipole moments. The polar ordering of molecules can be modified by an external electrostatic field along the director, as manifested by a E-field dependence of the SHG.

Polar ordering has never been observed in conventional liquid crystals in the nematic state; that is, molecules tend to have equal probabilities pointing in opposite directions along the director. Local anti-ferroelectric pairing of molecules is also known to exist in many liquid crystals consisting of strong polar molecules. This is due to the tendency of nearby dipoles to be anti-parallel to each other in order to reduce their dipole-dipole interactions [10]. In QLCs, however, nearby merocyanine molecules are separated by a few spiropyrans, which could screen the dipole-dipole interactions among merocyanines and allow a preferred polar alignment of these molecules in the electrostatic field. The spiropyrans themselves could also have a polar ordering with an axis presumably pointing opposite to that of the merocyanines.

3. Photo- and Thermochromic Liquid Crystal Polymers

Liquid crystal polyacrylates were prepared according to the scheme shown in the Fig. 2. The detailed synthesis and free radical polymerization of the monomers were described earlier [6]. Liquid crystal polysiloxanes were obtained via active ester-mesogenic copolymers, which were prepared by addition of mesogenic and hydroxysuccinimide-ester olefines to Si-H containing polymers [7] (Fig. 3).

Observation with a polarizing microscope revealed that the clearing points of the copolymers are lower the higher the content of spiropyran units in the copolymer. Typical phase transition temperatures and composition for polyacrylate and polysiloxane copolymers are given in Figs. 4 and 5.

3.1 Photo- and thermo-chromic behavior. The photochromic behavior of the copolymers depends strongly on the physical state of the materials (glassy, mesomorphic, isotropic) and the spiropyran - merocyanine equilibrium which exists at a given temperature.

In the fluid state, the copolymer films acquire a pink color due to the thermally formed merocyanine molecules. Irradiation of such a film with visible light brings about a pale yellow color ($\lambda_{max} \approx 370$ nm), which

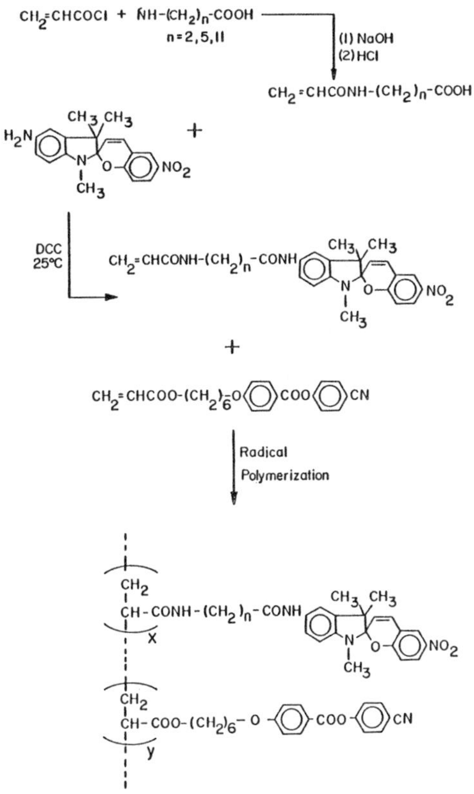

Figure 2. Reaction scheme for the synthesis of photochromic polyacrylates.

corresponds to the spiropyran absorption. If the yellow film is irradiated with u.v. light at temperatures at which the side chains are immobilized the characteristic blue color (λ_{max} 580 nm) of isolated merocyanine molecules is observed. For the polysiloxane copolymers, for example, this occurs at temperatures below –10 °C (Fig. 6). If the u.v. irradiation of a yellow film is performed at temperatures around and above the glass transition,

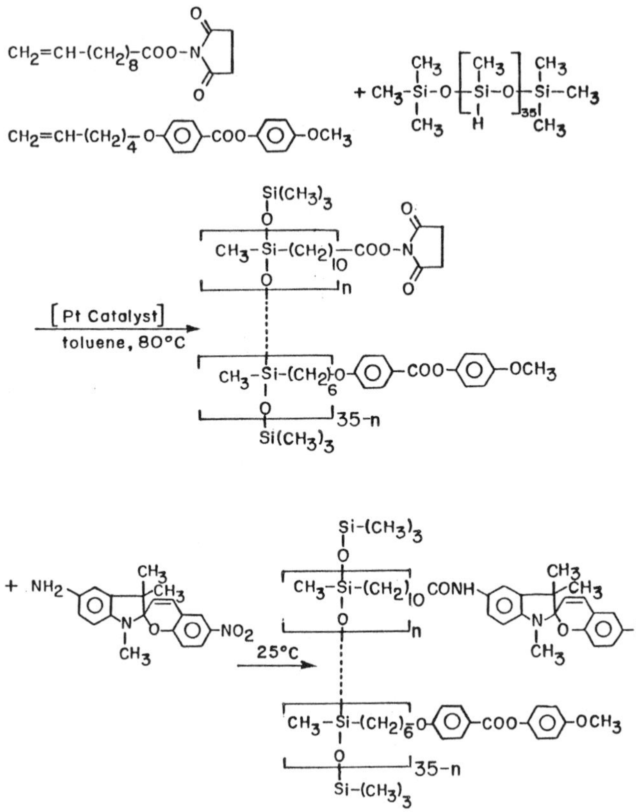

Figure 3. Synthetic route for preparation of photochromic polysiloxanes

physical crosslinking of the macromolecules occurs due to aggregation of the dye moieties. This is accompanied by appearance of a red color (λ_{max} 550 nm) The network formation is responsible for the appearance of a new rheo-optical effect described in part 3.2.

The yellow color can be restored by irradiation of blue or red films with visible light, ie, the photoinduced crosslinking of the macromolecules is reversible.

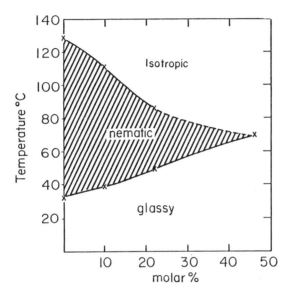

Figure 4. Phase behavior of the spiropyran containing polyacrylates with n=5 in Fig. 1

The mechanism for these transformations is summarized below:

The copolymers exhibit thermochromic properties, ie, change of color

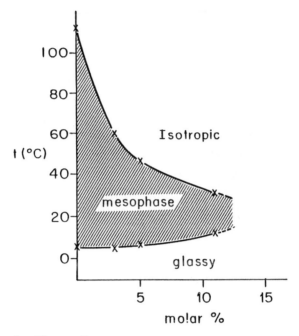

Figure 5. Phase diagram of the polysiloxane copolymers with spiropyran groups.

progressively with temperature. The electronic absorption spectra of the copolymer films show an increase of the merocyanine's absorption with temperature rise. The transition from amorphous to liquid crystalline phase is accompanied by aggregation of the merocyanine molecules in molecular stacks. The optical density does not change appreciable with temperature up to the clearing point. The transition from mesophase to isotropic phase coincides with an increase of non-aggregated merocyanine molecules. Probably, above the clearing point, formation of the network is completed and restricted segmental mobility hinders further merocyanine association.

3.2 Rheo-optical properties. A very remarkable feature of the isotropic films formed above the clearing point by the copolymers is their very strong transient translucence between cross polarizers when they are squeezed

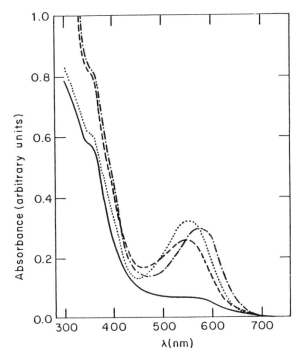

Figure 6. Absorption spectra of polysiloxane with 11% of spiropyran units: (—) yellow film at 25° C; (· · ·) red film at 25° C; (– · –) blue film at –20° C; (– – –) red film obtained by heating the blue film to 25° C.

between two glass slides or even lightly touched with the tip of a spatula.

The liquid crystal homopolymers, which do not contain spiropyran, do not exhibit this effect.

Usually such instant birefringence during mechanical disturbance is considered as an indication of homeotropic (orthogonal to solid surface) orientation of mesogenic molecules [12]. Therefore, one could conclude that instead of nematic-isotropic transition at the clearing point we observe a transformation of orientation of mesogenic groups of macromolecules from parallel to perpendicular to the surface.

To check this, we treated the glass surface with cremophor [13] and nylon [14] which promote the planar orientation of liquid crystals and with 1-dodecanol [13] for the homeotropic alignment. We found no effect of the surface on the "sparkling phenomenon". A decisive experiment was performed with droplets of the copolymers in an isotropic phase having a relatively small area connected with a surface. For these droplets again even a very gentle touch gave remarkable sparkling though in this case the effect of the surface must be insignificant. This suggests that the transient brightening is caused by at least partial restoration of liquid crystalline order, induced by mechanical disturbance.

The DSC measurements give endothermic peaks at temperatures that coincide with the microscopic observation of the clearing points. This confirms that the static birefringence disappearance relates to the mesophase-isotropic phase transition.

The rheo-optical properties of the copolymer films as a function of temperature were investigated with a parallel-disk type rheometer with transparent glass disks [8]. The results revealed that introduction of relatively small portion of the spiropyran groups causes a drastic increase in the viscosity of the polymers. This indicates that the interaction of the merocyanine groups, formed on heating or irradiation, gives rise to the aggregation of macromolecules in a network with high viscosity.

The formation of such a network by physical crosslinking of the macromolecules, due to dimerization of the merocyanine side groups, is responsible for the appearance of the strong dynamic birefringence above the clearing point.

Presumably, the more rigid structure of the network favors the preservation by macromolecules of the conformation acquired in the mesophase even above the clearing point. This makes the dynamic ordering easier.

4. Conclusion

We have presented a brief summary of the photo- and thermochromic properties of low molar mass molecules and polymers containing spiropyran and mesogenic groups. The results indicate that these materials exhibit a complex structure, which is determined by the superposition of the merocyanine interactions on the self assembling property of the mesogenic molecules.

The possibility of altering the mesophase by light opens a way to many scientific and practical applications [15]. In particular detailed investigations on the use of the materials as storage medium for optical information are currently being performed.

References

[1] Priestley, E.B., Wojtowicz, P.J., and Shen, P. *"Introduction to Liquid Crystals"*, Plenum, New York, 1975; Chapter 14.

[2] Bertelson, R.C. In *"Photochromism"*, G.H. Brown, Ed.; Wiley, New York, 1971.

[3] Cabrera, I., Shvartsman, F., Veinberg, O., and Krongauz, V. *Science*, **1984**, 226, 341.

[4] Shvartsman, F., and Krongauz, V. *Nature*, **1984**, 309, 608.; Shvartsman, F., Cabrera, I., Weis, A., Wachtel, E., and Krongauz, V. *J. Phys. Chem.*, **1985**, 89, 3941.

[5] Hsiung, H., Rasing, Th., Shen, Y.R., Shvartsman, F., Cabrera, I., and Krongauz, V. *J. Chem. Phys.*, in press.

[6] Cabrera, I., and Krongauz, V. *Macromolecules*, in press.

[7] Cabrera, I., Krongauz, V., and Ringsdorf, H. *Angew. Chem.*, in press.

[8] Cabrera, I., and Krongauz, V. *Nature*, **1987**, 326, 582.

[9] Shen Y.R. *"The Principles of Nonlinear Optics"*, Wiley and Sons, New York, 1984 Chapters 6,7.

[10] Chandrasekhar, S. *"Liquid Crystals"*, Cambridge University press,

Cambridge, 1977, p.83

[11] Eckhardt, H., Bose, A., and Krongauz, V. *Polymer*, in press.

[12] Kelker, H., and Hatz, R. *"Handbook of Liquid Crystals"*, Chemie, Berlin, 1980 Chapter 1.

[13] De Jeu, W. H. *"Physical Properties of Liquid Crystalline Materials"*, Gordon and Breach, New York, 1980, p. 23.

[14] Patel, J., Leslie, T., and Goodby, J. *Ferroelectrics,* **1984**, 59, 137.

[15] Attard, G. and Williams, G. *Nature,* **1987**, 326, 544.

NOVEL POLYCONDENSATION SYSTEMS FOR THE SYNTHESIS OF POLYAMIDE AND POLYESTER

Naoya OGATA

Department of Chemistry, Sophia University

7-1 Kioi-Cho, Choyoda-Ku, Tokyo 102,

JAPAN

ABSTRACT

Direct polycondensation for the synthesis of polyamide or polyester was successively carried out by using triphenyl phosphine (TPP) or triphenyl phosphine dichloride (TPPCl$_2$) under mild conditions. High molecular weight polyamide or polyester were obtained from the combination of aromatic diamines or bisphenols with dicarboxylic acids. Reaction conditions, including the amount of reagents and the concentration of monomers, solvents and acid acceptors, were investigated in terms of optimum yields and solution viscosities of resulting polymers.

Polymeric initiators having pendant triphenyl phosphine units were prepared and their activities for the initiation of polycondensation reactions were investigated. The principal advantage of these initiators having TPPCl$_2$ is that recovered triphenyl phosphine oxide can be regenerated to the reactive triphenyl phosphine dichloride by treating with phosgene or oxalyl chloride so that the recycling system can be established for the industrial application.

INTRODUCTION

Condensation polymers such as polyamide or polyester can be prepared under mild conditions by interfacial or solution polycondensation of a dicarboxylic acid with a diamine or a bisphenol in the
presence of inorganic or organic bases as acid
acceptors. Active diesters with an enhanced reactivity can also be used to prepare condensation
polymers under mild conditions.

Recently , convenient methods have been developed
for the direct polycondensation reactions of dicarboxylic acids with diamines or diols by using
various phosphorylating agents such as triphenyl
phosphite or triphenyl phosphine (1,2).   The direct
polycondensation reactions require a stoichiometric
amount of the phosphorylating agents to initiate
the direct polycondensation reaction.   Therefore,
the recycling system of the phosphorylating agents
has to be established in terms of industrial application of the direct polycondensation method.
When triphenyl phosphite or triphenyl phosphine are
used for the direct polycondensation, they transform into penta-valent phosphine oxide which has
to be reduced in order to regenerate these initiators.
It is known that triphenyl phosphine oxide is
quantitatively transformed into triphenyl phosphine
dichloride when it is treated with phosgene or
oxalyl chloride (3,4).   Therefore, when triphenyl
phosphine dichloride (TPPCl$_2$) could initiate the
direct polycondensation reactions, the recovered
triphenyl phosphine oxide is expected to be recon-
verted to the reactive TPPCl$_2$ by treating phosgene
or oxalyl chloride(5).

$$(C_6H_5)_3PO + COCl_2 \longrightarrow (C_6H_5)_3PCl_2 + CO_2$$

$$(C_6H_5)_3PO + (COCl)_2 \longrightarrow (C_6H_5)_3PCl_2 + CO + CO_2$$

The recovery and regeneration of triphenyl phosphine moiety becomes much easier when the triphenyl phosphine moiety is immobilized into a polymer. Moreover, the direct polycondensation reaction requires the co-existence of pyridine as an acid acceptor. When both triphenyl phosphine and pyridine moieties are immobilized into a polymer, it is expected that the direct polycondensation reaction could be carried out as a semi-continuous process, as illustrated as follows:

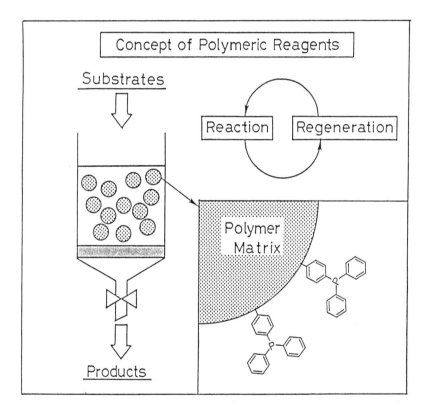

This paper deals with the direct polycondensation method using novel phosphorylating agents as an initiator for the synthesis of polyamide or polyester with an intention to make an approach to a semi-continuous process of the direct polycondensation.

EXPERIMENTAL

Triphenyl phosphine dichloride (TPPCl$_2$)system

Typical procedure for the synthesis of polyester is described as follows:

Polycondensation of Bisphenol A, Terephthalic Acid, and Isophthalic Acid

A 100-mL, four-necked flask equipped with a mechanical stirrer, a dropping funnel, a reflux condenser, and a nitrogen inlet was flushed with dry nitrogen gas and then charged with 20 mL of chlorobenzene and 8.94 g (0.032 mol) of triphenylphosphineoxide (TPPO). To this mixture was added dropwise during 10 min with stirring a solution of 4.10 g (0.032 mol) oxalylchloride in 10 mL of chlorobenzene. Gas evolved out of the solution. After the evolving gas was stopped, a mixture of 1.25 g (0.0075 mol) of isophthalic acid (IPA) and 1.25 g (0.0075 mol) of terephthalic acid (TPA) was added. The mixture was heated with stirring at a reflux temperature for 5 min, then cooled to room temperature. A portion of 3.42 g (0.015 mol) of 4,4'-bis(4-hydroxyphenyl) propane (BPA) was added and dissolved and finally a solution of 6.07 g (0.06 mol) triethylamine in 5 mL of pyridine was added with stirring at room temperature. After the solution was heated at a reflux temperature with stirring for 4 h, the flask was immersed in a water bath and 50 mL of chloroform was added to this solution. The solution was poured into 1000 mL of methanol and the precipitate was washed with hot methanol and dried for 18 h at 50°C in a vacuum.

Immobilized triphenyl phosphine (TPP) system

TPP moiety was immobilized by polymerizing 4-styryl diphenyl phosphine (4-SDP) or copolymerizing 4-SDP with 4-vinyl pyridine (4-VPy), using a conventional radical polymerization method as follows:

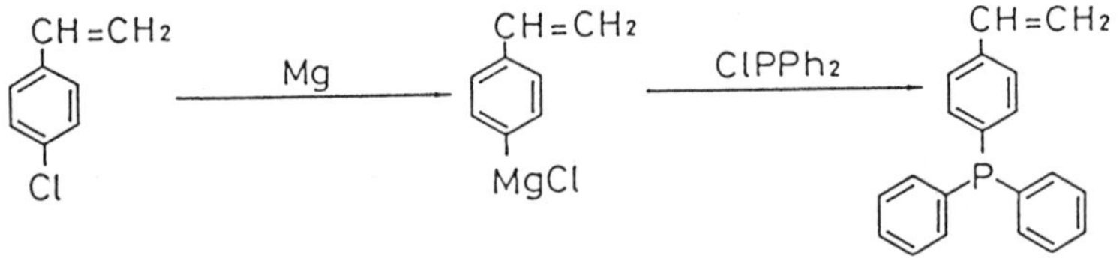

1

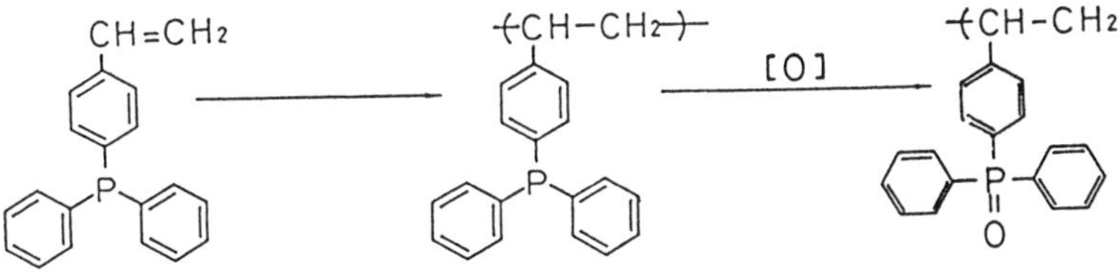

Poly 1   Poly 1 oxide

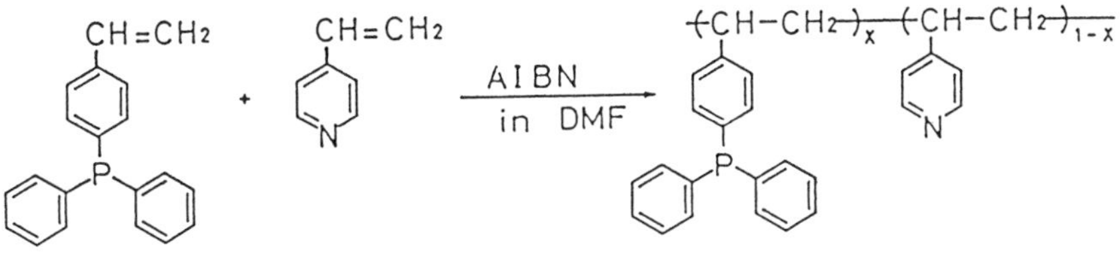

Poly(1-co-2)

The direct polycondensation method was essentially the same as monomeric TPP or TPPCl$_2$ systems.

RESULTS AND DISCUSSION

TPPCl$_2$ system for the synthesis of polyester

Table I summarizes results of the synthesis of aromatic polyester from various bisphenols, terephthalic acid (TPA) and isophthalic acid (IPA) by the direct polycondensation with TPPCl$_2$. In all cases, the polyester was obtained in quantitative yield and had a relatively high solution viscosity. The polyetser obtained from bisphenol A (BPA) and IPA/TPA (50/50) had the highest solution viscosity of η_{sp}/c = 1.66. This polyester and the polyester from TPA and BPA/resorcinol(RN) (50/50) were dissolved by chloroform, THF, DMF, or phenol. It could then be cast as a transparent film from the chloroform solution.

GPC measurement could be carried out by using THF as a solvent, with polystyrene as the standard polymer. Reults are summarized in Table II. These polyesters had molecular weight of couple ten thousands and M_w/M_n values were diversified.

TABLE I
Polycondensation of Various Bisphenols and IPA, TPA with TPPCl$_2$[a]

Monomer			Yield	
Acid	Bisphenol	Solvent	(%)	η_{sp}/C[b]
IPA	BPA	Chlorobenzene	99	0.63
IPA	BPA	Dichlorobenzene	98	1.00
TPA	BPA	Chlorobenzene	99	0.74
TPA	BPA	Dichlorobenzene	99	0.96
TPA	RN[d]	Chlorobenzene	86	0.36[c]
TPA	BPA/RN (50/50)	Chlorobenzene	96	0.60
IPA/TPA (50/50)	BPA	Chlorobenzene	96	1.66
IPA/TPA (50/50)	DDS[e]	Chlorobenzene	99	0.56

[a] Molar ratios of TPPO, (COCl)$_2$ to monomer: 1.07; total monomer concentration: 1.0 mol/dm^3; acid acceptor: (C$_2$H$_5$)$_3$N + pyridine; reflux temperature: 4 h.
[b] Measured in o-chlorophenol at 30°C.
[c] Measured in p-chlorophenol at 50°C.
[d] RN = Resorcinol.
[e] DDS = 4,4′-Dihydroxydiphenyl sulfone.

TABLE II
Result of GPC Measurements for Aromatic Polyesters

Composition	Acid acceptor	η_{sp}/C[a]	M_n[b]	M_w[b]	M_w/M_n
TPA, BPA/RN (50/50)	Triethylamine	0.60	11,200	28,100	2.51
IPA/TPA (50/50), BPA	Triethylamine	0.68	13,700	37,500	2.74
	Pyridine (Py)	1.26	10,900	66,300	6.08
	Triethylamine	1.42	37,800	86,400	2.29
	Triethylamine+Py	1.66	18,900	103,200	5.46
m-HBA/p-HBA (80/20)	Triethylamine	0.47	7,600	23,500	3.09

[a] Measured in o-chlorophenol at 30°C.
[b] Calculated from the poly(styrene) standard calibration curve.

Table II shows the effects of solvents and acid acceptors on the direct polycondensation reaction of BPA and IPA/TPA (50/50). The polyester was obtained in quantitative yield in such solvents as chloroform, chlorobenzene, dichlorobenzene, and nitrobenzene. In chlorobenzene, the polyesters with a high solution viscosity were obtained by using acid acceptors such as triethylamine or pyridine, while N-methylmorphorine and diethylaniline were not satisfactory. When triethylamine was added, reaction took place vigorously and the solution quickly became viscous. In contrast, when excess pyridine or the mixture of triethylamine and excess pyridine was used, a mild reaction took place.

Triphenyl phosphine requires the co-existence of hexachloroethane and pyridine for the direct polycondensation, while TPPCl$_2$ needs not the presence

of hexachloroethane and pyridine, and triethylamine
can be used as an acid acceptor.   Therefore, TPPCl$_2$
has an advantage for the practical use to initiate
the direct polycondensation reaction for the synthesis of polyamide or polyester.

TABLE III
Effect of Solvents and Acid Acceptors on the Polycondensation of BPA and
IPA/TPA (50/50)[a]

Solvent	Acid acceptor	Temperature (°C)	Time (h)	Yield (%)	η_{sp}/C[b]
Chloroform	Triethylamine	60	4	94	0.68
Chlorobenzene	Triethylamine	130	1	94	1.42
	N-Methylmorpholine	130	4	98	0.75
	Diethylamine	130	4	89	0.27
	Pyridine (Py)	130	4	99	1.26
	Triethylamine+Py	130	4	96	1.66
Dichlorobenzene	Triethylamine	180	1	99	0.86
Nitrobenzene	Triethylamine	130	1	100	0.49

[a] Molar ratios of TPPO, (COCl)$_2$ to monomer 1.07; total monomer concentration 1.0 mol/dm^3.

[b] Measured in o-chlorophenol at 30°C.

Figure 1 shows the effect of the amount of TPPO
and oxalyl chloride (COCl)$_2$ to form TPPCl$_2$ in situ,
on the solution viscosity of the polyester from
BPA and IPA/TPA (50/50).   The molar ratio of TPPO
and (COCl)$_2$ was found to have a significant influence on the solution viscosity of the resulting
polyester, reaching a maximum at the ratio of 1.07.

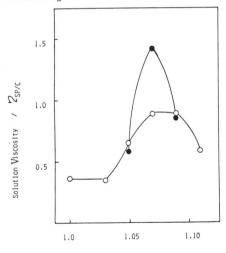

Fig. 1.   Effect of the amount of TPPO, (COCl)$_2$ on η_{sp}/C of BPA–IPA/TPA (50/50). IPA,
TPA: 0.25 mol/dm^3; BPA: 0.5 mol/dm^3; acid acceptor: (C$_2$H$_5$)$_3$N; solvent: chlorobenzene. ($\bigcirc$)
Room temp. (60 min); ($\bullet$) room temp. (10 min) + reflux (60 min).

Immobilized TPP system
_____

Immobilized TPP which was incorporated into a
polymer, was applied to the direct polycondensation
for the synthesis of polyamide and polyester.   The
initiator systems are called as the system (1) and
system (2), which are the combination of Poly-1
oxide and (COCl)$_2$,and the combination of Poly(1-co-
2) and hexachloroethane C$_2$Cl$_6$, respectively.

1.   Synthesis of polyamide

Synthesis of polyamide was carried out by using
both initiators of the systems (1) and (2), and
results are summarized in Table IV.   It is seen
in Table IV that these polymeric TPP systems could
initiate the direct polycondensation to form polyamide under mild conditions.   The system (2) which
was the combination of Poly(1-co-2) and C$_2$Cl$_6$ was
superior to the system (1) which was the combination of Poly-1 oxide and (COCl)$_2$, when solution
viscosities of the resulting polyamide were compared.

When the combination of Poly-1 and C$_2$Cl$_6$ was
used, pyridine was required for the initiation of
the direct polycondensation.   On the other hand,
the system (2) using Poly(1-co-2) and C$_2$Cl$_6$ did not
request the addition of pyridine for the direct
polycondensation as pyridine moiety was incorporated into the polymeric TPP.   The system (2) using
Poly(1-co-2) and C$_2$Cl$_6$ is advantageous from the
practical point of view since the direct polycondensation reaction proceeded in aprotic amide
solvents such as N-methyl pyrrolidone (NMP) or
tetramethyl urea (TMU) without adding pyridine as
an acid acceptor.

Resulting polymers were characterized as corresponding polyamides from infrared and elemental
analyses.   However, yields of the polyamides
frequently exceeded over 100% and elemental analysis of the resulting polyamide revealed an existence of a small amount of phosphorus atom (1 - 3
%).   Therefore, the resulting polyamides contained
a small amount of the polymeric TPP initiator.
The complete separation of the combined initiator
was very difficult by a simple extraction with a

Table IV Direct polycondensation using
polymeric TPP for the synthesis
of polyamides

System	Initiator Mole fraction of TPP	Monomer Kind	Concn. (mol/dm^3)	Solvent[d]	Acid acceptor	Polymer Yield(%)	ηsp/c[e]
(1)	1.0	TPA/APE[a]	0.15	NB	Py	26	0.26
	1.0	"	0.15	CB	Py	35	0.31
	1.0	4-ABA[b]	0.15	CB	TEA	16	0.30
	1.0	"	0.15	NB	Py	>100	0.38
(2)	$\underset{\sim}{1}$[c]	4-ABA	0.400	Py	Py	89	2.87
	1.0	"	0.042	"	"	>100	0.37
	1.0	"	0.083	"	"	>100	0.46
	1.0	"	0.160	"	"	>100	0.49
	1.0	"	0.167	"	"	>100	0.51
	1.0	"	0.417	"	"	>100	0.71
	1.0	"	0.833	"	"	>100	0.90
	1.0	"	0.250	NMP	None	>100	0.61
	0.877	"	0.250	"	"	>100	0.78
	0.788	"	0.250	"	"	>100	0.75
	0.428	"	0.250	"	"	43	0.66
	0.345	"	0.100	"	"	47	0.25
	0.343	"	0.05	"	"	0	-
	0.343	"	0.400	TMU	"	0	-
	0.154	"	0.100	NMP	"	trace	-
	0.154	"	0.0105	TMU	"	0	-
	0.094	"	0.0139	NMP	"	trace	-

a) TPA/APE=terephthalic acid/4,4'-diaminodiphenyl ether

b) 4-ABA=4-aminobenzoic acid

c) 1 = 4-SDP

d) NB=nitrobenzene, TEA=triethylamine, NMP=N-methyl pyrrolidone,
TMU=tetramethyl urea

e) Measured in H_2SO_4 at 30°C.

solvent and a hydrolytic cleavage of the combined
inititor might be required.

As seen in Table IV, both yields and solution
viscosities of the resulting polyamide decreased
with decreasing content of TPP in Poly(1-co-2) and
a favorable composition of Poly(1-co-2) was in the
range of 0.9-0.8/0.1-0.2 of the molar ratio of 4-
SDP and 4-VPy in the copolymers. There were no
direct relationships between yields and solution
viscosities of resulting polymers since the direct
polycondensation was not a simple step-wise polycondensation. The decrease in yields and solution
viscosities of resulting polymers with decreasing
content of TPP might be attributed to less interactions between TPP and pyridine moieties in
Poly(1-co-2) owing to an unfavorable conformation
of polymer chains to enhance the formation of an
active N-phosphnium complex.

It was found that Poly(1-co-2) could produce
polyamide in a quantitative yield with a solution
viscosity of about 0.7 when NMP was used as a
solvent.

2.   Synthesis of polyesters

Table V summarizes results of the direct polycondensation for the synthesis of various polyesters. It can be seen in Table V that polyesters
were produced in a good yield from terephthalic
acid (TPA)/bisphenol-A (BPA) or 4-hydroxybenzoic
acid (4-HBA) when the initiator of the system (2)
was applied to these monomers in various solvents.

Resulting polyesters were characterized by
infrared and elemental analyses which revealed the
corresponding structure. However, a small amount
of phosphorus atom was found in the resulting
polyesters. Solid-phase polycondensation of the
polyesters at 230°C yielded insoluble and infusible
polymers which suggested that a cross-linking
reaction took place as soon as they were heated.
Therefore, the polymeric initiator might be
attached on the resulting polyester and the
complete separation of the initiator was difficult.

Poly(1-co-2) which was recovered after the
direct polycondensation was applied again for the

Table V   Direct polycondensation using polymeric
TPP for the synthesis of polyester

Monomer[a]	Initiator P/N ratio (Poly(1-co-2))	Solvent[b]	Acid acceptor	Polymer Yield(%)	ηsp/c[c]
IPA/BPA	1.0/0	Py	Py	25	0.56
	1.0/0	CB	Py	27	0.26
	0.88/0.12	Py	Py	23	0.13
TPA/BPA	0.79/0.21	CB	None	100	0.31
	"	NB	"	94	0.56
	"	TCE	"	95	0.61
	"	CB/NMP(1/1)	"	99	0.61
	"	CB/DMAc(1/1)	"	99	0.40
	0.43/0.57	CB	None	65	0.32
	"	NB	"	50	0.25
	"	TCE	"	45	0.26
	"	NB	"	60	0.45
4-HBA	1.0/0	Py	Py	>100	insol.
	"	Py/ -Pico(1/1)	Py	16	"
	"	β-Pico	β-Pico	94	"
	"	β-Pico/TMU(1/1)	β-Pico	86	"
	"	NMP	None	10	"
	0.43/0.57	NMP	"	0	-

a)   IPA/BPA=isophthalic acid/bisphenol A
      TPA/BPA=terephthalic acid/bisphenol A
      4-HBA=4-hydorxybenzoic acid
b)   Py=pyridine, CB=monochlorobenzene, NB=nitrobenzene, TCE=
      tetrachloroethane, β-pico= β-picoline, TMU=tetramethyl urea,
      NMP=N-methyl pyrrolidone
c)   Measured in phenol/tetrachloroethane=1/1 at 30°C.

second run after it was isolated and treated with
(COCl)$_2$ so that phosphine oxide group was converted
to phosphine dichloride which could initiate the
polycondensation.   Yields and solution viscosities
of the resulting polyesters decreased down to about
half and the initiator activity of the recovered
Poly(1-co-2) decreased by the regeneration process.
However, the regeneration process was much easier
than the monomeric initiator and future remodifi-
cation of the regeneration process would make it
possible to establish a complete recycling of the
initiator.

REFERENCES

1.   N.Yamazaki and F.Higashi, J.Polym.Sci., Polym.
      Chem. Ed., 12, 2149(1974).

2.   G.Wu, H.Tanaka, K.Sanui, and N.Ogata, Polym.J.,
      14, 635 and 797(1982).

3.   R.Appel and W.Heinzelmann, German Patent
      1,192,205(1965).

4.   M.Masaki and K.Fukui, Chem. Lett., 151(1977).

5.   S.Kitayama, K.Sanui, and N.Ogata, J.Polym.Sci.,
      Polym. Chem. Ed., 22, 2705(1984).

Part V
High-Performance Polymers
from Flexible Macromolecules

1. HOW STRONG ARE CURRENT HIGH MODULUS HIGH STRENGTH
   POLYETHYLENE FIBERS?

Dusan C. Prevorsek

Allied-Signal Inc.
P.O. Box 1021R
Morristown, NJ 07960

Introduction

High tenacity-high modulus polyethylene (HTHMPE) fibers
are the latest entry into the field of high performance
fibers. Many years of research activities at the univer-
sities and industry resulted in an appreciable amount of
technical and patent literature [1-3]. The fibers are com-
merically available under the trade name of Spectra®-900
and Spectra®-1000. This latter grade exhibits in the form
of multifilament yarn, a strength of 3.0 GPa and modulus of
172 GPa.

Considering that this is an emerging technology it can
be anticipated that Spectra and other HTHMPE fibers will
change with time to better suit the broad market needs and
that new grades tailored for specific applications will be
developed. Potential for further increasing the strength
has also been demonstrated.

HTHMPE fibers are produced using very high molecular
weight linear polyethylene. The process yields highly
ordered fibrous structures having exceptionally high melting point relative to melt spun polyethylene fibers [4].
Based on the technical and patent literature data this process yields fibers of highest specific strength achieved to
date as a commerical product and specific modulus approaches to that of high modulus carbon fibers. The specific
properties of various "solution spun" fibers reported in
the literature compared with those of other reinforcing
fibers are shown in Figure 1 and Table 1.

Since the specific modulus and strength are the two
most important properties for a variety of applications it
can be anticipated that ultra high performance polyethylene fiber will soon penetrate a significant portion of
the market which can tolerate the relatively low melting
point and the low end use temperature of these fibers.
Potential use in ballistic armor has also been demonstrated
[10].

Fig. 1. Specific Strength and Modulus of Reinforcing Fibers

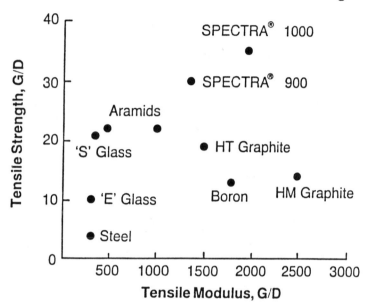

Table 1

Comparison of Mechanical Properties of Various High Performance Fibers.

Name	Unit	Spectra 900	Spectra 1000.	Aramid LM	Aramid HM	Carbon HS	Carbon HM	Glass S
1. Yarn Spec.	d/fil.	1200/118	650/120	1500/1000	1500/1000	1730/5100	1630/3000	
2. Fil. Diameter	μ	38	27	12	12	7	7	7
3. Density	g /cm³	0.97	0.97	1.44	1.44	1.81	1.81	2.50
	lb/ln³	0.035	0.035	0.052	0.052	0.065	0.065	0.090
4. Tensile Strength	ksi	375	435	406	406	443	350	665
	GPa	2.59	3.0	2.8	2.8	3.1	2.4	4.6
5. Tensile Modulus	Msi	17	25	9	18	33	55	13
	GPa	117	172	62	124	228	379	90
6. Tensile Elongation	%	3.5	2.7	3.6	2.8	1.2	0.6	5.4
7. Specific Tensile Strength	10^6in	10.7	12.4	7.8	7.8	6.8	5.4	7.4
8. Specific Tensile Modulus	10^8in	486	714	173	346	507	846	140

Parallel to the development of spinning technology for high molecular weight PE which yields fibers of exceptionally high strength and modulus, Prof. I.M. Ward of Leeds University developed a melt extrusion/ multiple stage drawing process for a much lower molecular weight polyethylene [5,6]. Those interested in composites made of this lower strength (1.0 - 1.5 GPa) and lower modulus (40 - 70 GPa) polyethylene fiber may obtain such information in references [7-9].

After these technological accomplishments have been achieved, it is appropriate to assess potential for further improvements in strength. It has been known since the early work of H. Mark that the maximum theoretical tensile strength of uniaxially oriented polymers is almost two orders of magnitude higher than that of polymeric fibers. Although a great deal has been written about the maximum achievable strength, it is still not settled what this limit in strength is. As shown below, the differences in the results depend greatly on the methods of calculations, and laboratories involved in advanced fiber research can neither assess on the basis of published data the merits of current technologies nor set realistic goals.

This study was undertaken to answer one of the most relevant questions of fiber science - where on an absolute scale of strength are current HTHMPE fibers? Is the technology sufficiently close to the theoretical limits in strength to consider the current processes as the last step in ultra strong fiber technology, or is the gap between current strength and theoretical potential so large that we should keep searching for new processes allowing another quantum jump in strength? The same question was addressed in 1972 by Perepelkin [11] whose results are frequently quoted and are still accepted as a yard-stick to estimate the quality of fibers. If we accept Perepelkin's data we have already reached with Spectra® fibers the "maximum attainable strength" of ~ 4.0 GPa (45 g/den) that the author places at ~ 1/5 of the theoretical fiber strength. (See Table 2)

Because of our long range commitments to fiber technology we had to reassess the validity of Perepelkin's conclusions along with those of other authors, and establish if required, our own criteria for technological limits in fiber strength. A recent study by Termonia et al at DuPont is particularly important in this respect[12].

Table 2. Theoretical and Maximal Achievable Strength
of PE Fibers (after K.E. Perepelkin)

Maximum Achievable Strength (GPa)		Theoretical Maximum Strength (GPa)
Molecular Approach (Morse)	Polymer Fracture (Zurkov)	1/3-1/6 Theoretical
18.6 - 19.6	19.5 - 22.9	2.9 - 4.9 (GPa) (33 - 55 g/den)

We also hoped that if the difference between theoretical and the achieved strength are larger than concluded by
previous authors, a comprehensive theoretical study may
provide clues to the cause.

Background

The low temperature, high rate of deformation modulus
of Spectra® fibers is so close to the theoretical limits
that extensive research to further increase the modulus is
not warranted. The measured crystalline modulus is within
the limits predicted by theories indicating that the
structure of a large volume fraction (~75%) of the fiber
is near perfect. This conclusion is corroborated by the
crystalline orientation function, crystalline density and
very high melting point of the fiber.

Although the degree of chain folding, in HTHMPE fibers
has not yet been established one experiment is worth
noting: the effect of draw ratio on thermal shrinkage. A
fully extended crystalline structure should be dimen-
sionally stable until its melting point. In less extended
structures, the inter and intra fibrillar tie molecules
start contracting below the melting point because they are
under tension and cannot be easily incorporated into perfect crystalline domains. The fibers, therefore, start
shrinking well below the crystalline melting point.

As the volume fraction of the tie molecules increases
with the draw ratio, so does the thermal shrinkage of melt
spun fibers. With ultra high strength PE, this is the case
only up to a certain threshold at high levels of draw
ratio. If this threshold in draw ratio is exceeded, the
shrinkage starts to decrease with further increases in the
draw ratio. This shows that at very high draw ratios it is
possible to achieve structures where the fraction of taut

tie molecules starts to decrease with increasing draw
ratio. This can be achieved by a complete chain extension
of a tie molecule and incorporation of the chain into a
nearly perfect crystal. The high melting point, perfection of the crystalline phase, fiber modulus, and the
thermal shrinkage data led us to conclude that Spectra®
fibers behave, at least in these respects, like extended
chain fibers.

Modulus and strength do not necessarily go hand in
hand. Modulus is a volume property where contribution of
all domains are averaged according to the schemes which
depend on the structural arrangements of these domains.
Conversely, strength, is governed on the weakest element
of the structure. The strength of heterogeneous structures, such as fibers falls therefore, far below (any
volume or weight) the average strength of various domains.

This statement is true of any volume or weight
averaging scheme.

Theoretical Strength of Fibers

General. The calculations of theoretical strength of
fibers involve two separate steps:

1. the strength of an individual molecule under stress
   and

2. the strength of an ensemble of perfectly aligned
   extended chain molecules

The first step is a problem of quantum mechanics. It
has been studied by several authors, and the results have
recently been reviewed and expanded by Crist et al [14].
It will be shown that, depending on the method of
approach, the predicted strength of a polyethylene molecule varies by a factor of four! The strength of the
ensemble of molecules, is a problem of fracture mechanics.
Here, the situation is even worse. In the range of intermediate molecular weights (Mn ~ 60,000) the predicted
strength differs by an incredible two orders of magnitude!

Let us begin with the strength of individual molecules
under stress. The problem is in the uncertainty about
decrease in the activation energy barrier to break a C-C
bond when the bond is under stress. Based on the experimental evidence it is generally accepted that the rate of
bond breakage V_i is given by:

$$V_i = \tau_o \exp [-(U - \beta\sigma)/kt]$$

where τ_o is thermal vibration frequency, T is the absolute
temperature, β is the activation volume and σ is the local
stress. (13)

The activation energy for unstrained C-C (i.e. disso-
cation energy) is ~ 80 Kcal/mol while the thermal fluc-
tuation approach to fracture yields a value of U $\approx$ 25
Kcal/mol. This indicates that under stress the energy
barrier is lowered to about 1/3 of the value in the
unstrained state.

The theoretical calculation of force strain relationship for ethylene repeat unit by Crist et al. [14]
Boudreaux, [15] and empirical Morse function yield the
plots shown in Figure 2.   The latter function was derived
assuming C-C bond energy D_o = 80 Kcal/mol and the initial
slope adjusted to yield a modulus of 405   GPa.

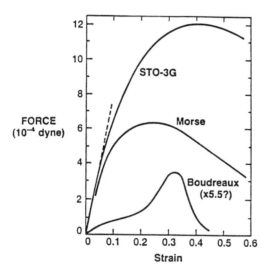

Figure 2. Force-strain curves for the ethylene repeat unit
          calculated for ab initio (STG-3G), semiempirical
          (Boudreaux, Ref. [15]), and empirical (Morse)
          potential function

For the semiempirical result of Boudreaux (1) there
seems to be an inconsistency noted by Crist et al.[14].   If
the reported maximum stress of 19 GPa is compared with the
reported modulus of 297 GPa, the force should be increased

by a factor of 5.5. In the absence of Boudreaux original calculations, we will try to resolve the uncertainty regarding the appropriate values of E and σ_{max} by using both sets of data.

The integration of force strain curves between the origin and maximum force yields the activation energies for bond breakage. These values were obtained by us using the figure in Reference [14]. These activation energies are listed in Table 3 along with corresponding values of strength σ^*, modulus E, and elongation at break ε_B.

Table 3

	σ^*(GPa)	E(GPa)	$\varepsilon_{B}\%$	ΔF (ergs/bond)
Crist	66	405	43	3.39×10^{-12}
Morse	34.6	405	20	1.015×10^{-12}
Boudreaux (1)	19.25	59	33	0.467×10^{-12}
Boudreaux (2)*	106	297	33	2.566×10^{-12}
Experimental				2.26×10^{-12}

*Adjusted following Crist's finding

With these data it is possible to estimate the strength of an ensemble of perfectly aligned molecules whose stress-strain behavior is represented by the stress-strain relationship shown in Figure 2. A treatment of this problem was recently published Termonia, Meakin and Smith [12]. Since we find their conclusions surprising, we decided that a more realistic study should be undertaken to examine the validity of their results, and revise, if required, the assessment of recent fiber technologies.

The most controversial conclusion of the study of Termonia et al. is the molecular weight dependence of strength. These authors find an excellent agreement between experimental and predicted values. A monodispers PE having $Mn = 10^6$ yields according to their calculations only one-half of the strength of the polymer having an infinite molecular weight. Our calculations predict for the same system a much faster increase in strength with molecular weight. We find that Mn of ~ 60,000 is sufficient to achieve 95% of the strength given by a polymer having infinite molecular weight!

This brings us to the key issue of this paper. If the results of Termonia et al are correct, the strength of Spectra® fibers is about 1/5 of theoretical which leaves very little room for improvement. Their data also suggest the most desirable approach to further enhance the properties is to increase the molecular weight of the polymer. Our calculations, on the other hand, suggest that molecular weight is not a major factor, and that the causes for low strength lie in the imperfect structure.

The conclusion regarding the molecular weight dependence on strength reported by Termonia attracted our attention. It is so far from an educated guess that it deserved another less biased examination. The following will illustrate the reasons for our doubts.

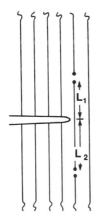

Assume that a crack propagating perpendicular to the parallel ensemble of extended chain molecules reaches an unbroken molecule. Under the increasing applied stress the molecule will be either pulled from the surrounding matrix of molecules or broken, depending on whether the energy to pullout is smaller or larger than the breaking strength of the molecule.

Schematic of
Crack Propagation

Since the pullout can occur on both sides, we have the chain breakage when both chain ends are sufficiently far from the fracture plane. The breakage will occur when

$$L_1 \, q_{coh} \gtrless Q_B \quad \text{and}$$
$$L_2 \, q_{coh} \gtrless Q_B$$

Here L_1 and L_2 are distances in repeat units from the fracture plane, q_{coh} is the cohesive energy per unit length, and Q_B is the energy to break L the primary bond.

This yields for the total length = $L^* = L_1 + L_2$ the relationship

$$L^* = 2Q_B/q_{coh}$$

Since the probability that both ends of the molecule
are equidistant from the fracture plane is 1/L, there will
be very little chain breakage at this point. Assuming A_B
= 45 and q_{coh} = 0.5, we shall then observe with a monodispers system the onset of chain breakage at ~ Mn = 2520!
When L = 10 L* the probability of chain breakage increases
to about .9. This simple consideration shows that with
extended chain uniaxial polyethylene where chain ends are
randomly distributed in the polymer matrix, the major
increase in strength should be observed between Mn = 2500
and 25,000, and increases beyond 25,000 should yield relatively insignificant increases in strength. More exact
calculations are presented below.

The Model

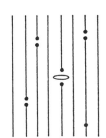

The strength of an ensemble of molecules will be calculated assuming that
the crack is circular and that it pro-
pagates perpendicularly across the specimen. At each molecule we will
consider the path of minimum energy
i.e., pullout vs chain breakage.

The free energy associated with the formation of such
a crack of radius r, is given by [17]

$$\Delta f_r = 2\pi\rho^2 \, \rho - \frac{8(1-\mu^2)r^3\sigma^2q^2}{3E} \tag{1}$$

In this expression ρ is the specific energy, defined
as the work required to form a unit area of surface in a
brittle material by formation of two planes each half a
unit in area; σ is the applied tensile stress, q is the
stress concentration factor and E is Young's modulus. The
curve of Δf_r vs r has a maximum for cracks with radius

$$r* = \frac{\rho E}{2(1-\mu^2)q^2\sigma^2}$$

As soon as the radius is exceeded, the crack becomes
unstable and the specimen fractures catastrophically.

If it is assumed that:

1.   growth of cracks is a thermally activated process,

2.   the time to rupture equals the time to form an unstable crack, and

3.   the free energy change associated with the formation of cracks can be expressed by Eq. (1),

one arrives at the following expression for the time to break t_b of a specimen under constant stress [16].:

$$t_b = (h/kTZV)$$

$$\exp\left\{(1/kT)\ [\Delta f_r^* + \Delta F^* - \frac{(vq^2\sigma^2)}{2E}\ ]\right\} \tag{2}$$

In this expression, k and h are the Boltzman and Planck constants, respectively. T is absolute temperature, Z is the concentration of crack nucleation sites, V is the volume of specimen, ΔF^* is the free energy of activation associated with the process of crack growth, v is the effective volume of the vacancy created by the breakage of a chain segment and

$$\Delta f_r^* = \frac{\pi^3 \rho^3 E^2}{6(1-\mu^2)^2 q^4 \sigma^4} \tag{3}$$

In proceeding to an expression for the time to failure in a tensile test, we apply the linear rule of cumulative damage [18] which under conditions of time dependent stress assumes the form

$$\int_0^{t^*} \frac{dt}{t_b[\sigma(t)]} = 1 \tag{4}$$

where t^* is the time to failure. The validity of Eq. (4) has been established both theoretically [19] and experimentally [20] for conditions where $\sigma(t)$ increases monotonically with time.

It follows from Eqs $(\underline{2})$, $(\underline{3})$, and $(\underline{4})$ that:

$$\frac{VZkT}{h} \exp\left(-\frac{\Delta F^*}{Kt}\right)$$

$$\int_0^{t^*} \exp\left\{\frac{1}{kT}\left[\frac{v\sigma^2(t)}{2E} - \frac{\pi^3\rho^3 E^2}{6(1\ \mu^2)^2\sigma^4(t)q^4}\right]\right\} dt = 1 \quad (5)$$

In a tensile test performed at a constant rate of increase of stress, $\sigma(t) = \sigma t$ where σ is rate of stressing and

$$\frac{VZkT}{h} \exp\left(-\frac{\Delta F^*}{kT}\right)$$

$$\int_0^{t^*} \exp\left\{\frac{1}{kT}\left[\frac{v\sigma^2 t^2}{2E} - \frac{\pi^3\rho^3 E^2}{6(1\ \mu^2)^2\sigma^4 t^4 q^4}\right]\right\} dt = 1 \quad (6)$$

The upper limit of the integrals in Eqs. $(\underline{3})$, $(\underline{5})$ and $(\underline{6})$, t*, is the time to failure in a tensile test. Breaking stress σ^* can be calculated from

$$\sigma^* = \sigma t^* \quad (7)$$

Since most of the authors studying the rupture of fibers apply in their analysis a procedure introduced by Zhurkow [21], it is desirable to compare the two approaches. Zhurkov uses the following expression for the time to rupture under stress

$$\tau = \tau_o \exp\ [U_o - \gamma\sigma/kT] \quad (8)$$

where k is Boltzmann's constant; U_o the activation energy of the process of mechanical polymer fracture, τ_o the period of thermal oscillation of atoms, σ the applied stress and γ the structure dependent proportionality constant which reflects the magnitude of the effect of the applied stress σ on the activation energy of bond rupture.

The applicability of Eq $(\underline{8})$, which relies on only one molecular property (bond energy) while all other structural factors are included in a vaguely defined parameter γ, is severely limited when the study of fracture is concerned with the effects of fiber morphology and molecular parameters other than bond energies. The main flaw of Zhurkov's approach lies in the fact that it fails to take into account that:

a. the breakdown is a result of the formation and propa-
    gation of cracks

b. the strength is molecular-weight dependent and,

c. the cohesive energy and modulus play an important role
    in the strength of fibers.

This author's treatment represents, on the other hand,
an attempt to develop a model of the breakdown which is
consistent with the phenomenological observations.
Furthermore, this model provides the possibility of
assessing quantitatively the magnitude and the relative
importance of the secondary effects (E and Mn) on
strength. A discussion of these factors is presented
below.

Effect of Molecular Weight on Fracture Surface Energy.
In our calculations the molecular weight dependence of
strength is derived from the effect of molecular weight on
fracture surface energy ρ.[22] The procedure to calculate
ρ is outlined in the Appendix, the results can be plotted
in the dimensionless form as reduced fracture energy (ρ_r)
versus reduced chain length (L_r). The reduced fracture
energy is defined by

$$\rho_r = \rho/\rho \text{ max}$$

where

$\rho \equiv$ fracture surface energy

ρ max $\equiv$ maximum fracture surface energy

$= NQ_b$

$N \equiv$ number of molecules per unit area

$Q_b \equiv$ breaking energy per chain

The reduced average chain length is defined by

$$L = L/L^*$$

where

$L \equiv$ number of average molecular weight in
repeat units

$L^* \equiv 2Q_b/q_c$

$q_c \equiv$ cohesive energy per repeat unit

L* is measured in number of repeat units and represents the critical chain length above which chain breakage
occurs. The value of L* (i.e. $2Q_b/q_c$) is obtained as
follows. In the case of an isolated molecule inserted a
distance L into a matrix of the same cohesive energy as an
ensemble of molecules, the energy to remove the molecule
is given by Lqc. In order for breakage of a molecule in
an ensemble to occur, it is necessary that both ends of
the molecule be located at a distance greater than L*/2
from the fracture plane. Thus, $L*/lq_c$ = Qb or $L* = 2Q_b/q_c$

A schematic representation of a possible fracture path
as assumed in the calculations of fracture surface energy
and theoretical strength of a random array of parallel molecules is shown in Figure 3.

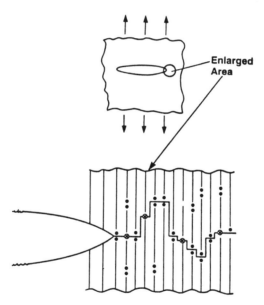

Figure 3. Schematic representation of a possible fracture
          path in a random array of parallel molecules as
          assumed in the calculations of fracture surface
          energies - fracture path

Figure 4 illustrates the dependence of reduced fracture energy surface (ρ_r) on the reduced average chain
length ($\langle L_r \rangle$) for the extended chain model in which a
"most probable" molecular weight distribution was
employed. The reduced fracture surface energy increases
monotonically with the reduced length and asymptotically

approaches unity. For this case, the fracture surface
energy and, hence, the strength are relatively independent
of cohesive energy if the molecular weight corresponds to
a chain length greater than 10L*. For example, suppose
the ratio of breaking energy to specific cohesive energy
is 25. Then for average chain lengths of say 500 we have
Lr = 10; i.e. we are at the flat portion of the curve.
Here a twofold increase in cohesive energy (i.e. a twofold
increase in ⟨Lr⟩) yields a negligible increase in fracture
surface energy. On the other hand, cohesive energy has a
strong effect on fracture surface energy in the low molecular range.

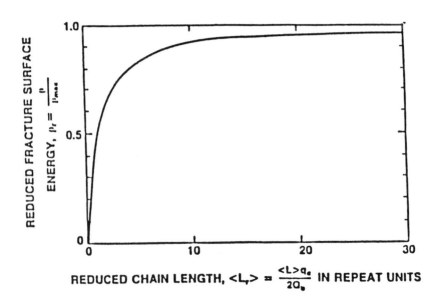

Figure 4.   Reduced fracture surface energy vs reduced
             chain length extended chain model

     The numerical values of parameters used in these
calculations are listed in Table 4.

Table 4. Numerical Values of Parameters Used in
Calculations of Theoretical Maximum Strength
of Uniaxial Extended Chain Polyethylene

Temperature T, °K	100
Poisson's ratio μ	0.4
Volume of specimen V, cm^3	3×10^{-5}
Concentration of nucleation sites Z, N cm^{-3}	10^{19}
Volume of an elementary vacancy v, cm^3	2×10^{-23}
Average distance between polymer chains ℓ, cm	4×10^{-8}
Stress concentration factor q	1
Stress rate σ, dyne $cm^{-2}sec^{-1}$	1×10^{6}
$\Delta F*$	3.13×10^{-12}

Using the activation energies and moduli in Table I,
our calculation yields the values of strengths and elongations at break listed in Table 5.

Table 5. Strength and Elongation at Break for Uniaxial
Extended Chain PE

	ρ	E*	ΔF	GPa	ε_b %
Crist	1730	3.0	3.39	72	24
Morse	519	3.0	1.02	26	20
Boudreaux (1)	239	2.9	0.47	14	46
Boudreaux (2)	1313	2.9	2.57	56	19
Experimental	1153	3.0	2.26	50	17

*Values of moduli are adjusted to yield the estimate
activation energy according to $\Delta F = \sigma^2/2E$. This is roughly
roughly the average of initial and secant modulus.

According to present calculations, the strength of a
perfect uniaxial PE falls between 14 and 72 GPa. Based on
our experimental data and remarks of Crist, we can safely
eliminate Boudreau (1) scheme. This restricts the range
to 26-72 GPa. Since the experimental activation energies,
and modulus agree very well with Boudreau (2) results, we
consider the theoretical maximum strength ~ 50-55 GPa the

most probable. All these schemes also predict very high
elongations at break ~ 20%. This is much higher than
observed with current high strength fiber (~ 2%).

With regard to strength at infinite molecular weight,
there is no principle difference between the findings of
this study and the results of Termonia et. al.

Unreconcilable differences exist, however, in the
effect of molecular weight on strength. Results of our
calculations are presented in Figure 5. Showing the plots
corresponding to moduli and activation energies obtained
by various authors. The main finding is that Mn = 60,000
is sufficient to achieve strength exceeding 95% of the
theoretical potential (i.e. Mn = ∞).

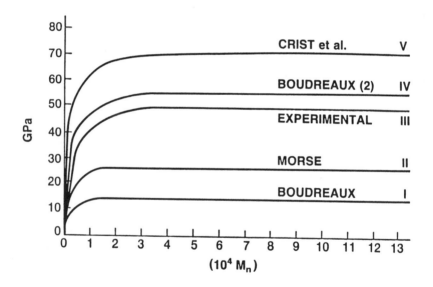

Figure 5. PE Extended chain model-random placement of chain
          ends; dependence of strength on molecular weight,
          modulus, and activation energies taken from stu-
          dies of various authors

In order to illustrate the difference between our results and Termonia's we show in Figure 6 the results of both studies on log-log plot. Note that at Mn = 6 x 10^4, Termonia shows a strength of ~ 5 GPa which is roughly 1/5 of the strength at Mn = ∞! Finally at Mn = 2000 Termonia had a strength of 0.4 GPa while our calculations assuming Crist modulus and activation energy yield 40 GPa!

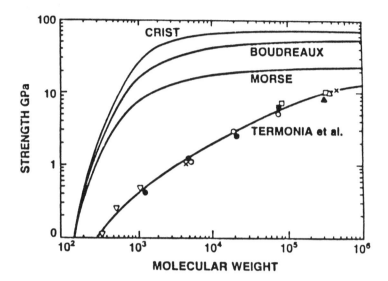

Figure 6. Dependence of the strength on the molecular weight for monodisperse polyethylene fibers. Crist, Bourdeaux and Morse, results of present work, Termonia et al from reference 2.

Before discussing these findings, we want to examine Termonia's and our results in terms of Flory's expression for the effect of Mn on strength. He derived the following relationship which was frequently supported by experimental results.

$$\sigma_B \ (\text{Mn}) = \sigma_{B \ \infty} - A \ M^{-1}$$

Here, σ_B (Mn) is the strength at a given Mn, $\sigma_{B\infty}$ is strength for infinite molecular weight and A is an empirical constant whose exact form depends on the failure mechanism and characteristics of the polymer.

In Figure 7 are shown the corresponding plots for present model and Termonia's calculation. Note that our

calculations reproduce the linear dependence very nicely.
The fit would be even better if we included in this plot
the decrease in modulus with decreasing molecular weight.
Termonia's results cannot be reconciled with Flory's
theory.

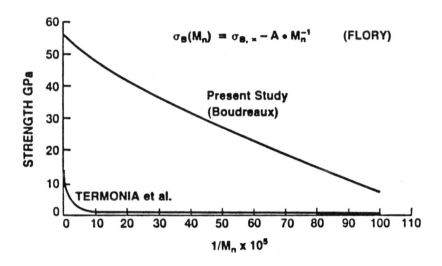

Figure 7. Extended chain models: strength vs 1/Mm

Nevertheless, this does not entirely settle this argument. According to a study of Smith, Lemstra and Pifjpers
the molecular weight dependence of gel spun PE follows a
power law

$$\sigma_B \approx Mn^{0.4}$$

which is reproduced nicely by Termonia's calculations but
not at all by ours.

Discussion

The predicted strength and elongation at break are
almost an order of magnitude higher than the experimental
values for exceptionally strong sections of PE fibers.
The predicted molecular weight dependence is so far from
the observed one that we must conclude that the theoretical model does not represent adeqately the structure of
current ultra high strength PE fiber.

Another purpose of this study was to rationalize on the basis of theoretical predictions, the main possible causes for the observed differences in strength and develop a research strategy to further increase the strength.

In comparing the theory to experimental data it should be noted that the calculations assume:

1.   the random distribution of chain ends and

2.   no segregation of molecular weight species and no disordered domains

Since the modulus of fibers can be reconciled using these assumptions, it is possible (if not very likely) that a large section of the microfil is relatively well represented by this model. The weak disordered domains, where the crack is initiated behave, however, very differently from the predictions based on this model.

Since the Termonia's results agree with observed molecular weight dependence, we can assume that the domains where the fracture is initiated have a much lower molecular weight than the fiber. It is possible that, in the process of drawing, the ends are rejected from the crystals and that the crystallization of the solution into the gel also introduces some fractionation of the sample.

The low elongation at break suggests that some of the domains are strained much more. About 20% local strain is required to start breaking the bonds. If a composite structure is assumed consistent with our microscopy and x-ray diffraction data, we can conclude that crystalline domains reinforce the weak elements of the amorphous matrix and that, because of very long L/D of crystallites, the strength of the weak domains is considerably below that of the fiber. This could be due to very low molecular weight of these domains and low orientation.

The following exercise yielding the lowest estimate of the molecular weight for weakest elements in structure is useful in this respect. Draw in Figure 5 a horizontal line from Termonia's to our plots. If this is done for $Mn = 10^5$ we intersect our curves at ~ $Mn\ 10^3$.

This means that a structure where the molecular weights are fractionated but the molecules are still perfectly oriented, extended, and chain ends spaced randomly would fit our calculation if the weakest domains had at polymer Mn of 10^5 a local Mn of 10^3.

In reality the Mn of the weakest domains must be much higher because of two additional factors; the tendency of chain ends to aggregate in one plane, and low orientation function of the amorphous domains. On both accounts, our model must regarded unrealistic.

In the process of solution spinning of PE, the dilute solution is cooled to form a crystalline gel. It is difficult to conceive that in this process the chain ends remain distributed randomly in spite of the efforts to stay away from thermodynamic equilibrium. For a short period of time the conditions in the spin way resemble those in fractionation of polymers by precipitation from dilute solutions. The point we want to make is that a small step from a complete randomness, towards segregation of domains of different molecular weights and aggregation of chain ends has a very large effect on fiber strength. The effect of the molecular weight has already been covered. The effect of agglomeration of chain ends can also be studied quantitatively by our computational methods. The doubling, tripling etc of chain ends reduces the fracture surface energy primarily at low molecular weight range. The effect decreases with increasing molecular weight. The net result is that our curves of strength are shifted towards that of Termonoia.

We could present quantitative data showing how much chain end aggregation and fractionation is required to make our strength curves match the relationship of Termonia or any other set of experimental data. However, this would be a useless exercise because fractionation and chain-end aggregation are not the only factors affecting the strength of HTHMPE fibers. Relatively low orientation of amorphous domains also contributes to the weakening of the fiber and we cannot establish the relative importance of these three factors.

While the limits in modulus have been sufficiently approached and it is doubtful whether a major step beyond current values is achievable, the strength is still far below the potential of a more uniform fiber structure. Most interesting, the elimination of very weak domains, which reach breaking extension while the macroscopic fiber strain is only ~ 2.5% would yield fibers whose elongation at break would increase with the draw ratio! This is because the individual PE molecules can be stretched before rupture almost 10 times more than current fibers.

This study also predicts that with more perfect fiber
structures it should be possible to achieve a threshold in
draw ratio beyond which both fiber strength and elongation
at break will start increasing with increasing draw ratio.
(See Figure 8)

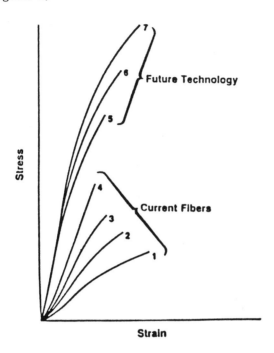

Figure 8.   A schematic representation of stress-strain
relationship for current and future fibers.
Draw ratio increases with increasing index
(1,2.....7)

References

[1] Prevorsek, D.C.; "Recent Advances in High Strength Fibers and Molecular Composites in Polymer Liquid Crystals". (A. Ciferri et al. eds.) 1982. Chapter 12.

[2] Pennings, A.J.; Menninger, K.E. in "Ultra High Modulus Polymers". (A. Ciferri and I. M. Ward, eds.) p. 117 Appl. Sci., London (1979).

[3] Smith, P.; Lemstra, P.J. J. Mater. Sci., 15, 505, (1980)

[4] Kavesh, S.; Prevorsek, D.C. U.S. Patents Nos. 4,413,110; 4,536,536.

[5] Cappaccio, G.; Ward, I.M. Polymer 15, 233, (1974).

[6] Cappaccio, G.; Gibson, A.G. Ward, I.M. in Ultra High Modulus Polymers (A. Ciferri and I. M. Ward, eds., Applied Science Publishers, London 1979. Chapter 1.

[7] Ladizesky, N.H.; Ward, I.M. Composites Science and Technology 26, 129 1986.

[8] Ladizesky, N.H.; Sitepu, M., Ward, I.M. Composites Science and Technology 26, 169, (1986).

[9] Ladizesky, N.H.; Ward, I.M. Composites Science and Technology 26, 199 (1986).

[10] Harpell, G.A.; Kavesh, S. Palley, I. Prevorsek D.C., U. S. Patents No. 4,403,012; 4,623,574.

[11] Perepelkin, K.E.; Angew, Makromol Chem. 22, 181 (1972)

[12] Tormonia, Y.; Meakin, P. Smith, P. Macromolecules, 18, 2246, 1985

[13] Zhurkov, S.N.; Vettegren, V.I. Korsukov, V.E. Novak, J.I. "Proceedings of the 2nd International Conference on Fracture, Brighton, Chapman and Hall, Ltd., London, 1969, p. 545 Zhurkow, S.N.; Korsukov, V.E. J. Polymer Sci., Polymer Phys. Ed. 12, 385, (1974)

[14] Crist, B.; Rafner, M.A. Brower, A.J. Sabin, R.J. J. Appl. Phys., 50, 6047 (1979).

[15] Boudreaux, D.P.; J. Polymer Sci., Polymer Phys. Ed., 11, 1285 (1973)

[16] Prevorsek, D.C.; J. Polymer Sci., Part A, 1, 993 (1963); J. Polymer Sci., Symp. No. 32 p. 343

[17] Sack, R.A.; Proc. Phys. Soc., (London), 58, 729 (1946)

[18] Miner, M.A.; J. Appl. Mech., Trans. ASME, 67, A159 (1945)

[19] Prevorsek, D.C.; Brooks, M.L. J. Appl. Polymer
Sci., 11, 925 (1965)

[20] Lyons, W.J.; Prevorsek, D.C. Text Res. J. 35, 1048
(1965)

[21] Zhurkov, S.N.; Tomashevsky, E.E. in Phys. Basis of
Yield and Fracture, Stickland, A.C.; (Ed) Inst.
Phys. & Phys. Soc., Conference Series No. 1, London
(1966), p. 200.

[22] Prevorsek, D.C.; Butler, R.H. Intern. J. Polymeric
Mater, 2, 167, 1973

[23] Smith, P.; Lemstra, P.J. Pijpers, J.P.L. J. Polym,.
Sci., Polym. Phys. Ed., 20, 2229 (1982)

DEVELOPMENT OF STRUCTURE IN PROCESSING POLYMERS COMPOSED OF FLEXIBLE VS. RIGID MOLECULES

Andrzej Ziabicki

Institute of Fundamental Technological Research,
Polish Academy of Sciences, Warsaw, Poland

Abstract

Manufacturing of high-performance, strong and rigid polymer materials requires that polymer molecules are oriented and crystallized into a structure of high degree of perfectness. Molecular orientation can be induced by flow of a fluid polymer (solution, melt, suspension) or deformation of a solid (plastic) material. Rigid, rod-like molecules and flexible, coiled-chain polymers behave differently, due to different relaxation times nad different stress-orientation characteristics. Crystallization, and crystal orientation distribution strongly depend on the conditions of processing and molecular orientation in the crystallizing material. The mechanisms of molecular orientation and crystallization of polymer materials in the conditions of processing have been discussed. Whereas highly oriented, fully crystalline structures can easily be produced in solutions of rigid molecules, solid-state processing seems to be necessary for obtaining similar results in flexible-chain polymers.

Introduction

The conditions of processing polymers with predetermined physical properties must be properly adjusted to chemical structure of the processed material and to the desired characteristics of the final product. *High performance* polymer materials with extremely high moduli and/or strengths should possess high degree of crystallinity, high molecular orientation, and crystal structure with high degree of perfectness, duly developed in the course of processing.

1

The first materials which satisfied such requirements were based on stiff-chain aromatic polymers (DuPont's *Kevlar*, US Air Force's *PPBT*) processed from liquid-crystalline solutions. Intensive studies aimed at finding new materials and manufacturing processes, resulted in the development of two groups of high-performance polymers. The first includes semi-stiff aromatic copolymers processed from random solutions and subsequently subjected to drawing (Teijin's *Technora*), the other one - ultra-high molecular weight, flexible-chain polymers, such as polyethylene (*Dyneema* from DSM, *Spectra* from Allied Signal), or polyvinyl alcohol. Processing of such materials involves so called *gel spinning* from low-concentration solutions, followed by *deep drawing* in the solid (gel) state.

Basing on simple theoretical models, we will analyze factors responsible for structure development in polymers composed of rigid-rod and flexible-coil molecules. The analysis shows why solution spinning is adequate for producing highly oriented structures in systems of rigid molecules, while solid-state processing is necessary for flexible chains.

Behavior of rigid rods and flexible coils in elongational flow

Uniaxial (multiaxial) extension provides a typical class of deformations in the formation of fibers, rods, bands, films, etc. When such deformation is applied to a fluid (polymer solution, melt, suspension), we are concerned with extensional flow. We will compare two asymptotic models of polymer fluids: a suspension of *rigid, elongated particles*, and a solution of *coiled chains* equivalent to ideally flexible, linear molecules. We will confine our analysis to *elongational flow* which provides a special (uniaxial and isochoric) case of extensional flow, typical for fiber spinning.

Elongated ellipsoids with half-axes $a \neq b = c$, volume v, and axial ratio $p = a/b > 1$, undergo *rotation* when subjected to elongational flow in a viscous medium. The rotation, controlled by the product of elongation rate, q, and the shape factor $R(p)$, tends to orient ellipsoids with their longer axes parallel to the direction of flow. This tendency is opposed by *random thermal motions* controlled by rotational diffusion coefficient, D_r [1,2]. In steady-state flow, equilibrium orientation distribution $w(\vartheta)$ is established [3]

$$w(\vartheta) = \text{const.} \exp[-qRsin^2 \vartheta/2D_r] \qquad (1)$$

ϑ denotes angle between symmetry axis of the particle and flow (fiber) axis. The *orientation factor* in such a flow reads

$$f_{or} = (3\langle\cos^2\vartheta\rangle-1)/2 =$$

$$= (2/15)(qR/D_r) + (1/315)(qR/D_r)^2 - \ldots \qquad (2)$$

According to the classical theory of suspensions, rotational diffusion coefficient is inversely proportional to the hydrodynamic friction coefficient and, consequently, to particle volume, v, and viscosity of the medium, η

$$D_r = kTZ(p)/v\eta; \qquad \tau_r = 1/6D_r \qquad (3)$$

where $Z(p)$ denotes another shape function, and τ_r is relaxation time for rotation. *Normal stress difference* for a suspension containing n particles in unit volume reads

$$\Delta p = p_{11}-p_{22} = (kTZx_2/vR)(qR/D_r)\cdot\varphi(p, qR/D_r) \qquad (4)$$

It can be noted that the resulting *orientation-stress coefficient*, $df_{or}/d\Delta p$, is proportional to particle volume, v, and inversely proportional to volume fraction of particles in the suspension, $x_2 = nv$

$$df_{or}/d\Delta p = const.(vR/x_2kTZ)[1 + \varphi_1(p, qR/D_r)] \qquad (5)$$

This means that the same stress will produce higher average orientation in a more dilute suspension, and one composed of larger and more asymmetric particles (axial ratio, p).

Orientation factor asymptotically approaches unity (ideal orientation, all particles parallel to flow axis) as stress increases to infinity. Figures 1. and 2. illustrate such relations [3]. Orientation and stress in the entire range of the variable qR/D_r, have been calculated from the closed-form distribution, eq. (1). Orientation factor is plotted vs. normal stress difference reduced by a constant reference volume, v_0. It is evident, that orientation is sensitive both to volume, v, of the suspended particles (Figure 1.), and their shape (axial ratio, p, Figure 2.).

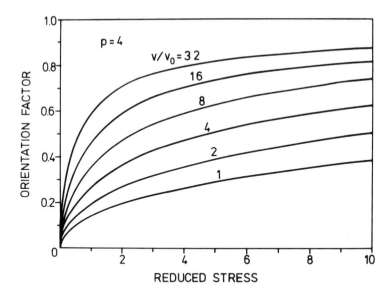

Figure 1. Orientation factor for a suspension of ellipsoids in the elongational flow vs. reduced normal stress difference, $\Delta p v_0 / x_2 kT$ [3]. p=4, reduced particle volume, v/v_0, indicated.

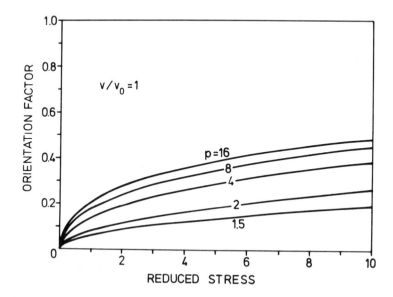

Figure 2. Orientation factor for a suspension of ellipsoids in the elongational flow vs. reduced normal stress difference, $\Delta p v_0 / x_2 kT$ [3]. $v/v_0=1$, axial ratio, p, indicated.

The behavior of flexible coils is different. Under extensional flow, the coils undergo *deformation* characterized by
the change of their end-to-end vectors, h. At the same time,
chain segments change their *orientations*. For a freely-jointed
chain composed of N segments of length l, segment orientation
factor results in the form [4]

$$f_{or} = (3/5)(h/Nl)^2 + (36/175)(h/Nl)^4 + \ldots \qquad (6)$$

where h/Nl denotes the degree of coil extension. n such coils
in unit volume of the system (solution, melt) produce normal
stress difference

$$\Delta p = (kTx_2/v_0)[3(h/Nl)^2 + (9/5)(h/Nl)^4 + \ldots] \qquad (7)$$

Since $x_2 = nv = nNv_0$, the stress is inversely proportional
to *segmental volume*, v_0, rather than volume of the entire
molecule, Nv_0. Consequently, orientation-stress coefficient
is independent of the molecular size, N, and contains only
volume of the segment

$$df_{or}/d\Delta p = (v_0/5kTx_2)[1 + \varphi_2(h/Nl)] \qquad (8)$$

The *average* orientation and stress in a system of flexible
coils subjected to steady-state elongational flow with deformation rate q, are functions of the parameter: $q\tau = qNl^2/3D_t$
[5,6], where τ and D_t denote, respectively, relaxation time
and translational diffusion coefficient of the entire macromolecule. The macroscopic orientation-stress coefficient is
proportional to segmental volume, v_0, and its dependence on
molecular size, N, is included in the parameter $q\tau$

$$df_{or}/d\Delta p = (v_0/5kTx_2)[1 + \varphi_3(q\tau)] \qquad (9)$$

In an uncrosslinked, fluid system, the average orientation
factor, f_{or}, and the average coil extension, h/Nl, approach
unity (full extension of all coils, ideal orientation of chain
segments), as the flow parameter, $q\tau$, and the related stress
difference, Δp, tend to infinity. The behavior of crosslinked
systems (networks) is different: crosslinks prevent some
chains from full extension, and the limiting orientation

factor is less than unity [6,7]. All these relations are shown
in Figure 3. Like in Figures 1. and 2., the orientation factor
is plotted vs. dimensionless stress reduced by a reference
volume, v_o, corresponding to a single segment.

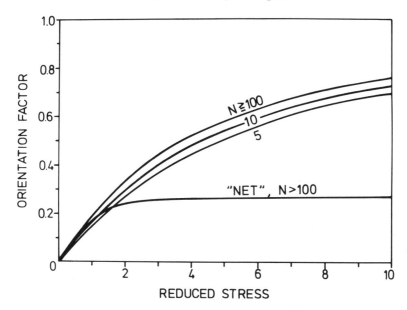

Figure 3. Segment orientation factor for a system of freely
jointed chains in elongational flow vs. reduced normal stress
difference, $\Delta p v_o / x_2 kT$ [6]. Number of chain segments, N, indicated. "Net" - relation for a crosslinked network.

As evident from Figure 3., systems of uncrosslinked chains
(solutions, melts) show only slight effect of chain length, N,
which disappears for N>100. At high stresses, f_{or} asymptotically approaches unity, corresponding to complete extension of
all chains and ideal orientation of all segments. On the other
hand, orientation of a crosslinked network (curve "Net" in
Figure 3.) levels off at $f_{or}= 0.22-0.25$. This is due to full
extension of chains oriented parallel to deformation axis,
combined with compression of perpendicular ones.

Several conclusions follow from the comparison of rigid,
and flexible molecules. Large, asymmetric, rigid particles can
be easily oriented in a fluid subjected to elongational flow.
On the other hand, effective extension and orientation of
flexible chains in a fluid state requires extremely high
stresses. Crosslinking of flexible chains excludes the possibility of reaching complete orientation.

Orientation of flexible chains in fluid-state
vs. solid-state processing

As follows from simple hydrodynamic considerations, the effectivity of orientation in a *fluid* state depends on two factors: intensity of flow (deformation rate), q, and diffusion rate, D (or relaxation time, τ). In a steady-state flow, equilibrium orientation distribution is controlled by the balance between the two competing effects described by the parameter q/D, or qτ. When flow is switched off, orientation relaxes with the rate inversely proportional to relaxation time. In the case of rigid particles, rotational relaxation time τ_r (eq. 3.) is proportional to particle volume, v, and effective viscosity of the fluid, η, which, in concentrated systems, is itself a function of v. Consequently, in a system of large, rigid molecules or their aggregates, relaxation is slow and orientation induced by flow can be maintained for a long period of time. For flexible, coiled chains, the translational relaxation time of the entire macromolecule, $\tau = N\ell^2/3D_t$, is also long, and strongly increases with the number of chain segments, N. However, relaxation of segmental orientation in such systems can be realized by short-range rotational motions of small segments rather than translation of the entire, large coil. The related (segmental) relaxation time is short, as controlled by small volume v_0. Therefore, in polymer fluids composed of flexible chains, orientation produced by flow is much less stable than in suspensions of large particles, and disappears shortly after cessation of flow.

The effectivity of producing orientation in flexible-chain systems and stability of this orientation can be improved by reduction of temperature and increase of viscosity, ultimately leading to solidification.

Processing of polymers in the *solid (plastic) state* is associated with very long (practically infinite) relaxation times. The mechanism of orientation is different to that active in fluids. The main role is played by the *extent of deformation* applied to the system, rather than *deformation rate*. In a steady-state, continuous extension (fiber spinning, drawing), the amount of deformation, e, is determined by the ratio of take-up and feeding velocities (V, V_0), while deformation rate is equal to velocity gradient along the drawing axis, x

$$e = \ln(V/V_0) \tag{10}$$

$$V = dx/dt; \quad q = de/dt = dV/dx \tag{11}$$

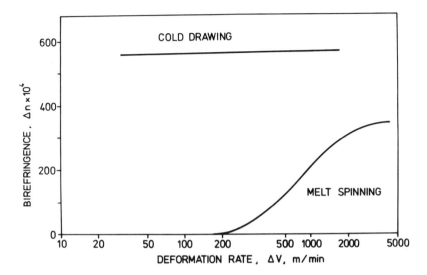

Figure 4. Birefringence of melt-spun [8] and cold-drawn [9] Nylon 6 fibers vs. deformation rate (velocity difference). Draw (spin-draw) ratio, V/V_0 = const.

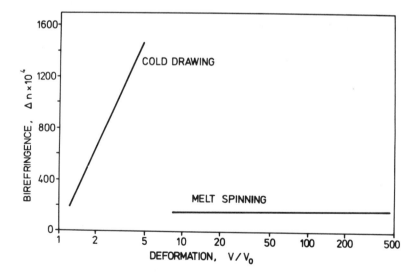

Figure 5. Birefringence of melt-spun [8] and cold-drawn [9] Nylon 6 fibers vs. deformation (velocity ratio). Spinning (drawing) velocity, V = const.

Figures 4. and 5. illustrate molecular orientation induced in Nylon 6 fibers by *melt spinning* (fluid processing)
and *cold drawing* (plastic deformation in a solid state). For
both processes optical birefringence (as a measure of orientation) is plotted vs. velocity difference, ΔV, as a characteristic of *deformation rate* (Figure 4.) and vs. velocity ratio,
V/V_0, related to *deformation*, e.

Orientation characteristics in Figures 4. and 5. indicate
fundamental differences between fluid, and solid-state processing. Orientation produced in melt-spinning is controlled by
spinning velocity, as expected for a flow-induced mechanism,
and hardly depends on spin-draw ratio V/V_0. On the contrary,
cold-drawing of solid fibers in a plastic state yields draw-
ratio-controlled orientation, insensitive to drawing rate.

Fluid flow and plastic deformation provide two extreme,
idealized mechanisms of orientation. In many systems, there
appears a mixture of deformation, and deformation-rate effects. The examples presented in Figures 4. and 5. seem to
illustrate pure mechanisms, though.

The relation between orientation and stress in solid-state
plastic processing is different to that characteristic of
fluids. Experimental data and theoretical models (e.g. the two
"limiting cases" analyzed by Kratky [10,11]) relate actual
orientation distribution of structural elements to their displacement and reorientation from original positions, caused by
macroscopic deformation of the sample. The rate with which
such deformation is performed, as well as molecular mobility,
temperature, viscosity, and other relaxation-related factors
play minor, if any, role. On the other hand, stress related to
plastic flow is a function of deformation rate which hardly
affects orientation. Therefore, high deformations and high
degrees of orientation can be achieved by solid-state processing without high stresses, if processing is slow, or plastic
viscosity reduced by swelling, plastification, etc. This is
exactly the way applied in producing ideally oriented fibers
from ultra-high-molecular-weight polyethylene. Melt processing
never yields comparable degrees of orientation.

Crosslinked polymers processed in the *rubberlike state*
behave differently. Although translational motions of the
network chains are suppressed by crosslinks, relaxation time
for segmental motions is short, and elastic equilibrium, similar to that for a system of uncrosslinked chains, is established. Orientation and stress are controlled by the extent of
deformation. The latter is limited by the presence of crosslinks, and full orientation cannot be reached (cf. Figure 3.)

Molecular orientation and crystallization

Molecular orientation plays specific role in the crystallization of materials composed of asymmetric molecules. The thermodynamic driving force for attachment of one kinetic element (molecule, chain segment) to a growing crystal, Δf, includes terms related to molecular deformation and orientation. For a polymer crystallizing in a uniaxial orienting field E_{or} (eg in elongational flow) it can be written [5,12]

$$\Delta f = \Delta f_0 + \delta f_{conf}(h; E_{or}) +$$

$$+ \text{const.}E_{or}\sin^2\vartheta - kT\ln[w(\vartheta; E_{or})/w_0] \qquad (12)$$

The first term is related to crystallization in an undeformed and unoriented system, free from external fields ($E_{or}=0$); δf_{conf} for flexible-chain polymers denotes change of the conformational free energy in the process of attachment, a function of the end-to-end distance, h; it is absent in crystallization of rigid molecules. The $E_{or}\sin^2\vartheta$ term describes change in the energy of interaction with the orienting field, experienced by the crystallizing unit. The last, logarithmic term, is related to the change of entropy of the amorphous phase on the loss of one crystallized unit with orientation ϑ. w and w_0 denote, respectively, the actual, and reference orientation distribution of the crystallizing units (molecules, segments). In systems subjected to elongational flow, intensity of the orienting field, E_{or}, can be identified with the flow field potential (qR/D_r, or $q\tau$) appearing in the theory of flow orientation.

The fact that free energy of crystallization depends on orientation of the growing crystals, has several practical consequences. At $E_{or}\neq0$, critical crystallization temperature, nucleation and growth rates, as well as other crystallization characteristics, become functions of the orientation angle, ϑ. In systems subjected to elongation, molecules (segments) oriented along the fiber axis crystallize at higher temperatures and faster than others [5,12].

Figure 6. presents critical crystallization temperature, T_{cr}, as a function of orientation angle, ϑ, for systems subjected to elongational flow with a sine-square potential [12]. It is evident that crystals formed at the highest temperature have perfect orientation ($\vartheta=0$). At lower temperatures, also less oriented crystals can appear, but their formation is slower, the more so, the wider is orientation angle, ϑ.

Orientation-induced crystallization is *selective*: differently oriented crystals are produced in different temperatures, with different speeds, and in different amounts. This rule applies to crystallization from any asymmetric (orientable) kinetic elements (molecules, chain segments). One of the consequences of such crystallization is very narrow crystal orientation distribution.

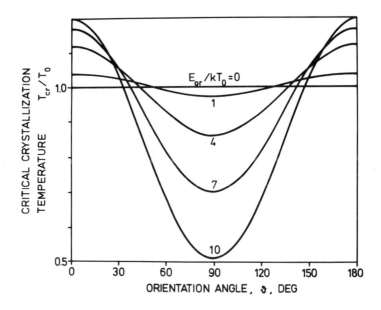

Figure 6. Critical crystallization temperature in elongational flow vs. orientation angle, ϑ [12]. Reduced intensity of the orienting field, E_{or}/kT_0, indicated.

Conformational effects in crystallization of flexible-chain polymers have been recognized long ago; it has been noted that partial crystallization of a polymer chain in a stretched, crosslinked sample can lead to recoiling of the uncrystallized portion of the chain, reduction of its free energy, and release of molecular tension [14, 15]. Conformational free energy of highly stretched chains (large $h/N\ell$ ratio) can stimulate crystallization, increasing *crystallization rates*, and critical *crystallization temperatures*. Stretched rubbers provide a classical example [14, 15]. Figure 7. presents crystallization half-periods of oriented, glassy polyethylene terephthalate [16,17] as a function of molecular orientation. Acceleration of the overall crystallization rate (inversely proportional to crystallization half-period) reaches several orders of magnitude, and is more pronounced in higher temperatures.

The data presented in Figure 7. concern low–temperature crystallization region, typical for mild annealing. Much stronger kinetic effects have been observed in melt processing (e.g. in high–speed fiber spinning).

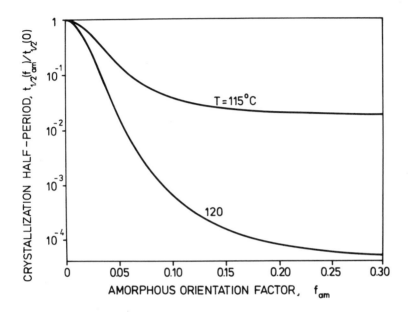

Figure 7. Reduced crystallization half–periods for glassy polyethylene terephthalate vs. amorphous orientation factor. Crystallization temperatures indicated. "120°C" – data by Smith and Steward [16]; "115°C" – data by Alfonso et al. [17].

It seems natural to expect that conformational distribution should affect *morphology of crystallization,* i.e. the way in which polymer chains are arranged in crystals and uncrystallized regions. No satisfactory theory of such effects is available, but simple qualitative considerations seem to point that in the conditions of high orientation and stress, *folded–chain* morphology should be gradually replaced by *extended–chain* crystals [5, 18]. In the manufacture of high–performance materials from flexible–chain polymers, the latter are most desired. Melt processing does not offer such morphology. Crystallization in a solid state under stress provides a promising alternative. Structures obtained in gel–spinning and drawing of ultra–high–molecular weight polyethylene fibers seem to confirm this point.

Crystal orientation vs. amorphous orientation

Although in many flexible-chain polymer materials it is *molecular orientation* what affects mechanical properties, *crystal orientation* plays also some role, especially in high-performance polymers with ultimate moduli and tenacities. For such materials, processing efforts aim at high crystallinity, high perfection of crystals, and high degree of crystal orientation. The latter characteristic strongly depends on the conditions in which crystallization was performed.

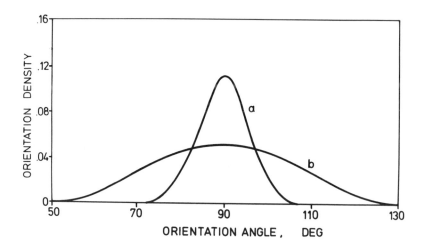

Figure 8. Crystal orientation distributions in crosslinked polychloroprene [19]. a. First oriented, then crystallized; b. First crystallized, then oriented.

Figure 8. presents X-Ray-determined crystal orientation distributions for two samples of crosslinked polychloroprene (Neoprene W) [19]. Sample "a" was stretched above critical crystallization temperature, and then crystallized in the stretched (oriented) state. Sample "b" was first crystallized in an unoriented state, and then oriented by stretching at room temperature. *Oriented crystallization* effective in the stretched sample, evidently leads to a much sharper distribution of orientations than *deformation-controlled rotation* of previously formed crystals. Moreover, in a sample crystallized from unoriented polymer, one can expect more defective, and more folded-chain crystals, introduced in the process of drawing. All this suggests oriented crystallization as a preferable way of producing crystalline textures with high degree of orientation and perfection.

As follows from the principles of orientation-induced crystallization [5,12,13], the average degree of crystal orientation is much higher than the average orientation of structural elements from which crystals have been formed. This fact has been noted long ago by Krigbaum and Roe [20]. Formation of crystals from oriented molecules, or chain segments can also reduce molecular orientation in the amorphous phase. At least two different mechanisms of such reduction can be imagined.

The first mechanism, effective in systems of *constrained* (crosslinked) *flexible chains*, is related to conformational changes in the crystallizing polymer. Consider a chain composed of N segments of length ℓ, with an end-to-end vector h. When N_1 chain segments loose their freedom of motion and are transformed into a rigid crystal characterized by a vector h_1, the $(N-N_1)$ segments remaining in the amorphous phase, form a reduced chain with an end-to-end vector $h_2 = h - h_1$.

In the Gaussian approximation of chain statistics, the conformational contribution to the free energy of crystallization per crystallized segment, (δf_{conf} in eq. 12) reads

$$\delta f_{conf} = (3kT/2N_1 \ell^2)[(h-h_1)^2/(N-N_1) - h^2/N] \qquad (13)$$

and the corresponding change in the orientation factor (eq. 6)

$$\Delta f_{or} = (3/5\ell^2)[(h-h_1)^2/(N-N_1)^2 - h^2/N^2] \qquad (14)$$

Dependently on orientation and dimensions of the produced crystal (vector h_1), molecular orientation of the amorphous phase may be increased or reduced by crystallization. The original concepts of stress-induced crystallization [14, 15] implied formation of bundle-like crystals oriented along highly extended end-to-end vector h. In such a special case, crystallization reduces conformational free energy of the system; molecular orientation and stress in the uncrystallized material decrease. However, more general analysis of the problem [18] shows that, dependently on temperature, deformation, and molecular structure, crystallization of crosslinked polymers can lead to different type of crystals, including folded-chain structures which increase, rather than reduce, stress and molecular orientation in the uncrystallized phase.

Orientation-differentiated kinetics of crystallization provides the other mechanism of crystallization-controlled orientation changes [5, 12, 13, 21]. Consider a polymer crystallizing in an external orienting field (e.g. in elongational flow). When crystallization progresses, crystallizing units leave the amorphous phase and join crystals. The number of units which contribute to amorphous orientation is reduced, and that contributing to crystal orientation increased by the same amount. At the same time, crystallizing units and crystals perform rotational motions controlled by the orienting field and rotational diffusion. These combined effects can be described by simultaneous equations for orientation distribution in the amorphous, $w_{am}(\vartheta,t)$, and crystalline phase, $w_{cr}(\vartheta,t)$

$$\partial w_{am}/\partial t - D_{r,1}[\nabla^2 w_{am} + \nabla(w_{am}\nabla U_1/kT)] =$$

$$= [\dot{x}/(1-x)]\cdot[w_{am} - j(\vartheta)] \qquad (15)$$

$$\partial w_{cr}/\partial t - D_{r,cr}[\nabla^2 w_{cr} + \nabla(w_{cr}\nabla U_{cr}/kT)] =$$

$$= - (\dot{x}/x)\cdot[w_{cr} - j(\vartheta)] \qquad (16)$$

$D_{r,1}$, $D_{r,cr}$ denote, respectively, rotational diffusion coefficients for single kinetic units and crystals, x – crystallinity, $\dot{x}$ – overall crystallization rate, and $j(\vartheta)$ – normalized distribution of the transition rates for differently oriented kinetic units. The Laplacian terms describe rotational diffusion controlled by the coefficients D_r. The terms including potentials $U_1(\vartheta)$, $U_{cr}(\vartheta)$, are related to the action of the orienting field. In the case of elongational flow, U_1, U_{cr} assume the form $(qRkT/2D_r)\cdot\sin^2\vartheta$ (cf. eq. 1) with appropriate values of D_r and R. The differential operators concern the space of orientations, in which conservation of the kinetic units is considered [5].

Two asymptotic cases can be considered. In the *absence of rotational mobility* ($D_r=0$), variation of the distributions $w(\vartheta)$ is controlled by the consumption/production term, proportional to crystallization rate, $\dot{x}$. Such a special solution was discussed in ref. [5] in connection with high-speed melt spinning. The other extreme concerns systems with *infinite mobility*. At $D_r=\infty$ orientation distributions reduce to the

equilibrium Boltzmann form, $\exp(-U/kT)$, with the potentials
appearing in eqs. (15) and (16). Such a distribution was discussed above (eq. 1) when molecular orientation in fluid-state
processing was considered.

Jarecki [22] obtained solutions of the orientation equations (15), (16), for crystallization either very fast
($\dot{x} \gg D_{r,1}$), or very slow compared to rotational diffusion
($\dot{x} \ll D_{r,1}$). The results are presented in Figure 9.

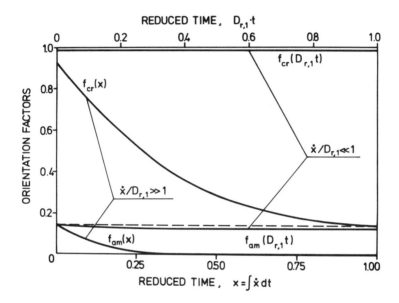

Figure 9.   Variation of crystal, and amorphous orientation
factors in the course of oriented crystallization [22]. Fast
crystallization: $\dot{x} \gg D_{r,1}$, orientation plotted vs. degree of
crystallinity, x; slow crystallization: $\dot{x} \ll D_{r,1}$, orientation
plotted vs. reduced time, $D_{r,1}t$.

In the first case, well oriented (= fast crystallizing)
segments in the amorphous phase are exhausted first; this
gradually reduces amorphous orientation factor to zero, and
the average weighted orientation of crystals to the original
orientation of the amorphous phase. In the other, diffusion-
controlled case, the units consumed in crystallization are
reoriented by the external field, and orientation levels approach equilibrium. In the calculations [22] crystal mobility
was neglected, as small compared with that of single kinetic
units ($D_{r,cr} \ll D_{r,1}$).

Processing of Liquid-Crystalline vs. Random Solutions

The fact, that first successful attempts of obtaining high-performance fibers from rigid molecules (*Kevlar*) involved processing of *liquid-crystalline solutions* [23], raises question about the importance of ordered domains in orientation and crystallization. It has already been mentioned that fibers with comparable characteristics have recently been obtained from semi-stiff molecules spun from unordered, random solutions and subsequently subjected to drawing (*Technora* from Teijin) [24]. So, liquid-crystalline order in the processed fluid does not provide a necessary condition for ultimate structure and properties.

First, it should be noted that liquid crystalline order in polymer fluids (solutions, melts) is a direct consequence of *molecular rigidity and symmetry*. The more rigid, and more elongated (flattened) are molecules, the lower is critical concentration, and the higher critical temperature for the appearance of such order.

The first difference between ordered and random fluids is their behavior in flow. As shown above, the effectivity of flow orientation and stability of oriented structures is controlled by *volume of a rotating unit*. If we assume that in liquid-crystalline fluids the role of such units is played by *ordered domains*, and in random solutions by *single molecules, or their segments*, longer relaxation times and higher orientation-stress coefficients will easily account for high spin-induced orientation in *Kevlar* or *PPBT*, and for the necessity of applying an additional treatment for semi-stiff *Technora*.

The other difference concerns crystallization. Crystallization in a non-liquid-crystalline fluid with an orientation distribution of crystallizing units, is controlled by the degree of orientation. The higher is average orientation (e.g. one induced by flow), the higher crystallization temperature, and crystallization rate [12, 13]. On the other hand, *intradomain crystallization* in an ordered fluid is controlled by *intradomain orientation*, i.e. so called "liquid-crystal order parameter", rather than by macroscopic orientation characteristics of the entire fluid [13]. Therefore high crystallizability is an internal property of liquid-crystalline solutions (melts), hardly influenced by flow, or stress.

The above two factors can account for high orientation and crystallinity produced in processing of liquid crystalline fluids, without the necessity of drawing or annealing.

High-Speed Melt Spinning

Formation of fibers with take-up speeds reaching 6000 10000 m/min (50% of the velocity of sound) creates extremal conditions for melt processing. Although attempts of producing in this way high-performance materials (comparable with *Kevlar*, or gel-spun polyethylene) seem to have failed, the process has found many applications The process has several peculiarities, especially when structure development is concerned, and seems to show limits of melt processing of flexible-chain polymers.

Maximum elongation rates involved in high-speed spinning, q, reach 10^3–10^4 sec^{-1}; normal stress difference, Δp, of the order of 10^8 dynes/cm^2 is only by a factor of ten lower than stress required for plastic, solid-state processing. Crystallinity and orientation are developed within a short section of the spinline, during a period as short as 100 microseconds. And yet, mechanical properties (tenacity, modulus), as well as texture of as-spun fibers, lie below the level of traditional fibers (melt-spun and drawn), to say nothing about modern high-performance materials [25].

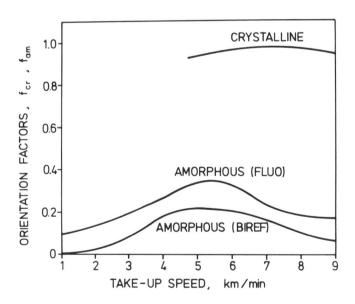

Figure 10. Orientation characteristics of as-spun PET fibers vs. take-up speed [5]. f_{cr} – crystal orientation factor from WAXS; f_{am} – amorphous orientation factors based on polarized fluorescence, and resolved birefringence measurements.

Figures 10. and 11. illustrate crystalline structure of polyethylene terephthalate (PET) fibers, as a function of spinning speed [5]. PET is a slowly crystallizing material, easily quenched into an amorphous glass. It is widely used for textile and industrial fibers, and provides the most important polymer for high-speed spinning.

Orientation characteristics shown in Figure 10. indicate two points. In the range where crystals exist (above 4000 m/min), *crystal orientation* is close to unity, and only slightly affected by spinning speed. Two independent characteristics of *amorphous orientation* – one based on polarized fluorescence, another one on optical birefringence resolved into crystalline and amorphous contributions – show much lower values, and exhibit a maximum near V = 5500 m/min.

Degree of crystallinity (Figure 11.) was characterized by density and the Lorentz–Lorenz ratio, LR, proportional to optical density and calculated from the average refractive index, n_{iso}

$$LR = (n^2-1)/(n^2+2); \quad n = n_{iso} = \tfrac{1}{3}(n_{\parallel} + 2n_{\perp}) \qquad (17)$$

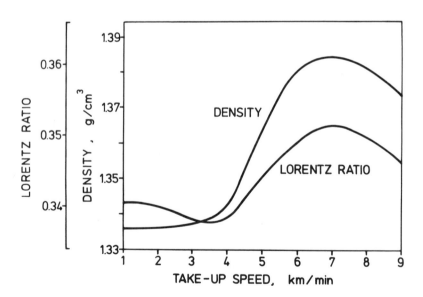

Figure 11. Crystallinity characteristics of as-spun PET fibers vs. take-up speed [5]. Density measured in a density gradient column; Lorentz-Lorenz ratio, based on refractive indices.

Crystallinity does not develop in slow spinning: glassy, amorphous fibers are formed, until spinning speed reaches 3000 – 4000 m/min and molecular orientation climbs to a level (0.1 – 0.2) effectively stimulating crystallization. Maximum of crystallinity (ca 40% in the absolute scale) is observed around 7000 m/min, followed by reduction at higher speeds (Figure 11.).

High crystal orientation factors accompanying moderate amorphous orientation (Figure 10.) are, doubtless, due to *selective crystallization* in the oriented melt. The reason for the appearance of *orientation maxima* is less obvious. Tensile stress monotonically increases with spinning speed, and yet amorphous orientation definitly drops down. Ziabicki and Jarecki [5] tried to explain this effect by the "consumption/production" mechanism, according to which high crystallization rates appearing above some critical speed, reduce rather than increase amorphous orientation. An alternative explanation could be based on *heterogeneity* of a partly crystallized system, in which stress would be supported by a hypothetic "network" of extended chains, filled with relaxed chains released by crystallization.

The *maxima of crystallinity* observed in Figure 11. can be understood, if one considers two opposite effects of spinning speed: an increase of orientation (stress), leading to higher *rate of crystallization*, and an increase in the *rate of quenching* (due to enhanced heat transfer) which reduces *time available for crystallization*. The idea of such maxima, first suggested by Nakamura et al. [26] was discussed in detail by the present author in ref. [27].

Little is known about *morphological structure* of high-speed-spun fibers. Small-angle X-Ray scattering shows that lamellar (folded-chain) structure seems to be less pronounced than in unoriented-crystallized and drawn fibers. On the other hand, little, if any, evidence is available about the existence of extended-chain crystals.

High-speed-spun fibers differ in many points from fibers manufactured in a conventional two-step process. Relatively low amorphous orientation, non-uniform, "skin-core" structure, some porosity appearing at very high speeds, as well as the maximum-characteristic behavior evident in Figures 10. and 11. do not qualify high-speed melt processing for manufacturing of polymer materials with ultimate structures and properties.

Closing remarks

I have tried to discuss factors responsible for development of highly oriented crystalline structure in polymers, to explain differences between solid-state and fluid-state processing, as well as the characteristic features of processing polymers composed of rigid vs. flexible molecules. It seems that *molecular mobility and relaxation* plays the crucial role. Long (rotational) relaxation times of rigid molecules and liquid-crystalline domains make possible effective processing in the fluid state, at moderate speeds and moderate stresses. Molecular motions in flexible-chain polymers characterized by short (segmental) relaxation times make fluid-state processing ineffective, as requiring extremely high stresses. Slow processing in a solid (plastic) state at reasonably high stresses seems to offer better perspectives.

References

[1] Jeffery, G.B. Proc. Roy. Soc. (London) 1922, A102, 161.
[2] Takserman-Krozer, R.; Ziabicki, A. J. Polymer Sci. 1963, A-1, 491.
[3] Ziabicki, A.; Jarecki, L. Rheol. Acta, in press.
[4] Kuhn, W.; Grün, F. Kolloid Z. 1942, 101, 248.
[5] Ziabicki, A.; Jarecki, L. in: ref. [25], p. 225.
[6] Ziabicki, A.; Jarecki, L. Colloid and Polymer Sci. 1986, 264, 343.
[7] Treloar, L.R.G.; Riding, G. Proc. Roy. Soc. (London), 1979, A369, 261.
[8] Ziabicki, A.; Kedzierska, K. J. Appl. Polymer Sci. 1962, 6, 111; Ziabicki, A. in: "Applied Fibre Science", vol.3; Happey, F., Ed.; Academic Press: London, 1979, p. 235.
[9] Arakawa, S. in: "Formation of Fibers and Development of Their Structure", vol.1.; Kagaku Dojin: Tokyo-Kyoto, 1969, p. 287.
[10] Kratky, O.; Platzek, P. Kolloid Z. 1939, 88, 78.
[11] Kratky, O. Kolloid Z. 1933, 64, 213.
[12] Ziabicki, A. J. Chem.Phys. 1986, 85, 3042.
[13] Ziabicki, A.; Jarecki, L. Inst. Fund. Technol. Res. Rep. 1982, 1.
[14] Flory, P.J. J. Chem.Phys. 1947, 15, 397.
[15] Kim, H.G.; Mandelkern, L. J. Polymer Sci. A-2 1968, 6, 181.
[16] Smith, F.S.; Steward, R.D. Polymer 1974, 15, 283.
[17] Alfonso, G.C.; Verdona, M.P; Wasiak, A. Polymer 1978, 19, 711.
[18] Kosc, M.; Ziabicki, A. Macromolecules 1982, 15, 1507; Kosc, M. Inst. Fund. Technol. Res. Rep. 1981, 12.

[19] Andrews, E.H.; Reeve, B. J. Mater. Sci. 1971, 6, 547.

[20] Krigbaum, W.R.; Roe, R.J. J. Polymer Sci. A-2 1964, 2, 4391.

[21] Ziabicki, A.; Jarecki, L. Colloid and Polymer Sci. 1978, 256, 232.

[22] Jarecki, L. Inst. Fund. Technol. Res. Rep. 1984, 32.

[23] Du Pont de Nemours USP 3,671,542; USP 3,767,756.

[24] Ozawa, S. (Teijin); "Aramid Copolymer Fibers Related to HM-50", paper presented at the ACS Symposium "Approaches to Property Limits in Polymers", Scanticon-Princeton, August 11-13, 1986.

[25] Ziabicki, A.; Kawai, H., Eds. "High-Speed Fiber Spinning" Interscience: New York, 1985.

[26] Nakamura, K.; Watanabe, T.; Amano, T. J. Appl. Polymer Sci. 1972, 16, 1077.

[27] Ziabicki, A. in: ref. [25], p. 21.

DIACETYLENE BASED SINGLE CRYSTAL FIBERS

Robert J. Young

Department of Polymer Science and Technology
UMIST
Manchester, M60 1QD, UK

Abstract

The development of polymer single crystals
as reinforcing fibers for applications such
as in polymer-matrix composites is reviewed
It is demonstrated that polydiacetylene
single crystal fibers made by the solid-
state polymerization of single crystal
monomers have high degrees of strength
and stiffness and are inherently resistant
to creep. It is also shown that investigations upon these materials have led
significant improvements in our understanding of structure/property relationships in polymer crystals and the micromechanics of fiber reinforcement. Some
preliminary results upon the mechanical
properties of composites consisting of
aligned polydiacetylene single crystal
fibers in an epoxy resin matrix are
described.

Introduction

Most of the effort in obtaining highly-oriented polymer
samples has in the past been concentrated upon trying to un-
coil and align long polymer molecules in solid state, melt or
solution. A variety of techniques have been employed with
conventional polymers such as polyethylene such as ultra-
drawing [1-31], high degrees of extrusion [3-5] and gel or
solution spinning [6,7]. New rigid-rod polymers have also
been developed from which highly oriented fibres can be prepared by the spinning of liquid crystalline solutions [8-10].
Although significant improvements and developments have taken
place over recent years and samples with high values of

modulus and strength have been prepared, perfect molecular
alignment as in single crystals is never achieved using such
approaches since defects such as chain-ends, chain-folds,
loop and entanglements are invariably trapped in the structures.

The technique of solid-state polymerization has the ability to produce polymer single crystal fibres in which the
molecules are perfectly aligned and the fibers have their
theoretical values of stiffness and strength. The technique
is essentially very simple. Single crystals of monomer molecules are prepared using conventional methods such as vapour
phase deposition or precipitation from dilute solution.
Alignment of the monomer molecules is present in the monomer
crystals. The transformation from monomer to polymer takes
place through a solid-state polymerization reaction by a rearrangement of bonding within the crystals but without any
appreciable molecular movement. The alignment of the monomer
molecules is therefore transferred to a uniaxial orientation
of the polymer molecules. Any of the normal problems encountered with aligning polymer molecules in the melt or
solution, such as entanglements are therefore avoided. In
this way polymer molecules can be incorporated into fibers
without passing through the melt or solution stage which
invariably produces polycrystalline samples and imperfections
in the fibers.

Even though the solid-state polymerization technique is
simple and attractive it suffers from several drawbacks. As
yet, the fibers which are produced relatively slowly are only
made as short strands with maximum lengths of no more than
5-10cm. Also the only really successful single crystal fiber
system is based on the solid-state polymerization of certain
substituted diacetylenes [11-13] and it cannot be applied
widely to all polymer systems. Nevertheless, these substituted diacetylene polymer fibers are found to have promising
and useful physical properties characteristic of one-dimensional solids.

Solid-State Polymerization

The solid-state reactions which are relevant to the
preparation of polymer single crystal fibers are termed
either 'topochemical' or 'topotactic' depending upon the detailed nature of the reaction. These two terms are used to
describe reactions which can take place in organic crystals
such as single crystals of fibre-forming monomers. In such
crystals the molecules are generally separated sufficiently

far that they are unable to react. However, if their is
sufficient mobility that the molecules are able to come within about 0.3nm of each other by diffusion or rotation then a
solid-state reaction may occur. Examples of such systems in
which monomer single crystals can undergo solid-state polymerization have been given by Wegner [11].

The most important reactions for the formation of good
polymer single crystals are topochemical reactions whereby as
Wegner [11] pointed out: "there is a direct transition from
the monomer molecules to polymer chains without destruction
of the crystal lattice and without the formation of non-crystalline intermediates". Hence, the behaviour is rather like
a martensitic transformation [14]. The centers of gravity of
the molecules do not move significantly from their lattice
site and the reaction takes place within the parent crystal.
The three-dimensional order of the monomer lattice is transferred directly to the polymer lattice with the result that
the polymer formed must have a well-ordered extended-chain
morphology. This behaviour was first reported by Wegner [12]
in 1969 who produced virtually defect-free extended-chain
single crystals of substituted polydiacetylenes from single
crystal monomers. In the intervening years a large number of
different substituted diacetylene have been produced and a
number of examples are shown in Table 1 along with the abbreviations that will be used to describe them in this chapter.

Table 1. Chemical formulae and abbreviations for the
         diacetylene derivatives described in the
         text with general formula: $R-C{\equiv}C-C{\equiv}C-R$

R	Chemical Name	Abbreviation
$-CH_2OCONHC_2H_5$	2,4-Hexadiyne-1,6-diol -bis(ethyl urethane)	EUHD
$-CH_2OCONHC_6H_5$	2,4Hexadiyne-1,6-diol -bis(phenyl urethane)	PUHD
$-CH_2N$	2,4Hexadiyne-1,6- -di(N-carbazol)	DCHD
$-CH_2OSO_2C_6H_4CH_3$	2,4Hexadiyne-1,6-diol -bis(p-toluene sulphonate)	TSHD

Topochemical polymerization reactions can proceed in either a homogeneous or heterogeneous manner. In the homogeneous case the polymer molecules form randomly in the monomer matrix. The other possibility is that polymerization is heterogeneous starting at specific sites in the monomer crystal such as at defects. If further chain growth takes place at the preexisting polymer nuclei there can be a tendency for the polymers formed to be polycrystalline. In the case of polydiacetylenes both homogeneous and heterogeneous polymerizations have been reported. For example, in the case of the toluene sulphonate derivative (TSHD) polymerized using synchotron radiation the reaction has been observed to take place homogeneously throughout a crystal [15] whereas when TSHD monomer is polymerized thermally the polymerization reaction is found to start preferentially at defects [16]. However, in both cases the polymer forms as a solid solution with monomer and the single crystal morphology is directly transferred from the monomer to polymer leading to poly TSHD crystals having a high degree of perfection.

The behaviour of other diacetylene derivatives can be more complex such as in the case of fiber-forming diacetylenes such as the carbazolyl (DCHD) [17,18] and ethyl urethane (EUHD) [19-21] derivatives. In both cases thermal polymerization produces polycrystalline samples which are not as crystallographically perfect as the single crystals produced when polymerized using γ-rays at room temperature. It has been suggested [17] that in the case of DHCD this may be due to the tendency for phase separation to be favoured kinetically at the higher temperatures used for the thermal polymerization.

Since high modulus fiber-forming diacetylenes are insoluble once polymerized it is impossible to determine the molar mass or degree of polymerization of the polymers. The fibers behave as if the molar mass is effectively infinite although recent work upon soluble diacetylene derivatives with long aliphatic side groups [22] has suggested that solid state polymerization leads to degrees of polymerization of the order of 1000-1500. It is debatable, however, as to how relevant this result is to the high modulus derivatives with short side-groups.

Structure

Polydiacetylene single crystals can be obtained essentially in two crystal forms, either as lozenges or as fibers.

The morphology is controlled by the conditions under which the monomer is crystallized from solution although the exact reasons why a particular morphology is obtained are not really understood. TSHD (Table 1) is normally only found in the form of lozenges when crystallized from most solvents [13] whereas DCHD is usually obtained as fibers [17], the aspect ratio of which depends upon the solvent, solution concentration and crystallization temperature. In contrast EUHD can be obtained in three crystal forms one   of which can undergo solid-state polymerization to give fully-polymerized single crystal fibers [19]. Good polymerizable fibers of PUHD (Table 1) are obtained when it is crystallized from dioxane/water mixtures and dioxane molecules trapped interstitially in the structure [23]. In each of these cases the morphology is retained when the monomer crystals are transformed into polymer by the solid-state polymerization reaction.

A particularly convenient way of investigating the structure of polydiacetylenes is by transmission electron microscopy. Monomer crystals which are sufficiently thin (~100nm) to allow penetration of the electron beam can be readily produced by allowing a droplet of dilute monomer solution to evaporate on a carbon support film on an electron microscope grid [24]. The monomer can then be polymerized by heating or most conveniently by exposure to the electron beam in the microscope.

The crystal structures of several polydiacetylenes have been determined to a high degree of accuracy using X-ray diffraction methods [25-27]. This can be contrasted with conventional polymers for which the crystal structures [28] determined using oriented polycrystalline samples are much less accurately known. The unique single crystal nature of polydiacetylenes has enabled the positions of the atoms in the unit cells and bond angles to be measured to a high degree of precision. Many diacetylene monomers and polymers have monoclinic $P2_1/c$ crystal structures (indexing the chain direction unconventionally as b). This detailed knowledge of the crystal structures is invaluable for the correlation of physical properties with structure in these materials.

Although X-ray diffraction is by far the most accurate method of determining crystal structures considerable extra information can be obtained using electron microscopy with the ability of obtaining high magnification images of the structure and performing simultaneous electron diffraction. One problem normally encountered with the study of organic materials in the electron microscope is that they are prone to radiation damage whereby the crystal structure is destroyed

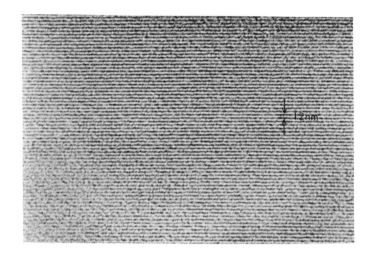

Figure 1. High magnification, high resolution transmission
electron micrograph of a fibrous crystal of polyDCHD showing
lattice planes with a spacing of 1.2nm parallel to the chain
direction.

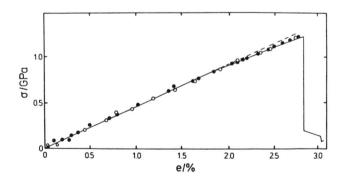

Figure 2. Stress/strain curve showing the elastic deformation
of a polyDCHD single crystal fiber.

by exposure to the electron beam [29]. It has been found
that most polydiacetylenes are relatively stable but that
polyDCHD is particularly outstanding in its ability to resist
damage in the electron beam being at least 20 times more
stable than polyethylene [30], the most widely-studied crystalline polymer. This high stability has allowed detailed
investigations to be made of the structure of polyDCHD at
high magnification. In particular it has been possible to
image the crystal lattice directly in the microscope [20,31]
as is shown in Figure 1 where lattice planes parallel to the
chain direction are imaged. All these observations indicate
that there is perfect molecular alignment and a high degree
of crystal perfection in the fibers. This can be contrasted
with the highly defective polycrystalline structures found by
similar electron microscope observation upon polymer fibers
produced using techniques such as mechanical orientation [32]
and solvent spinning of liquid crystalline solutions of rigid
rod polymers [33]. This difference in structure is reflected
directly in the differences in the mechanical properties
between the single crystal fibers and fibers produced using
other techniques as will be shown later.

    The presence of defects is known to affect the mechanical
behaviour of crystals of most materials and polymer single
crystals are no exception. Surface defects such as steps are
known to have an important effect upon the fracture behaviour
of polydiacetylene single crystal fibers and so the geometry
of such defects has been examined in detail [18].

                        Elastic Deformation

        The preparation of single crystals of polydiacetylenes
with centimeter dimensions has enabled the stress/strain behaviour of polymer single crystals to be determined for the
first time using conventional mechanical testing methods. In
1974 Baugham, Gleiter and Sendfeld [23] demonstrated that
fiber-like single crystals of polyPUHD (Table 1) could be
deformed elastically to strains of over 3%. The crystals
were found to have values of Young's moduli in the chain
direction of the order 45GPa. Such high degrees of stiffness have also been found for other polydiacetylenes with
Young's modulus values of 45GPa being determined for polyDCHD
[18] and 62GPa for polyEUHD [21].

    A typical stress/strain curve for a polyDCHD single
crystal fiber is given in Figure 2. The curve is linear up
to a strain of about 1.8% and there is a slight decrease in
slope above this strain until fracture occurs at a strain of

about 2.8%. Loading and unloading take place along the same path [18] indicating a lack of hysteresis in the deformation. The fracture strain is found to depend upon the fiber diameter, decreasing as the diameter is increased [18].

The stress/strain curve for a polymer single crystal in Figure 2 gives a clear indication of how deformation takes place on the molecular level as is corresponds essentially to a molecular stress/strain curve. Baughman et al [23] suggested that the slight deviation from Hooke's law above about 2% strain, which is clearly not a yield process, might be due to the anharmonic part of the interaction potential between neighbouring atoms on the polymer chain. Since the polydiacetylene fibers are highly-perfect polymer single crystals the deformation directly involves the stretching and bending of bonds along the polymer backbone. This has been confirmed using Resonance Raman Spectroscopy [20,34,35] where it has been shown that the frequencies of the C-C, C=C and C$\equiv$C stretching modes in polydiacetylenes depend upon the deformation of the crystals, decreasing with applied strain. An example of the variation of the frequency of the C$\equiv$C stretching mode for polyDCHD as a function of applied strain is shown in Figure 3. The consequent reduction in force constants is one of the factors leading to the reduction in the slope of the stress/strain curves at high strains [18,21,23].

Baughman and coworkers [23] pointed out that the levels of chain direction modulus displayed by poldiacetylene single crystals were extremely high when account was taken of the high cross-sectional area, A, of the molecular chains due to the relatively large sidegroups on the polydiacetylene molecules. It is also found that for a particular diacetylene derivative the Young's modulus depends upon degree of conversion from monomer to polymer and the method of polymerization. There is a linear increase in the modulus with the volume fraction of polymer in the crystals for the thermal polymerization of EUHD whereas the moduli of the crystals produced by γ-ray polymerization are somewhat lower [21]. Although it is known that the thermally-polymerized crystals are crystallographically less-perfect it is thought that the high doses of radiation needed to polymerize EUHD damage the crystals and so reduce their moduli.

Although it is the Young's modulus in the chain direction that is of the greatest importance for engineering applications, it must be remembered that a large number of elastic constants are needed to fully describe the elastic behaviour of a crystal [36]. Polydiacetylene single crystals usually possess monoclinic symmetry and therefore have 13 elastic constants.

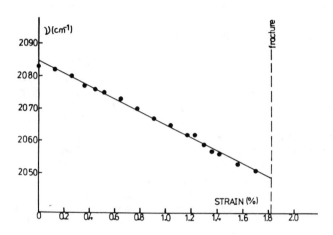

Figure 3. Change in C≡C triple bond Raman stretching freqency with applied strain for a polyDCHD single crystal fiber.

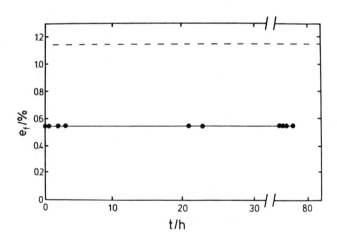

Figure 4. Variation of the fiber strain with time for a poly DCHD fiber held at constant stress corresponding to 50% of its fracture strain (given by the dashed line).

There have been several attempts to calculate the 9 elastic constants for orthorhombic polyethylene [37] but as yet there have been no similar calculation for a polydiacetylene. The calculation ought to be feasible as the crystal structures and atomic positions of several polydiacetylenes are known to a high degree of accuracy [25-27]. However, there are also a large number of different bonds in each unit cell making such a calculation extremely complex.

The modulus of polyTSHD crystals deformed in the chain direction has been calculated by Batchelder and Bloor [34] by employing the method of Treloar [38]. They used force constants measured from Raman Spectroscopy and estimated a Young's modulus for polyTSHD crystals in the chain direction to be 50GPa which is in good agreement with the value of 45GPa determined for polyDCHD [18] which has similar unit cell dimensions. The method of Treloar is relatively simple but it cannot be extended to determine the full set of elastic constants since it only takes into account the covalent bonding along the polymer backbone [38].

It has been possible to measure some of the 13 elastic constants for polydiacetylene single crystals using Brillouin scattering [39,40] and sound velocity measurements [41]. Leyrer, Wegner and Wettling [39] employed Brillouin scattering to determine 6 of the elastic constants of the monomer and 3 for the polymer of TSHD at room temperature. Rehwald, Vonlanthen and Meyer [41] showed that sound velocity measurements enabled 9 of the elastic constants to be determined for TSHD monomers and polymers over a wide range of temperature, although the errors involved in some of the determinations were rather large. It is found that on the whole, the agreement between the elastic constants determined by the different techniques is relatively good.

Plastic Deformation

The restrictions of not breaking covalent bonds during deformation means that polymer crystals are only capable of undergoing a limited amount of plastic deformation [42]. In polycrystalline samples of conventional polymers much of the deformation is taken up by the non-crystalline amorphous phase [42]. Since there is not amorphous material in single crystals of polymers such as polydiacetylenes then it is found that their ability to undergo plastic deformation is everely restricted and so they are highly resistant to creep [18,21]. However, polydiacetylenes are found to be capable of undergoing some limited plastic deformation through twinning [43,44].

One of the most striking aspects of the mechanical properties of polydiacetylene single crystals is that it is not
possible to measure any time-dependent deformation (or creep)
when crystals are deformed in tension parallel to the chain
direction [18,21]. This behaviour is demonstrated in Figure
4 for a polyDCHD single crystal held at constant stress at
room temperature, and preliminary measurements have indicated
that creep could not be detected during deformation at temperatures of up to at least 100°C [18].

Creep and time-dependent deformation are normally a
serious problem in the use of high-modulus polymer fibers such
as polyethylene [45] in engineering applications. Such oriented fibers produced by drawing or spinning contain a high
density of defects such as chain-ends, loops and entanglements.
These allow the translation of molecules parallel to the chain
direction during deformation which leads to creep. Polydiacetylene single crystal fibers contain perfectly aligned
long polymer molecules (cf Figure 1) and so there is no mechanism whereby creep can take place even at high temperatures.
In addition, polydiacetylenes tend to degrade rather than
melt and this should be contrasted with polyethylene which melts
at about 140°C [36] and so has only very limited high-temperature applications.

Studies upon the deformation of polydiacetylene single
crystals [43,44] have demonstrated that polymer single crystals
twin when deformed in compression parallel to the chain direction. This type of deformation leads to the formation of
twins which involve the molecules kinking over at a well-defined
angle [43,44]. Figure 5 shows a schematic diagram of a twin
in a polyDCHD [46] fiber with the corresponding molecular
displacements involved. The process can be differentiated
from the formation of kink-bands which are found in other high
modulus fibers [47] since the material within the twinned
region has the same crystal structure as the undeformed
crystal. There is also a mirror image orientation relationship between the deformed and undeformed regions [43] which is
not the case with kink bands [47].

The ability of polymer single crystal fibers to undergo
chain twinning has important consequences upon their use in
composites. For example, it is possible to tie knots in poly-
DCHD fibers [18]. This process involves a high degree of
deformation but the crystals are able to cope with the high
strains by undergoing chain twinning on the inside surface
which is subjected to compression in the chain direction, but
the crystals remain relatively intact even when pulled into a
tight knot [18]. This behaviour has implications for the use

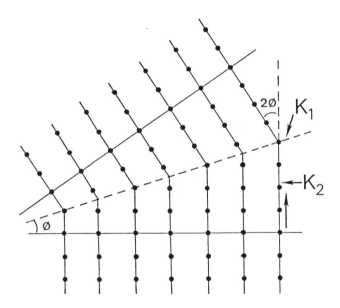

Figure 5. Schematic diagram of a twin in a polyDCHD single crystal. The deformation takes place by the polymer molecules bending at a well-defined angle across the twin boundary (given by the dashed line). It is analogous to the formation of kink bands in other polymer fibres.

of such fibers in composites as their ability to absorb strain
by twinning means that the fibers should not break up and undergo 'fiber attrition' as easily as other fibers such as those
of glass and carbon.

Fracture

Polydiacetylene single crystals readily undergo cleavage
parallel to the chain direction reflecting the relative strength
of covalent bonding compared with the van der Waals bonding
between the polymer molecules. In polydiacetylenes this
cleavage takes place preferentially on certain crystallog-
graphic planes. However, the most interesting aspect of the
fracture behaviour of polymer single crystal fibers is the
high strength they can exhibit when deformed parallel to the
chain direction.

Investigations into the fracture behaviour of polydiacetylene single crystal fibers has revealed a strong dependence of the fracture stress, σ_f, upon the fiber diameter,
d [18,21,23]. This is demonstrated for polyDCHD fibers in
Figure 6 and it is similar to the size-dependence reported
earlier for inorganic high-strength fibers [48] where the
dependence of σ_f upon d was found to follow a relation of the
form

$$\sigma_f \; \alpha \; 1/d \qquad\qquad (1)$$

This behaviour was thought to be due to the presence of surface
defects which give rise to a stress concentration when the
fibers are deformed. Since the size of the defects in the
inorganic fibers was found to scale with the fiber diameter
the size dependence could be predicted [49]. However, more
detailed examination of the data in Figure 6 has shown that
equation (1) is not accurately obeyed for polydiacetylene
single crystal fibers. In addition, theoretical calculations
have sown that a different relationship is expected [18,21].
When the data in Figure 6 were replotted in the form of a
log/log plot it was found [18] that the dependence of σ_f upon
d was given more accurately by

$$\sigma_f \; \alpha \; 1/d^{0.55} \qquad\qquad (2)$$

This is precisely the dependence predicted from the theoretical
considerations and detailed analysis has allowed the thoretical
strength of polyDCHD crystals to be calculated as 3±1GPa [18].

Frank [50] pointed out several years ago that polymer

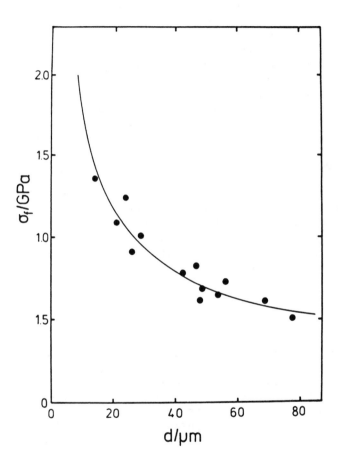

Figure 6. Dependence of the fracture strength of polyDCHD
fibers upon the fiber diameter.

molecules can have very high values of strength when deformed
parallel to the direction of their molecular chains. The
determination of the theoretical strength of polydiacetylene
single crystal fibers allows the strength of individual molecules to be estimated. From the knowledge of the crystal
structure of polyDCHD [27], it can be shown that each molecule
supports a cross-sectional area, A of about $1nm^2$. The theoretical strength of 3GPa therefore corresponds to a force to
break molecules of about 3nN and a fracture strain of the
order of 6-8%. It is of interest to compare this with the
theoretically-calculated values of strengths of covalently-
bonded polymer molecules [48,51,52]. Kelly [48] has estimated the strength of a polyethylene molecule as about 6nN but
this is though to be rather high. Kausch [52] has shown that
a covalently-bonded polymer molecule should be broken by a
force of about 3nN, which is identical to the value determined
for polyDCHD. A molecular strength of this magnitude corresponds to a fracture stress of the order of 20GPa for poly
ethylene single crystals where the area supported by each
molecule is considerably smaller than for polyDCHD. However,
it has not yet been possible to make single crystals of polyethylene with macroscopic dimensions and so even highly-oriented polyethylene fibers are found to have strength values
only of the order of 4GPa [3,5] which is well below the theoretical strength.

Composites

Composites produced by incorporating high-modulus fibers
in a brittle matrix such as an epoxy resin can have outstanding
mechanical properties [53]. Recent examples have included
composites produced with high-modulus polyethylene fibers [54]
and aromatic polyamide fibers [55]. Polydiacetylene single
crystal fibers also offer considerable promise as reinforcing
fibers in polymeric matrices because they have properties
such as high stiffness, high strength, low creep, good thermal
stability and low density.

Recent investigations [46,56,57] into the behaviour of
polydiacetylene single crystal fibers in epoxy resin matrices
have shown that not only do such composites have promising
mechanical properties but that important fundamental details
of the mechanisms of fiber reinforcement can also be revealed
from their study.

It was pointed earlier that the vibration frequencies of
certain main-chain Raman active modes were found to change with
the level of applied strain [20,34,35]. In particular it was

found that the C≡C triple bond stretching frequency changes by
the order of 20cm⁻¹ for 1% of strain [34] as is shown in
Figure 3. This property can be used to determine the strain
in a polydiacetylene fiber subjected to any general state of
stress. The strain can be measured to a high degree of spatial
resolution and accuracy as beam diameters of as small as ∿10µm
can be used and changes in frequency can be determined to ∿1cm⁻¹.
The fiber therefore, behaves as though it has an internal
molecular strain gauge and the Raman technique has been recently
been employed to monitor the point-to-point variation in strain
in polydiacetylene fibers in epoxy composites [46,56,57].

    The first case investigated was that of a single short
fiber in a matrix subjected to an overall strain which is a
classical problem in the theory of fiber reinforcement [53,58].
Model specimens were fabricated. When the specimen is deformed
the matrix strain is measured using a strain gauge adhered to
the surface of the specimen and the point-to-point variation
in strain is monitored using Raman spectroscopy. Typical
results [56] are shown in Figure 7 where the fiber strain is
plotted as a function of position along the fiber for different
levels of applied matrix strain and it was found that the
results agreed moderately well with theoretical predictions
of Cox [58]. For example, it can be seen that at higher
levels of matrix strain thefiber strain rises from the end to
a constant value along the length of the fiber and then falls
off at the other end as predicted by the theory. These regions
of rise and fall of strain are known as the 'critical length'
[53] and it was also demonstrated [56] that the critical
length was proportional to the fiber diameter. Although this
result was expected from the theoretical analyses [56,58]
this was the first time it has been demonstrated by direct
experiment for any composite system.

    Some preliminary measurements have been made of the me-
chanical properties of composites consisting of aligned poly-
DCHD single crystal fibers (∿10mm long) in an epoxy resin
matrix [57]. It is found that the mechanical properties of
polyDCHD/epoxy composites depend strongly upon the fiber volume
fraction and this is demonstrated in Figure 8 where a plot is
given of the Young's modulus, E, as a function of V_f. It can
be seen from Figure 8 that the stiffness of the composite
increases as the volume fraction of fibers increases. The
experimental points fall between the Voigt (uniform strain)
and Reuss (uniform stress) lines [53]. For a uniaxially-
aligned composite sample it would be expected that the data
should fall close to the Voigt line and it is thought [57]
that the short-fall in modulus is due to a combination of
fiber misalignment and fiber-twinning due to resin shrinkage.

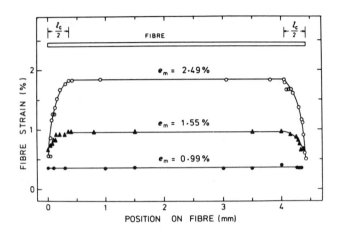

Figure 7. Variation of fiber strain, e_f with position along the fiber for different levels of matrix strain, e_m for a polyDCHD/epoxy single fiber composite.

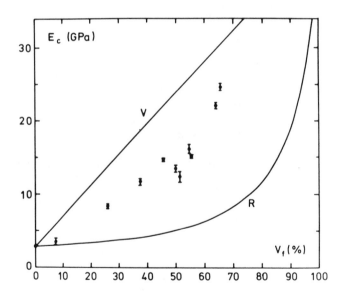

Figure 8. Variation of the Young's modulus of a polyDCHD/epoxy composite with fiber volume fraction, V_f. The lines show the Voigt (V) and Reuss (R) average.

It is also found that there is a general increase in the strength of the composites with increasing fiber volume fraction. However, at low values of V_f the strength is below that of the pure resin until sufficient fibers are present to produce reinforcement [48] and at high values of V_f (>60%) the strength falls off because their is not enough resin to wet the fibers [57].

Conclusions

The high degree of molecular alignment that is found in polymer single crystal fibers means that such materials can display high levels of stiffness and strength and are inherently resistant to creep. As a consequence of this it has been shown that composites produced by incorporating polydiacetylene single crystals fibers in an epoxy resin matrix have promising mechanical properties.

As well as enabling the development of a new type of reinforcing fiber the study of these polymer single crystals fibers has led to significant advances of our understanding of the structure/property relationships in polymers. Examination of the structure of these materials using electron microscopy has enabled a detailed study of defects such as dislocations to be made and has also led to the development of high-resolution techniques allowing molecular detail to be seen at a level hitherto unobtainable with polymers. The characterization of surface defects such as steps has enabled estimates to be made of the theoretical strength of fibers and hence the strength of individual molecules. The observation that the frequencies of the Raman-active main-chain stretching modes are a strong function of externally applied strain has allowed the fibers to be used to follow the micromechanics of fiber reinforcement in composites in detail.

As well as leading to the development of a new class of reinforcing fiber these studies upon polydiatetylene single crystal fibers have helped to open up new areas of research in more conventional polymer systems. For example, there has recently been an upsurge of interest in the application of high-resolution transmission electron microscopy to a wide range of polymer crystals. Also, it has been found that the Raman strain measurement echnique is applicable to other types of high modulus polymer fibers such as aromatic polyamides [59] and this will allow detailed studies to be made of the micromechanis of reinforcement of composite systems reinforced with these fibers.

Acknowledgements

Most of the work described above was carried out at Queen Mary College and the author is grateful to Professor D. Bloor and Dr. D N. Batchelder for introducing him to poly-diacetylenes and for their continued encouragement and help. He would also like to thank Mr. D. Ando and Mr. I.F. Chalmers for their help with the preparation of monomers. Finally he must extend his gratitude to Dr. R.T. Read, Dr. C. Galiotis, Dr. P.H.J. Yeung and Dr. I.M. Robinson who performed the bulk of the experimental work.

References

[1] Andrews, J.M.; Ward, I.M. J.Mater. Sci., 1970, 5, 411.
[2] Capaccio, G.; Ward, I.M. Polym. Eng. Sci. 1975 , 15, 219.
[3] Capaccio, G.; Gibson, G.; Ward, I.M. in 'Ultra-High Modulus Polymers', Ed. Ciferri, A.; and Ward, I.M., Applied Science, London, 1979, p.1.
[4] Zachariades, A.E.; Mead, W.T.; Porter, R.S., in 'Ultra-High Modulus Polymers', Ed. Ciferri, A. and Ward, I.M., Applied Science, London, 1979.
[5] Mead, W.T.; Desper, C.R.; Porter, R.S., J. Polym. Sci., Polym. Phys. Ed., 1979,17, 859.
[6] Lemstra, P.J., "private communication".
[7] Pennings, A.J.; Meihuizen, K.E., in 'Ultra-High Modulus Polymers', Ed. Ciferri, A. and Ward, I.M., Applied Science, London, 1979, p.117.
[8] Schaefgen, J.R.; Bair, T.I.; Ballou, J.W.; Kwolek, S.L.; Morgan, P.W.; Panar, M.; Zimmerman, J., in 'Ultra-High Modulus Polymers', Ed. Ciferri A. and Ward, I.M., Applied Science, London, 1979, p.173.
[9] Wolfe, J.F.; Loo, B.H.; Arnold, F.E., Macromolecules 1981 14, 915.
[10] Allen, S.R.; Farris, R.J.; Thomas, E.L., J. Mater. Sci. 1985 20 2727.
[11] Wegner, G,;Pure. Appl. Chem., 1977 49 443.
[12] Wegner, G.; Naturforsch, Z., 1969 24b 84.
[13] Bloor, D.; Kaski, L.; Stevens, G.C.; Preston, F.H.; Ando, D.J., J. Mater. Sci., 1975 10 1678.
[14] Kelly, A.; Groves, G.W., 'Crystallography and Crystal Defects', Longman, London, 1970.
[15] Dudley, M.; Sherwood, J.M.; Bloor, D.; Ando, D., J. Mater. Sci. Lett., 1982 1, 479.
[16] Schermann, W.; Williams, J.O.; Thomas, J.M.; Wegner, G., J. Polym. Sci., Polym. Phys. Ed., 1975 13 753.
[17] Yee, K.C.; Chance, R.R., J. Polym. Sci., Polym. Phys. Ed., 1978, 16, 431.

[18] Galiotis, C.; Read, R.T.; Yeung, P.H.J.; Chalmers, I.F.; Bloor, D., J. Polym.Sci., Polym.Phys. Ed. 1984, 22, 1589.
[19] Galiotis, C.; Young, R.J.; Ando, D.J.; Bloor, D., Makromol. Chem. 1983, 184 1083.
[20] Galiotis, C.; Young, R.J.; Batchelder, D.N., J. Polym. Sci., Polym. Phys. Ed., 1983 21, 2483.
[21] Galiotis, C.; Young, R.J., Polymer, 1983, 24, 1023.
[22] Wenz, G.; Wegner, G., Makromol. Chem., Rapid Comm., 1982, 3, 231.
[23] Baughman, R.H.; Gleiter, H.; Sendfield, N., J. Polym. Sci., Polym. Phys. Ed., 1975, 13, 1871.
[24] Read, R.T.; Young, R.J., J. Mater. Sci., 1979, 14, 1968.
[25] Kobelt, D.; Paulus, E.F., Acta. Cryst., 1974, B30 232.
[26] Hadicke, E.; Mez, H.C.; Krauch, C.H.; Wegner, G.; Kaiser, J., Angew. Chem., 1971, 83, 253.
[27] Apgar, P.A.; Yee, K.C., Acta. Cryst., 1978, B34 957.
[28] Wunderlich, B.; 'Macromolecular Physics' 1973, Vol. 1, Academic Press, London.
[29] Grubb, D.T.; J. Mater. Sci., 1974, 9 , 1715.
[30] Read, R.T.; Young, R.J., J. Mater. Sci., 1984, 19, 327.
[31] Read, R.T.; Young, R.J., J. Mater. Sci., 1981, 16, 2922.
[32] Frye, C.J.; Ward, I.M.; Dobb, M.G.; Johnson, D.J., J. Polym. Sci.,Polym. Phys. Ed. 1982, 20, 1677.
[33] Dobb, M.G.; Johnson, D.J.; Saville, B.P., Phil. Trans. Roy. Soc. Lond., 1980, A294, 483.
[34] Batchelder, D.N.; Bloor, D., J. Polym. Sci., Polym. Phys. Ed., 1979, 17, 569.
[35] Galiotis, C.; Ph.D. Thesis, University of London, 1982.
[36] Young, R.J., 'Introduction to Polymers', Chapman and Hall, London, 1981.
[37] Odajima, A.; Maeda, T., J. Polym. Sci., 1966, C15, 55.
[38] Treloar, L.R.G., Polymer, 1960, 1, 95.
[39] Leyrer, R.J.; Wegner, G.; Wettling, W., Ber. Bunsenges. Phys. Chem. 1979, 82, 697.
[40] Enkelmann, V.; Leyrer, R.J.; Schleier, G.; Wegner, G., J. Mater. Sci., 1980, 15, 168.
[41] Rehwald, W.; Vonlanthen, A.; Meyer, W.; Phys. Stat. Sol. (a) 1983, 75, 219.
[42] Bowden, P.B.; Young, R.J., J. Mater. Sci., 1974, 9, 2034.
[43] Young, R.J.; Bloor, D.; Batchelder, D.N.; Hubble, C.L., J. Mater. Sci., 1978, 13, 62.
[44] Young, R.J.; Dulniak, R.; Batchelder, D.N.; Bloor, D., J. Polym. Sci., Polym. Phys. Ed., 1979, 17, 1325.
[45] Wilding, M.A.; Ward, I.M., Polymer, 1978, 19, 969.
[46] Robinson, I.M.; Yeung, P.H.J.; Galiotis, C.; Young, R.J.; Batchelder, D.N., J. Mater. Sci., 1986, 21, 3440.
[47] Dobb, M.G.; Johnson, D.J.; Saville, B.P., Polymer, 1981, 22, 960.
[48] Kelly, A. 'Strong Solids', Clarendon Press, Oxford, 1966.

[49] Marsh, D.M.; in 'Fracture in Solids', Eds. Drucker, D.C.;
     Gilman, J.J., Interscience, New York, 1963.
[50] Frank, F.C., Proc. Roy. Soc. 1970 A319, 127.
[51] Kinloch, A.J.; Young, R.J., 'Fracture Behaviour of
     Polymers', Applied Science, London, 1983.
[52] Kausch, H.H.; 'Polymer Fracture', Springer-Verlag,
     Berlin, 1978.
[53] Hull, D, 'An Introduction to Composite Materials',
     Cambridge University Press, 1981.
[54] Ladizesky, N.H.; Ward, I.M., paper presented at 6th
     International Conference, 'Yield, Deformation and
     Fracture of Polymers', Cambridge, Plastics and Rubber
     Institute, 1985.
[55] Greenwood, J.H.; Rose, P.G.;, J. Mater. Sci., 1974, 9, 1809.
[56] Galiotis, C.; Yeung, P.H.J.; Young, R.J.; Batchelder, D.N.,
     J. Mater. Sci., 1984, 19, 3640.
[57] Robinson, I.M.; Galiotis, C.; Young, R.J.; Batchelder,
     D.N., to be published.
[58] Cox, H.L.; Brit. J. Appl. Phys., 1952, 3, 72.

[59] Galiotis, C.; Robinson, I.M.; Young, R.J.; Smith, B.E.J.;
     Batchelder, D.N., Polymer Comm., 1985, 26, 354.

ULTIMATE BEHAVIOR OF ADVANCED POLYMERIC FIBERS AND WHISKERS: A STOCHASTIC STUDY OF SIZE EFFECTS

H. D. Wagner

Materials Research Department
The Weizmann Institute of Science
Rehovot 76100 (Israel)

Summary

The ultimate mechanical properties of materials, such as the (short-term) strength under a linearly increasing load history or the (long-term) lifetime under a constant load history, are well known to be dependent on the volume of the tested specimen.

Current approaches to the effects of size on ultimate properties of crystalline and semi-crystalline polymers, and single crystals, are discussed, and an approach based on statistical/stochastic methods to model size effects in polymers is presented. This is illustrated through an analysis of experimental data for ultra–high molecular weight polyethylene (UHMWPE) fibers, Kevlar 29 and 49 fibers, and polydiacetylene whiskers. Results are presented according to different schemes based on the Weibull distribution for failure strength, and maximum likelihood estimation (MLE) is used to determine the values of the Weibull parameters in each case, as well as to test which scheme is the most appropriate for each material.

Previous authors studied this problem by means of models based on linear elastic fracture mechanics (LEFM). Our approach is shown to yield analytical predictions which are at least as accurate as LEFM schemes, and to possess the advantage of possible simultaneous interpretation of both length and diameter effects. In addition, the proposed scheme provides some interesting information regarding the type of flaw population yielding to failure in a given polymeric material.

1. Introduction

A fundamental aim of any theory of material failure is to assess the relationship between material geometry and microstructure, on the one hand, and some significant macroscopic variable which describes the ultimate mechanical behavior of the material, on the other hand. One such macroscopic variable is the breaking strength under a linearly increasing load history. Although not a true material constant like the fracture toughness (K_{Ic}), for instance, the strength is nevertheless widely used in science and engineering, mainly because of convenience and historical reasons. Under a constant load history, the fracture behavior of a material is assessed by measuring the time elapsed between the instant of loading and the time of failure, or specimen lifetime. There is considerable interest in assessing the link between the strength and the lifetime of a solid, and its microstructural and geometrical characteristics.

As observed for many decades, strength is a parameter which is a function of the volume of the specimen under test, a fact which is true for most types of materials (metals, ceramics, polymers, and their composites). It is generally – but not always – observed that larger specimens fail at lower stresses and this size effect is apparently stronger for inherently brittle materials than for tough, ductile ones. A much stronger size effect is predicted for the lifetimes of brittle specimens, but there seems to be no experimental measurement whatsoever in the literature of a size effect in material lifetime. Throughout this paper we are concerned mainly by the effect of material size on strength.

The purpose of the present study is to discuss the current approaches to the effect of size on the ultimate behavior of crystalline and semi-crystalline polymers and to propose an approach based on stochastic methods to model size effects in polymers.

2. Size Effect: Definition and Significance

Since Griffith's work on soda lime silica glass [1] and Peirce's study on cotton yarns [2], it is known that many brittle materials exhibit a higher mechanical strength when the volume of the tested specimen is smaller. This effect was apparently discovered by Leonardo da Vinci as he tested iron wires, as reported by Hertzberg [3]. Another early mention of the size effect is due to Chaplin [4].

An excellent introduction to size effect concepts can be found in a 1958 article by Irwin [5], and an extensive survey of the literature on size effects was performed by Harter [6].

Since we essentially deal with fibers, the size effect will be defined as a longitudinal size effect or transversal size effect if the effects of fiber length or fiber diameter, respectively, are analyzed. There are many important aspects which motivate the study of the effect of specimen size upon fracturing:

(1) A great deal of experimental work is underway, these days, to produce very high strength polymers in fibrous or whisker form. If significant enough, the transversal size effect in such materials might well provide a complementary avenue for improved strength properties of fibrous polymers by obtaining finer filaments.

(2) If a probabilistic interpretation to failure is adopted, the effect of size upon fracture stress provides an indication of the amount of expected scatter due to the distribution of flaws in the specimens and, in addition, some important information on the nature and characteristics of the flaw population can be obtained. This will be clarified below.

(3) In certain cases, if significant enough, the effect of size on mechanical strength or lifetime can be utilized in a way similar to the effect of temperature, or voltage etc., thus providing a new way of performing accelerated testing. (Curiously, this seems to have been unnoticed so far.). Changing the specimen size provides a way to generate a "generalized stress" (in the sense of Mann et al. [7]).

(4) Laboratory scale specimens are advantageous from a point of view of cost and time savings. When dealing with large structures, one must be aware that small scale laboratory specimens may yield different data if a size effect exists, and it is thus essential to know the amount of change in the data as a function of size.

We now analyze the lifetime and strength of polymeric filaments using probabilistic concepts and here we particularly emphasize the aspects related to the effects of size on failure parameters, using recent relevant data published in the literature.

3. Theoretical Background

3.1 Weakest Link Rule

The weakest link concept is a simple and powerful statistical tool utilized for the first time by Peirce [2] to predict the mean and variance of the failure strength of long cotton yarns, assumed to be made up of shorter lengths whose statistical characteristics were known. The application of weakest link principles to material failure problems has been the subject of several excellent early studies by Epstein [8-10].

The weakest link theory is closely related to the statistical theory of extreme values [7,11]. Its fundamental principle is as follows: it is assumed that in an assembly of n small elements linked together in a chain-like structure fracture occurs when the weakest element fails. If each element possesses a probability of failure $F(x)$ under a stress increase from 0 to x and if the strengths of the elements are independent, identically distributed random variables, then the strength distribution of the chain, $F_n(x)$, is given by the distribution of the smallest order statistic:

$$F_n(x) = 1 - [1 - F(x)]^n \qquad (1)$$

Clearly, as the number of elements n increases, the probability of failure of the chain increases. This is the size effect for materials possessing a chain-like structure. The same concept is applicable to the lifetime under a constant stress history (rather than strength under a linearly increasing stress history) of a chain-like structure made up of smaller elements. From the above expression, the size effect is seen to depend on the type of distribution $F(x)$ for the single element. Based on various failure probability distributions, it can be explicitely expressed in terms of mean strength (or lifetime) as a function of specimen size [8-10].

3.2 Weibull/Weakest Link Model

Assuming that $F(x)$ is the well-known two-parameter Weibull distribution, given by

$$F(x) = 1 - exp\left\{-(x/a)^b\right\} \qquad (2)$$

where a is the scale parameter and b is the shape parameter, we obtain the strength distribution for a specimen made up of n elements, using the weakest link rule:

$$F_n(x) = 1 - exp\left\{-n\left(x/a\right)^b\right\}$$ (3)

which again is a Weibull distribution (this is the self-reproducing property of the Weibull distribution) with the same shape parameter b, but with a scale parameter equal to:

$$a' = a\Omega^{-1/b}$$ (4)

This is the effect of size on the scale parameter and thus on the mean (or median) strength. Note that in equation 4 we have presented the size effect for the chain-like material in terms of elemental volumes instead of elemental lengths, and Ω is thus the fiber volume. The effect of size on strength is a function of b, the shape parameter. Strength experiments with fibers for composites give typically $2 \leq b \leq 12$, and for lifetime experiments $b \leq 1$ is typical. Thus size effects in lifetime are generally much more pronounced than size effects in strength. Using the Weibull model written in terms of the volume of the elements rather than in terms of the number of elements allows us to interpret the weakest link rule in a slightly different, more general way. Indeed, the rule now relates to two- or three-dimensional structures which are not necessarily physically chain-like systems, but still fail at their weakest flaw. (A balloon is a simple example of such a two-dimensional structure).

3.3 Maximum Likelihood Estimation and Plotting Procedure

We have utilized two variants of the Weibull distribution, as follows:

(i) If it assumed that the critical defect population is located primarily in the fiber surface region, then the Weibull cdf is given by

$$F_s(x) = 1 - exp\left\{-DL\left(x/a\right)^b\right\}$$ (5)

where D is the dimensionless fiber diameter and L is the dimensionless length.

(ii) If critical defects are distributed throughout the fiber bulk, then the Weibull cdf adopted is:

$$F_v(x) = 1 - exp\left\{-D^2L\left(x/a\right)^b\right\}$$ (6)

Given these expressions, the shape and scale parameters were computed as follows, using maximum likelihood estimation. First each strength datum x_i is multiplied by $D_i^{1/b}$ (for the surface flaw model) or by $D_i^{2/b}$ (for the volume flaw model). This is equivalent to a change of variable and a new strength datum x_i' (for the surface model) or x_i'' (for the volume model) is obtained. However, since b is unknown *a priori*, a first approximation must be used, then x_i' (or x_i'') obtained, b recomputed, etc., until a final value of b and a final set x_i' (or x_i'') are obtained. Next, the new data sets x_i' (for the surface model) and x_i'' (for the volume model) are ordered, and Weibull plots performed. Finally a decision is made regarding the best model, based on the best fit to the data.

4. Data Analysis and Discussion

The experimental data used concern ultra high molecular weight polyethylene (UHMWPE) fibers [12,13] and Kevlar 29 and 49 fibers [14], all highly oriented polymeric materials, as well as results obtained with polydiacetylene (PDA) whiskers [15-18]. The conventional Weibull distribution as well as the surface/Weibull and volume/Weibull distributions were fitted to all data sets.

A full discussion of the results obtained using various Weibull modelling approaches and maximum likelihood estimation of the parameters is presented elsewhere [19].

Several modelling schemes for the size effect for strength are presented in Table 1. These are either empirical, or LEFM (fracture mechanics)-based, or statistics-based schemes. The approach adopted here, based on the Weibull probability distribution for failure, does not apply equally well to all the data analysed, probably because the polymers reviewed are quite different from each other from a structural/molecular viewpoint. For example, failure in fibrillar polymers such as Kevlar and, perhaps, UHMWPE proceeds by a splitting/fibrillation pull-out mechanism [20,21], whereas PDA crystals fail by a molecular cleavage mechanism. A comparison between the LEFM- and stochastic- based approaches was performed by checking various dependences of strength on diameter as obtained from different models (see Table 1). Regression lines of σ *versus* D^{-1}, of σ^{-1} *versus* $D^{1/2}$, and of $log(\sigma)$ *versus* $log(V)$ were obtained for all data sets, and the coefficient of correlation computed, as reported in Table 2. From the data in the Table, it appears that, for PDA crystals, the size effect predicted by the Weibull model is more appropriate than the size effect predicted by the LEFM-based model, whereas for UHMWPE the reverse is true, although the differences between both predictions are not very large. The model proposed by Baughman *et al.* [15], i.e. a linear dependence between σ and D^{-1} [or a slope of -1 on a $log(\sigma)$ *versus* $log(D)$ plot], is not well obeyed for most data sets.

Table 1. Various Models for Strength-Size Dependence.

Model	Source	Size Dependence of Strength*	Cylindrical Fibre: Superficial Flaw Distribution	Volumetric Flaw Distribution
Empirical	Baughman *et al.* (Ref. 15)	$\sigma \propto D^{-1}$	–	–
Based on Fracture Mechanics Concepts	Smook *et al.* (Ref. 12)	$\sigma^{-1} \propto D^{1/2}$	–	–
	Galiotis *et al.* (Ref. 16)	$\sigma^{-1} \propto D^{1/2}$	–	–
	Weibull distribution	$\tilde{\sigma} \propto V^{-1/b}$	$D^{-1/b}L^{-1/b}$	$D^{-2/b}L^{-1/b}$
Based on Statistical Concepts	WSGT distribution**	$\tilde{\sigma} \propto V^{-\alpha/b}$	$D^{-\alpha/b}L^{-\alpha/b}$	$D^{-2\alpha/b}L^{-\alpha/b}$
	Uniform (rectangular) distribution, Epstein (Ref. 8)	$\tilde{\sigma}$ independent of V	–	–

* V = specimen volume, D = specimen diameter, L = specimen length. $\tilde{\sigma}$ is the most probable value, or mode.

** WSGT: Watson-Smith-Gutans-Tamuzs, see references 22 and 23.

<u>Table 2.</u> Correlation coefficient (r^2) from linearized fits of UHMWPE
and polydiacetylene data using LEFM and Weibull strength-
diameter dependences.

	r^2 (from linear regression)			
Model	UHMWPE (Smook *et al.* [12])	Set # 4	Polydiacetylene* Set # 5	(Baughman *et al.* [15])
$\sigma^{-1} \propto D^{1/2}$ (LEFM)	0.93	0.72	0.89	0.72
$log\tilde{\sigma} \propto log D$ (Weibull)	0.91 (slope=-0.46)	0.79 (slope=-0.61)	0.90 (slope=-0.55)	0.76 (slope=-0.83)

* Sets 4 and 5 are taken from Refs. [15–18].

Some anomalous behavior exists regarding the Weibull approach
for polymeric (and other) fibers: Indeed, there is some unconclusive-
ness regarding the "true" value of the shape parameter of the distribution since Weibull plots of strength data at a given gauge length
(or diameter) yield a value for this parameter which is significantly
different from that obtained from $log(strength)$ *versus* $log(length)$
[or $log(diameter)$] plots. This is shown elsewhere (Ref. 19). A possible solution to this problem was suggested simultaneously by Watson
and Smith [22] and Gutans and Tamuzs [23], using a new probability
distribution, but this opens the way for more questions, particularly
regarding the physical interpretation of this new distribution.

In conclusion, Weibull-based models seem to provide a satisfactory explanation for transversal size effects in polymeric fibers, (although some problems exist regarding the "true" value of the shape
parameter obtained, as fully discussed elsewhere [19]). The longitudinal size effect is also well interpreted with this model, as seen for
Kevlar fibers [14]. However, since none of the available literature
data included <u>both</u> length and diameter effects (for a given polymer), more experimental work is needed to assess the correctness of
any modelling scheme for size effects.

Acknowledgements

The author is a recipient of the J. and A. Laniado Career Development Chair in Industrial and Energy Research. This work was supported by the Levi Eshkol Fund of the National Council for Research and Development (Ministry of Science and Development, Israel). The author expresses his gratitude to A.J. Pennings, P. Schwartz, H. Gleiter and R.J. Young for permission to use their experimental data. Sincere thanks are due to S.L. Phoenix, R.L. Smith and D.R. Cox for many critical discussions.

References

[1] Griffith, A.A., Philos. Trans. of the Royal Soc. (London), 1921, A221, 163.

[2] Peirce, F.T.S., J. Text. Inst., 1926, 17, 355.

[3] Hertzberg, R.W., Deformation and Fracture Mechanics of Engineering Materials, 1976, John Wiley & Sons, p. 235.

[4] Chaplin, W.S., Van Nostrand's Engineering Magazine, 1880, 23, 441.

[5] Irwin, G.R., in Encyclopedia of Physics; Flügge, S. Editor, Vol. 6, Springer-Verlag, Berlin, 1958, 551.

[6] Harter, H.L., AFFDL–TR–77–11, 1977.

[7] Mann, N.R., Schafer, R.E., Singpurwalla, N.D., Methods for Statistical Analysis of Reliability and Life Data, J. Wiley & Sons, 1974, Chapter 9.

[8] Epstein, B., J. Appl. Phys., 1948, 19, 141.

[9] Epstein, B., J. Amer. Stat. Soc., 1948, 43, 403.

[10] Epstein, B., Technometrics, 1960, 2(1), 27.

[11] Gumbel, E.J., Nat. Bur. Standards, Appl. Math. Series, 1954 33.

[12] Smook, J., Hamersma, W., Pennings, A.J., J. Mater. Sci., 1984, 19, 1359.

[13] Schwartz, P., Netravali, A., Sembach, S., Text. Res. J., 1986, 56, 502-508.

[14] Wagner, H.D., Phoenix, S.L., Schwartz, P., J. Comp. Mater. 1984, 18, 312.

[15] Baughman, R.H., Gleiter, H., Sendfeld, N., J. Polym. Sci. (Polym. Physics Ed.), <u>1975</u>, 13, 1871.

[16] Galiotis, C., Young, R.J., Polymer, <u>1983</u>, 24, 1023.

[17] Galiotis, C., Young, R.J., Batchelder, D.N., J. Mat. Sci. Letters, <u>1983</u>, 2, 263.

[18] Galiotis, C., Read, R.T., Yeung, P.H.J., Young, R.J., Chalmers, I.F., Bloor, D., J. Polym. Sci. (Polymer Physics), <u>1984</u>, 22, 1589.

[19] Wagner, H.D., <u>1987</u>, submitted manuscript.

[20] Wagner, H.D., J. Mater. Sci. Lett. <u>1986</u>, 5, 229.

[21] Wagner, H.D., J. Mater. Sci. Lett. <u>1986</u>, 5, 439.

[22] Watson, A.S., Smith, R.L., J. Mater. Sci., <u>1985</u>, 20, 3260.

[23] Gutans, J., Tamuzs, V., Mech. of Comp. Mater. (in Russian), <u>1984</u>, 20(6), 1107.

Part VI
Polymer Networks

INTERPENETRATING POLYMER NETWORKS

L. H. Sperling, J. H. An, M. C. O. Chang, and D. A. Thomas*

Polymer Science & Engineering Program, Dept. of Chemical
   Engineering, Materials Research Center, Coxe Lab #32
   Lehigh University, Bethlehem, PA   18015 USA

*Dept. of Materials Science and Engineering

Abstract

An interpenetrating polymer network,
IPN, contains two polymers in network form.
In general, the networks may be crosslinked
by chemical or physical bonds, and may be
formed simultaneously or sequentially. Each
mode of synthesis yields unique morphologies
and physical properties. This paper will
emphasize sequential IPN's with covalent
crosslinks. Evidence is presented that
some interpenetrating polymer networks have
two continuous phases. Recent experiments
on cross-polybutadiene-inter-cross-
polystyrene sequential IPN's show that constant diameter interconnected cylindrical
structures of polystyrene are formed,
suggesting a modified form of spinodal decomposition as the major mechanism of phase
separation during polymerization. IPN's
suitable for sound and vibration damping
exhibit a microheterogeneous type of morphology, yielding broad loss modulus-
temperature peaks. The area under the linear
loss modulus-temperature curve, LA, was found
to depend on the chemical structure of the
homopolymer, statistical copolymer, or IPN
in a predictable way, leading to a group
contribution analysis.

Introduction

Interpenetrating polymer networks, IPN's, are defined
in their broadest sense as an intimate mixture of two or
more polymers in network form [1,2]. They constitute a

branch of multicomponent polymer materials along with polymer blends, blocks, grafts, and AB-crosslinked polymers. These may be classified into two groups in terms of type of bonding. Polymer blends, grafts, and blocks are ordinarily thermoplastic, able to flow and dissolve. AB-crosslinked copolymers, IPN's, and semi-IPN's (one polymer crosslinked) are ordinarily thermoset. An exception is the case where physical crosslinks (block copolymers, ionomers, etc.) are used, rather than covalent crosslinks.

Most multicomponent systems undergo phase separation, yielding many varieties of morphology according to their thermodynamic and kinetic environments. Studies of their morphological features have been central in the research of multicomponent systems, because domain sizes, shapes, and interfacial bonding characteristics determine key mechanical properties along with the glass transition temperature, molecular weight, etc. A proper understanding of these features often allows synergistic behavior to be developed. Table I summarizes features important in the development of morphology in IPN's. Of course, most of these features are important for other multicomponent polymer materials.

Ideally, IPN's can be synthesized by either swelling the first crosslinked polymer with the second monomer and crosslinker, followed by in-situ polymerization of the second component (sequential IPN's) or by polymerizing a pair of monomers and crosslinkers at the same time through different, non-interfering reaction paths, forming simultaneous interpenetrating networks, SIN's. In fact, many variations of these ideas exist in both the scientific and the patent literature [1,2].

Important features of IPN synthesis include: (a) More crosslinks in both polymers limit phase separation and reduce domain size. (b) Polymerization is accompanied simultaneously by phase separation. (c) The final product is often thermoset, restricting diffusional motions. Consequently, the morphology of IPN's often stabilizes at a quasi-equilibrium state [3] determined by a balance among the several kinetic and thermodynamic factors, Table I.

In many of the early studies of IPN's, interest centered on the characterization of morphology and thermo-mechanical behavior of fully polymerized IPN's [4-7], particularly interrelationships among composition, crosslinking level, and temperature. The first electron

Table I. Factors Affecting IPN Domain Morphology.

Chemical Characteristics of Constituents
• Crosslink density
• Crosslinking rate

Physical Properties
• Miscibility
• Chain mobility
• Interfacial tension
• Temperature

Preparation Process
• Sequential or simultaneous
• Composition control
• Synthesizing pressure

microscopy study of morphology by Curtius et al. [4] on cross-polybutadiene-inter-cross-polystyrene (PB/PS) sequential IPN's showed well-formed domains of polystyrene of about 600-800 A. This was followed by more detailed electron microscopy studies [5,6,8,9]. The decrease of the domain sizes with increasing crosslink density was recognized by Yeo et al. [10,11], Donatelli, et al. [12], and Michel, et al. [13] who derived a series of equations for predicting domain sizes. Recently, scattering techniques such as small-angle X-ray scattering (SAXS) [14-16] and small-angle neutron scattering (SANS) showed that these colloid sized domains form most probably by a modified spinodal decomposition mechanism [3,17-19].

The different modes of phase separation during IPN synthesis provide a range of opportunities for the application of IPN's. First, one can choose broad ranges of polymer pairs regardless of reaction mechanism (addition or condensation) and miscibility level, providing a spectrum of properties. Secondly, the domain size may be controlled, in significant measure, in a given pair by the crosslinking level of each component. There are many applications for IPN's, some of which have already become commercial, and others which are still in the research or development stages [1].

One of the applications to be discussed herein is sound and vibration damping. When polymers are vibrated in the vicinity of their glass transition, the time required to complete an average coordinated movement of the chain

segments approximates the length of time of the
measurement. At the glass transition conditions, involving both temperature and frequency effects, this
coordinated molecular motion converts a maximum fraction
of mechanical work into heat.

Vibration damping is important for reducing noise and
also fatigue failure in aircraft, automobiles, and
machinery. The common damping materials are homopolymers
or copolymers, which exhibit efficient damping only over
about a 20 to 30°C range centered around the glass
transition temperature [20]. The acoustic spectrum spans
from 20 Hz to 20,000 Hz, three decades of frequencies. For
many polymers, the time-temperature equivalence is 6 to 7°C
per decade of frequency, or about 20°C for the acoustic
range. Thus, homopolymers or statistical copolymers often
suffice to damp the full acoustic range at one temperature.
However, many systems undergo large temperature variations,
causing the homopolymer damping materials to become
inefficient. For example, an aircraft with a homopolymer
loss maximum situated at 25°C would have low damping
efficiency at 0°C or -20°C, commonly encountered at high
altitudes. One solution to the problem lies in selecting
multicomponent polymer systems with controlled degrees of
partial miscibility. With such systems, damping can be
made to take place over the entire temperature range
between the glass transition of the two homopolymers.

Many materials, such as mechanical blends and grafts,
have been explored to achieve damping over a wide
temperature [2,21-31]. As will be developed below, the
linear loss modulus, E", will be used throughout for
improved quantitative characterization. Materials
consisting of incompatible polymer blends or grafts with
widely separated transition temperatures exhibit two
regions of high damping separated by an intermediate region
of low damping [29-31], see Figure 1. On the other hand,
semi-miscible (or microheterogeneous) multicomponent
polymer systems which have extensive but incomplete
molecular mixing show a rather broad transition [2,21-28],
see Figure 1. One method to broaden the damping range is
through the use of interpenetrating polymer networks
[1,32-55]. In IPN's, the introduction of crosslinks into
both polymers restricts the domain size and enhances the
degree of molecular mixing, often forming a microheterogeneous morphology. In some ways, IPN crosslinks behave
like block copolymer junctions, since both mechanisms
restrict domain size to the level of molecular dimensions.

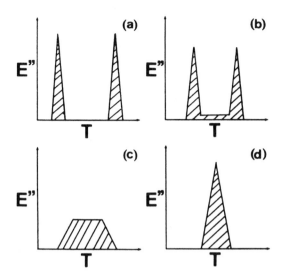

<u>Figure 1.</u> Schematic diagram for loss modulus vs. temp-
erature behavior of polymer I/polymer II with
different miscibility, (a) totally immiscible
polymer pairs, (b) polymer blends with limited
miscibility, (c) microheterogeneous system,
(d) miscible system. The weight fraction of
each component is 50% for all these system.

The application of IPN's in acoustic damping dates
back to the 1970's work by Sperling and Thomas [32-35].
Their IPN systems were primarily based on constrained layer
damping. This consists of a two-layer system with an IPN
layer under a stiff constraining layer. A shearing effect
occurs in the damping layer along with flexuring and extension motions as the composite panel vibrates, increasing
the amount of energy dissipated per cycle. These latex
IPN's were synthesized by first preparing a seed latex of
one crosslinked polymer then introducing a second monomer
and crosslinker, and initiator, but no new soap. Thus,
each latex particle consisted of a micro-IPN system.

This review will examine the recent works in this
laboratory as well as others, with emphasis placed on
measurement of domain size with SANS, phase separation in
IPN's, and the research efforts in the field of sound and
vibration damping [36-55].

Phase Dimensions Via SANS

One of the new instruments useful in characterizing multicomponent polymer materials such as IPN's is small-angle neutron scattering, SANS. Such studies yield domain sizes and interfacial surface area, and aid in the determination of the phase separation mechanisms [5].

Even though TEM and SEM played major roles in the study of IPN morphological features, there are various shortcomings, such as staining artifacts, difficulties in sample preparation for very rubbery materials, and the two-dimensional viewing limit for the former. Recently, various scattering techniques have been applied to measure the phase dimensions of IPN's via statistical treatment. The principles of neutron scattering theory as applied to the phase separated materials have been described in a number of papers and review articles [56-59].

For a two-phased system where domains are randomly dispersed, Debye [60,61] showed that the scattered intensity can be expressed in terms of a correlation function, $\gamma(r)$, in a simple exponential form.

$$\gamma(r) = \exp\left(-r/a\right) \tag{1}$$

where the quantity a represents the correlation distance defining the size of the heterogeneities in local fluctuations of scattering length density at a distance r apart.

The specific interfacial surface area, S_{sp}, defined as the ratio of interfacial surface area, A, to the volume, V, is calculated from the value of correlation distance obtained by plotting the scattered intensity in the Debye fashion [18,61].

$$S_{sp} = \frac{A}{V} = \frac{4\phi_1(1-\phi_1)}{a} = \frac{4\phi_1 \cdot \phi_2}{a} \tag{2}$$

where ϕ_1, ϕ_2 represent the volume fraction of each network, respectively.

The transverse length across the domains, the average distance of straight lines drawn across a domain, is given by [62,63]

$$\ell_1 = \frac{a}{(1-\phi_1)} \tag{3a}$$

$$\ell_2 = \frac{a}{(1-\phi_2)} \tag{3b}$$

It must be pointed out that the above equations do not require any assumption of domain shape in their derivation.

Fernandez et al. [17] carried out SANS analysis on fully polymerized PB/PS IPN's, semi-IPN's, and chemical blends. The specific interfacial surface area was shown to increase with increasing crosslink density in the order of chemical blends, semi-II IPN's, semi-I IPN's, and full IPN's, as expected from many earlier studies. Its value varies from 20 to 200 m^2/gm, in the range of true colloids.

So far, most of the experimental studies have been limited either to fully polymerized or a high plastic content samples. The interrelationships among synthetic detail, morphology, and mechanical behavior showed that the phase-separated nature of these materials was important in obtaining high strength products. In order to understand the domain formation process, an investigation of the intermediate stages before formation of the final morphology is required. There are several different ways to prepare such intermediate materials [3,64,65], see Figure 2. The characteristic domain dimensions of PB/PS IPN's prepared by these routes are compared in Figures 3 and 4 [3,18,66].

In Figure 3, the transverse length of the polystyrene domains increases steadily during the polymerization of monomer II, showing a more rapid increase later in the polymerization. Specific interfacial surface area (Figure 4) does not increase monotonously with PS content. Rather, it shows a maximum at near the midrange of PS content depending on the synthetic detail.

Phase Separation in Sequential IPN's

Phase separation in a polymer I/polymer II/monomer III polymerizing system at constant temperature follows the schematic ternary phase diagram shown in Figure 5. First, the two components are miscible. Then the binodal line and the spinodal (solid line) is crossed, yielding the possibilities of two types of phase separation; nucleation and growth, and spinodal decomposition.

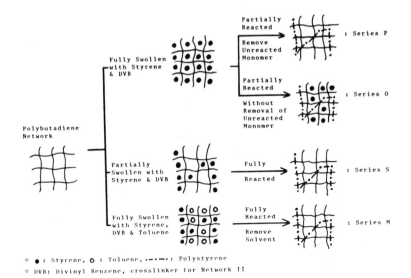

Figure 2. Typical ways preparing partially formed cross-polybutadiene-inter-cross-polystyrene IPN's.

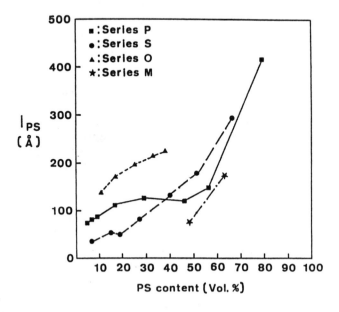

Figure 3. Increase of the polystyrene transverse length with polystyrene content [66].

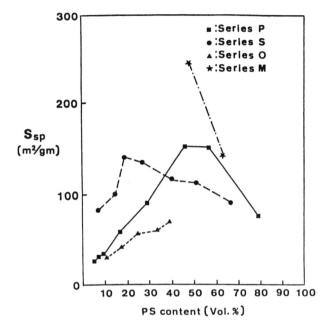

Figure 4. The specific interfacial surface area goes
through a broad maximum as the polystyrene
content increases [66].

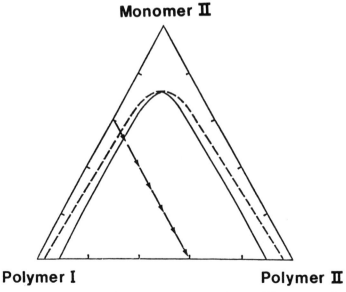

Figure 5. Schematic three component phase diagram, showing the composition locus as monomer II is
polymerized [66].

Phase separation in the polymerizing system polymer I/ polymer II/monomer II has been investigated by a number of people [66-69,74], see Table II. Most recently Lipatov et al [68,69] emphasized spinodal composition in semi-II IPN's.

When Fernandez et al. studied partially polymerized PB/PS systems, evaporating off the remaining styrene (series P) systems (see Fig. 2), SANS results yielded peaks in the scattering patterns [65]. Early attempts at understanding these peaks were troublesome due to the multiple causes for such peaks. Later, similar peaks were observed in another set of samples, which were partially swollen, then fully reacted (series S) by An et al. [66]. In this case, shoulders instead of distinct maxima were observed (Figure 6). In both cases, such maxima or shoulders appeared at mid-range compositions, then were followed by smoothly decreasing curves on further polymerization of monomer II.

At the same time, however, other workers [67-74] were making contributions which would permit a preliminary understanding of such peaks in terms of spinodal decomposition. There are two known mechanisms of phase separation, nucleation and growth [75], and spinodal decomposition [76], Figure 7. Nucleation and growth is associated with metastability, implying the existence of an energy barrier and the occurrence of large composition fluctuations. Domains of a minimum size, so-called "critical nuclei", are a necessary condition. Spinodal decomposition, on the other hand, refers to phase separation under unstable conditions in which the energy barrier is negligible, so even small composition fluctuations increase in amplitude.

For nucleation and growth, the intensity, I, of light scattered at a particular low angle is given by

$$I = n \cdot t^2 \tag{4}$$

where t is the time, and n is a constant. The scattering relationships for spinodal decomposition are more subtle. Following Cahn [76] and Nishi et al. [77], the logarithm of intensity is proportional to time:

$$\ln I(k,t) \quad \alpha \quad R(k)t \tag{5}$$

Table II. Ternary Phase Diagams for Polymer I/Polymer II/Monomer II.

Group	Institute	System	Ref.
G. Reiss et al.	Ecole Nationale Superieure de Chimie de Mulhouse	Polybutadiene/ Polystyrene/Styrene blocks and blends	70
S. L. Rosen et al.	Carnegie Mellon University	Polybutadiene/ Polystyrene/Styrene graft copolymers	71
E. V. Thompson et al.	Univ. of Maine at Ontario	Poly(methyl methacrylate)/ Polystyrene/Styrene blend	72
Y. S. Lipatov et al.	Academy of Science of the Ukranian SSR	Poly(n-butyl acrylate)/ Polystyrene/Styrene, Semi-II IPN	69
D. J. Walsh et al.	Imperial College	Poly(n-butyl acrylate)/ Poly(vinyl chloride)/ Vinyl chloride blend	73
M. T. Purvis et al.	Rohm and Haas Co.	Polyurethane/ Poly(methyl methacrylate)/ Methyl methacrylate blend	74
L. H. Sperling et al.	Lehigh University	Polybutadiene/Polystyrene/ Styrene full IPN	66

L.H. Sperling et al.

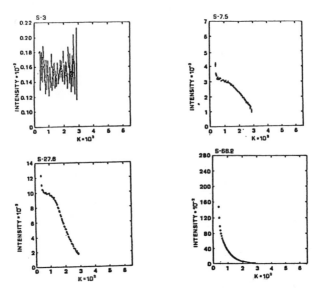

Figure 6. Scattering patterns of series S, cross-polybutadiene-inter-cross-polystyrene IPN's, showing peaks at intermediate composition [66].

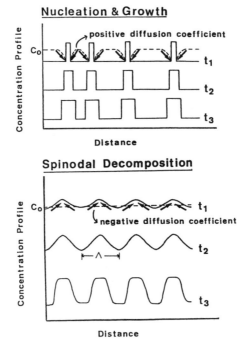

Figure 7. Mechanisms of phase separation.

where k is the wave vector and R(k) is the amplification factor at angle k. The wave vector of spinodal decomposition can be determined from the wavenumber at which the maximum in the scattering occurs. This wavelength, Λ, is given by

$$\Lambda = 2\pi / k_m \tag{6}$$

The main features of spinodally decomposed systems can be summarized as follows: (1) the logarithm of the scattered intensity increases linearly with time, (2) the presence of an interference intensity maximum, and (3) the formation of a uniform and highly interconnected structure. The rather constant value of Λ during phase separation, with increasing amplitude of the waves, is also illustrated in Figure 7.

Lipatov carried out the first experiments [69] on IPN's with an attempt to separate out the extent of spinodal decomposition from nucleation and growth. He studied a semi-II system of poly(butyl methacrylate) and polystyrene, the latter crosslinked with divinyl benzene. During the initial stages of reaction, the light scattering intensity was found to follow a logarithmic relation based on equation (5). A characteristic wavelength of 1-1.5 microns was found for these materials. Most importantly, they found negative diffusion coefficients which are characteristic of spinodal decomposition.

In the case of PB/PS IPN's studied by SANS [66], a "wavelength" of 600 A for the repeat unit was calculated from the angular position of the peak using equation (6). A preliminary light scattering study [78] on the above PB/PS IPN's was found to obey the logarithmic relationship proposed by Cahn [76] based on equation (5), yielding a wavelength of 1.5 microns and a negative diffusion constant. A single SANS experiment was also shown to yield a negative diffusion constant [66] and a 540 A characteristic dimension. It must be pointed out that SANS and light scattering might measure different types of structure, such differences reflecting the 2000X difference in wavelength.

Binder and Frisch [67] constructed free energy functions for spinodal decomposition in IPN's. They predict initial domain wavelengths which are both larger and smaller than the typical distances between network crosslinks. In the latter case, they anticipate a coarsening of the structure until the domain size becomes comparable to the distances between crosslinks. This, indeed, is what An et al. [66] found, because a wavelength of 600 A is of the same order of magnitude on the distance between crosslinks in a network containing about 1% crosslinker.

The type of spinodal decomposition encountered in IPN formation differs from the classical temperature quench in the sense that a composition change constitutes the driving force, at a fixed temperature. In this case, the rate of composition change is deeply involved in the phase separation process [70], which severely limits the applicability of current spinodal theory. In fact, Binder and Frisch [67] assume the polymerization rate is rapid enough to limit the phase separation. On the contrary, Lipatov et al. [69] kept the rate of polymerization to a minimum during the phase separation process. These systems resemble polymer I/polymer II/cosolvent materials [79].

The collage of electron micrographs shown in Figure 8 illustrates a probable transition from a nucleation and growth phase separation mechanism to a spinodal decomposition (see Figure 5). The 7% PS samples shows more or less spherical PS domains, characteristics of nucleation and growth. However, a few domains appear to have some elliptical characteristics. When the PS content increases up to midrange, the shape of the PS domains becomes more obviously elliptical, suggestive of cylindrical cross-sections taken at an angle. As the PS content increases further, the domains seem to form highly interconnected structures typical of the spinodal decomposition mechanism.

The co-occurrence of nucleation and growth and spinodal decomposition was just observed in a temperature quench of poly(2,6-dimethyl-1,4-phenylene oxide)/toluene/caprolactam system [80-81]. The typical morphology formed by nucleation and growth was observed when the quench temperature was slightly above the spinodal boundary. Conversely, if the quench temperature was somewhat lower than the spinodal boundary, they observed interconnected structures as well as small droplets. In the case of IPN's, the presence of crosslinks seems to favor spinodal decomposition more than for linear blends, considering that high molecular weight components generally have a narrower

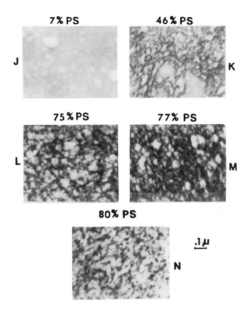

Figure 8. Collage of serpies P transmission electron
         micrographs. The morphology changes from
         spheres to cylinders as monomer II polymer-
         izes. The morphology is highly suggestive
         of spinodal decomposition [66].

gap between the binodal and spinodal boundaries. The
relatively high viscosity of the nascent IPN acts as an
unfavorable factor to nucleation and growth, which requires
a longer mobility range and is known to be a slow process.

    In summary, direct evidence for spinodal decomposition
in IPN's includes modulus-composition studies showing dual
phase continuity, the cylindrical, interconnected form of
the phases shown by TEM and SEM, and the maxima or
shoulders in the SANS scattering patterns. Figure 9
illustrates a model of interconnected cylinders with
occasional spheres, drawn from the sum total of the
observation on this system. The interconnected irregular
cylinders are a result of the spinodal decomposition, while
the more or less spherical domains may be reminiscent of
the initial nucleation and growth stage, or a coarsening of
the structure with breakup of the cylindrical structures
during the latter stages of polymerization.

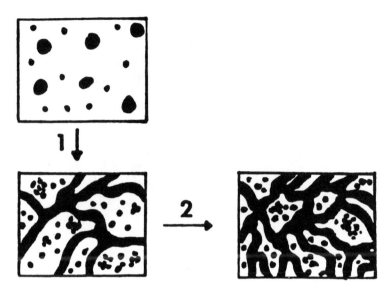

Figure 9. Model of nucleation and growth in IPN's,
         followed by spinodal decomposition [66].
         Note interconnected cylinder formation.

Sound and Vibration Damping with IPN's

Over the years, a number of research teams studied the
damping characteristics of IPN's [1,32-55], see Table III.
Klempner and co-workers [53] studied cross-polyurethane-
inter-cross-epoxy IPN foams. By an impedence tube
absorption method, they found that these IPN foams showed
superior sound absorption at all frequencies when compared
to pure polyurethane foams. Wong and Williams [23]
characterized the tan δ vs. temperature behavior of
polyurethane-inter-cross-epoxy 70/30 semi-II-simultaneous
interpenetrating polymer networks, noting one very broad
peak ranging from 273 to 373°K. Hourston and Zia [44]
prepared semi-simultaneous IPN's based on a polyurethane-
inter-cross-poly[ester-graft-(styrene-stat-methyl
acrylate)]. These compositions were found to be microheterogeneous, having a very broad glass transition
temperature range, for example, see Figure 10. Foster
et al. [38] made a cross-poly(n-butyl acrylate)-inter-
cross-poly(n-butyl methacrylate) 50/50 sequential IPN
filled by 6 vol-% graphite. They found that graphite-
filler increased the absolute value of tan δ and the area
under linear loss modulus-temperature plot.

Table III. Research Teams in the Area of Sound
and Vibration Damping with IPN's

Group	Institute	IPN's studied	Ref.
Sperling and Thomas	Lehigh University	Full, semi-, and filled IPN's based on acrylic polymers.	32-42
Hourston	University of Lancaster	1. polyurethane/ acrylic polymer based on full and semi-SIN's.	45-48
		2. poly(vinyl isobutyl ether)/poly(methyl acrylate)	43
		3. polyurethane/ modified polyester based semi-II-SIN's.	44
		4. latex IPN's based on acrylic polymers.	49-50
Fox	Naval Research Laboratory	Polyurethane/poly(n-butyl acrylate-stat-n-butyl methacrylate) based IPN's and SIN's.	52
Klempner and Frisch	University of Detroit	Polyurethane/ epoxy based SIN's.	53
Williams	University of Toronto	Polyurethane/epoxy based semi-II-SIN's.	23
Meyer	Louis Pasteur University	Polyurethane/ poly(methyl methacrylate) semi- and full SIN's.	54,55

*IPN's : sequential interpenetrating polymer networks,
SIN's : simultaneous interpenetrating polymer networks,
FULL : both components in IPN's were crosslinked,
Semi-I (or -II) : component 1 (or 2) in IPN's was cross-linked.

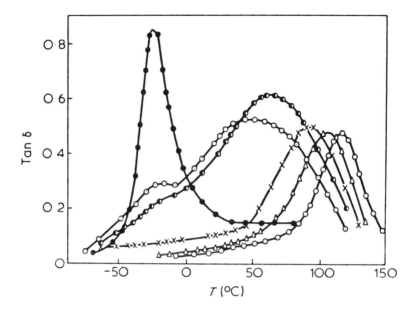

<u>Figure 10.</u> Tan δ vs. temperature curves for polyurethane
(●), modified polyester (O), and the semi-
SIN's containing 70 wt-% (⊖), 60 wt-% (◑),
50 wt-% (X), and 30 wt-% (△), polyurethane
[44].

LA and Group Contribution Analysis of Damping

The microheterogeneous morphology in IPN's was first
observed by₀Huelck et al. [5], see Figure 11. Domains of
about 100 A are observed. The domains in 11b are
substantially composed of interfacial material, pure phase
being absent. Each tiny area has a different composition,
its glass transition appearing at a different temperature.
The result is a broad damping peak spanning the range
between the two homopolymer transitions. By contrast, the
cellular structure in Figure 11a yields two peaks in the
loss modulus-temperature pattern.

It is well known that at a single temperature or
frequency, the energy dissipated in a complete cycle is
proportional to the loss modulus [82], i.e., loss modulus
measuring the conversion of mechanical energy into thermal
energy. Thus,

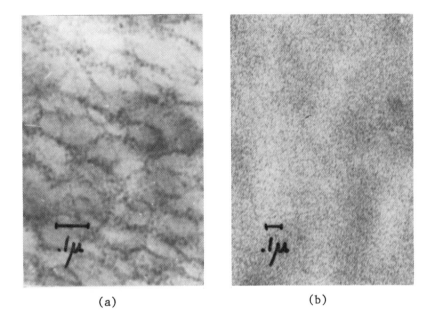

(a)   (b)

Figure 11. Morphology of sequential IPN's [5]. (a)
          Cross-poly(ethyl acrylate)-inter-cross-
          polystyrene, showing typical celluarl struc-
          ture and a fine structure within the cell
          walls. (b) Cross-poly(ethyl acrylate)-
          inter-cross-poly(methyl methacrylate), show-
          ing fine domains of 100 A level. PEA phase
          stained with OsO$_4$.

$$E_d = \pi \, E'' \, \gamma_0^2 \tag{7}$$

where E_d is the energy dissipated in a full cycle and γ_0
represents the amplitude of the experiment. Hence, the
quantity most suitable for the characterization of the
damping efficiency of a extensional damping material is the
loss modulus.

Using acrylic and methacrylic based IPN's, Fradkin
et al. [37] determined that the fundamental measure of
damping is the area under the linear loss modulus vs.
temperature, see Figure 12. The word linear is emphasized
because most workers traditionally use logarithmic loss
modulus vs. temperature plots. Although decrosslinking and
annealing treatments significantly modified the shape of
the loss modulus-temperature curves as a result of

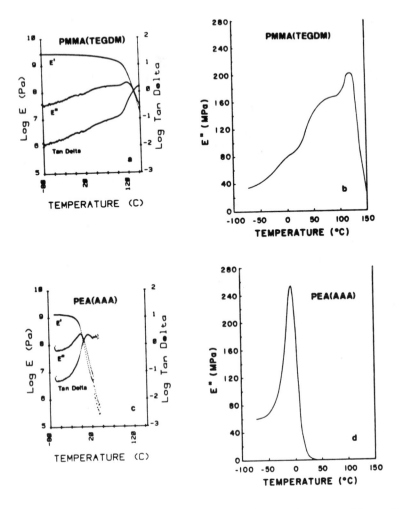

Figure 12. (a) and (c), Dynamic mechanical properties of
crosslinked poly(methyl methacrylate) and
poly(ethyl acrylate), respectively. (b) and
(d), corresponding linear E''-temperature
plots [37].

molecular and phase rearrangements, no effect on the area
under the curves was noted. Chang et al. [39-42]
introduced background corrections into the Rheovibron
curves by using a low damping aluminum plate, improving the
reproducibility of the dynamic mechanical experiments. For
example, Figure 13 [41,42] shows the linear loss modulus
vs. temperature plot for the cross-poly(methyl acrylate)-

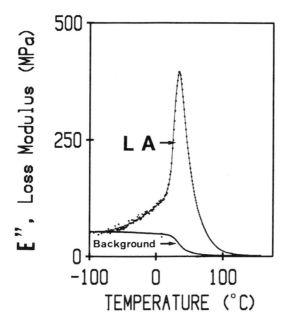

Figure 13. LA determination for a PMA/PEMA IPN, showing the background correction [42].

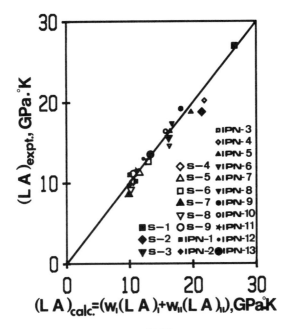

Figure 14. Additivity of LA's [42].

inter-cross-poly(ethyl methacrylate) 55/45 IPN, with the
background loss subtracted from the observed data. Thus,
the area LA (loss area) can be determined. Via the study
on various IPN's and statistical copolymers based on
acrylic, methacrylic, and styrenic polymers, Chang et al.
[41,42] found that area obeys a simple linear mixing rule
within experimental error,

$$LA = w_I * (LA)_I + w_{II} * (LA)_{II} \tag{8}$$

where w_I and w_{II} are the weight fractions of components I
and II in IPN's or statistical copolymers, respectively.
Calculated vs. experimental values are shown in Figure 14.

Chang et al. [39-42] found that the quantity LA is a
molecular characteristic, governed by the structure of the
polymer. This phenomenon relates to that which is already
known about infrared spectroscopy, where the area under an
absorption peak is quantitatively proportional to the
absorbing moiety present. The damping properties depend on
the nature of each moiety, independent of environments of
the mers.

In a similar vein, de Gennes [83] postulated the
reptation model of motion in chains. This model shows the
modes of conversion of mechanical vibration into chain
motion, i.e., heat.

On the basis of the above concept, Chang et al.
[41,42] utilized the group contribution method of Small
[84] and Hoy [85] to develop a quantitative method of
predicting LA. The following equation expresses the
formulation analytically:

$$LA = \sum_{i=1}^{n} \frac{(LA)_i * M_i}{M} = \sum_{i=1}^{n} \frac{G_i}{M} \tag{9}$$

where M_i is the molecular weight of the ith group in the
mer, M is the mer molecular weight, $(LA)_i$ is the loss area
contributed by the ith group, G_i is the molar loss constant
for the ith group, and n represents the number of moieties
in the mer. From equation [9] and LA's for various
polymers, the LA values and molar loss constants for about
twenty different moieties was obtained [41,42], shown in
Table IV. With Table IV one can select polymers with

Table IV. Group Contribution to LA

Group	Group Location	$(LA)_i$ Gpa·°K	G_i (Gpa·°K) (g/mole)
$\begin{array}{c} H \;\; H \\ \| \;\; \| \\ -C-C- \\ \| \;\; \| \\ H \end{array}$	1	3.4	91.8
$\begin{array}{c} O \\ \| \\ -C-O-(\;)-H \end{array}$	2	20.8	936
$\begin{array}{c} O \\ \| \\ -O-C-(\;)-H \end{array}$	2	20.1	905
$\begin{array}{c} O \\ \| \\ -C-OH \end{array}$	2	20.8	936
$-\langle\bigcirc\rangle$	2	11.9	916
$-CH_3$	2	11.0	165
$-OCH_3$	2	21.7	674
$-C\equiv N$	2	23.2	603
$-O-H$	2	4.7	80
$-Cl$	2	9.2	327
$\langle\;\rangle$	3	3.5	287
$\langle\bigcirc\rangle$	3	2.2	166
$\begin{array}{c} CH_3 \\ \| \\ -CH_2-CH-CH_3 \end{array}$	3	-1.7	-98.2
$\begin{array}{c} \| \\ -CH- \end{array}$	3,4	0.5	7.0
$-CH_2-$	3	-3.0	-42.0
$-C\equiv N$	3	14.5	377
$-Cl$	3	15.7	556

(1) Backbone. (2) Side group attached to backbone directly. (3) Side group not attached to backbone. (4) Value derived from isobutyl side group.

desired capability; thus optimum damping materials can be
obtained more systematically. For sound and vibration
damping, IPN's with a high but controlled LA are desirable.
LA values should prove useful in quantitative mechanical
analyses.

Chang et al. [42] noted that the value of the storage
modulus can affect the LA value. Hence, modulus of the
dynamic mechanical measurement should be calibrated in
order to obtain LA values comparable to Table IV. The LA's
shown in Table IV were collected using a storage modulus of
a polystyrene equal to $2.85 * 10^{10}$ dyne/cm^2 (25°C, 110 Hz)
[42].

LA values were calculated from various literature
papers, to obtain a quantitative comparison. The
background loss which the original workers [35,49,54,86-89]
did not apply was neglected, but loss curves of homopolymers
were made to be symmetrically centered around the glass
transition. LA's of polymethacrylate polymers including
β-relaxation were from previous work [40]. After the
calculations, it can be concluded that most of the
predictions are satisfied within ± 15% error noting that
the modulus should be calibrated before calculating LA's,
see Table V. In this table, the correction factor is the
factor for multipling the storage modulus in order to
obtain LA values corresponding to those in Table IV.

Conclusions

In at least some of the IPN systems described above,
evidence indicates that spinodal decomposition is
important. While no such evidence is yet available for the
microheterogeneous morphologies of interest to damping,
certain features suggest it might be important:

(1) As with spinodal decomposition, both phases seem
to be continuous.

(2) Available electron micrographs show structures
compatible with the notion of cylindrical structures.

If one considers a three-dimensional temperature-
composition phase diagram, (Fig. 5 with a temperature axis)
it may be that microheterogeneous morphologies form just
inside the spinodal. The area under the E"-temperature is

Table V.   Correction Factor for LA's

Group	Ref.	Polymers	Glassy Storage Modulus, Gpa	Corr. factor
Foster, et al.	38	PnBA, PnBMA, and IPN's based on acryl-ic polymers	3.05 (PnBMA, glassy state)	1.0
Widmaier, et al.	87	PS, PnBA, and IPN's based on PS/PnBA	2.75 (PS, room temperature)	ca. 1.0
Hourston, et al.	49	IPN's based on acrylic polymers	For example, 3.0, (cross-PiBa-inter-cross-PEMA 50/50 IPN, glassy state)	ca. 1.0
Oberst	86	Poly(vinyl acetate)	2.5 (PVA, glassy state)	ca. 1.0
Hermant, et al.	54	PMMA	3.3 (PMMA, room temperature)	1.1
Lorenz, et al.	35	IPN's based on acrylic polymers	For example, 1.6 (cross-poly(EMA-stat-EA)-inter-cross-poly(nBA-stat-EA) (21-stat-9)/(49-stat-21-IPN, glassy state)	1.8
Gillham, et al.	88	PMMA	4.2 (PMMA, room temperature)	0.9
Weiss, et al.	89	PS	1.0 (PS, room temperature)	2.85

constant, governed by the contribution of the individual moieties making up the material. A group contribution analysis can be used for fundamental or predictive purposes for homopolymers, statistical copolymers, or IPN's.

Acknowledgments

The authors wish to thank the National Science Foundation, Grant No. DMR-8405053, Polymers program and Office of Naval Research, Contract No. N00014-84-K-0508 for financial support.

References

[1] (a) Sperling, L.H. "Interpenetrating Polymer Networks"; Plenum: New York, 1981. (b) Sperling, L.H. In "Multicomponent Polymer Materials"; Paul, D.R.; Sperling, L.H., Eds.; American Chemical Society: Washington, D.C., 1986.

[2] Klempner, D.; Frisch, K.C. "Polymer Alloys III"; Plenum: New York, 1983.

[3] An, J.H.; Fernandez, A.M.; Wignall, G.D.; Sperling, L.H. Polym. Mat. Sci. Eng. 1987, 56, 541.

[4] Curtius, A.J.; Covitch, M.J.; Thomas, D.A.; Sperling, L.H. Polym. Eng. Sci. 1972, 12(2), 101.

[5] Huelck, V.; Thomas, D.A.; Sperling, L.H. Macromolecules 1972, 5(4), 340.

[6] Huelck, V.; Thomas, D.A.; Sperling, L.H. Macromolecules 1972, 5(4), 348.

[7] Klempner, D; Frisch, H.L.; Frisch, K.C. J. Polym. Sci., Part A-2 1970, 8, 921.

[8] Donatelli, A.A.; Sperling, L.H.; Thomas, D.A. Macromolecules 1976, 9(4), 671.

[9] Donatelli, A.A.; Sperling, L.H.; Thomas, D.A. Macromolecules 1976, 9(4), 676.

[10] Yeo, J.K.; Sperling, L.H.; Thomas, D.A. Polym. Eng. Sci. 1982, 22(3), 190.

[11] Yeo, J.K.; Sperling, L.H.; Thomas, D.A. Polymer 1983, 24, 307.

[12] Donatelli, A.A.; Sperling, L.H.; Thomas, D.A. J. Appl. Polym. Sci. 1977, 21, 1189.

[13] Michel, J.; Hargest, S.C.; Sperling, L.H. J. Appl. Polym. Sci. 1981, 26, 743.

[14] Shilov, V.V.; Lipatov, Yu.S.; Karbanova, L.V.; Sergeeva, C.M. J. Polym. Sci. Polym. Chem. Ed. 1979, 17, 3083.

[15] Lipatov, Yu.S.; Shilov, V.V.; Bogdanovitch, V.A.; Karbanova, L.V.; Sergeeva, L.M. Polym. Sci. USSR 1980, 1492.

[16] Lipatov, Yu.S.; Shilov, V.V.; Bogdanovtich, V.A.; Karbanova, L.V.; Sergeeva, L.M. J. Polym. Sci. Polym. Phys. Ed. 1987, 25, 43.

[17] Fernandez, A.M.; Wignall, G.D.; Sperling, L.H. In "Multicomponent Polymer Materials"; Paul, D.R.; Sperling, L.H., Eds.; Adv. Chem. Ser. No. 211, 1986.

[18] An, J.H.; Fernandez, A.M.; Sperling, L.H.; Macromolecules 1987, 20, 191.

[19] McGarey, B.; Richards, R.W.; Polymer 1986, 27, 1315.

[20] Alkonis, J.J.; MacKnight, W.J. "Introduction to Polymer Viscoelasticity"; 2nd Ed.; Wiley-Interscience: New York, 1983.

[21] Chen, A.C.; Williams, H.L. J. Appl. Polym. Sci. 1976, 20, 3387.

[22] Chen, A.C.; Williams, H.L. J. Appl. Polym. Sci. 1976, 20, 3403.

[23] Wong, D.T.H.; Williams, H.L. J. Appl. Polym. Sci. 1983, 28, 2187.

[24] Hourston, D.J.; Hughes, I.D. J. Appl. Polym. Sci. 1977, 21, 3099.

[25] Hourston, D.J.; Hughes, I.D. J. Appl. Polym. Sci. 1981, 26, 3487.

[26] Bandyopadhyay, P.K.; Shaw, M.T. J. Vinyl Technol. 1982, 4(4), 142.

[27] Woo, E.M.; Barlow, J.W.; Paul, D.R. Polymer 1985, 26, 763.

[28] Woo, E.M.; Barlow, J.W.; Paul, D.R. J. Appl. Polym. Sci. 1986, 32, 3889.

[29] Woo, E.M.; Barlow, J.W.; Paul, D.R. J. Appl. Polym. Sci. 1985, 30, 4243.

[30] Keskkula, H.; Paul, D.R. J. Appl. Polym. Sci 1986, 31, 1189.

[31] Margaritis, A.G.; Kalfoglou, N.K. Polymer 1987, 28, 497.

[32] Sperling, L.H.; Chiu, T.W.; Gramlich, R.G.; Thomas, D.A. J. Paint Technol. 1974, 46, 47.

[33] Grates, J.A.; Thomas, D.A.; Hickey, E.C.; Sperling, L.H. J. Appl. Polym. Sci. 1975, 19, 1731.

[34] Sperling, L.H.; Thomas, D.A. U.S. Patent 3,833,404, 1974.

[35] Lorenz, J.E.; Thomas, D.A.; Sperling, L.H. In "Emulsion Polymerization"; Piirma, I.; Gardon, J.L., Eds.; ACS Symp. Series No. 24, American Chemical Society: Washington, D.C., 1976, Chapter 20.

[36] Fradkin, D.G.; Foster, J.N.; Sperling, L.H.; Thomas, D.A. Polym. Eng. Sci. 1986, 26, 730.

[37] Fradkin, D.G.; Foster, J.N.; Sperling, L.H.; Thomas, D.A. Rubber Chem. Technol. 1986, 59, 255.

[38] Foster, J.N.; Sperling, L.H.; Thomas, D.A. J. Appl. Polym. Sci. 1987, 33, 2637.

[39] Chang, M.C.O.; Thomas, D.A.; Sperling, L.H. Polym. Mat. Sci. Eng. 1986, 55, 350.

[40] Chang, M.C.O.; Thomas, D.A.; Sperling, L.H. J. Appl. Polym. Sci. 1987, 34, 409.

[41] Chang, M.C.O.; Thomas, D.A.; Sperling, L.H. submitted, Polymer Preprints, New Orleans, 1987.

[42] Chang, M.C.O.; Thomas, D.A.; Sperling, L.H. J. Polym. Sci. Polym. Phys. Ed., submitted.

[43] Hourston, D.J.; McCluskey, J.A. Polymer 1979, 20, 1573.

[44] Hourston, D.J.; Zia, Y. Polymer 1979, 20, 1497.

[45] Hourston, D.J.; Zia, Y. J. Appl. Polym. Sci. 1983, 28, 2139, 3475, 3849; 1984, 29, 629, 2951, 2963.

[46] Hourston, D.J.; McCluskey, J.A. J. Appl. Polym. Sci 1986, 31, 645.

[47] Hourston, D.J.; Huson, M.G.; McCluskey, J.A. J. Appl. Polym. Sci. 1986, 31, 709.

[48] Hourston, D.J.; Huson, M.G.; McCluskey, J.A. J. Appl. Polym. Sci. 1986, 32, 3881.

[49] Hourston, D.J.; Satgurunathan, R. J. Appl. Polym. Sci. 1984, 29, 2969.

[50] Hourston, D.J.; Satgurunathan, R. J. Appl. Polym. Sci 1986, 31, 1955.

[51] Hourston, D.J.; Saturunathan, R.; Varma, H. J. Appl. Polym. Sci. 1987, 33, 215.

[52] Fox, R. B.; Binter, J.L.; Hinkle, J.A.; Carter W. Polym. Eng. Sci. 1985, 25, 157.

[53] Klempner, D.; Wang, C.L.; Ashtiani, M.; Frisch, K.C. J. Appl. Polym. Sci 1986, 32, 4197.

[54] Hermant, I.; Damyanidu, M.; Meyer, G.C. Polymer 1983, 24, 1419.

[55] Morin, A.; Djomo, H.; Meyer, G.C. Polym. Eng. Sci. 1983, 23, 394.

[56] Wignall, G.D.; Child, H.R.; Samuels, R.J. Polymer 1982, 23, 957.

[57] Caulfield, D.; Yao, Y.F.; Ullman, R. "X-Ray and Electron Methods of Analysis"; Plenum: New York, 1968.

[58] Higgins, J.S. J. Appl. Cryst. 1978, 11, 346.

[59] Sperling, L.H. Polym. Eng. Sci. 1984, 24(1), 1.

[60] Debye, P.; Bueche, A.M. J. Appl. Phys. 1949, 20, 518.

[61] Debye, P; Anderson, H.R.; Brumberger, R.J. Appl. Phys. 1957, 28, 649.

[62] Kriste, R.; Porod, G. Kolloid-Z 1962, 184, 1.

[63] Miffelbach, P.; Porod, G. Kolloid-Z 1965, 202, 40.

[64] Adachi, H.; Kotaka, T. Polymer J. 1982, 14(5), 3791.

[65] Fernandez, A.M. Ph.D. thesis, Lehigh University, 1984.

[66] An, J.H.; Fernandez, A.M.; Wignall, G.D.; Sperling, L.H. Polymer, submitted.

[67] Binder, K.; Frisch, H.L. J. Chem. Phys. 1984, 81(4), 2126.

[68] Lipatov, Yu.S.; Shilov, V.V.; Gomza, J.P.; Kovernik, G.P.; Grigor'eva, O.P.; Sergeyeva, L.M. Makromol. Chem. 1984, 185, 347.

[69] Lipatov, Yu.S.; Grigor'eva, O.P.; Kovernik, G.P.; Shilov, V.V.; Sergeyeva, L.M. Makromol. Chem. 1985, 186, 1401.

[70] Graillard, P.; Ossenbach-Sauter, M.; Riess, G. In "Polymer Compatibility and Incompatibility"; Solc, K., Eds.; MMI Press: New York, 1981.

[71] Ludwico, W.A.; Rosen, S.L. Polym Sci. Technol. 1977, 10, 401.

[72] Parent, R.D.; Thompson, E.V. In "Multiphase Polymers"; Cooper, S.L.; Ester, G.M., Eds.; Adv. in Chem. Series No. 176, American Chemical Society: Washington, D.C., 1979, Chapter 20.

[73] Walsh, D.J.; Sham, C.K. Polymer 1984, 25, 1023.

[74] Purvis, M.T.; Yanai, H.S. In "Polymer Compatibility and Incompatibility"; Solc, K. Eds.; MMI Press: Harwood, Switzerland, 1982, p. 261.

[75] Volmer, M.; Weber, A. Z. Phys. Chem. 1925, 119, 277.

[76] Cahn, J.W. J. Chem. Phys. 1965, 42(1), 93.

[77] Nishi, T.; Wang, T.T.; Kwei, T.K. Macromolecules 1975, 8, 227.

[78] An, J.H.; Traubert, C. S.; Sperling, L.H., to be published.

[79] Inoue, T.; Ougizawa, T.; Yasuda, O.; Miyasaka, K. Macromolecules 1985, 18, 57.

[80] van Emmerik, P.T.; Smolder, C.A. J. Polym. Sci., Part C 1972, 38, 73.

[81] van Emmerik, P.T.; Smolder, C.A.; Greymaier, W. Europ. Polym. J. 1973, 8, 309.

[82] Tschoegel, N.W. "The Theory of Linear Viscoelastic Behavior"; Academic Press: New York, 1980.

[83] deGennes, P.G. J. Chem. Phys. 1972, 55, 572.

[84] Small, P.A.   J. Appl. Chem. 1953, 3, 71.

[85] Hoy, K.L.   J. Paint Technol. 1970, 46, 76.

[86] Oberst, H.   Koll.-Z., Z. Polymere 1967, 216-217, 64.

[87] Widmaier, J.M.; Sperling, L.H.   J. Appl. Polym. Sci. 1982, 27, 3513.

[88] Gillham, J.K.; Staduicki, S.J.; Hazony, Y.   J. Appl. Polym. Sci. 1977, 21, 401.

[89] Weiss, R.A.; Lefelar, J.A.   Polymer 1986, 27, 3.

WATER-SWELLABLE POLYMERIC NETWORKS:KINETICS OF SWELLING

S.J. Candau, A. Peters, F. Schosseler[*]

Laboratoire de Spectrométrie et d'Imagerie Ultrasonores
Unité Associée au C.N.R.S.
Université Louis Pasteur
4, rue Blaise Pascal
67070 Strasbourg Cedex
France

[*] Institut Charles Sadron (CRM-EAHP)
6, rue Boussingault
67083 Strasbourg Cedex
France

Abstract

Networks composed of covalently linked polymer chains are of great interest, not only for their unique physical and chemical behaviour, but also for their possible technological application, as switches, actuaters for robots, artificial muscles, water reservoirs. For all these applications, the understanding of the equilibrium, and dynamic properties of water-swellable networks is of fundamental importance. In this paper, the current theories for the dynamic fluctuations and the kinetics of swelling of the networks are reviewed. Recent results obtained by dynamic light scattering and kinetics of swelling on polyacrylamide and polyacrylic acid gels are presented. The cooperative diffusion of the gels is investigated as a function of the ionization degree, the crosslinks density of the network and the salt content of the diluent. Also the effect of polydispersity of gel particles on the kinetics of swelling is discussed.

Introduction

Water swellable polymeric networks are employed in an increasingly large number of applications. In chemistry and

biochemistry, they are widely used in the analytic methods of
chromatography and electrophoresis that allow a separation of
molecules according to the speed with which they percolate
through the pores of the gel. During the last five years, ionic hydrogels have received considerable interest from both
academic and applied research laboratories. These materials
possess a very high retention capacity (up to 1000 times the
volume of the dry polymer in pure water) which makes them useful for new industrial applications as water reservoirs [1].
For these applications, the samples are supplied in the form
of powders of spherical beads of small size ($\sim$ 100 μm). Ionic
gels are also known to undergo volume-phase transitions when
external conditions such as temperature, solvent composition
or osmotic pressure, change [2-8]. These phenomena can be used
to develop new technological applications of gels as switches,
actuaters, memories or display units.

The understanding of the kinetics of volume change of gels
is of primary importance for all applications considered. The
kinetics of swelling of spherical neutral gels have been studied both theoretically and experimentally by Tanaka and Fillmore who showed that the macroscopic swelling behavior was
described by a diffusion equation with a diffusion coefficient
D given by [9] :

$$D = M_{os}/f = (K_{os} + 4\mu/3)/f \qquad (1)$$

where M_{os}, K_{os} and μ are the osmotic longitudinal modulus, the
osmotic compressional modulus and the shear modulus of the gel
respectively and f is the friction coefficient describing the
viscous interaction between the polymer and the solvent. When
a partly swollen gel is immersed in an excess of its own solvent, it swells to reach a new equilibrium volume. For spherical gels, the evolution of the radius during the swelling process is described by an infinite sum of exponential decays
with characteristic times proportional to the square of the
final equilibrium radius a. Modes with the smallest relaxation
times vanish rapidly and the final step of the swelling process is described by a single exponential relaxation. The following expression of the characteristic time τ has been derived by Tanaka and Fillmore in the limit case where the shear
modulus μ is negligible compared to the compressibility modulus K :

$$\tau = a^2/\pi^2 D \qquad (2)$$

If the above condition is not fulfilled, the relaxation
time is still proportional to a^2/D but with a different proportionality constant [10].

The diffusion equation describing the swelling process also applies to the thermal fluctuations of polymer network observed by dynamic laser light scattering spectroscopy. In
this case, the relaxation time of the local concentration
fluctuations with wavevector K is given by [11] :

$$T_r = (DK^2)^{-1} \qquad (3)$$

Equations (2) and (3) indicate that the relaxation time
associated with a volume change of a gel is determined by a
characteristic length such as the gel radius or the wavelength
of fluctuations and the diffusion coefficient. The square dependence on the size has been experimentally verified both in
macroscopic kinetics and microscopic dynamics. Values of D
obtained with the two methods were found to be in good agreement [10] [11]. As a matter of fact, the two methods of determination of D, kinetics of swelling and laser light scattering
spectroscopy, are complementary. For highly swollen samples,
the scattered signal is very weak so that light scattering
experiments are difficult to perform and kinetics of swelling
of macroscopic sample are more adapted. The same is true for
small spherical beads of gels. On the other hand, large, moderately swollen samples can be conveniently studied by quasi-
elastic light scattering.

For neutral gels, the coefficient D depends both on the
microscopic structure of the network and on the polymer-sol-
vent interactions. It was shown that D follows simple scaling
laws as a function of the equilibrium swelling ratio of the
gel [12-14].

For ionized gels, the swelling process is more complex because of electrostatic interactions [15]. Ricka and Tanaka
have shown that the swelling of weakly charged ionized gels
could be interpreted quantitatively from the Donnan theory
in which the ionic forces depend only on the ionic composition of the solvent and on the concentration of fixed ionizable groups in the gels [8]. This amounts to add in the osmotic compressibility K_{os} a supplementary contribution due to
the swelling pressure of free ions in the gels. Quasi-elastic
light scattering experiments performed on hydrolysed polyacrylamide gels confirmed the validity of this assumption
[16]. The case of strongly ionized gels is far more difficult
to handle theoretically as the effect of interactions between
the ions within the gel cannot be neglected. However, preliminary experiments suggested that the swelling kinetics could
still be described by an equation of diffusion as in the case
of neutral gels [1].

The purpose of the present study is twofold :
- First to extend the theory of the kinetics of swelling
to cylindrical gel samples that are often more practical for
experiments.
- Second to investigate the kinetics of swelling of powders of small spherical beads, commercially used as water reservoirs, as a function of the crosslinking degree, the ionization degree and the salt content of the diluent.

Also, some general, model-independent, polydispersity effects relevant for the swelling of an assembly of spheres are
discussed. These effects are of great importance for potential
applications since they control the overall kinetics of swelling.

Theory

Kinetics of swelling. The equation of motion of a gel network as given by Tanaka, Hocker and Benedek is [11] :

$$\frac{\partial \underset{\sim}{u}}{\partial t} = \frac{M_{os}}{f} \, grad \, (div \, \underset{\sim}{u}) + \frac{\mu}{f} \, \Delta \, \underset{\sim}{u} \qquad (4)$$

where Δ denotes the Laplacian and $\underset{\sim}{u}(\underset{\sim}{r},t)$ is the displacement
vector that represents the displacement of a point in the net-
work from its final equilibrium location after the gel is fully swollen. Under this definition $u = 0$ at $t = \infty$.

We consider a cylindrical gel of infinite length. To derive
the characteristic swelling time we apply the theoretical
treatment that has been already used for spherical gels [3]
[10]. The equation of motion for the radial deformation $u(\rho,t)$
is obtained from (1) and is given by :

$$\frac{\partial u}{\partial t} = D \, (1 + \frac{\mu}{M_{os}})(- \frac{1}{\rho^2} \, u + \frac{1}{\rho} \frac{\partial u}{\partial \rho} + \frac{\partial^2 u}{\partial \rho^2}) \qquad (5)$$

As the axis of the cylinder is not displaced during the
swelling process, one has the condition

$$u(o,t) = 0 \quad at \, all \, t \qquad (6)$$

The boundary condition on the surface of the gel is obtai-
ned by cancelling the stress $\sigma_{\rho\rho}$ normal to the lateral surface for $\rho = b$

$$\sigma_{\rho\rho} = M_{os} \left[\frac{\partial u}{\partial \rho} + \frac{1}{\rho} (1 - \frac{2\mu}{M_{os}}) u \right] = 0 \quad (t \to \infty, \rho = b) \tag{7}$$

where b is the final radius of the gel cylinder in equilibrium with the surrounding fluid. This condition is assumed to hold at finite time since we are primarily interested in the later stage of the swelling process, that is the range of large t.

The initial condition for the displacement vector is :

$$\frac{u(\rho,o)}{u(b,o)} = \frac{\rho}{b} \tag{8}$$

The spatial and temporal variation of the displacement vector is obtained by solving Eq.(5), using the conditions (6) and (7). This yields :

$$u(\rho,t) = \frac{1}{\sum 1/\lambda_n'^2} \sum e^{-t/\tau_n'} \frac{J_1(\lambda_n'\rho/b)}{\lambda_n'^2 J_1(\lambda_n')} \tag{9}$$

where

$$\lambda_n' = b \left[\tau_n'D(1 + \mu/M_{os}) \right]^{-1/2}$$

is the nth root of the equation :

$$\frac{J_0(\lambda')}{J_1(\lambda')} = \frac{2\mu}{M_{os}} \frac{1}{\lambda'} \tag{10}$$

$J_0(\lambda')$ and $J_1(\lambda')$ are the Bessel functions of order 0 and 1 respectively.

We consider the first term of the expansion (9) which becomes dominant over the higher order terms at large t, that is, at the last stage of swelling. This term is given by :

$$- \ln \frac{u(b,t)}{u(b,o)} = \frac{t}{\tau_1'} - \ln \frac{\lambda_1'^{-2}}{\lambda_1'^{-2} + 0.077} \tag{11}$$

The above relation predicts a linear dependence of $-\ln[u(b,t)/u(b,o)]$ with time. Both the time constant τ_1' and the intercept depend on the ratio μ/M_{os} through the parameters $\lambda_1', \lambda_2' \ldots \lambda_n'$. The analysis of the roots of Eq.(10) shows that the longest relaxation time τ_1' is nearly independent on μ/M_{os}

contrary to the case of the spherical samples [10] and can be approximated by

$$\tau_1 = \frac{b^2}{\lambda_1'^2 D} \simeq \frac{b^2}{5.826\ D} \tag{12}$$

The expression (11) is similar to that previously derived for spherical samples of radius a [10] :

$$-\ln \frac{u(a,t)}{u(a,o)} \simeq \frac{t}{(\tau_1)_s} - B(X_n) \tag{13}$$

where $B(X_n)$ is a function of X_1, X_2 ... X_n, roots of the equation :

$$\tan X = \frac{(4\mu/M_{os})X}{(4\mu/M_{os})-X^2} \tag{14}$$

and

$$(\tau_1)_s = \frac{a^2}{DX_1^2} \tag{15}$$

In the limit $\mu/M_{os} = 0$,

$$(\tau_1)_s = \frac{a^2}{\pi^2 D} \tag{16}$$

In Fig.1 are reported the variations of $-\ln[u(b,t)/u(b,o)]$ calculated from Eq.(11) for the two extreme values of the ratio μ/M_{os} (0 and 0.5). The shift of the two corresponding curves is rather small and the experimental accuracy doesn't allow to estimate μ/M_{os} from the extrapolate.

Quasi-elastic light scattering. The diffusion coefficient of a gel can also be determined from the characteristic time of concentration fluctuations of the polymer network [11]. These concentration fluctuations are experimentally observed using quasi-elastic light scattering. The autocorrelation function of the intensity of light scattered from a gel is given by :

$$\frac{\langle \Delta I(t+t')\ \Delta I(t')\rangle}{\langle \Delta I^2(t')\rangle} = \exp(-2DK^2 t) \tag{17}$$

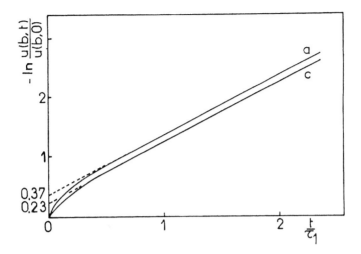

Figure 1. Variations of $-\ln[u(b,t)/u(b,o)]$ with the scaled time t/τ_1 calculated from Eq.(11) : (a) $\mu/M_{os} = 0$; (c) μ/M_{os} = 0.5.

where $\Delta I(t)$ is the fluctuating part of the scattered light intensity at time t, < > denotes the time average over t'. The magnitude of the scattering vector K is given by :

$$K = 4\pi n \sin(\theta/2)/\lambda \qquad (18)$$

where θ is the scattering angle, λ is the wavelength of the incident light in a vacuum and n is the index of refraction of the scattering medium.

Experimental Section

Materials. The polyacrylamide gels investigated here were prepared by a standard redox reaction employing ammonium per-sulfate and tetramethyl-ethylene-diamine (TEMED). The concen-trations used were 5 g of recrystallized acrylamide monomer, 0,133 g of N,N'-methylene bisacrylamide, 40 mg of ammonium persulfate, and 400 µl of TEMED dissolved in water to make a total volume of 100 µl. After thorough mixing, the prepara-tions are poured through a small aperture in spherical or cy-lindrical glass moulds. Once the gelation performed, the moulds are broken and the gel samples are transferred into a cell containing an excess of water. This time is taken to be zero $(t = 0)$ for the kinetics experiments. The concentration of polymer in the gel is 0.051 g cm^{-3}. At the swelling equilibrium it is : 0.033 ± 0.004 g cm^{-3}.

Ionized gels are acrylic acid-sodium acrylate copolymers. The samples were provided by Norsolor Company. They are obtained through an inverse suspension process. In this technique, the aqueous phase, containing an hydrophilic monomer, is dispersed in an organic phase, such as an alicyclic or aliphatic hydrocarbon.

The first step of the synthesis is the neutralization of acrylic acid by an aqueous sodium hydroxide solution. The degree of neutralization determines the ionization degree. A crosslinking agent and an initiator are then dissolved in the aqueous phase.

The inverse suspension is obtained by mixing the organic solution of a non-ionic surfactant with the aqueous solution of partially neutralized acrylic acid salt. The temperature is then raised and polymerization reaction is carried out for several hours. The samples are then dried and are ready for characterization. Figure 2 shows an electronic microscopy photograph of the spheres which are obtained in the final step.

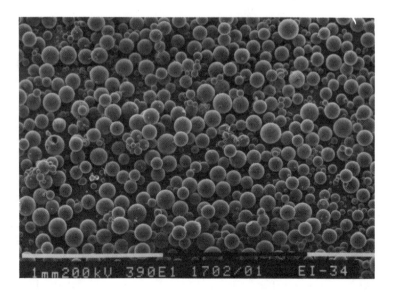

Figure 2. Electron microscopy photograph of spherical beads.

Quasi-elastic light scattering experiments. In order to obtain an independent determination of the diffusion coefficient D we have measured the average decay rate $\bar{\Gamma} = 2DK^2$ of the autocorrelation function of light scattered from cylindrical polyacrylamide gels with the same composition as in the swelling experiments. The dynamic light scattering spectrophotometer operated in a photon-counting mode using a Spectra Physics argon-ion laser ($\lambda = 488$ nm) and a Malvern correlator.

Measurements of swelling of a single sphere of ionized gel. The solvent used to swell the polymer spheres is freshly distilled water with pH around 6.5 at room temperature. Solutions containing salt are prepared with deshydrated sodium chloride.

The experimental procedure is the following. One gel particle is placed on the stage of a profile projector (magnification 50 to 200 x). Displacements of the stage are controlled through micrometric screws. This allows a precise measurement ($\pm$ 1μm) of the initial diameter   of the spheres. Deviations from spherical shapes have been found to be typically of less than a few per cent. We then deposit a drop of solvent on the sphere and follow the gel swelling. In fact, swelling is very fast, a typical duration time to reach the maximum swelling being about a few minutes. Photographs of the screen of the profile projector are taken at regular time intervals. Measurements of the diameter of the spheres are performed on the negatives.

A typical swelling process can be described as follow. During the first step, solvent penetrates the sphere from the periphery. There is a gradient of polymer concentration in the sphere which looks inhomogeneous and non-spherical. In the second step, solvent has reached the sphere center, polymer concentration is homogeneous, the contrast between pure solvent and gel is constant and the gel is spherical : rearrangement of the network becomes the essential swelling process. We take this moment as the time origin because theory is made for originally homogeneous gels. However data analysis does not depend on this time origin. Swelling becomes slower and slower until the sphere reaches its maximum swelling radius. We observe then a plateau where the swelling is constant as a function of time. The swelling curve thus obtained is then fitted to a single exponential providing a measurement of $(\tau_1)_s$.

Results and Discussion

Swelling of cylindrical polyacrylamide gels. Measurements
of τ_1 for a series of cylindrical gels with different values
of the ratio h/b were made according to the procedure described above. Using the experimental values of τ_1 it is then possible to draw a universal plot of $-\ln[u(b,t)/u(b,o)]$ as a function of the scaled time t/τ_1 for different cylinders. For

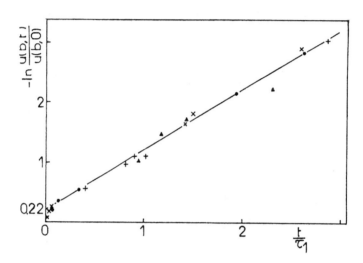

Figure 3. Universal plot of $-\ln[u(b,t)/u(b,o)]$ as a function
of the scaled time t/τ_1 for four cylindrical gels with final
length/radius ratioes h/b = 4.08 (▲) , h/b = 5.6 (●), h/b =
7.4 (+), h/b = 13.6 (x).

$\tau/t_1 > 0.25$, the data fit a straight line with an extrapolate
0.22. The latter value corresponds closely to the condition
$\mu/M_{os} = 0.5$. Indeed a value of 0.4 was previously obtained
from combined measurements of shear modulus and intensity of
scattered light [14]. The difference between the two values
is largely accounted for by the experimental accuracy. From
the experimental values of b and τ_1, it is possible, using
Eq.(12) to calculate the diffusion coefficient D. As a matter
of fact Eq.(12) is valid only in the limit h/b ≫ 1 . In Fig.4
is plotted the variation of D as determined from Eq.(12), as
a function of h/b. It can be seen that for h/b > 3, the values
of D agree within experimental accuracy with the dynamic light-
scattering data and with the results of the kinetics of swelling of spherical samples.

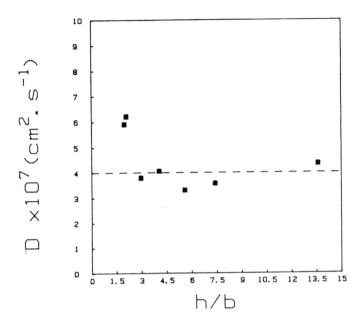

$D \times 10^7 (cm^2 . s^{-1})$

h/b

<u>Figure 4</u>. Variation of the apparent diffusion coefficient de-
termined from the experimental value of τ_1 and using Eq.(12),
versus h/b for gel cylinders. The dashed horizontal line corresponds to the value obtained by Quasi-Elastic-Light-Scatte-
ring. The value obtained from kinetics of swelling experiments
in spherical gels is $D = 4.1 \pm 0.3 \times 10^{-7}$ cm^2 s^{-1}.

Kinetics of swelling of small beads of ionized gels

a) <u>Qualitative picture</u>. We now turn to swelling experiments
performed on strongly ionized gels. In this case we don't know
the detailed form of the free energy. However, we can have a
primitive guess about the parameters which are important in
the swelling process.

As for neutral gels the restoring force is the elasticity
of the network chains. We can expect that variations of the
crosslinking degree will give the same general trend both for
neutral and ionized gels, that is increasing the crosslinking
degree will decrease the swelling ratio and increase the diffusion coefficient.

The main qualitative effects of the introduction of electrostatic interactions are also easily understood. Increasing
the ionization degree will increase both the swelling ratio
and the diffusion coefficient. The gel expands to minimize

the repulsive interactions between the charges fixed on the chains. Similarly the amplitude of the thermal concentration fluctuations decreases and these fluctuations are damped very rapidly. The chains adopt a local rod-like conformation on distances which are shorter than the Debye-Hückel screening length.

In the presence of a large amount of added salt, the Debye-Hückel length decreases to the value of the mean distance between ionic groups, so that the ionic gel behaves like a neutral one.

In samples with high degree of neutralization we expect Manning's condensation to reduce the number of effectively ionized groups, even when added salt is present.

b) <u>Single gel bead behavior</u>. The results obtained through the swelling kinetics of single gel beads are in agreement with the preceding qualitative picture.

We consider first the influence of crosslinking degree. Figure 5 shows, for different salt concentrations, the variation of the diffusion coefficient D versus the swelling ratio Q, for gels with the same neutralization degree (80 %) but

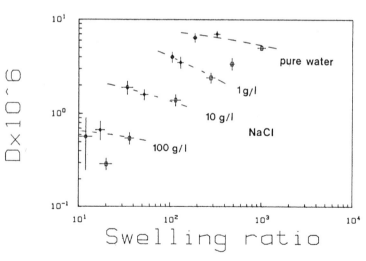

<u>Figure 5</u>. Variation of diffusion coefficient with swelling ratio at different salt concentrations and different crosslinking degrees : □) low, C_{NaCl} (g/l) = 0, 0.317, 1, 10, 100, 300 . ◆) intermediate, C_{NaCl} (g/l) = 0, 1, 10, 100. o) high, C_{NaCl} (g/l) = 0, 1, 10, 100. The neutralization degree is 80% for all samples.

with different crosslinking degrees. Data points move to the
upper left of the Q-D plane as the crosslinking degree is increased. It is to be noted that, for a given salt concentration, the crosslinking degree has an effect mainly on the
swelling ratio while the diffusion coefficient remains nearly
unaffected. This suggests that for high neutralization degrees,
the diffusion coefficient is dominated by electrostatic effects, i.e. by local rigidity of the chains.

On the other hand, the variation of the salt concentration
affects drastically both Q and D. For high salt concentration
the measured values of D and Q are of the same order of magnitude than those measured on neutral gels with a similar crosslinking degree.

We consider now the influence of neutralization degree.
On Figures 6 and 7, we have plotted the variation of respec-

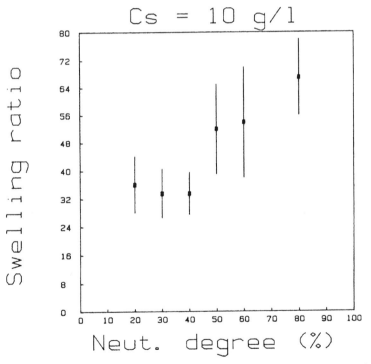

Figure 6. Variation of the swelling ratio of gels with different neutralization degrees and same crosslinking degree.The
salt concentration is kept constant to C_{NaCl} = 10 g/l.

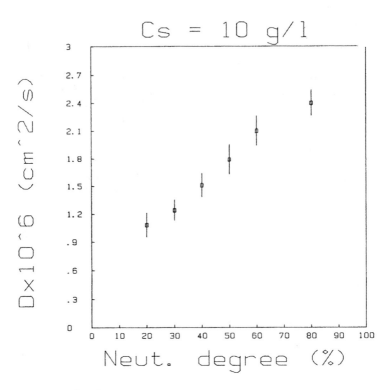

Figure 7. Variation of the diffusion coefficient for the samples of Fig.6 in the same conditions.

tively the swelling ratio and the diffusion coefficient with the neutralization degree for gels with the same crosslinking degree. The salt concentration has been kept constant to 10g/l, a typical concentration of interest in usual applications.

We observe that, for this high salt concentration, the swelling ratio and the diffusion coefficient still increase up to a neutralization degree of 80 %.

It can be noticed that the large variation in the amount of ionizable groups is responsible for relatively small variations (a factor 2) in D and Q. This can be compared with the big effects of varying salt concentration (Fig.5). These results suggest that Manning's condensation decreases the amount of effectively ionized groups.

c) <u>Swelling behavior of gel beads</u>. In usual practical applications of strongly ionized gels two relevant parameters are the retention capacity (swelling ratio Q) and the absorption rate. These two quantities are strongly dependent on the microscopic structure of the gel, i.e. crosslinking degree and neutralization degree. The absorption rate depends on the latter parameters through the diffusion coefficient D. However D only characterizes the intrinsic capacity of the gel to swell at a fast rate.

The effective swelling time is of the order of magnitude of the characteristic time τ_1 (see equation 16) and thus depends strongly on the gel dimensions at swelling equilibrium. But samples synthesized through inverse suspension reactions usually have a sphere size distribution with a finite width. This has some important consequences for the macroscopic swelling behavior of the samples.

As an example we consider an hypothetic sample consisting of a mixture of 99 per cent of spheres with initial radius r and one per cent of spheres with initial radius 5r. Assuming the same microscopic structure, the characteristic swelling time for the large spheres is 25 times longer than for the small spheres (equation 16). Moreover the amount of water retained in the small fraction of large spheres is more than the half of the total amount absorbed by this particular sample. Thus a small fraction of large spheres can dominate the swelling behavior of the whole sample.

In practice sphere size distribution are smooth functions and the effects are not so dramatic. However, due to the very particular averages involved in the swelling process, it is not possible to find a mean characteristic sphere size in the sample which would adequately describe its swelling behavior.

Figure 8 shows a real sphere size distribution measured on a typical sample. Using this distribution we can calculate various mean radii for the gels. We find $<r>_0 = 51.8$ μm, $<r>_1 = 64.8$ μm, $<r>_2 = 71.8$ μm where $<r>_n$ is defined as :

$$<r>_n = \frac{\sum_i P(r_i) r_i^{n+1}}{\sum_i P(r_i) r_i^n} .$$

As a first approximation, to simulate the macroscopic swelling behavior, we consider only the slowest relaxation modes in the swelling equation and we calculate as a function of time t the amount of solvent absorbed by unit of mass of the sample.

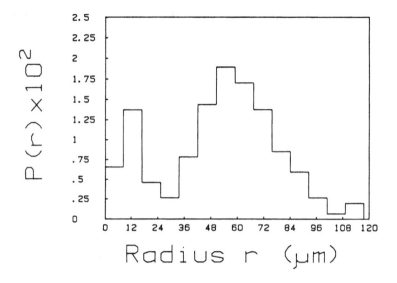

Figure 8. Typical sphere size distribution measured on a sample obtained by inverse suspension polymerization; $P(r)$ is the fraction number of spheres with radius r.

If we replace the true distribution by an idealized population of mean radius $<r>_1$, $A(t)$ is simply given by :

$$A(t) = Q(1-e^{-t/\bar{\tau}})^3 - 1$$

where $\bar{\tau} = \dfrac{Q^{2/3}<r>_1^2}{\pi^2 D}$. The result is shown as a dashed curve on Figure 9.

If we consider now the true distribution $P(r_i)$, $A(t)$ has a more complicated form :

$$A(t) = Q \frac{\sum\limits_i P(r_i)\, r_i^3(1-e^{-t/\tau_i})^3}{\sum\limits_i P(r_i)r_i^3} - 1$$

where $\tau_i = \dfrac{Q^{2/3}\, r_i^2}{\pi^2 D}$.

The obtained swelling curve is the continuous curve on Figure 9. The two curves are significantly different. This means that some attention should be paid, in the design of

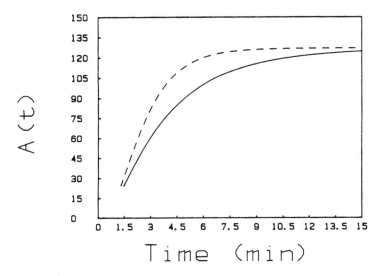

Figure 9. Time evolution of the amount of solvent absorbed by unit of mass of a sample constituted by a collection of gel beads. Continuous line : real size distribution (Fig.3). Dashed line : uniform sphere size equal to the mean radius $<r>_1$ of the distribution in Fig.3.

commercial samples, not only to Q and D values but also to the sphere size distribution.

At a first sight, we would expect samples with the highest Q and D values (i.e. high neutralization degree and small crosslinking degree) and the smaller sphere sizes to be the better ones. This is probably true with salted water when the salt is a mono-monovalent one. However, the introduction of even a small quantity of divalent ions can affect considerably the swelling behavior of the gels on long time scale. Deswelling can occur because of the complexation of the ionized sites by the divalent ions [7]. Preliminary results on the kinetics of this process suggest that the above conclusions have to be corrected in usual situations where divalent ions are present [17].

Conclusion

The results presented in this paper show that the kinetics of swelling of cylindrical or spherical gels can be used to obtain information about the microscopic structure of the gel through its diffusion coefficient. This coefficient depends strongly on the ionization and the crosslinking degree of the network. We have shown that the macroscopic swelling of a collection of spheres depends on both the microscopic structure of the gel and the sphere size distribution. To optimize this swelling behavior, both parameters can be varied. This leads to numerous possibilities to design industrial products with well defined swelling properties.

Acknowledgements

This work was supported by Norsolor Company. The authors thank Dr. C.Crétenot and Dr. P.Mallo for kindly providing the ionized gel samples and for stimulating discussions. They thank M. Najeh for technical assistance in the kinetics of swelling experiments.

References

[1] Schosseler, F.; Mallo, P.; Crétenot, C.; Candau, S.J.
     to be published in Journ. of Disp. Science and Techn.
[2] Tanaka, T. Phys. Rev. Lett. 1978, 40, 820.
[3] Tanaka, T.; Fillmore, D.J.; Sun, S.T.; Nishio, I.; Swis-
     lov, G.; Shah, A. Phys. Rev. Lett. 1980, 45, 1636.
[4] Tanaka, T.; Nishio, J.; Sun, S.T.; Ueno-Nishio, S. Scien-
     ce 1982, 218, 467.
[5] Ilavsky, M. Macromolecules 1982, 15, 782.
[6] Hirokawa, T.; Tanaka, T.; Sato, E. Macromolecules 1985,
     18, 2782.
[7] Rička, J.; Tanaka, T. Macromolecules 1985, 18, 83.
[8] Rička, J.; Tanaka, T. Macromolecules 1984, 17, 2916.
[9] Tanaka, T.; Fillmore, D.J. J. Chem. Phys. 1979, 70, 1214.
[10] Peters, A.; Candau, S.J. Macromolecules 1986, 19, 1952.
[11] Tanaka, T.; Hocker, L.; Benedek, G.B. J. Chem. Phys.
     1973, 59, 5151.
[12] Munch, J.P.; Candau, S.J.; Hild, G.; Herz, J. J. Phys.
     (Paris) 1977, 38, 1499.
[13] Candau, S.J.; Bastide, J.; Delsanti, M. Adv. in Polym.
     Sci. 1982, 44, 27.
[14] Geissler, E.; Hecht, A.M. Macromolecules 1980, 13, 1276.
[15] Helfferich, F. "Ion Exchange", Mc Graw Hill,New York,1962.
[16] Mallo, P.; Cohen, C.; Candau, S.J. Polymer Comm. 1985,
     26, 232.
[17] Candau, S.J.; Najeh, M.; Schosseler, F. to be published.

HYDROPHILIC–HYDROPHOBIC IPN MEMBRANES FOR

PERVAPORATION OF ETHANOL/WATER MIXTURE

Jae Heung Lee and Sung Chul Kim

Department of Chemical Engineering,
Korea Advanced Institute of Science
and Technology, Seoul, 131, Korea

Abstract

Interpenetrating polymer network (IPN) membranes of the hydrophilic polyurethane (PU) and the hydrophobic polystyrene (PS) were prepared by simultaneous polymerization methods under atmospheric pressure and high pressure up to 5000 kg/cm^2. The pervaporation of water/ethanol mixture was carried out to evaluate the effects of membrane thickness, PS composition, concentration of feed solution, temperature, and synthesis pressure on the pervaporation rate and the permselectivity. The IPN membranes of PU and PS showed high permeability and good water selectivity.

Introduction

Pervaporation is defined as a membrane technique to separate a liquid mixture by partially vaporizing it through a non-porous permselective membrane. This pervaporation process is potentially useful in fields where conventional distillation techniques are difficult to separate a liquid mixture, such as the fractionation of close-boiling components, azeotropic mixtures, isomeric mixtures, or heat-sensitive mixtures [1,2].

Several attempts have been made to improve the permeability, the selectivity, and the mechanical properties of a polymer membrane [3,4]. In general, the hydrophobic polymers exhibit good mechanical strength and high permselectivity, but low flux in most cases of the membrane application of water mixture, while the hydrophilic polymers reveal reverse characteristics. The combination of the hydrophilic and the hydrophobic polymers appears as a separation medium of water mixture. However, these polymer blends exhibit a large degree of phase separation due to the incompatibility of the component polymers and poor mechanical properties owing to the poor adhesion at the phase boundaries. Thus block or graft copolymers of the hydrophilic-hydrophobic constituents have mainly been treated because they reveal naturally microphase-separated structures [5,6]. Besides block or graft copolymers, the interpenetrating polymer networks, which are defined as a mixture of two or more crosslinked polymer networks which have partial or total physical interlocking between them [7,8], can be used to combine the hydrophobic and the hydrophilic components because the phase separation between the component polymers is restricted due to the interlocking between them. The resulting IPN membranes will exhibit the improved membrane characteristics. Moreover, it was reported that the IPN materials synthesized at high pressure showed the dispersed domains below about 100 $\overset{\circ}{A}$ in size, which resulted from both the increased compatibility and the reduced phase separation rate [9,10].

In the present investigation, the IPN membranes of the hydrophilic polyurethane and the hydrophobic polystyrene were prepared. The effect of the concentration of feed solution, temperature and the synthesis pressure on the permeability and the selectivity of the binary water-ethanol mixture was studied.

Experimental

Membrane preparation. The hydrophilic-hydrophobic IPN membranes were prepared by polymerizing the PU prepolymer mixture (the prepolymer which was synthesized by reacting 1 equiv of polyethylene glycol (mol wt: 1000) and 2 equiv of hexamethylene diisocyanate at 55 $^\circ$C; trimethylol propane; 0.01 wt% dibutyltindilaurate) and the styrene monomer mixture (93.33 wt% styrene monomer; 5.67 wt% divinylbenzene; 1 wt% benzoyl peroxide) simultaneously. The membranes were demolded from a mold of two glass plates between which a polyethylene terephthalate (PET) film spacer was placed, after polymerized at 80 $^\circ$C for one day and post-cured at 120 $^\circ$C for 4 hr. The thickness of the membrane was about 75 μ. The membranes were also prepared under high pressure up to 5000 kg/cm^2. A diagram of a mold used for the preparation of the membranes under high pressure was shown in Figure 1. The high pressure membranes were prepared by the same procedure as described above.

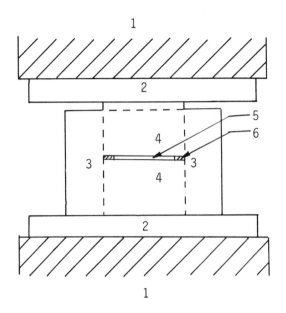

1. Press   4. Mold
2. Hot plate   5. Membrane
3. Guide   6. PET spacer

Figure 1. Apparatus for the preparation of
          a membrane under high pressure.

In this experiment, the membranes were coded as follows.
For the membrane of U50S50, the letters U and S denoted the
polyurethane and the polystyrene, respectively and the Arabic
number denoted the composition in wt%.

Scanning electron micrograph (SEM).   SEMs were obtained on
a scanning electron microscope (Hitachi; S-510). The SEMs of
the fractured membrane surface were obtained after a
vacuum-dried membrane was fractured in liquid nitrogen, and
the fractured surface of the membrane was metallized with
gold.

Swelling ratio.   The equilibrium swelling ratio was
calculated by the following equation.

$$\text{Swelling ratio}(\%) = \frac{W_s - W_d}{W_d}$$

where W_s is the weight of swollen sample and W_d is the weight of dry sample.

Ethanol composition in the membrane. The ethanol composition in the membrane was measured by gas chromatography after the solvent in the swollen membrane was allowed to evaporate under reduced pressure of below 2 torr at 80 °C and collected in a cold trap.

Pervaporation. The pervaporation apparatus was shown in Figure 2. It was consisted of a pervaporation cell which held the membrane and liquid to be permeated, traps to condense and collect the permeant vapor, a vacuum controller, a manometer and a vacuum pump. The downstream pressure was maintained at about 1 torr.

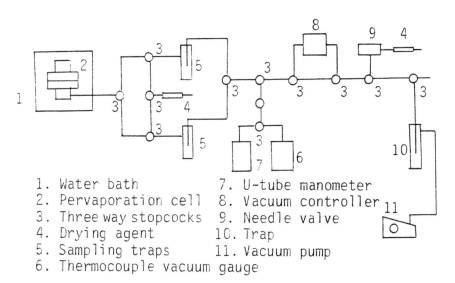

1. Water bath
2. Pervaporation cell
3. Three way stopcocks
4. Drying agent
5. Sampling traps
6. Thermocouple vacuum gauge
7. U-tube manometer
8. Vacuum controller
9. Needle valve
10. Trap
11. Vacuum pump

Figure 2. Pervaporation apparatus.

The area of the membrane in contact with the feed solution was about 16 cm^2. The permeate was collected in a trap cooled with liquid nitrogen. The flux was determined by weighing the permeate and the composition of the permeate was analyzed by gas chromatography (Hewlett Packard; 5840A, equipped with a 6 ft long Porapak Q column). The pervaporation rate, P, was defined as

$$P = \frac{w \cdot 1}{a \cdot t}$$

where w, 1, a and t denoted the permeate weight, the membrane thickness, the membrane contact area and time, respectively. The separation factor of water (α_w) (water permeated preferentially through the IPN membranes) was defined as

$$\alpha_w = \frac{P_W / P_E}{F_W / F_E}$$

where subscripts W and E represented water and ethanol, respectively, and P_W/P_E and F_W/F_E represented the concentration ratios of the permeate and the feed solution, respectively.

Results and Discussion

Effect of membrane thickness.   Figure 3 showed the effect of the membrane thickness on the pervaporation of the water/ethanol azeotropic mixture (ethanol, 95.6 wt%) at 25 ^{o}C.

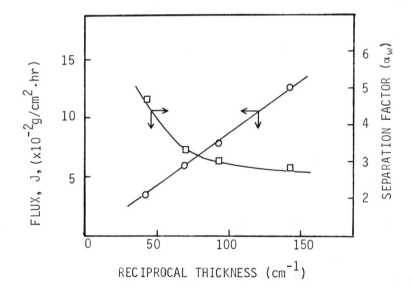

Figure 3.   Effect of membrane thickness on flux and separation factor (α_w) of the IPN membranes of U50S50

The flux, J, was inversely proportional to the thickness, 1, of the membrane according to Fick's law of diffusion. A line passing approximately through the origin was shown in Figure 3. Thus we could approximate the pervaporation rate, P, (unit, g cm/cm^2 hr) by flux, J, (unit, g/cm^2 hr) multiplied by the swollen membrane thickness, 1 (unit, cm). However, the effect of the membrane thickness on the separation factor revealed rather complex changes. As the thickness of the membrane increased, the pathways through which the penetrant molecules passed, became longer. Thus differences in diffusivity of water and ethanol in the membrane could play a significant role in determining the selectivity. As shown in Figure 3, the separation factor was nearly constant when the membrane thickness was below about 150 μ and increased abruptly when above 150 μ, which implied that the selectivity was strongly influenced by the differences in diffusivity of water and ethanol in the membrane.

Effect of PS composition. Figure 4 showed the scanning electron micrographs of the IPN membranes synthesized with varying PS composition. SEMs revealed the dispersed PS domains for the IPN membrane of U75S25 (Figure 4(a)). As the composition of PS was increased, the domain sizes and the continuity of PS phase increased.

Figure 5 showed the effect of the PS composition on the pervaporation rate of the IPN membranes. Feed composition was the azeotropic mixture of water/ethanol, and temperature was 25 °C. The pervaporation rate showed lower values than those calculated by simple additivity rule. This was because the PS phase in the PU-PS SIN membrane acted as a filler which did not allow the permeant to pass through and suppressed the plasticization of the membrane.

Figure 6 showed the relation of selectivity and PS composition. It showed an increase in the selectivity and a decrease in the product of the pervaporation rate and the separation factor (P x α_w) with an increase in the PS composition.

Effect of feed composition. Figure 7 showed the effect of the feed composition on the pervaporation rate and the equilibrium swelling ration of the IPN membranes of U50S50. The equilibrium swelling ratio showed a maximum at 60-70 wt% ethanol (EtOH) composition in feed solution. The change of the pervaporation rate seemed to reflect that of the equilibrium swelling ratio.

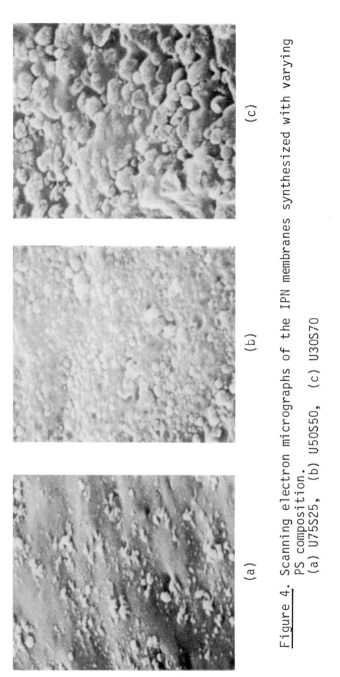

Figure 4. Scanning electron micrographs of the IPN membranes synthesized with varying PS composition. (a) U75S25, (b) U50S50, (c) U30S70

Jae Heung Lee and Sung Chul Kim

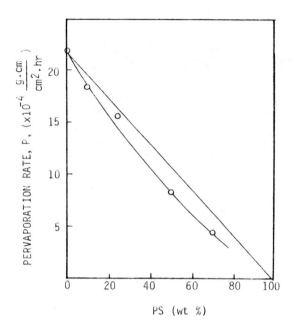

Figure 5. Effect of PS composition on pervaporation rate of the IPN membranes.

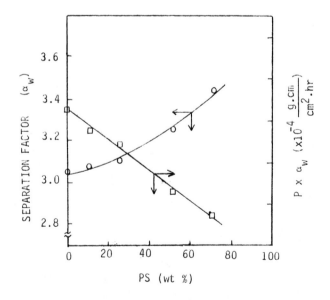

Figure 6. Effect of PS composition on separation factor and $P \times \alpha_w$ of the IPN membrane at 25 °C.

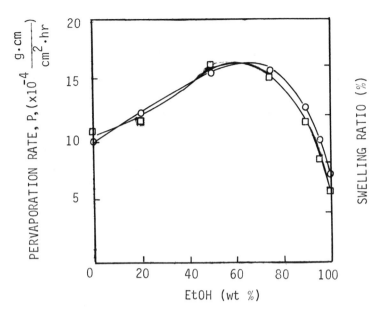

<u>Figure 7.</u> Effect of ethanol concentration on pervaporation
rate and equilibrium swelling ratio of the IPN
membrane of U50S50 at 25 °C.

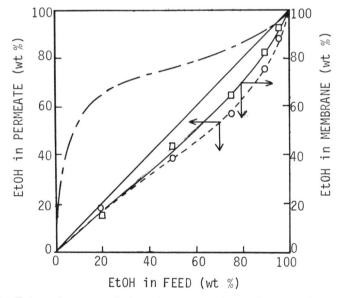

<u>Figure 8.</u> Ethanol composition in permeate and in membrane vs.
ethanol concentration in feed for the IPN membrane
of U50S50 (dashed line (——— - ———): vapor-liquid
equilibrium curve of water/ethanol mixture)

Figure 8 showed the plot of the ethanol composition in the permeate and in the membrane as a function of ethanol composition in feed. To compare this membrane process with the distillation process, the liquid-vapor equilibrium diagram for water/ethanol mixture at atmospheric pressure was plotted on the same figure. As shown in this figure, simple distillation process showed a tendency of higher ethanol separation while the membrane separation process revealed water selectivity. The IPN membranes also had to some extent an affinity toward water as shown in Figure 8. By considering that the permeability was a product of solubility and diffusivity, it might be concluded that the higher water selectivity resulted from the differences in both solubility toward the membrane and diffusivity in the membrane.

Effect of temperature. For the azeotropic feed composition of water/ethanol, the pervaporation rate, the equilibrium swelling ratio, and the water selectivity were measured over a temperature range from 25 °C to 65 °C for the IPN membranes of U50S50 (Figures 9 and 10).

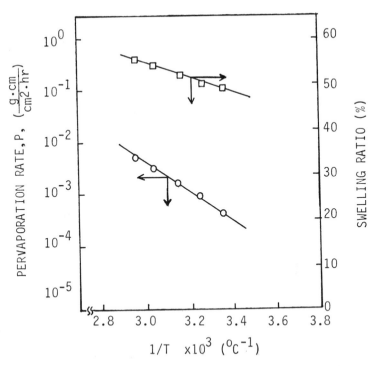

Figure 9. Effect of temperature on pervaporation rate and equilibrium swelling ratio of the IPN membrane of U50S50.

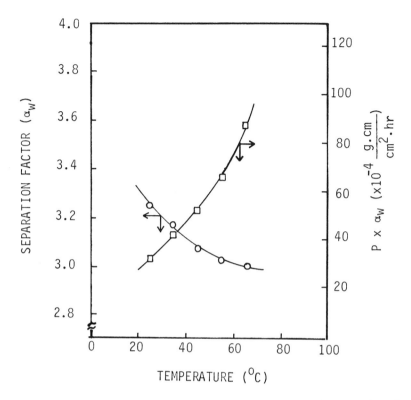

<u>Figure 10.</u> Effect of temperature on separation factor (α_w)
and P x α_w of the IPN membrane of U50S50.

Pervaporation rate increased with an Arrhenius type behavior
while the selectivity of water reduced from 3.25 at 25 °C to
2.99 at 65 °C. The equilibrium swelling ratio increased
slightly (from 48% to 55%) with increasing the temperature.
The activation energy of pervaporation rate was 6.34
kcal/mole. The increase in the pervaporation rate might be
presumed to be mainly due to the increase in diffusivity with
increasing temperature if we considered the slightly increase
in the equilibrium swelling ratio. The decrease in the
selectivity could be attributed to an increase in the thermal
motion of the polymer chains which allowed either of the
diffusing molecules to pass more readily. The P x α_w also
increased sharply with increasing temperature.

<u>Effect of synthesis pressure</u>. It was reported in the
previous paper [11] that the compatibility between the
constituents in the IPN of the hydrophilic PU and the
hydrophobic PS increased due both to the decrease in Gibbs

1 μ

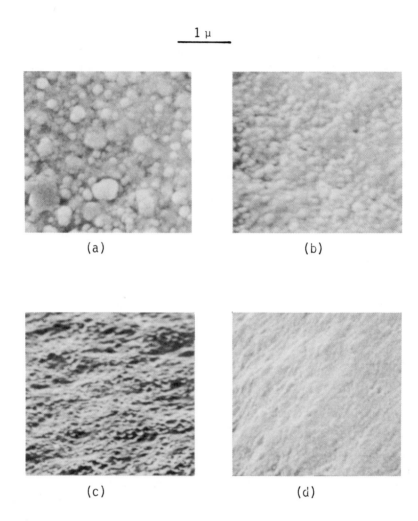

(a)   (b)

(c)   (d)

Figure 11. Scanning electron micrographs of the IPN
           membranes of U50S50 synthesized with varying
           pressure.
           (a) atmospheric, (b) 1000 kg/cm^2
           (c) 2500 kg/cm^2, (d) 5000 kg/cm^2

free energy of mixing and to the reduced chain mobility to phase separate when the IPN's were synthesized at high pressure.

Figure 11 showed the scanning electron micrographs of the fractured surface of the IPN membranes of U50S50 synthesized with varying pressure from atmospheric pressure up to 5000 kg/cm². The IPN membranes synthesized at atmospheric pressure revealed the PS dispersed domains with a few thousand Å in size. As the synthesis pressure was increased, the size of the PS domains decreased to about 500 Å at 5000 kg/cm².

Figure 12 showed the water selectivity of the IPN membranes synthesized with varying pressures. The water selectivity increased with increasing the ethanol composition. As the synthesis pressure was increased, the water selectivity also increased due to the decreased plasticizing effect (ie, the decreased swelling ratio) which resulted from the increases both in the number of the hydrophobic PS domains(acted as physical crosslinks) and in the continuity of PS phase.

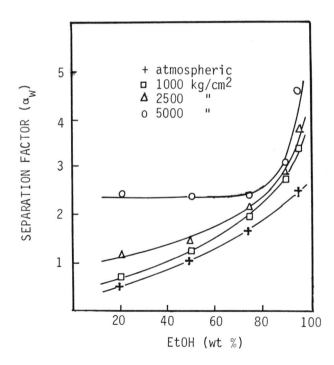

Figure 12. Effect of synthesis pressure on separation factor of the IPN membrane of U50S50 at 25 °C.

Thus it was possible to control the permeation characteristics by varying the morphology of the hydrophilic–hydrophobic IPN membranes.

References

[1] Binning, R.C.; Lee, R.J.; Jenings, J.F.; Martin, E.C. Ind. Eng. Chem. 1961, 53, 45.
[2] Long, R.B. I & EC Fundamentals 1965, 4, 445.
[3] Aptel, P.; Cuny, J.; Jozefowicz, J.; Morel, G.; Neel,J. J. Appl. Polym. Sci. 1972, 16, 1061
[4] Yoshikawa, M.; Yokoi, H. ; Sanui, K.; Ogata, N.J. J. Polym. Sci. Polym. Chem. Ed. 1984, 22, 2159
[5] Pusch, W.; Walch, A. Angew. Chem. Int. Ed. Engl. 1982, 21, 660
[6] Tamisugi, H.; Kotaka, T. Polymer J. 1985, 17, 499
[7] Sperling, L.H., "Interpenetrating Polymer Networks and Related Materials"; Plenum Press: New York, 1981
[8] Kim, S.C.; Klempner, D.; Frisch, K.C.; Radigan, W.; Frisch, H.L. Macromolecules 1977, 9, 253
[9] Lee, D.S.; Kim, S.C. Macromolecules 1984, 17, 268
[10] Lee, D.S.; Kim, S.C. Macromolecules 1984, 17, 2193
[11] Lee, J.H.; Kim, S.C. Macromolecules 1986, 19, 644

KINETIC STUDIES ON INTERPENETRATING POLYMER NETWORKS FORMATION

S.R. Jin , J.M. Widmaier and G.C. Meyer

Institut Charles Sadron (EAHP-CRM)
4, rue Boussingault
67000 Strasbourg
France

Introduction

Interpenetrating polymer networks (IPNs) present an attractive solution to the blending of two polymers (1). The mutual entanglement prevents any further phase separation once both constituents have been crosslinked, and therefore, no subsequent evolution in properties has to be feared. However, IPNs are quite complicated systems and only few of the usual investigation methods for polymers are suitable for their study. As a consequence, general structure - property relationships have not yet been established, and only their synthesis and some application oriented properties are usually reported in the literature (1). Attempts have been made to modelize the formation of IPNs (2), but the results merely apply to the system under investigation and may not be generalized. On the other hand, the chemical aspects of IPN formation, i.e. the kinetics, the viscosity and the compatibility changes during polymerization, etc., have been rarely investigated. Nevertheless, it is important to understand how one network is formed in the presence of the other or of its precursors. Once the chemical system chosen, the factors which govern the polymerization are directly responsible of the resulting morphology and properties of the material.

Amongst the various IPNs under investigation in our laboratory (3), we have chosen a very simple

polyurethane - poly(methyl methacrylate) system for
a in depth study of the processes involved in the
network formation leading to IPNs. The elastomeric
polyurethane (PUR) consists of an aromatic
triisocyanate combined with a poly(oxypropylene)
glycol. The rigid phase (PAc) is formed by methyl
methacrylate copolymerized with a trimethacrylate.
After the mixing of all the reagents, PUR is
prepared first at room temperature, followed by the
acrylic constituent at 60°C. We have adopted the
denomination of "in situ sequential" IPNs for such
materials, which emphasizes that all the reagents
are introduced simultaneously in the reaction
vessel, but that the networks are formed in a
sequential mode.

In this paper, we report on the formation of the
acrylic phase in the presence of the already formed
polyurethane network. The influence of various
parameters on the kinetics is examined : the PUR
content, the amount of acrylic crosslinker and the
temperature of the reaction medium. The main
results have been obtained through Fourier
transform infra red (FTIR) spectroscopy.

                          Experimental

PUR - PAc IPNs were directly synthesized in the
IR cell following the basic procedure described in
the first paper of this series (4). The elastomeric
PUR network was prepared by reacting an aromatic
triisocyanate, Desmodur L (Bayer AG), with a
poly(oxypropylene) glycol (ARCO Chemical),
molecular weight = 2000 g/mol; the catalyst was
stannous octoate (Goldschmidt). The rigid PAc
network was obtained by radical copolymerization of
methyl methacrylate (Merck) and trimethylolpropane
trimethacrylate, TRIM (Degussa), in the presence of
2,2'- azobisisobutyronitrile, AIBN. The mixture of
all reagents was injected in a cell formed by two
sodium chloride plates separated by a 20 um thick
gasket. The cell, which was sealed afterwards, was
fixed into a Specac heating chamber. The infra red
spectra were obtained on a Nicolet, model 60SX FTIR
spectrometer by averaging 32 consecutive scans
with a resolution of 2 cm^{-1}. The sampling interval
was 1 min during most of the reaction. Reaction
conversion was calculated from the change of the

normalized   absorbance. The   variation   of   the
isocyanate peak (2275 cm^{-1}) was followed during the
PUR formation at room temperature. When the conversion   ratio   was   over   90%   (see   Figure   1),   the
temperature   was   raised   to   initiate   the   radical
copolymerization.   The C=C peak   at 1639 cm^{-1},   not
overlapping with   its   neighbours,   was   used   to
calculate the PAc conversion.

Results and discussion

In   a previous paper (4),   we have checked that
FTIR   spectroscopy   is   an   appropriate   method   to
follow the monomer to polymer conversion,   and that
the   Beer   - Lambert   law   remains   valid   in   the
concentration   and temperature ranges used in   this
work.   Typical   conversion versus time curves,   P =
f(t),   for PUR and PAc are shown in Figure   1.   The
rate   of polymerization, R_p,   is deduced from   the
slope   at   each point of such curves.   For all   the
parameters influencing the kinetics of the   acrylic
system i.e.   polyurethane content (%PUR),   crosslink
density   (%TRIM),   temperature (T),   two series   of
curves,   P   = f(t) and R_p = f(P) were drawn,   which
allows to discuss the main features of the   kinetic
process.

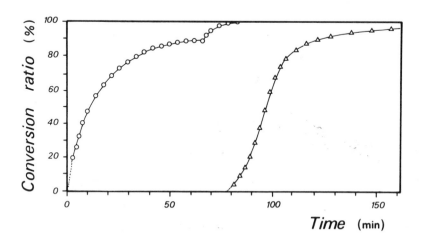

<u>Figure 1.</u> Conversion profiles for an IPN (25/75):
PUR (o), PAc (Δ).

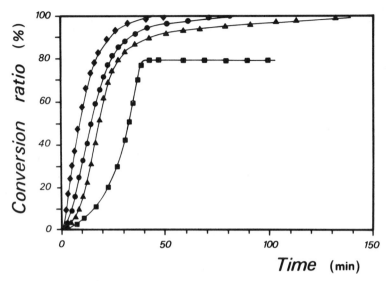

Figure 2. Conversion profiles for methyl methacrylate with TRIM copolymerization at various PUR network contents : (■): 0%, (▲): 25%, (●): 34%, (◆): 66%.

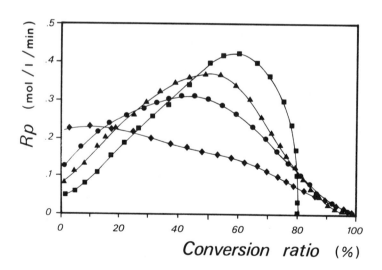

Figure 3. Rate profiles for methyl methacrylate with TRIM polymerization at various PUR network contents :(■): 0%, (▲): 25%, (●): 34%, (◆): 66%.

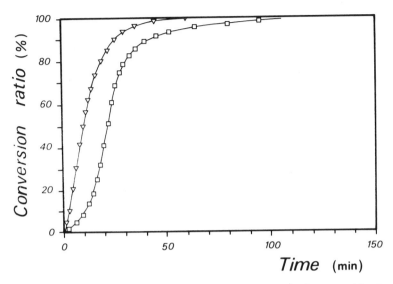

Figure 4. Conversion profiles for methyl methacrylate polymerization for a 34/66 IPN, 7.5% TRIM (∇) and for a 34/66 semi-IPN, 0% TRIM (□).

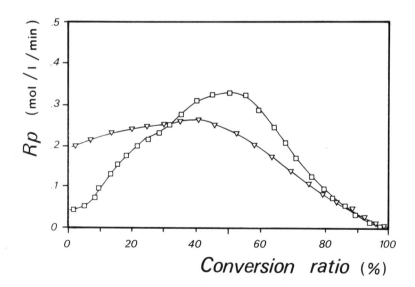

Figure 5. Rate profiles for methyl methacrylate polymerization for a 34/66 IPN, 7.5% TRIM (∇) and for a 34/66 semi-IPN, 0% TRIM (□).

Figure 2 and Figure 3 show the influence of the polyurethane content. With increasing PUR from 0 to 66% by weight, the rate and the final degree of conversion both increase. At a 66% PUR content, P = 90% after 20 min; without PUR, only a 15% conversion ratio is reached after the same time. The initial polymerization rate is about five times higher at 66% PUR than for the PAc system in bulk. Also, the R_p = f(P) plots are very different: with a low elastomer content, the curves show a pronounced maximum, and total conversion is not obtained. On the other hand, in the 34/66 IPN, the curve is flattened and the conversion is 100%. Increasing the PUR content therefore accelerates the polymerization and leads to a higher degree of conversion of the acrylic monomers.

The effect of the acrylic crosslinker on the kinetics of the second network is described by Figure 4 and Figure 5 for a 34/66 IPN. As the %TRIM is varied from 0 to 7.5, the conversion curves are shifted leftwards, showing an acceleration of the polymerization process; the initial rate is increased, but the maximum rate is lowered. Adding more crosslinker therefore causes autoacceleration at a smaller conversion and has a similar effect as the increase of PUR content: both parameters induce an earlier onset of the Trommsdorff effect with a higher initial conversion rate.

Table 1. PAC final conversion ratio as a function of polyurethane content and amount of crosslinker.

% PUR % TRIM	0	15	25	34
0	86	88	92	100
5	81	97	100	100

Experimental conditions:
temperature: 60°C, AIBN: 1 wt-%

The effect on final conversion is also somehow the same (Table 1). As already shown, the system tends towards completion with more polyurethane. Adding TRIM to the acrylic phase allows a complete polymerization at even lower PUR content, around 25%. Only in the absence of PUR, the final conversion is lower for the crosslinked PAc than for the linear PMMA (5). This observation shall be discussed later.

The effect of temperature is given in Figure 6 and Figure 7 for IPNs containing 0%, 34% and 66% PUR. Similar results are obtained for the other compositions. The temperatures under consideration correspond to our usual experimental conditions: all are below the glass transition temperature of the rigid phase (about 105°C). As may be expected, the rate of propagation and the degree of conversion increase when the temperature of the reaction medium is raised. The change in the rate profiles with the concentration of polyurethane is significant: a higher elastomer content shifts the maximum rate towards smaller conversion ratios.

The P versus t and R_p versus P curves shown above resemble those of a radical polymerization of methyl methacrylate in bulk (6,7): after an initiation period, the Trommsdorff effect causes an important viscosity raise of the reaction medium, and the polymerization is first accelerated, then the propagation finally slows down to zero. At this point, the conversion of the monomer may not be complete, depending upon the polymerization temperature (8,9).

In Figure 8, the copolymerization of methyl methacrylate with TRIM in the presence of 34% PUR is compared with that in different other reaction media, in bulk and in solution. The rate profile with PUR presents a maximum which is more damped than those usually encountered in bulk polymerization (6). It seems that the polyurethane behaves rather as a diluent towards the acrylic polymerization; like with ethylacetate or the more viscous polyol, a final conversion of 100% is reached. The presence of the elastomeric network prevents the reaction medium from attaining the glassy state exactly as would do a solvent.

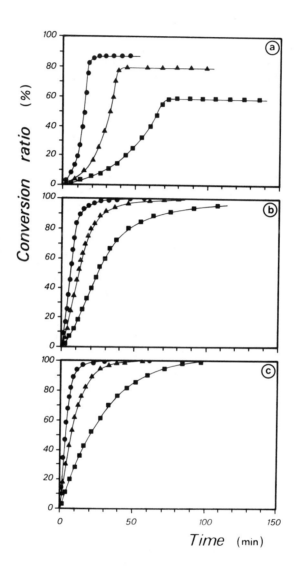

<u>Figure   6.</u>   Influence of temperature   on   PAc
formation. Conversion profiles in the presence of :
a) 0% PUR, b) 34% PUR and c) 66% PUR. Temperatures :
        (■): 50°C, (▲): 60°C, (●): 70°C.

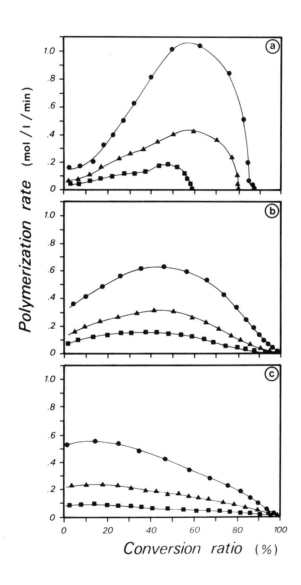

<u>Figure 7.</u> Influence of temperature on PAc formation. Rate profiles in the presence of : a) 0% PUR, b) 34% PUR and c) 66% PUR. Temperatures : (■): 50°C, (▲): 60°C, (●): 70°C.

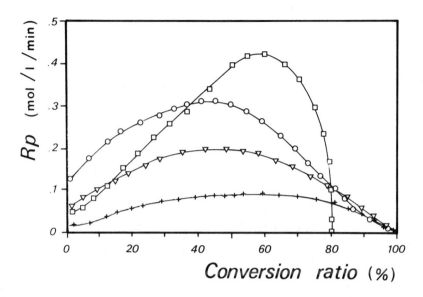

<u>Figure 8.</u> Rate profiles for methyl methacrylate
with TRIM copolymerization in bulk (□); in solution
  ethyl acetate (+), polyol (∇); in presence of
                 PUR network (○).

     As a matter of fact, the reaction medium changes
from the beginning of the polymerization process to
its   end: first it consists of PUR swollen by the
acrylic monomers, than, as the conversion proceeds,
the   concentration in PAc increases at the   expense
of the monomers. Without the presence of PUR,   a
limiting conversion would be observed at a   given
polymer/monomer   ratio,   depending   on   the
polymerization temperature. As   this is not the
case, it follows that at least from this point on,
polyurethane   is   necessary   as   a   diluent   for
obtaining a 100% conversion.

     Dynamic   mechanical   investigations   have   shown
(10)   that phase separation exists in the   present
type   of PUR/PAc in situ sequential IPNs,   even
though some improved miscibility is   observed. If
the   polymerization process leads to   large   glassy
domains,   it   will be necessary that at least   some
radicals   are   on   the   domain   border   or   in   an
interphase containing both PUR and PAc in order   to
be   reached   by   the   remaining   monomer
molecules.   The   presence   of   radicals   on   the

elastomeric phase may be excluded, as no grafting on PUR through transfer reactions has been detected for the present system (3). Finally, it is also possible that the combining of PUR and PAc by interpenetration causes a high mutual dispersion of the phases so that all the domains are small enough to allow monomer diffusion towards a radical, until its total consumption. Further work is necessary to settle this point, but the main result is that phase separation does not hinder the complete monomer to polymer conversion.

Another factor to be considered in order to explain the experimental results is the viscosity of the reaction medium: the initial rate increases as the reaction medium becomes more viscous, i.e. when going from the solvent polymerization to the polymerization in the presence of PUR, and the onset of the gel effect also begins at smaller conversion ratios (Figure 8). In that way, the PUR behaves like a viscous medium in which the acrylic polymerization proceeds. Raising the temperature shows the classical accelerating effect and needs no further comments. Of course, a temperature of the reaction medium which would be above the glass transition temperature of the rigid constituent would allow a complete conversion of the monomers even without PUR. From the temperature data, the activation energy of the PAc system was calculated (5) : the value found is around 19 kcal/mol whatever the conversion ratio for crosslinked PAc as well as for all IPNs; this means that these parameters do not influence the activation process of the acrylic monomers.

Our results concerning the crosslinking of the hard phase, show that the presence of TRIM interfers with that of PUR. In a classical radical crosslinking reaction in bulk (11), adding the crosslinker produces an earlier onset of the Trommsdorff effect, as well as earlier maximum rate and final stop of polymerization. According to Miller and Macosko (12) the crosslinker takes part in the polymerization in the very first steps of conversion, causing the viscosity to raise and therefore an earlier gelation of the reaction medium. The same situation is observed in the PUR/PAc IPNs, except that the value of the final conversion increases as soon as polyurethane is present (Table 1): the elastomer maintains the medium beyond the glassy state, and the reaction can proceed to completion. Therefore, both PUR and TRIM produce an earlier Trommsdorff effect, with a higher initial rate; but the role of PUR as a diluent allows a more even reaction rate and a final conversion ratio closer to 100%: in some way, the elastomer counterbalances the influence of the crosslinker.

Conclusion

Polyurethane has two effects on the formation of the acrylic phase in PUR/PAc in situ sequential IPNs. It confers a high viscosity to the reaction medium from the very beginning of the polymerization process, inducing a high initial rate and an early gelation effect. On the other hand, polyurethane acts as a diluent which keeps the glass transition temperature of the reaction medium below the glass transition temperature of the rigid phase, allowing therefore a complete monomer to polymer conversion.

References

(1) Sperling, L.H. in "Interpenetrating Polymer Networks and Related Materials" Plemum Press, New York, 1981.

(2) Widmaier, J.M., Yeo, J.K. and Sperling, L.H. Colloid Polym. Sci. 1982, 260, 678.

(3) Djomo, H., Morin, A., Damyanidu, M. and Meyer, G.C. Polymer 1983, 24, 65.

(4) Jin, S.R. and Meyer, G.C. Polymer 1986, 27, 593.

(5) Jin, S.R. Thesis, Louis Pasteur University, Strasbourg, 1986.

(6) Hayden, P. and Melville, H. J. Polym. Sci. 1960, 43, 201.

(7) Balke, S.T. and Hamielec, A.E. J. Appl. Polym. Sci. 1973, 17, 905.

(8) Marten, F.L. and Hamielec, A.E. J. Appl.
    Polym. Sci. 1982, 27, 489.

(9) Friis, N. and Hamielec, A.E. Am. Chem. Soc.,
    Symp. Ser. 1976, 24, 82.

(10) Hermant, I., Damyanidu, M. and Meyer, G.C.
     Polymer 1983, 24, 1419.

(11) Hayden, P. and Melville, H. J. Polym. Sci.
     1960, 43, 215.

(12) Miller, D.R. and Macosko, C.W. Macromolecules
     1976, 9, 206.

Part VII
Polymer Blends

Degradation of Blends of Poly(vinyl acetate) and Poly(vinylidene Fluoride)

Ling-Yu-He*, Eli M. Pearce, and T. K. Kwei
Polytechnic University
Brooklyn, New York 11201

Although studies of polymer degradation abound in the literature, the degradation characteristics of polymer mixtures have been the subject of only a few investigations.[1-8] In a heterogeneous blend of polybutadiene and polystyrene, the unsaturation in polybutadiene was found to accelerate the rate of oxidation of polystyrene.[1] On the other hand, blends of poly(acrylic acid) and poly(ethylene oxide) containing 30 to 70% of the latter exhibited better thermal stability than either component polymer.[4] Thus, the degradation of a blend is not necessarily a superposition of the processes of its components. We wish to report here some preliminary findings on the oxidation of mixtures of poly(vinyl acetate), PVAc, and poly(vinylidene fluoride), PVDF. Poly(vinyl acetate) is more susceptible to thermal and oxidative degradation than PVDF, but the two polymers are miscible over the entire range of composition.[9] The purpose of this study is to investigate whether the presence of a stable polymer, PVDF, in the mixture imparts a beneficial effect on the oxidative stability of PVAc.

*South China Institute of Technology,
Guangzhou, China

Poly(vinyl acetate) from Aldrich Co., with a
reported Mn of 150,000, and PVDF from Pennualt
Corp. (Kynar 721, Mn=80,000) were dissolved in
dimethylacetamide containing 1% acetic anhydride.
Solution cast films, about 0.5 mm in thickness,
were dried at room temperature under a nitrogen
atmosphere and then at 100°C in a vacuum oven
for at least two days. The glass transition
temperatures of the blends and the melting
points of PVDF in the mixtures were determined
by differential scanning calorimetry with a
DuPont thermal analyzer (Model 1090), using a
heating rate of 10°C/min. A single Tg was
found for each blend and the value lies between
the Tg of PVAc (37°C) and that of PVDF (-40°C).
(Although the assignment of glass transition of
PVDF is still a contentious issue, [10-13] our
experimental data are compatible with a low Tg
value). The melting point of PVDF was determined
to be 170°C. It decreases as the PVAc content
of the blend increases. For a mixture containing
82% PVAc the melting point is 150°C. The magnitude of melting point depression is in agreement
with literature values.[5]

Our first set of oxidation experiments were
carried out at 120°C in oxygen. At these conditions, PVAc suffers a loss in weight of about
20% after 200 hours while PVDF remains unchanged.
The weight loss data for the blends are therefore
calculated on the basis of PVAc content in the
mixture and are shown in Fig.1. The weight loss
decreases to 11% when the blend contains 21%
PVDF and to 4% for a mixture containing 40% PVDF.
It was thought at first that the reduction in
weight loss was caused by the crystalline domains
of PVDF which might have decreased the rate of
oxygen diffusion or the rate of chain propagation
reaction. Therefore, additional experiments were
carried out at 200°C, above the melting point of
PVDF. The weight of PVDF remained unchanged after
200 hr at 200°C in oxygen but the weight loss for
PVAc was 68%. The value decreased to 53% for
blend containing 21% PVDF and to 33% for blend

containing 40% PVDF, (Fig. 2). The same trend
found in our weight loss data above and below
the melting point of PVDF suggests that factors
other than the crystallinity of PVDF are responsible for the observed decrease in weight loss.
Experimen carried out in air at 200°C yielded
similar results. (Fig. 3)

The volatile products of oxidation at 200°C
were collected in a dry ice-acetone trap and
analyzed by gas chromatography. The major product was acetic acid but minor amounts of formal-
dehyde aned water were also identified. The
amount of acetic acid, also expressed as percent
of PVAc, again decreases as the PVAc content decreases in the blend (Fig. 4). As a point of
reference, the theoretical amount of acetic acid
evolution is 69.8% for PVAc, if all the acetate
groups are liberated as acetic acid.

The soluble material in the oxidized polymer
was extracted by chloroform in a Soxhlet extrac-
tor, evaporated to dryness, and weighed. The
weight of the crosslinked fraction was computed
as the difference of the weight of the chloroform
insoluble material and that of PVDF in the blend.
The ratio of the sol to gel fraction is also dependent on the composition of the blend, as shown
in (Fig. 5).

Our experimental results, although preliminary
at this stage, point to a profound effect of the
presence of PVDF on the oxidation of PVAc. A
mechanistic explanation of this effect can not
be offered at present. But we speculate that
the interaction between the two polymers involving the carbonyl group of PVAc may be responsible
for our observations.   In our Fourier Transform
Infrared Spectroscopy studies, shifts in the
carbonyl absorption frequency of about 4-7 cm^{-1}
were observed in the blends, in agreement with
similar studies of PVDF-poly(methyl methacrylate)
mixtures.[14]

Acknowledgment. This work was supported by
a grant from the National Science Foundation
(Materials Research Group Grant DMR 8508084).

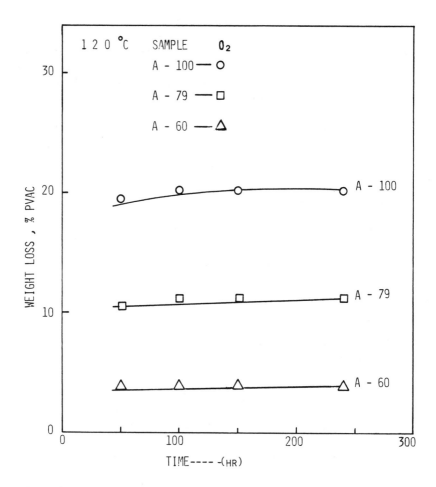

Fig. 1.

Weight Loss Measurement of Thermo-oxidation
of PVAC, PVDF and their blends in oxygen,
at 120°.

A-100 PVAC
A-79 PVAC - 79%
A-60 PVAC - 60%

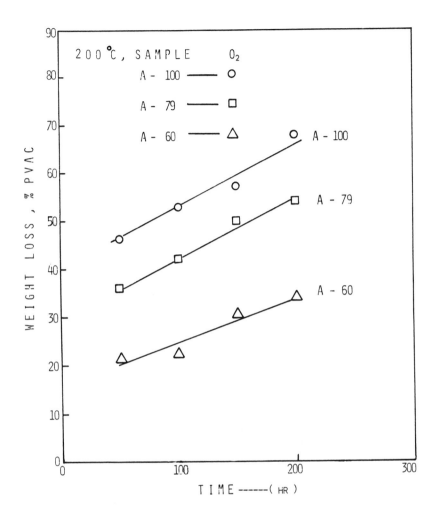

Fig. 2.

Weight Loss Measurement of Thermo-oxidation
of PVAC, PVDF, and their blends, in oxygen
at 200°.

    A-100 PVAC
    A- 79 PVAC - 79%
    A- 60 PVAC - 60%

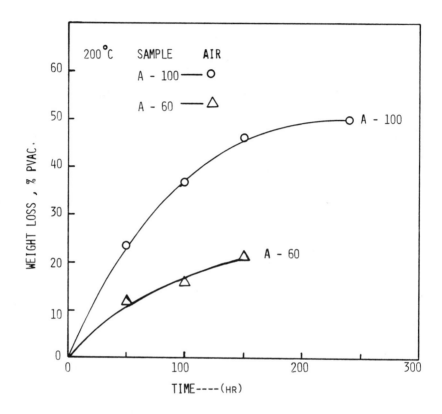

Fig. 3.

   Weight Loss Measurement of Thermo-oxidation
   of PVAC,PVDF and their blends, in air,
   at 200°.

                    A - 100   PVAC
                    A - 60   PVAC - 60%

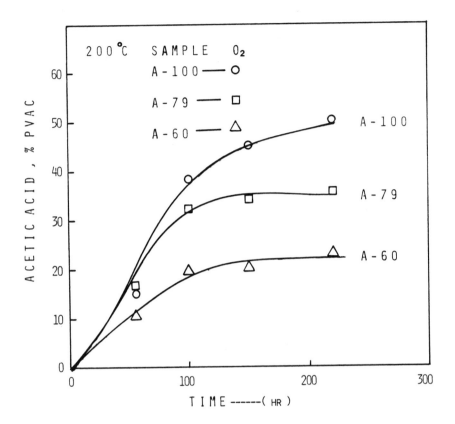

Fig. 4

Acetic acid evolution in Thermo-oxidation
of PVAC, PVDF and their blends, in oxygen,
at 200°.

A - 100 PVAC,
A - 79 PVAC - 79%
A - 60 PVAC - 60%

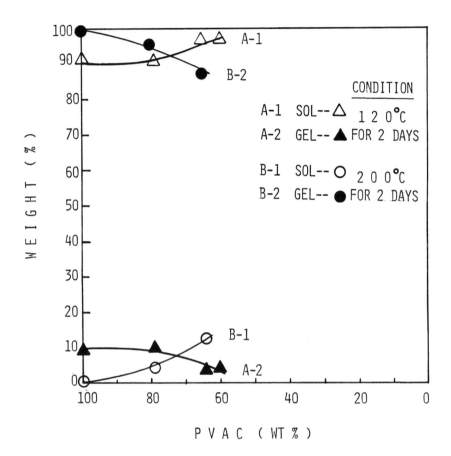

Fig. 5.

   Sol-Gel Weight fraction analysis of Thermo-
   oxidation of PVAC, PVDF, and their blends.

References

1. G. Scott in "Stabilization and Degradation of Polymers", Adv. in Chem. Ser., No. 169, D. L. Allara   and W. L. Hawkins, Eds., Amer. Chem. Soc.(1978).
2. K. Naito and T. K. Kwei, J. Polym. Sci., Polym. Chem. Ed. 17, 2935 (1979).
3. J. Jachowicz, M. Kryszewski, and M. Mucha, Macromolecules, 17, 1315 (1984).
4. K. L. Smith, A. E. Winslow, and D. E. Peterson, Ind.Eng.Chem. 51, 1361, (1959).
5. C. Sadrmohaghegh, G. Scott, and E. Setoudeh, Polymer Degradation and Stability, 3, 469 (1981).
6. G. Scott and M. Tahan, Polymer (London) 13, 981 (1977).
7. A. Ghaffar, C. Sadrmohaghegh, and G. Scott, Polymer Degradation and Stability, 3, 341 (1981).
8. G. Scott and Setoudeh, Polymer (London), 18, 901, (1982).
9. D. R. Paul, J. W. Barlow, R. E. Bernstein, and D. C. Wahrmund. Polym. Eng. Sci., 18, 1225 (1978).
10. J. S. Noland, N.N.C. Hsu, R. Saxon, and J. M. Schmitt,Adv. Chem. Ser., No. 9, Amer. Chem. Soc., (1971).
11. R. L. Imken, D. R. Paul, and J. W. Barlow. Polym. Eng. Sci.,16, 593, (1976).
12. T. K. Kwei, G. D. Patterson, and T. T. Wang, Macromolecules, 9, 780 (1976).
13. B. Hahn, J. Wendroff, and D. Y. Yoon, Macromolecules, 18,718 (1985).
14 M. M. Coleman and J. Varian, D. F. Varnell, and P. C. Painter. J. Polym. Sci.,Polym. Lett. Ed. 15, 745 (1977).

STUDY OF ETHYLENE-PROPYLENE
BASED POLYMER BLENDS BY MECHANICAL SPECTROSCOPY

C. Jourdan , J.Y. Cavaillé
GEMPPM, INSA Bât. 502
69621 Villeurbanne Cedex

M. Glotin, J.F. Pierson
ATOCHEM, CERDATO
27470 Serquigny

ABSTRACT

Mechanical spectroscopy is used to study various blends of polypropylene, high density polyethylene and ethylene-propylene copolymers.

An extension of the Halpin-Kardos model to viscoelastic measurements allows for a description of the microscopic organization of the various phases in the blends. It is shown that the cristallite thickness of the PP matrix is influenced by the addition of HDPE while the addition of EP copolymer modifies the amorphous regions.

The use of a Kerner-Dickie type model allows for a description of the macroscopic behavior of the multiphase systems (modulus) and brings information on the volume fraction of inclusions.
Results show that in a ternary blend, HDPE is mainly included inside EPR particles. This is confirmed by electron microscopy observations.

1. INTRODUCTION

The aim of the present work is to show how dynamic mechanical measurements can be used as a supplementary technique to get informations on the morphology of blends at different scales. It is well known that this type of technique can provide information on molecular movements since they are responsible for some relaxation processes [1] ; for instance figure 1 shows the results in G' and tan δ obtained on a ternary polypropylene/high density polyethylene/ethylene propylene copolymer blend (PP/HDPE/EPR) at 1 Hz as a function of temperature, indicating several relaxation processes : one at

150 K is related to the amorphous phase of PE, two others
around 225 K and 275 K correspond to the glass transitions of
EPR and PP respectively and the last one at 300 K is related
to the cristalline phase [2].

However, as the macroscopic behaviour depends upon (i) the
behavior of each phase, and (ii) the way they are arranged
in the material (morphology), it is necessary to use a systematic method to separate these two aspects.

In the following, we give two types of informations related on one hand to each phase (microscopic scale ∿ 1 nm) and
on the other hand to the morphology (semi microscopic scale
100 nm) : it requires the help of two types of theoretical
approaches depending on the scale at which information is needed. Thus physical models for the deformation of homogeneous
phases (amorphous and crystalline phases) are used to get
structural parameters, such as crystallite thickness... ; on
the other hand quantitative informations are obtained from
experimental results with help of mechanical models.

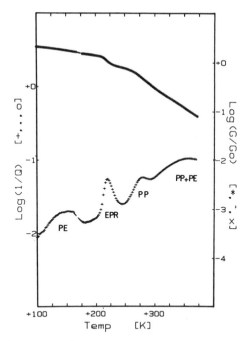

Figure 1. Spectrum in temperature of a PP.PE.EPR blend.

2. MATERIALS AND TECHNIQUES

Systems

Four systems have been used for this study :
System A : is a pure polypropylene homopolymer. M_w = 260 000.

System B : is a mixture of A with 6 % in weight of a 50/50 ethylene-propylene copolymer.
System C : is a mixture of B with 4 % in weight of HDPE
System D : is the same system as C, but with a higher EPR content (18 % in weight).

As observed in phase contrast optical microscopy, samples B, C, D contain a small amount (< 15 % for system C) of spherical dispersion in a PP matrix.

Samples were taken from injection molded plates (100 x 100 x 2 mm), always in the same central area to avoid anisotropy effects. Preliminary isothermal heat treatment (24 hours at 393 K) was applied to each sample.

Techniques

Dynamic mechanical measurements : The apparatus used in this study has been developed in our laboratory (G.E.M.P.P.M). It is an inverted forced oscillation pendulum. The real (G') and imaginary (G") parts of the dynamic shear modulus and the internal friction tan δ (G"/G') are measured either at a fixed frequency in a temperature range from 100 K to 400 K (isochronal measurements) or at a fixed temperature, between 10^{-4} and 5 Hz (isothermal measurements). The behavior has been verified to be independent of the applied stress or strain (in the range of 10^{-5}). The limit of detectability for tan δ is about 10^{-3}. When this value is reached, precision and reproductibility are better than 1 %. All the measurements were performed below the heat treatment temperature, so that the morphology remains constant during the test.

Optical microscopy : Samples 2 to 3 μm in thickness were obtained by cooling from the melt (475 K) at a constant rate of 0.2 K/min, and observed in transmission by phase contrast.

Scanning electron microscopy : Cryofractured samples were observed after gold metallization. Samples were also submitted to hot heptane for 1 hour to selectively extract EP copolymers. Then, they were dried under vacuum and gold metallized.

3. RESULTS AND INTERPRETATION

3.1. <u>System A (pure polypropylene)</u>. Figure 2 shows the results obtained on pure PP versus temperature at 1 Hz, show-ing the two main relaxation processes (β at 275 K and α at 340 K). The β relaxation is doubtless associated with the glass transition of amorphous PP. The origin of the α relaxa-tion is much more controversial in the literature. We have suggested somewhere else [4] that it is due to movements of defects in the crystalline phase along the chain axis. Move-ments are thermally activated and follow an Arrhenius law. For an accurate study of both relaxations, corresponding respecti-vely to the crystalline and the amorphous phases, the beha-vior of each phase has been separated with the help of a me-chanical model, considering the material as being composite. In this approach, the amorphous phase constitutes the matrix reinforced by the crystalline entities. The Halpin-Kardos mo-del [5], first developped for multilayer fiber composites, has been chosen because it can take into account some specifi-cities of our systems such as geometry of crystallites and anisotropy effect.

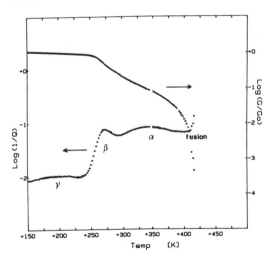

<u>Figure 2</u>. Modulus and internal friction at 1 Hz for PP

We have modified this model in order to apply it to visco-elastic experimental results ; then the approach can be sum-marized by the following equation :

$$G_{sc} = f(G_a, G_c, X_c, \xi)$$

G_{sc}, G_a, G_c : complex moduli of the semi-crystalline mate-

rial, of the amorphous phase, of the crystalline phase.
X_c: cristallinity ratio, ξ form factor of inclusion. Two of
theses parameters are unknown, namely G_a and G_c.

We have considered G_c as constant in the temperature range of the β relaxation process (its value is available in
the literature [6]). This allows us to extract the amorphous
modulus variations corresponding to the β relaxation alone.

Two main results have to be recalled here, related to the
behavior of each phase [3,4] : (i) the amorphous behavior
is strongly influenced by the presence of the crystalline
phase. This is equivalent to a crosslinked polymer with a
high rubbery plateau (10^7 Pa) and low value of tan δ max
(~ 1). (ii) Inversely, in the case of the α relaxation, correlated movements in the amorphous phase are required in order
to accomodate local strain when defects are moving inside the
crystalline phase. A parameter "k_α" has been introduced to
describe this effect in the amorphous phase involved for the
α process. Its value is directly obtained from experimental
results as drawn in a complex plane plot (G'' versus G', see
figure 3a). If ω is the value of the angle between the real
axis and the experimental curve at the origin, $k_\alpha = 2\pi/CO$.

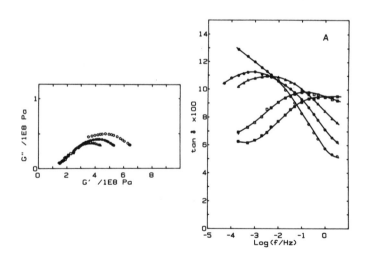

Figure 3. Sample A : isothermal measurements
a) Cole-Cole diagram   b) Internal friction
Temperatures : ▼: 310 K ; o : 321 K ; ●: 332 K ; Δ: 342.5 K
▲: 353 K ; □ : 363.4 K ; ■: 374 K ; ◊: 384.5 K

The mean relaxation time τ_{max}, which can be evaluated from the position of tan $_{max}$ at a given temperature (see figure 3 b) is thus related (i) to the thickness "1" of the crystallites and to the diffusion coefficient D (both connected to the time necessary for a defect to move from its position to the limiting surface of the crystallite) and (ii) to this parameter k_α. Thus it has been shown that calculations lead to the following set of equations :

$$\tau_{max} = to^{\,(k_\alpha - 1)/k_\alpha} - \left| \frac{1^2}{10\,D} \right|^{\,1/k_\alpha} \qquad (1)$$

with $D = D_o ex\ (-\Delta H/RT)$

to : mean time involved in correlated movements occuring in the amorphous phase (to $\sim 10^{-5}$s)
1 : thickness of crystallite
D : diffusion coefficient for defect movements in the crystalline phase.
k_α : correlation parameter ($0 < k_\alpha < 1$)
ΔH : activation energy involved in the diffusion of defects.
It has been shown that the apparent activation energy ΔH_α observed fot the α process is related to ΔH by $\Delta H_\alpha = \Delta H/k_\alpha$.
Thus, crystalline lamellar thickness appears to be a very important parameter on the observed α relaxation process.

3.2. <u>Crystalline phase of PP in blends</u>. Differential scanning calorimetry has been performed on the different samples and results are shown in figure 4.

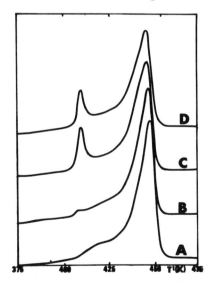

Figure 4. D.S.C. recordings

Samples were heated at a rate of 10 K/min and flow endotherms were recorded. From the area under the peaks, the cristallinity ratio in PE and PP can be evaluated when the enthalpy of fusion is known. The value for PP has been calibrated
from X-Ray measurements on pure PP (sample A), the
value for PE is available in the literature [7]. Crystallinity values are given in table 1. The global crystallinity
of different blends was also obtained from X-Ray measurements
and the results are in good agreement. The results show that
there is only a slight change in the crystallinity ratio (no
change in crystallographic structure is detected from the study of X Ray recordings).

Table 1

	A	B	C	D
X_c of PP	61.6	61.6	57.3	54
X_c of PE	-	4	71	69

On the contrary, optical observations show that for identical
cooling conditions the average diameter of spherulites is dependent upon the blend (see figure 5) :

A and B $\sim$ 80 μm,   C and D $\sim$ 20 μm

Dynamic mechanical measurements lead to results shown in
figure 6 to 8 for samples B, C, D. Frequencies (F_{max}) at
which tan δ is maximum are reported as a function of inversed temperature for each blend and pure PP on figure 9. It
appears that F_{max} (or τ_{max}) presents the same temperature dependence for (i) A and B, and (ii) C and D. Slopes of the
Arrhenius curves are identical in both cases leading to the
same apparent energy ΔH_A. D has been evaluated from data
available in the literature, D $\sim$ 10^8 m/s 8 . Although the
uncertainty in the parameter k_α ($k_\alpha \sim$ 0.18) is large, it is
possible using equation (1) to calculate the crystallite
thickness and furthermore to compare the results obtained for
each blend. Thus we have evaluated the crystallite thickness
to be 10 % lower for sample C and D. This result is consistent
with observations carried out by optical microscopy (figure 5)
: they show that the diameter of spherulites in samples containing HDPE (i.e. C and D) is divided by a factor 4 when compared to that of samples without HDPE (ie. A and B). This
phenomenon has been assigned to an additional heterogeneous
nucleation effect due to HDPE for which the crystallization

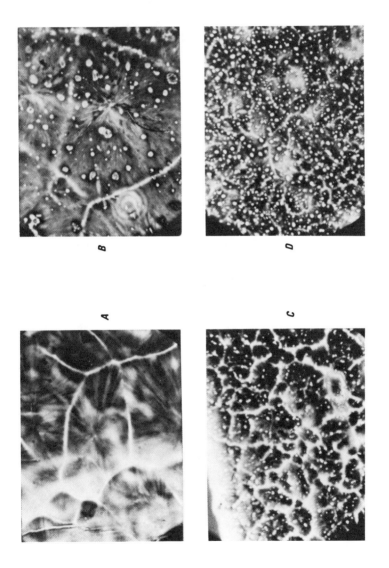

Figure 5. Phase contrast optical micrographs of slowly cooled thin films of systems A, B, C and D.

C. Jourdan et al.

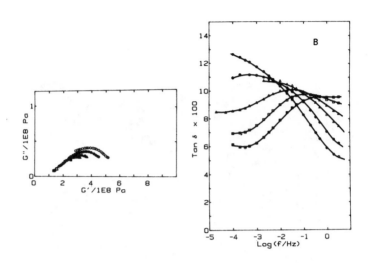

Figure 6. Sample B, isothermal measurements
Cole-Cole diagram and internal friction

Temperatures : ▼: 310 K ; o: 321 K ; ●: 332 K ; Δ: 342.5 K
▲: 353 K ; ▫: 362.4 K ; ◼: 374 K ; ◊: 384.5 K

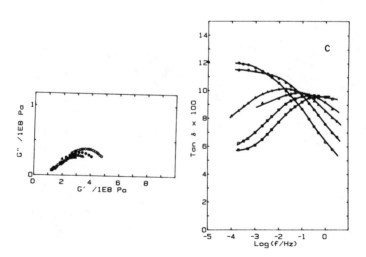

Figure 7. Sample C, isothermal measurements
Cole-Cole diagram and internal friction
(Temperatures and symbols same as in figure 6)

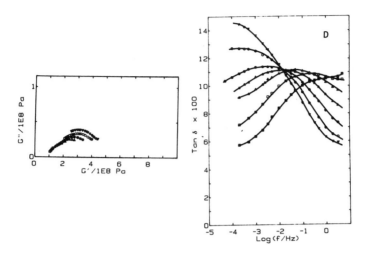

Figure 8. Sample D, isothermal measurements :
Cole-Cole diagram and internal friction
(Temperatures and symbols same as figure 6)

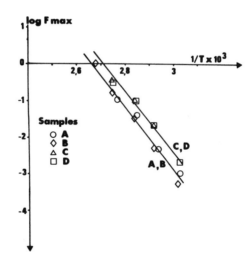

Figure 9. Maximum of tan δ as a function of inversed temperature

rate is higher than for pure PP (under the cooling conditions used here) [9]. Although no direct determination of lamellar thickness has been done, it is reasonable to think that its change should be connected to change in spherulite diameter.

3.3. <u>Amorphous phase of PP</u>. Figure 10 shows the temperature dependance of tan δ and G' in the β relaxation range for each blend and pure PP ; following the method mentionned in § 3.1, the behavior of pure amorphous PP has been determined and is given on figure 11. Two main effects have to be noted :

3.3.1. The value of the rubbery modulus decreases from A to B and from B to D, ie. it is modified when the amount of EPR increases, but is not modified by HDPE (B to C).

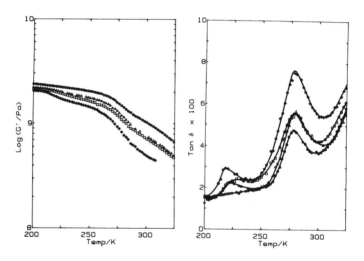

Figure 10. Samples A, B, C, D : G' and tan δ

● : A ; Δ : B ; □ : C ; ◊ : D

The level of the rubbery modulus is representative of (i) the density of "crosslink nodes", considering that the amorphous phase is crosslinked by the crystalline phase, (ii) for a given density of nodes of the state of tension of chains. We can imagine that the presence of liquid inclusions of EPR when PP crystallizes, modifies the morphology of chains (longer chainloops, smaller tension ...).

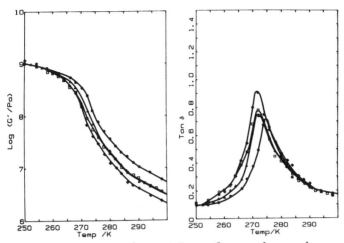

<u>Figure 11.</u> Dynamic modulus of amorphous phase
in Tg region symbols as in figure 10

3.3.2. A slight shift of the tan δ peak towards lower temperatures (∿ 5 K) is observed when EPR is added to system A ;
this could be due to a very large distribution in propylene
content among EPR chains : the higher the propylene content,
the more compatible it is with PP, probably leading to a
slight plastification effect of amorphous PP.

3.4. <u>Morphology of the blends.</u> Optical microscopy has revealed the presence of spherical inclusions. This observation
is also confirmed by scanning electron microscopy (see figure
12).

The description of a composite system as a function of the
properties of the constituants is a complex problem, which
finds generally only approximative answers. In the case of
spherical inclusions in a matrix, the Kerner equation [10] is
well adapted for a content lower than 25 % such as in our systems. Dickie [11] has successfully applied this equation to
viscoelastic data. For the shear modulus, the equation is :

$$\frac{G}{G_m} = \frac{(1 - v)\ G_m + (\alpha + v)\ G_i}{(1 + \alpha v)\ G_m + \alpha(1 - v)\ G_i} \qquad \alpha = \frac{2(4 - 5\ \nu_m)}{7 - 5\ \nu_m} \qquad (2)$$

Where G_m is the complex modulus of the matrix

   G_i is the complex modulus of the inclusion

   G is the complex modulus of the composite

   ν_m is the Poisson coefficient of the matrix

If the modulus of inclusions can be neglected (rubbery inclusions in a glassy matrix), equation (3) becomes :

$$G = \frac{1 - v}{1 + \alpha v} G_m$$

Thus, the modulus of the composite depends only on the volume fraction of inclusions (and not on the nature of inclusions) and is proportionnal to the modulus of the matrix. Then, it is possible to deduce the volume fraction of inclusion from the ratio $\frac{G}{G_m}$ measured at a temperature intermediate between those of the two relaxations corresponding to the glass transition of EPR and amorphous PP respectively. A temperature of 250 K has been chosen and results are :

	A	B	C
G/GM	1	0.867	0.804
V(%)	0	7.5	11.5

The volume fraction V for system B is in good agreement with the composition of system B. For system C, the result is unexpected, because HDPE remains rigid at this temperature and should not contribute to the decrease in modulus. A possible interpretation is that the HDPE is localized inside the EPR and thus increases the apparent volume of inclusions. This has been confirmed by scanning electron microscopy : for system B, holes are observed after heptane extraction, which has removed EPR inclusions. For system C the same phenomenom occurs but some free insoluble particles of material still remain, out of or inside the holes. As they appear only in system C and not in B, it is reasonable to assume that they are HDPE particles previously located inside the EPR inclusions.

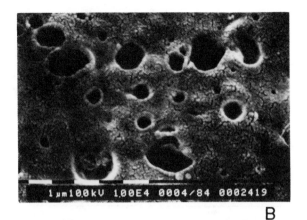

B

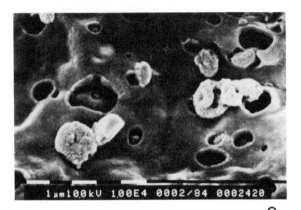

C

Figure 12. Electron micrographs of fracture surfaces of
samples B and C, EP copolymers have
been selectively removed by hot heptane

Calculations bases on the Kerner-Dickie equation require
the knowledge of the mechanical behavior of each constituant,
so experiments on pure EPR were performed. Predicted and experimental results are compared on figure 13 in the case of system B : agreement between both is rather good considering the
small variations discussed in this section.

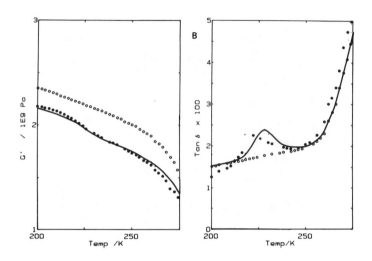

Figure 13. Mechanical modelization of system B
Experimental results : O : A ; ● : B
Continuous line : calculations based on model

4. CONCLUSION

As it is shown above, the use of mechanical spectroscopy
can lead mainly to two different types of information on heterogeneous polymer blends ; on one hand, it is possible to determine how each phase is modified, while on the other, the
morphology of the blend can be quantitatively deduced. In the
case of the blends considered here, the main conclusions can
be summarized as follows :

– The crystalline lamellar thickness decreases by addition
of HDPE (systems A and C). Amorphous PP is plasticized by the
addition of EPR (systems A and B).

– EPR forms a quasi immiscible phase (systems B and C).
HDPE is mainly included in the EPR phase (systems C and D).

The results have been confirmed by optical and electron
microscopy.

Since the study of such materials is very complex, it is
interesting to get informations from as many experimental methods as possible : the use of this spectroscopic method
should be considered extensively.

References

[1] Ferry, J.D. ; "Viscoelastic Properties of Polymers" ;
     J. Wiley : New York, 1980
[2] Yeh, P.L. ; Birley, W ; Hemsley, D.A Polymer 1985, 26,
     1155.
[3] Jourdan, C. Thesis : Lyon(Fr), 1987.
[4] Jourdan, C. ; Cavaillé, J.Y ; Perez, J. Submitted for
     publication.
[5] Halpin, J.C ; Kardos, J.L. ; J. Appl. Phys. 1972, 43,
     2235.
[6] Sawatari, C. ; Matsuo, M. ; Macromolecules 1986 19, 2653.
[7] Mandelkern, L. ; Allou, A.L. ; Gopalan, M. J. Appl. Phys.
     1968, 72, 309.
[8] Renecker, D.H. ; Mazur, J. Polymer 1982, 23, 401.
[9] Teh, J.W. J. Appl. Polym. Sci. 1983, 28, 605.
[10] Kerner, E.H. Proc. Phys. Sco. 1956, 69 B, 808.
[11] Dickie, R.A. J. Appl. Polym. Sci. 1973, 17, 45.

THERMAL PROPERTIES OF MIXTURES OF POLYCARBONATE WITH SEMIFLEXIBLE, THERMOTROPIC COPOLYESTERS

Robert Kosfeld, M. Heß, K. Friedrich

FB-6, Physical Chemistry
University-GH-Duisburg, D-4100 Duisburg
Federal Republic of Germany

We present here some results of our recent investigations on the thermic properties found in blends prepared from poly(carbonate) based on bisphenol-A (PC) and copolyesters (COP) from poly(ethyleneterephthalate) (PET) and p-hydroxi-benzoate (PHB).

Blends prepared from coil molecules as, e.g. PC, and rigid rod type polymers, e.g. COP, are of increasing interest with respect to creating polymers with outstanding physical properties (1). Semiflexible polymers may allow a more detailed adaptibility due to their range of variation of stiffness. A better miscibility with coil shaped polymers may be attainable varying the content of flexible parts in the chain. This concept means an alternative way to the compatibilization by introduction of sidechains onto rigid rods (2).

Of special interest are those polymers which exhibit thermotropic mesophases due to mesogenic groups incorporated into the main chain.

One promising material for this type of liquid crystalline polymer is the previously mentioned COP, first described by Jackson and Kuhfuss (3). This copolyester allows a high variation in composition and shows a liquid crystalline nematic phase above about 30 mol % PHB. The optimum in mechanical properties seems to be about 60 mol % PHB. The material is easy to prepare and commercially available.

We have prepared a series of compositions from about 10 mol % PHB up to about 60 mol % PHB. Preparation, characterization and further details of analysis are given in (4).

By means of special techniques it is possible to create different types of morphologies: optically clear and turbid ones (4). The turbid material showed a microscopically visible phase separation.

The dependence of $c_p(T)$ of blends of different composition was examined by Differential Scanning Calorimetry (DSC) using a Perkin Elmer DSC-2.

It was found that the blends exhibited two separated glass transition processes. Detailed analysis showed that the PC-phase is capable of absorbing some COP while the COP-phase strictly excludes the coil shaped molecules as predicted by Flory (5).

Table 1. Glass temperatures for the copolyester component (T_{g1}) and for the polycarbonate component (T_{g2})

COP content	Morphology	T_{g1}	T_{g2}
100%	–	345	–
0%	–	–	422
25%	turbid	–[a]	416
50%	turbid	346	415
75%	turbid	346	413
25%	clear	–[a]	416
50%	clear	348	412
75%	clear	348	408

[a] not detectable

The glass transition temperature (T_{g1}) of COP remains almost unaltered with increasing PC content, while the T_g of PC is shifted to lower values depending on the amount of COP in the blend.

With respect to the reactions probably occuring in technical processes during manufacturing, we had a look on annealing at constant temperature in the melt under protective gas.

All annealing experiments were immediately executed in the DSC apparatus in the melt at 543 K.

Within a time of about 10 minutes the clear foils showed a decreasing of T_{g2}, while T_{g1} increased, coinciding in one single transition temperature. On further annealing a slow and comparably small increase of this joint transition was observed. In comparison with it the turbid films showed the same effects but with a kinetic which was about 3 times slower.

In polarisation microscopy this phenomenon is accompanied by a quick decrease of birefringence resulting in a dark object under crossed polarizers. This phenomenon has been observed in further examples much more pronounced (6).

Our interpretation of these findings is that a transesterification process occurs which first results in a block formation.

Besides the compatibilizing effect of end blocks built up in the early stage of reaction, further annealing leads to a random block copolymer with COP and PC blocks showing a T_g behaviour depending on the composition according to (7). Further annealing leads to a rearrangement to a quite new distribution of the polymer constituents, then.

In those cases, where there was a macroscopic phase separation observable, reactions were slower compared to those samples with a submicrophase separation which requires a larger reaction surface and a shorter time to pass through the phase. Both samples, nevertheless, resulted in materials with the same thermal properties.

The interpretation is in accordance with the results published by Devaux et al. (8-11) found in the system poly(butyleneterephthalate)/PC and by ^{13}C-NMR investigations done in our laboratory the results of which are summarized in the following.

Up to now it was not possible to achieve a complete analysis of what happens during annealing. Due to the great amount of new chemical bonds being created the whole NMR resonance scheme is subjected to alterations especially in the region of aromatic carbons.

It is more informative to restrict first analysis on the carbonyl sequence of PC at 152.4 ppm. The exchange of a neighbouring bisphenol-A unit for an ethyleneglycol unit is proved by taking rise of a new resonance at 151.6 ppm and a decrease of the one first mentioned.

Acknowledgements

The authors are greatfully indepted the Deutsche For-schungsgemeinschaft for financial support. We want to thank Prof. W. Borchard, University of Duisburg, FB-6, Angewandte Physikalische Chemie, and Prof. Witold Brostow, Drexel University, Philadelphia, for helpful discussions.

References

(1)   Prevorsek, D.C. in "Polymer Liquid Crystals"; Ciferri, A.; Kriegbaum, W.R.; Meyer, R.B., Eds.; Academic Press: New York, 1982.

(2)   Ballauff, M. Macromol. 1986, 19, 1366.

(3)   Jackson, W.J.; Kuhfuss, H.F. J. Polym. Sci., Polym. Chem. Ed. 1976, 14, 2043.

(4)   Kosfeld, R.; Hess, M.; Friedrich, K. Materials Chemistry and Physics, in press.

(5)   Flory, P.J. Macromol. 1978, 11, 1138

(6)   Friedrich, K.; Hess, M.; Kosfeld, R. Mol. Cryst. Liquid Cryst., in press.

(7)   MacKnight, W.J.; Karasz, F.E. in "Polymer Blends,Vol.1"; Paul, D.R.; Newman, S., Eds.; Academic Press: New York, 1978, Chapter 5.

(8)   Devaux, J.; Godard, P.; Mercier, J.P. J. Polym. Sci., Polym. Phys. Ed. 1982, 20, 1875.

(9)   Devaux, J.; Godard, P.; Mercier, J.P.; Touillaux, R.; Dereppe, J.M. ibid 1982, 20, 1881.

(10) Devaux, J.; Godard, P.; Mercier, J.P. ibid 1982, 20, 1885.

(11) Devaux, J.; Godard, P.; Mercier, J.P. ibid 1982, 20, 1901.

NEW ENGINEERING THERMOPLASTIC POLYMER BLENDS
MISCIBILITY BEHAVIOUR, PHASE MORPHOLOGY
AND DEFORMATION MODES

G. Groeninckx and D. Deveen

Catholic University of Leuven, Dept. of Chemistry,
Celestijnenlaan 200F, 3030 Heverlee, Belgium

D. Dufour, P. Keating and J. Pierre

Monsanto Company, Technical Centre,
1348 Louvain-La-Neuve, Belgium

Synopsis

Thermoplastic polymer blends of PVC, a rubber containing modifier and a random terpolymer comprising mainly styrene and maleic anhydride (S-MA) have been studied with respect to their miscibility behaviour, phase structure and toughness properties.

Blends of the rubber containing modifier with the random terpolymer (hereafter called Cadon®) are completely miscible. This miscible blend system was found to be partially miscible with PVC giving rise to a PVC rich phase and a Cadon rich phase. The distribution and composition of the partially miscible phases depend strongly on blend composition.

Tensile stress-strain experiments performed on these complex blend systems indicate an increase of the yield stress, elongation at rupture and toughness with increasing PVC content. Crazing and shear yielding have been observed as plastic deformation mechanisms. While notched PVC shows brittle behaviour under impact conditions, alloys of PVC and Cadon exhibit a significant improvement in impact resistance.

1. INTRODUCTION

The actual interest in the development of new multi-component polymeric materials based on mixtures of polymers arises from their growing technological importance in engineering, electrical, electronic, biomedical and space applications.

The industrial development of new polymers is expected to be rather limited in the future, and blending of existing polymers will probably be the most effective way to design new polymeric materials with the desired combination of properties. In the last few years, the scientific literature on this subject has expanded very rapidly (1-5); it clearly appears now that the potential for new polymeric materials via alloying is very large.

In this area, however, we still have a long way to go as far as the fundamental understanding and prediction of the miscibility behaviour and structure-property relations of polymer blends is concerned (5). The degree of miscibility of the components in a blend is of crucial importance with respect to its morphology and physical properties.

Most of the research work carried out in this field has been mainly devoted to completely miscible and completely immiscible blend systems (2,4). Only limited data exist in the literature about the morphological characterization and properties of partially miscible polymer blends.

The systems considered in this paper are blends of a rubber containing modifier with a styrene-maleic anhydride terpolymer and PVC. These systems consist of three glassy components exhibiting partial miscibility and one non-miscible rubbery component.

The miscibility behaviour, morphology, mechanical properties and deformation modes of these complex blend systems will be investigated.

2. EXPERIMENTAL

2.1. Materials and Blend Preparation

The thermoplastic polymers used in this blend study are :
PVC ($\bar{M}_w$ = 215 000, $\bar{M}_n$ = 66 000), a rubber containing
modifier and a random terpolymer mainly based on styrene
(69 wt.%) and maleic anhydride (25 wt.%). The rubber
containing modifier used contains both large (1.5 μm) and
small (0.17 μm) rubber particles; the polybutadiene content
is approximately 31 wt.% and acrylonitrile content is
30 wt.%.

Blends of the rubber containing modifier with the
terpolymer are commercially available under the name Cadon®
(Monsanto Europe SA); the Cadon composition used here is
Cadon 330. This blend exhibits a two-phase morphology
consisting of a miscible matrix phase in which grafted
polybutadiene particles are dispersed. Cadon can thus be
considered as a rubber-modified glassy two-component matrix.

Blends of PVC with Cadon have been prepared over a wide
range of compositions by intensive melt-mixing using a
Banbury mixer. They were subsequently compression moulded
at 185° C (3 minutes) into sheets of 3 mm thickness from
which test samples were cut.

2.2. Physical Testing Methods

Tensile stress-strain experiments were performed using
an Instron machine at a cross-head speed of 1 and 100 mm/min.

For the quantitative investigation of the plastic
deformation modes, a special device has been designed to
measure the thickness change (lateral contraction) of the
samples during the tensile tests (6). The preformed neck-
length (dumbell-shape) of the tensile samples was only 3 mm
in order to be able to follow locally the thickness change
during elongation.

Storage and loss moduli, as a function of temperature,
were determined by dynamic mechanical analysis using a Dupont
1090 DMA apparatus at a heating rate of 5° C/min. The
notched Izod impact strength and the heat deflection
temperature of the samples were measured according to
standard ASTM specifications.

The phase morphology of the blends was examined with a transmission electron microscope Jeol 100 U.

3.   RESULTS AND DISCUSSION

3.1.   Miscibility and Phase Distribution in PVC/Cadon Blends

The degree of miscibility of the different components in the blends has been evaluated by studying their glass-transition behaviour using DMA.

Figure 1 compares the dynamic loss modulus-temperature curves of PVC, Cadon and their blends at different composi-tions. For the blends, two peaks can be observed above 70° C which correspond to two glass-transition regions. The Tg's of these blends are located between those of the base compo-nents (Tg of PVC : 87° C, Tg of the miscible matrix of Cadon : 139° C), which indicates partial miscibility, i.e., a two phase system with a PVC rich phase and a Cadon rich phase.

Figure 2 represents the temperature dependence of the storage modulus. From this figure, interesting information can be deduced with respect to the distribution of the partially miscible phases as a function of the blend composition. If one considers the value of the storage modulus at 110° C, one can observe that the blend modulus increases only slightly from 0 to 60 wt.% Cadon; in these blends the soft PVC rich phase forms the continuous matrix in which the rigid Cadon rich domains are dispersed. The pronounced increase of the blend modulus between 60 and 80 wt.% Cadon indicates that phase inversion is occuring, and as a consequence, the Cadon rich phase is then forming the continuous phase. In this composition range, the mechanical response of the blend at 110° C becomes more and more dominated by the rigid Cadon rich phase, as this phase increases in continuity throughout the sample.

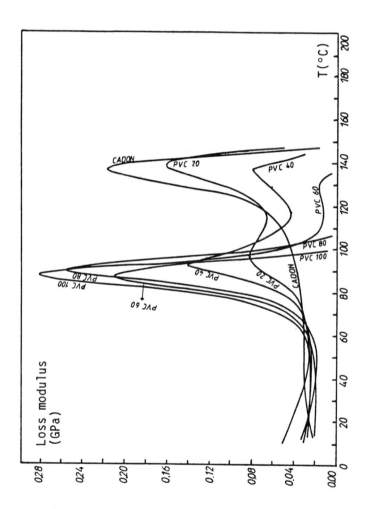

Figure 1   Temperature dependence of the tensile loss modulus of PVC/Cadon blends

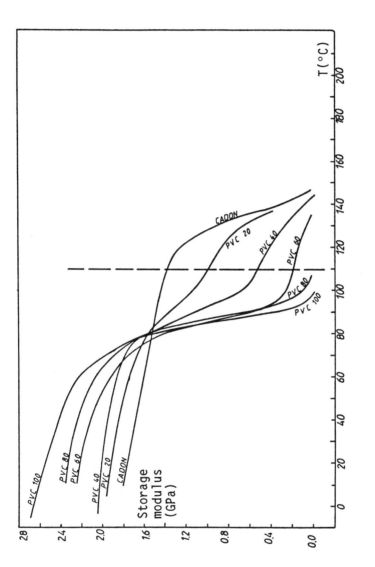

<u>Figure 2</u>   Temperature dependence of the tensile storage modulus of PVC/Cadon blends

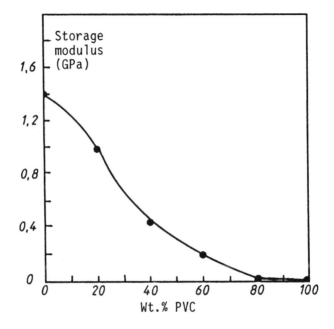

Composition dependence of the storage modulus at
            110°C for PVC/Cadon blends

    In figure 3, the storage modulus measured at 110°C is
plotted against the blend composition;   the S-shaped
dependence of the storage modulus on blend composition
clearly indicates the presence of two phases with the phase
inversion occurring in the range of 20 to 40 wt.% PVC.

    Electron microscopic observations on ultramicrotome
sections confirm the conclusion from the DMA-analysis that
the PVC rich phase is already forming the continuous matrix
phase at a concentration of only 40 wt.%.

3.2.   Mechanical Behavior and Deformation Mechanisms of
        PVC/Cadon Blends

    The tensile stress-strain behavior of the PVC/Cadon blends
has been studied at room temperature at different strain
rates.

    The stress-strain curves are represented in Figures 4 and
5 for strain rates of 1 mm/min and 100 mm/min, respectively.
As can be seen, PVC, Cadon and their blends clearly exhibit
plastic deformation. The yield stress and elongation at

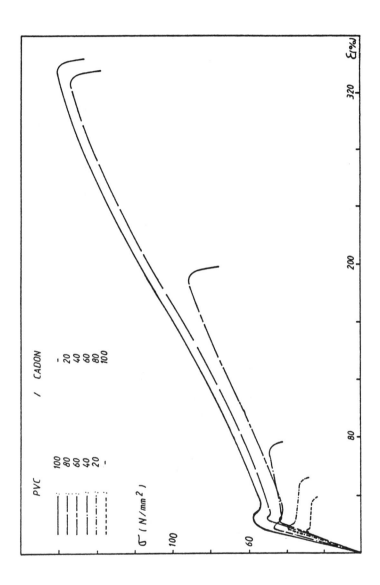

Figure 4   Stress-strain curves of PVC/Cadon blends at 20°C   (strain rate: 1 mm/min)

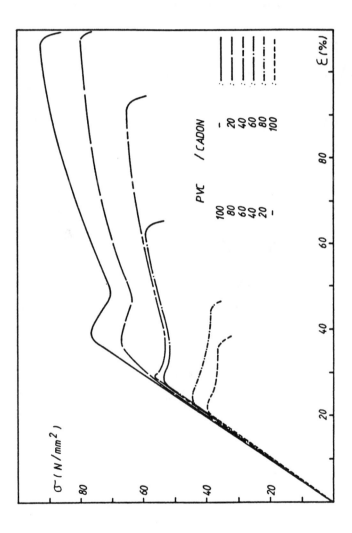

Figure 5   Stress-strain curves of PVC/Cadon blends at 20°C   (strain rate: 100 mm/min)

rupture of the blends increase markedly with increasing PVC content. The tensile toughness, as deduced from the area under the stress-strain curves, also rises significantly with increasing PVC content in the blends. The higher fracture strain of the 20/80 and 40/60 PVC/Cadon blends compared to Cadon must be attributed to the good interfacial adhesion between the glassy PVC phase and the glassy miscible matrix of Cadon, as a result of their partial miscibility.

The elongation at fracture is almost independent of the applied strain rate within the range considered for Cadon as well as for the 20/80 and 40/60 PVC/Cadon blends with a continuous Cadon rich phase. For the 60/40 and 80/20 blends with a PVC rich matrix phase, and for pure PVC, the elongation at rupture is much lower at higher strain rates. The tensile toughness follows nearly the same trend as the elongation at rupture with increasing strain rate.

The impact behaviour of the different materials has also been investigated. The notched Izod impact strength as a function of blend composition is given in Figure 6. As can be seen, the Izod impact value increases strongly with increasing Cadon content for the blends where the PVC rich phase forms the matrix phase.

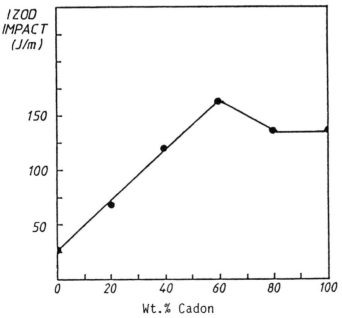

Figure 6   Notched Izod impact strength vs. composition for PVC/Cadon blends at 20° C

This shows that the dispersed rubber particles within Cadon
play a very important role in the toughening mechanism of
these blends at very high deformation rates. Cadon can
thus be used as an effective impact modifier for PVC; PVC
becomes brittle when notched, particularly under impact
conditions. Above the phase inversion point, where the
Cadon rich phase forms the continuous matrix, the Izod
impact strength remains nearly constant.

Vicat softening temperatures of the PVC/Cadon blends
were determined in 1 kg and 5 kg-tests as a function of
of blend composition (Figure 7). As can be seen, the blends
show a marked increase of the Vicat point compared to PVC.
For 50/50 PVC/Cadon blend, an increase of the Vicat
point of 15° C in a 5 kg-test and 30° C in a 1 kg-test has
been observed.

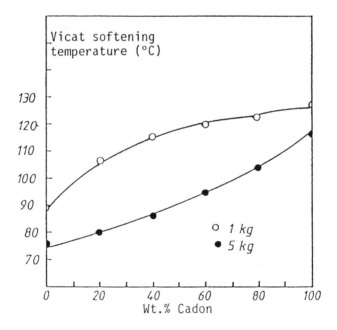

Figure 7   Vicat softening temperature vs. Cadon content
            in PVC/Cadon blends

The mechanisms of plastic deformation of PVC, Cadon and their blends during the tensile experiments will now be examined.

In glassy polymers, two different deformation mechanisms have been identified : crazing and shear yielding (7,8). Crazing is a tensile yielding process accompanied by molecular orientation in the stretching direction and extensive formation of microvoids (9). Crazes occur in a direction perpendicular to the tensile stress axis and cause an increase in the sample volume. Shear yielding involves elongation by shearing under a certain angle with regard to the tensile stress direction, with no change in sample volume.

Measurements of longitudinal and lateral strains can be used to evaluate quantitatively the contribution of crazing and shear yielding to the extension of the samples (7,10). This analysis is based on the assumption that shear deformation occurs at constant volume, so that any volume increase has to be attributed to crazing.

Pure PVC deforms completely by shear yielding while Cadon deforms predominantly by crazing (about 90 %). Cadon is much more sensitive to crazing than the rubber containing modifier used here (about 70 % crazing) because the deformation mechanism of the terpolymer is 100 % crazing.

The fraction f_c of the total deformation of the samples studied due to crazing is given in Figure 8 as a function of the PVC content in the blends, for tensile strain rates of 1 and 100 mm/min, respectively.

The plastic deformation of the PVC/Cadon blends is partly by crazing and partly by shear processes, and the fraction of the deformation of the blends due to crazing decreases with increasing PVC content. At a strain rate of 100 mm/min, a lower amount of crazing is observed than at 1 mm/min. The curves in Figure 8 exhibit a sigmoidal shape as a result of the partial miscibility in the blends.

Transmission electron microscopic observations performed on tensile samples of the blends with co-continuous matrix phases indicate craze initiation and propagation in the Cadon rich phase and craze termination at the ductile PVC rich phase, together with extensive deformation of the matrix phases and of the dispersed rubber particles. Similar observations have recently been made by Mendelson in blends of ABS/Styrene-maleic anhydride-methyl methacrylate terpolymer/polycarbonate of bisphenol A (11).

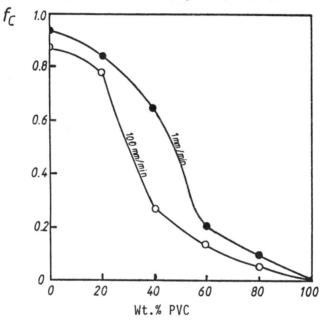

Figure 8   Relative amount of crazing f_c vs. PVC content in PVC/Cadon blends

The PVC/Cadon blends described in this paper present an excellent combination of properties : the incorporation of Cadon in PVC during intensive melt-mixing results in partially miscible systems which show a substantial improvement in practical heat distortion temperature and impact strength compared to PVC. PVC/Cadon alloys are thus very interesting from a practical point of view.

The combination of crazing and shear yielding in PVC/Cadon blends make them very attractive for the systematic study of the kinetics of plastic deformation during tensile, creep and impact experiments.

ACKNOWLEDGEMENTS

The authors wish to thank Mr. D. Maes and Mr. G. Defieuw for the many discussions during the experimental work and for the electron microscopic investigations.

REFERENCES

(1)   Paul, D.R.; Newman, S. (Eds.) In "Polymers Blends"; Academic Press : New York, 1978, vols. I and II.

(2)   Olabisi, O.; Robeson, L.M.; Shaw, M.T. In "Polymer-Polymer Miscibility"; Academic Press : New York, 1979

(3)   Klempner, D.; Frisch, K.C. (Eds.) In "Polymers Alloys"; Plenum Press : New York, 1980, vol. I.

(4)   Solc, K. (Ed.) In "Polymer Compatibility and Incompatibility : Principles and Practices", Harwood Acad. Publ., New York, MMI Press Symp. Series, 1982, vol. 2.

(5)   Walsh, D.J.; Higgins, J.S.; Maconnachie, A. (Eds.) In "Polymer Blends and Mixtures", Martinus Nijhoff Publishers, Dordrecht, Nato Asi Series, 1985.

(6)   Maes, D.; Groeninckx, G.; Ravenstijn, J.; Aerts, L.; in press.

(7)   Bucknall, C.B. In "Toughened Plastics", Applied Science, London, 1977.

(8)   Kambour, R.P., J. Polym. Sci., Macromol. Rev., 1973, 7, 1.

(9)   Maes, D.; Groeninckx, G.; Ravenstijn, J.; Aerts, L., Polymer Bulletin, 1986, 16, 363.

(10) Groeninckx, G.; Chandra, S.; Berghmans, H.; Smets, G. In "Multiphase Polymers", Adv. in Chem. Ser., ACS, Washington D.C., 1979, 176, 337.

(11) Mendelson, R.A., J. Polym. Sci., Phys. Ed., 1985, 23, 1975.

Part VIII
Composites and Adhesives

STUDIES ON MODIFIED BISMALEIMIDE RESINS

I.K.Varma, Sangita, S.P.Gupta and D.S.Varma

Centre for Materials Science & Technology and
Department of Textile Technology,
Indian Institute of Technology,
Hauz Khas, NEW DELHI-110016, INDIA

Bismaleimide based thermoset resins are very attractive for high temperature applications and are increasingly used as matrix resins for advanced fibre-reinforced composites. Attempts have been made in the past to improve processability and fracture toughness of bismaleimide resins by nucleophilic addition reaction to maleimido double bonds, coreaction with reactive diluents and blending with oligomers and elastomers. Modifications of bismaleimide resins with nonstoichiometric amounts of aromatic/aliphatic diamines (chain extended bismaleimides), some allyl type toughners and epoxy resins have been described in literature. The present studies were undertaken to investigate thermal behaviour of chain extended 4,4'-bis-(maleimidophenyl) methane (BM) or 4,4'-bis(maleimidophenyl) ether (BE) with epoxy resin containing an amino type hardner or a diallyl compound 4,4'-diallyl 2,2'-dimethoxy diphenyl sebacate (AEg). Chain extension of BM or BE resin was done with 4,4'-diaminophenyl methane (M) or 4,4'-diamino diphenyl ether (E) in 1:0.3 molar ratios. Diglycidyl ether of bisphenol A containing a stoichiometric amount of tris-(m-aminophenyl) phosphine oxide (TAP) [5,10,30 & 50% (w/w)] was solution blended with BE-E resin and curing characteristic & thermal stability of cured products was evaluated. Similar blends were also prepared using BM-M and AEg which was prepared by reaction of eugenol with sebacoyl chloride. A significant change in the curing exotherm was observed on blending of BE-E/BM-M resins. Isothermal curing of these resin blends was done in the temperature range of 200-220°C for several hours. A reduction in initial decomposition temperature (IDT) was observed on blending. A decrease in char yield at 800°C in BM-M resin was observed on increasing AEg content whereas an increase in char yield in BE-E resin was observed on increasing epoxy-TAP content from 10-50%.

Introduction

Epoxy resins have been used in the past as matrix resins for the fabrication of carbon-fibre reinforced composites. However, these resins show only limited temperature capability and poor performance in hot-wet environment. Bismaleimide resins, on the other hand, exhibit excellent retention of properties in hot-wet environment and are very attractive candidates for advanced fibre-reinforced composites suitable for high temperature applications.

Bismaleimide (BM) resins, however, are brittle in nature due to high crosslink density and aromatic structure of the backbone. Attempts have been made in the past to improve processability and fracture toughness of BM resins by nucleophilic reaction to maleimido double bond. Resin formulations based on BM and nucleophiles such as diamines [1-6], dihydrazide [7], aminoacid hydrazide [8] and bisphenols [9] have been developed. Copolymerisation of maleimido double bond with vinylic compounds helps in improving the processability of BM resins. Recently a class of new propenyl-terminated sulfone-ether oligomer [10] has been developed which is a useful reactive comonomer for bismaleimide resins. Triallyl cyanurate [11], triallyl isocyanurate [12], vinyl ester [13,14], o-diallyl phthalate [15], o,o-diallyl bisphenol A [16] and allyl type toughners [17] have also been used for this purpose. o,o-diallyl bisphenol A in combination with 4,4'-bis(maleimidophenyl) methane and bis(allylphenyl) type liquid modifiers have been found to be very effective toughening modifiers for bismaleimide resins [18]. These allyl benzene derivatives undergo a linear chain extension reaction with bismaleimides via a ene-type addition reaction thus leading to tough cured resins. Prepregs with good tack and drape could be obtained by using these compounds as reactive diluents. Bismaleimide resin formulations with improved properties have been developed by blending with elastomers [19] and epoxy resins [20].

It would be of interest to investigate the curing characteristics and thermal behaviour of bismaleimide resin blends containing reactive diluents. Two such formulations were investigated in the present work. A novel diallyl modifier derived from 4-allyl, 2-methoxy phenol (eugenol) and sebacoyl chloride was prepared and used as modifier for chain extended 4,4'-bis(maleimido phenyl) methane (BM). Chain extended BM resin was obtained by reacting 1 mole of

BM with 0.3 mole of 4,4'-diamino diphenyl methane. The BM-M resin thus obtained was blended with 4,4'-diallyl, 2,2'-dimethoxy diphenyl sebacate (AEg),(0, 5, 10, 15, 20, 25, 35, 45%, w/w). The second resin formulation was obtained by blending bismaleimide resin with diglycidyl ether of bis-phenol A containing stoichiometric amount of tris(m-amino-phenyl) phosphine oxide (TAP) as hardner. 4,4'-bis-(maleimido phenyl) ether (1 mole), chain extended with 4,4'-diaminodiphenyl ether (0.3 mole) (BE-E resin) was blended with 5,10,30 and 50% (w/w) of epoxy-TAP resin for this purpose. Curing characteristics and thermal behaviour of cured resins was evaluated by using DSC and TG techniques.

Experimental

Materials. Diglycidyl ether of bisphenol A (Hindustan Ciba-Geigy), 4,4'-diaminodiphenyl methane (Fluka), 4,4'-diaminodiphenyl ether (Fluka) were used as such. Synthesis of 4,4'-bis (maleimidophenyl) methane (BM) and 4,4'-bis-(maleimidophenyl) ether (BE) was carried out according to procedure described elsewhere [4,21].

Chain extension reaction of BM with 4,4'-diamino diphenyl methane (M) and BE with 4,4'-diamino diphenyl ether (E) was carried out by refluxing an acetone solution of bismaleimide with amine (1:0.3 molar ratio) for 3-4 h until a homogeneous solution was obtained. Acetone was removed under vacuum at a temperature of 50-55°C using a rotary evaporator and yellow shiny powder of chain extended bismaleimide was obtained.

Tris (m-amino phenyl) phosphine oxide (mp 258-260°C) was prepared from triphenyl phosphine oxide (Fluka) by nitration and subsequent reduction with Pd-C/hydrazine hydrate [22].

Bis(4-allyl, 2-methoxy phenyl) sebacate (AEg) was syn-thesized by reacting 2.0 mole of 4-allyl, 2-methoxy phenol (eugenol) (Fluka) with 10% aq. NaOH solution. Sebacoyl chloride (1 mole)(BDH) was then added with constant stirring. The reaction mixture was stirred at room tempera-ture for 15-20 min.The resulting solution was then poured in excess of water. The precipitates thus obtained were filtered and dried at 50°C in an air oven. Crystallization of AEg was done in chloroform (mp 76.3°C, yield 70%).

Solution blending of BM-M resin with AEg was done in chloroform and BE-E resin with epoxy-TAP was done in MEK. The solvent was later removed under reduced pressure/air oven.

Curing Studies. The curing of BM-M/AEg blends was done at 200°C for 2h in an air oven. BE-E/epoxy-TAP blends were cured at 200°C for 4 h. Post curing at 220°C for 4h was also done for BE-E/epoxy-TAP blends.

Percent weight loss was noted down during curing of various resin blends. The extent of crosslinking was determined by solubility measurements in DMF. The cured resin was boiled in DMF for five minutes and the insoluble fraction was collected on filter paper, washed with acetone and dried. The % solubility was thus estimated

$$\% \text{ solubility} = \frac{w_0 - w_1}{w_0} \times 100$$

Where w_0 = initial weight of polymer and
        w_1 = weight of insoluble material

Characterization. The purity of AEg, bismaleimides and chain extended bismaleimides was checked using High Performance Liquid Chromatography (HPLC) technique. A Du Pont HPLC having Zorbax-ODS column was used. The elemental composition of AEg, was determined using a Perkin Elmer 240 C elemental analyser. IR spectra of eugenol and AEg in KBr pellets were recorded on a 5 DX Nicolet FTIR spectrophotometer. H-NMR of eugenol and AEg were recorded in CDCl$_3$ using a Jeol JNM-FX 100 FT-NMR spectrometer.

A Du Pont 1090 thermal analyser was used to evaluate the thermal behaviour of these resins. The DSC measurements were done using a 910 DSC module in static air at a heating rate of 10°C/ min. Thermogravimetric analysis was carried out on a 951 TG module in nitrogen atmosphere (flow rate 100 ml/min). Sample size of 11$\pm$2 mg and a heating rate of 10°C/min was used.

Results and Discussion:

The purity of various samples was checked by HPLC using
chloroform as an eluent, Zorbax-ODS column and UV detector.
A flow rate of 1 cm³/min, pressure, 57-58 bar and temperature of 30°C was used. In AEg, a single peak was observed
at a retention time of 1.9 min indicating thereby, that the
product was pure and comprised of single component. For BM
and BE resins, a single sharp peak with retention time of
2.1 and 1.8 min respectively was observed. In BM-M and BE-E
resins, two peaks were observed, one peak was identified as
unreacted BM or BE resin on the basis of retention time
while the retention time for second peak was around 3.5 min
and is due to chain extended resin. Thus BM-M and BE-E
resins comprised of two components-one the unreacted bismaleimide while the other one is the product of nucleophilic addition of diamine with maleimido double bonds. Such a
mixture is expected on the basis of molar ratios of resins
(1 mole) and diamines (0.3 mole) taken for chain extension.

The   results of elemental analysis of AEg agreed well
with the theoretical values, and were found to be C=71.5%,
H=7.64%. Theortical values were   C=72.8%, H=7.69%. In the
IR spectrum of AEg, characteristic absorption bands for
ester were observed at 1760±10(νC=O) and 1210±10 cm⁻¹ (C-O-

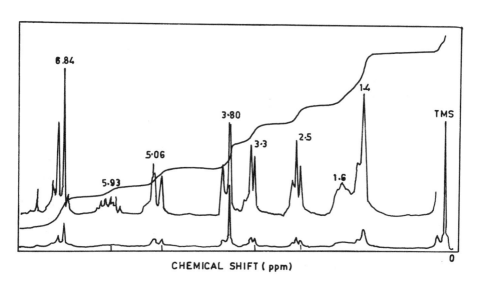

CHEMICAL SHIFT (ppm)

<u>Figure 1.</u>   ¹H NMR of Bis(4-allyl,   2-methoxy phenyl)
          sebacate (AEg).

C-stretch). The structure of AEg was also confirmed by ^{1}H-NMR spectroscopy. In the PMR spectrum of eugenol, characteristic proton signals of benzylic $-CH_2-$ and methoxy group were observed at 3.25, and 3.8 ppm respectively. The two protons of vinylic $-CH_2-$ group of eugenol are magnetically not equivalent and also are coupled with $-CH=$ proton, therefore, the signal split into multiplet centred at 5.06 ppm. The trans coupling constant of the order of 16 cps was observed. The signal due to C-H proton also appeared as a multiplet centred at 5.93 ppm due to coupling with benzylic-CH_2- and vinylic $-CH_2-$ groups. The proton signal for -OH was observed at 5.66 ppm. In AEg the signal due to -OH proton disappeared (Fig-1) and additional signal due to (CH_2) protons appeared as multiplet centred at 1.4 ppm. Signal due to $-CH_2-CO-$protons was at 2.5 ppm and due to $-\underline{CH}_2-CH_2CO-$ at $1.\overline{60}$ ppm.

Curing Studies. The details of various resin formulations prepared are given in Table I. The resin designations given for these samples in this Table have been used throughout the text. Isothermal curing of various resin formulations was done at 200°C for different time intervals. In BE-E/epoxy-TAP resins, postcuring was also done at 220°C for 4h. Percentage weight loss was noted down at different time intervals during this isothermal curing and these results are summarized in Table-I. In BM-M/AEg blends approx. 1-4.5% weight loss was observed by heating at 200°C for 1h. The weight loss increased as AEg content increased in the blends. Heating for additional 1h did not lead to additional weight loss. In BE-E/epoxy-TAP systems 5-6% weight loss was observed after heating for 4h at 200°C when low weight percent of epoxy-TAP was present (BEET and BEET resins). However, on increasing the percentage of epoxy-TAP in these blends (> 30%), higher weight loss was observed during isothermal curing (BEET$_3$ and BEET$_4$).

The results of solubility of cured resins in DMF are given in Table-II. In BM-M/AEg blends, percentage solubility increased on increasing AEg content. This may be due to unreacted AEg in these blends. BE-E resin was found to have 9.65 % solubility in DMF after curing at 200°C whereas 89-94 % solubility was observed for blends having low weight percent epoxy-TAP resin. Blends containing 30-50 % epoxy-TAP were found to have solubility in the range of 29-30 % after curing for 4h at 200°C. Hence further curing of these samples was done at 220°C for 4h. Such a treatment resulted in an increase in crosslinking and a consequent reduction in solubility (Table II).

Table-I   Results   of   Percentage Weight Loss on Curring   of Bismaleimide   resin blends at 200°C   for   differnt time intervals.

Resin System	Wt % of diluent	Resin designation	Wt loss(%) after curing		
			1 h	2 h	4 h
BM-M/AEg	5	BMMA$_1$	2.62	2.64	-
	10	BMMA$_2$	1.20	1.20	-
	15	BMMA$_3$	2.60	2.67	-
	20	BMMA$_4$	3.8	3.80	-
	25	BMMA$_5$	4 .2	4.24	-
	35	BMMA$_6$	4.40	4.40	-
	45	BMMA$_7$	3.90	3.96	-
	100	AEg	23.0	-	-
BE-E/Epoxy -TAP	0	BE-E	-	-	2.35 (1.63)
	5	BEET$_1$	4.1	4.58	5.74 (1.06)
	10	BEET$_2$	4.08	4.49	5.43 (1.10)
	30	BEET$_3$	5.9	6.71	8.26 (2.05)
	50	BEET$_4$	7.38	8.16	9.98 (2.44)

Figures in parenthesis indicate the weight loss observed on further postcuring of resin samples at 220°C for 4h.

Table-II Results of percentage solubilty (in DMF) on
          Curing of various samples for 2h/4h

S.No.	Resin designation	% Solubility after 2h	Resin designation	% Solubility after 4h
1	$BMMA_1$	8.56	BE-E	19.46 (9.65)
2	$BMMA_2$	8.76	$BEET_1$	93.46 (36.74)
3	$BMMA_3$	18.29	$BEET_2$	89.4 (24.27)
4	$BMMA_4$	25.58	$BEET_3$	29.89 (14.4)
5	$BMMA_5$	27.52	$BEET_4$	29.48 (13.64)
6	$BMMA_6$	44.59	-	-
7	$BMMA_7$	34.00	-	-

Figures in parenthesis indicate % solubility in DMF after
post curing at 220°C for 4h

    In the DSC traces of resin samples endothermic transi-
tion due to melting and exothermic transition due to curing
were observed. These DSC traces were characterized by
noting down the temperatures of these transitions. These
included temperature of endothermic peak position (T_m),
exothermic peak position (T_{exo}). Temperature of onset of
curing reaction (T_1) (determined by extrapolation) and for
completion of reaction T_2 were also determined. Heat of
polymerization (ΔH) was obtained from the area under the
exothermic transitions.

    In BM-M resin (Fig.2a) no endotherm due to melting was
observed. A broad exotherm with peak positions at 136.7 and
283.6°C was observed in the temp. range of 70-330°C. Since

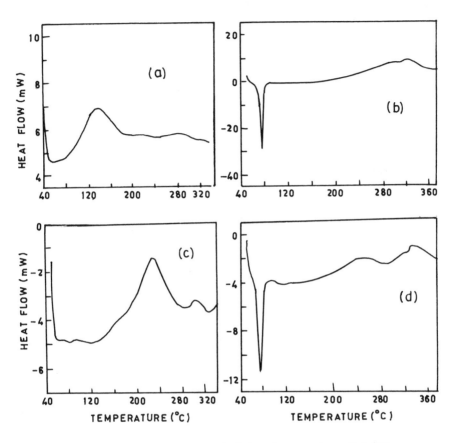

<u>Figure 2.</u>   DSC trace of (a) BM-M resin   (b) AEg
           (c) BMMA$_3$ and   (d) BMMA$_7$

the BM-M resin was a two component system therefore, the
two peaks correspond to curing of BM resin at 283°C and BM-
M resin at 137°C. In BE-E resin, an endotherm due to
melting at 90°C was observed. A broad bimodal exotherm in
the temperature range of 118-320°C with peak positions at
162.7 and 270°C was observed.

Addition of 5-15% AEg to BM-M resulted in an increase in
heat of curing ΔH (from 54.5 to 109.0 J/g). However further
increase in the AEg content from 15 to 45%, significantly
reduced the heat of curing (from 109.0 to 59.7 J/g).
Addition of 5-10% AEg also changed the bimodal character of
exothermic peak.

In BMMA$_4$ - BMMA$_7$ resins, endothermic transition associated with melting of AEg was observed at 76°C. These
DSC results thus indicate the presence of unreacted AEg in
blends containing higher weight % of AEg (Fig.2)(Table-III)

<u>Table-III</u>   Effect   of AEg on curing characteristics of BM-M
           resin.

Resin designation	T$_m$ (°C)	T$_1$ (°C)	Texo (°C)	T$_2$ (°C)	ΔH J/g
BM-M	-	93.0	136.0	200	49.0
		263.0	283.6	330	5.39
BMMA$_2$	-	181.8	215.8	310	86.6
BMMA$_3$	-	177.3	228.1	330	109.0
			301.1		
BMMA$_4$	67.0	183.6	228.3	328	101.6
			302.7		
BMMA$_5$	70.0	189.8	235.4	330	81.0
			308.0		
BMMA$_6$	76.0	189.8	234.9	375.0	43.8
		302.7	318.0		
BMMA$_7$	75.3	202.0	242.7	370	30.2
		316.9	332.9		
AEg	76.3	275.4	320.0	375.0	297.0

Addition of 5% epoxy-TAP resin to BE-E resulted in a
decrease in ΔH from 36.9 J/g to 7.74 J/g. Further increase
in the epoxy-TAP content resulted in disappearance of
exotherm.

The relative thermal stability of various resins was assessed by comparing the initial decomposition temperture (IDT), the temperature of maximum rate of weight loss (T_{max}) and final decomposition temprature (FDT). IDT & FDT were obtained by extrapolation (Fig. 3). Percent char yield (Y_c) at 800°C was also determined. These results are summarized in Table IV. A significant reduction in IDT was observed on increaisng the AEg content from 5-45%. An increase in T_{max} was found on increasing the content of AEg. Blends of BM-M and AEg (5%) showed a char yield of 47.5% . Addition of AEg to BM-M resulted in a decrease in char yield. A two step decomposition was observed in blends of BM-M/AEg with 25 and 45% AEg. this may be due to the presence of unreacted AEg in these blends.

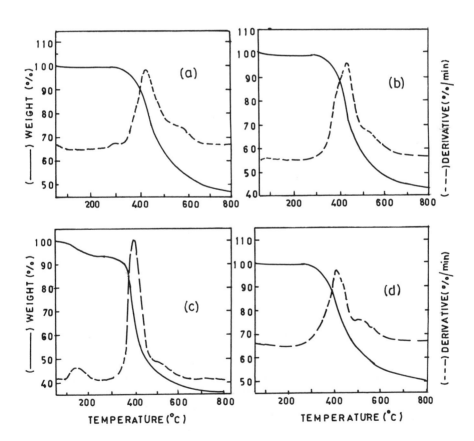

Figure 3. TG traces of cured resins (a) BMMA$_1$ (b) BEET$_2$ (d) BEET$_4$ (c) BEET$_4$ (uncured)

<u>Table-IV</u> Results of TG studies of Cured bismaleimide
resins formulation (curing of BM-M/AEg resins was
done for 2h at 200°C and BE-E/Epoxy-TAP resins
was done for 4h at 200°C and 4h at 220°C).

Resin designation	IDT (°C)	T_{max} (°C)	FDT (°C)	Y_c (%)
$BMMA_1$	375.5	424.0	512.0	47.5
$BMMA_3$	357.7	437.0	524.6	41.5
$BMMA_5$	350.1	315.0 445.7	529.7	39.5
$BMMA_7$	290.0	330.7 448.5	448.5 495.8	29.5
BE-E	387.7	413.6	499.0	42.2
$BEET_1$	373.0	431.7	503.8	43.0
$BEET_2$	369.8	429.9	500.1	43.0
$BEET_3$	358.0	428.8	519.7	51.0
$BEET_4$	356.9	413.5	491.5	50.5
	(370.8)	(409.6)	(460.4)	(35.5)

Figures in parenthesis indicate the results of TG studies
of uncured $BEET_4$.

In BE-E/epoxy-TAP blends, a systematic decrease in IDT
was observed on increasing the Epoxy-TAP content. All these
cured blends were found to be stable upto 350°C (Table-IV).
A reduction in T_{max} ($\backsim 18$°C) was noticed when epoxy-TAP
content was increased from 5-50%. Char yield at 800°C was
not significantly affected on adding low weight % epoxy-TAP
($BEET_1$ and $BEET_2$) but 5% increase in char yield was found
on increasing epoxy-TAP content from 10-50% (Table-IV).

In uncured resin having 50% epoxy-TAP, $\backsim$ 6% weight loss
was observed in the temperature range of 100-200°C (Fig.3)

whereas no weight loss in this temperature range was noticed in cured resins (BEET$_4$).

Conclusions

These results thus indicate that curing characteristics of bismaleimide resins are significantly altered in presence of allyl/expoxy type modifiers. Addition of bis-(4-allyl, 2-methoxy phenyl) sebacate upto 20% results in a decrease in Texo values indicating thereby copolymerisation of allyl group with maleimide double bonds. Further increase in AEg did not influence the curing characteristics. This was primarily due to unreacted AEg in these blends. Thermal stability of these blends was only marginally affected by adding AEg upto 20%. Further increase in AEg however, resulted in a decrease in IDT and char yield. Addition of low weight percent of epoxy-TAP to bismaleimides did not affect the char yield whereas crosslinking and percent char yield were increased on adding 30-50% epoxy-TAP resin.

References

[1]   Bergain, M.; Combet, A.; Grosjean, P. U.S.Pat. 1971, 3,562,223.

[2]   Crivello, J.V. J. Polym. Sci. Polym. Chem. Ed., 1973, 11, 1185.

[3]   Gherasim, M.G.; Zugravescu, I. Eur. Polym. J. 1978, 14, 985.

[4]   Varma, I.K.; Sangita; Varma, D.S. J. Appl. Polym. Sci. 1983, 28, 191.

[5]   Varma, I.K.; Sangita; Varma, D.S. J. Appl. Polym. Sci. Polym. Chem. Ed. 1984, 22, 1419.

[6]   Varma, I.K.; Sangita; Varma, D.S. J. Appl. Polym. Sci. 1984, 29, 2807.

[7]   Asahara, T. JAP. 1972,12, 14745; Chem. Abstr. 1972, 77, 152864n.

[8]   Stenzenberger, H.D. U.S.Pat. 1980, 4,211,861.

[9]   Renner, A.; Forgo, J.; Hoffmann, W.; Ramsteiner, K. Helvetia Chimica Acta. 1978, 61, 4.

[10] Stenzerberger, H.D.; Konig, P.; Herzog, M.; Romer, W.; Canning, M.S.; Pierce, S. Int. SAMPE Tech. Conf. 1986, 18, 500.

[11] Varma, I.K.; Gupta, S.P.; Varma, D.S. J. Appl. Polym. Sci. 1987, 33, 151.

[12] Harruff, P.W. Nat. SAMPE Tech. Conf. 1979, 11, 1.
[13] Varma, I.K.; Choudhary, M.S.; Rao, B.S.; Sangita; Varma, D.S. J. Macromol. Sci. Chem. Ed. 1984, A21, 793.
[14] Varma, I.K.; Sangita; Varma, D.S. J. Appl. Polym. Sci. 1984, 23, 1885.
[15] Stenzenberger, H.D.; Herzog, M.; Romer, W.; Scheib-lich, R.; Pierce, S.; Canning, M. 29th Nat. SAMPE Symp. 1984, 29, 1043.
[16] King, J.J.; Choudhary, M.; Zahir, S. 29th Nat. SAMPE Symp. 1984, 29, 392.
[17] Stenzenberger, H.D.; Romer, W.; Herzog, M.; Pierce, S.; Canning, M.; Fear, K. 31st Nat. SAMPE Symp. 1986, 31, 920.
[18] Carduner, K.R.; Chattha, M.S. Polym. Mat. Sci. & Engg. 1987, 56, 660.
[19] Varma, I.K.; Fohlen; G.M.; Parker, J.A.; Varma, D.S. in "Polyimides: Synthesis, Characterization and Applications" (Vol.2)Ed. K.L.Mittal, Plenum Press, NY 1984, p. 683.
[20] Landman, D. Proceed. 28th Nat. SAMPE Symp. 1983, 740.

CHEMICAL MODIFICATION OF KEVLAR^R FIBER SURFACES FOR IMPROVED ADHESION IN COMPOSITES

G.C. Tesoro, R. Benrashid

Polytechnic University
Brooklyn, New York 11201

L. Rebenfeld, Umesh Gaur

Textile Research Institute
Princeton, New Jersey 08542

Abstract

Surface-controlled reactions have been carried out on Kevlar^R fibers to introduce functional groups designed to interact with matrix resins and improve adhesion in fiber-reinforced composites. The effects of reaction environments and conditions on the mechanical properties and surface topography of modified fibers have been investigated. For Kevlar^R fibers modified by nitration and by subsequent reduction of nitro groups to amino groups, significant concentrations of functional groups have been introduced with minor strength loss and significant improvement in the interfacial shear strength with an epoxy resin, as evaluated by the microbond technique. The results provide new insights concerning the effect of these functional groups on adhesion in epoxy composites.

Introduction

The increasing importance of aramid fibers (particularly Kevlar) as reinforcing elements in advanced composites stimulates continuing research on chemical modifications of fiber surfaces for the purpose of enhancing fiber/matrix interactions and improving adhesion in composites. Promising results have been reported for the chemical modification of Kevlar surfaces by plasma treatments [1] [2]. Attempts

to provide surface amine sites by hydrolytic scission of
amide groups [3], and by reaction sequences introducing
flexible, reactive pendant groups on the fiber surface [4]
have also been reported. An exploratory study of the surface-limited nitration of Kevlar fibers followed by reduction of nitro groups [5] has shown that significant concentrations of these functional groups could be introduced on
the fiber surface without major loss in fiber strength, and
that improvements in adhesion in epoxy composites could be
realized by these reactions. An in-depth investigation of
this promising approach is the subject of the present paper.

The essential requirement for evaluating the effects of
systematically varied surface modifications on interfacial
bond strength for fiber samples prepared in small amounts
(about one gram) has been met by employing a new experimental technique for direct measurement of interfacial adhesion between fiber and remain matrix which has been developed at Textile Research Institute [6] and termed "microbond." The technique involves application of microdroplets
of suitable resin onto a single fiber, followed by measurement of the force required to pull out or debond the fiber
from the droplet. The main advantage of this modified pull-
out technique is that small diameter (ca 10µm) fibers, commonly used in fiber-reinforced composites, can be evaluated
successfully.

The Microbond Method

The commonly used single fiber pull-out test involves
measuring the force, in a direction parallel to the fiber
axis, to "pull out" a single fiber from a pool of cured
resin in which it is partially embedded. However, if the
embedded length is greater than a critical value, ℓ_c, tensile failure will occur in the nonembedded portion of the
fiber. This critical fiber length is given by

$$\ell_c = \frac{\sigma d}{4\tau}$$

where   d = fiber diameter,
          σ = ultimate fiber tensile stress at break, and
          τ = interfacial bond strength (shear stress).

For very fine fibers such as those associated with engineering composites (glass, carbon, and aramids), the
critical fiber length becomes extremely small. These
fibers can be expected to have diameters of ca 10µm, tensile

strengths ranging from 2000-4000 MPa, and average interfacial shear strengths with typical resins of 10-70 MPa.
Therefore their critical fiber lengths fall in the range of
0.07-1.0 mm. Such very short embedded lengths are difficult
to attain experimentally, and are hard to handle and process.
In addition, meniscus rise around the circumference due to
wetting (Figure 1) can increase the embedded length enough
so that the critical fiber length is exceeded and the fiber
fails prior to pull-out. Another potential problem is that
the thinner resin coating in the meniscus region may rupture prior to debonding, leaving behind a crown or resin
cone on the surface of the fiber (also shown in Figure 1).

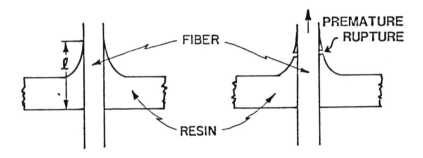

Figure 1. Extended embedded length ℓ produced by
meniscus, and potential rupture of meniscus before debonding.

Although this complex failure process does not invalidate
the subsequent shear strength measurement, it does require
the inspection of every specimen after debonding to determine the true embedded length. Furthermore, this cohesive
fracture suggests that the load recorded during pull-out
may include the load to fracture the resin cone.

The TRI microbond pull-out test involves the deposition
of a small amount of resin onto the surface of a fiber in
the form of one or more droplets which form concentrically
around the fiber in the shape of ellipsoids. After appropriate hardening or curing, the fiber diameter and the droplet dimensions are measured with the aid of an optical
microscope, the size of the droplet along the fiber axis
determining the embedded length.

The fiber specimen is pulled out of the droplet on an
Instron tensile tester using a special device to grip the
droplet bonded to the fiber (Figure 2). This device consists of two adjustable plates that form a slip or microvise that is placed on the crosshead of the Instron.

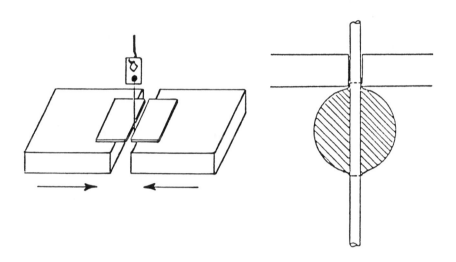

Figure 2. Arrangement for shear debonding and enlarged
schematic of a resin droplet on a fiber under the shearing
plates.

The plates are positioned just above the droplet and the
slit is narrowed symmetrically until the plates just make
contact with the fiber (Figure 2). As the crosshead moves
downward, an initial frictional force between the fiber and
the shearing plates is registered, indicating that the slit
is just touching the fiber and the droplet has little chance
of slipping through; this frictional force is reproducible
and can be adjusted to the same value for each specimen. As
the shearing plates continue to move downward, they encounter
the droplet and exert a downward force on it which is recorded. At present, two droplets are usually deposited on
each fiber specimen a certain distance apart. After the
lower droplet has been sheared, the slit is opened and the
plates moved up and positioned above the second droplet.
Using this procedure, two pull-out results can be obtained
from one fiber specimen.

It is important to be able to distinguish between successful shear debonding and droplet slipping or fiber breakage.
The three types of force traces obtained, illustrated in
Figure 3, accomplish this clearly.   When the droplet is
sheared under a constant rate of strain, the force curve
rises steeply, from some initial frictional force, to a
peak and then drops abruptly to another frictional level.
This curve represents shear debonding and subsequent downward displacement of the droplet along the fiber.   The second type of curve observed also rises steeply but does not
exhibit a sharp peak.   Instead the force continues to rise
slowly until it drops down to another frictional level.
This type of curve is attributed to droplet slippage in the
slit and is disregarded.   The third curve in Figure 3 is
initially similar to the previous traces except that the
force peak is followed by a sharp drop to zero force without
a subsequent frictional force.   This type of curve is characteristic of fiber tensile failure outside the droplet region and is not used in the calculation of interfacial bond
strength.

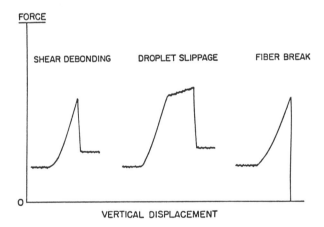

Figure 3. Force traces for the three possible results
of a shear test.

The interfacial shear stress, τ, is calculated using the
equation:

$$\tau = \frac{F}{A} = \frac{F}{\pi d \ell}$$

where F = measured debonding force,
    A = interfacial area,
    d = fiber diameter, and
    ℓ = embedded length.

The droplet size is an important consideration in the
microbond method. The general relationship used to calculate shear strength (τ) implies that the applied force is
distributed uniformly and equally to all elements of the
interface. This is not likely to be happening during microbond shearing, nor in any other version of a pull-out test.
Force applied to a small portion (the area of contact between the microvise and the droplet) of a finite elastic
body (the droplet) would not be evenly distributed but
should decrease with distance from the point of application
(Saint-Venant's Principle).

Since the droplets in these experiments are very small,
the variation in applied stress along the embedded length
might be expected to have only a minor effect on the bond
strength measurement. However, because of this small size,
embedment length can be varied by a factor of five for some
fiber/resin systems. (This is not possible with macro pull-
out experiments.) Figure 4 shows pull-out load versus embedded length data collected for a large set of Kevlar 49/
Epon 828 specimens. Because of the high tenacity of the
fiber, it was possible to use large drops without having the
bond shearing process pre-empted by fiber rupture. The
points in Figure 4 show a distinct trend away from linearity
and toward a reduced slope (ie, apparently lower bond
strengths).

The results in Figure 4 suggest that the applied shearing
force is not being distributed uniformly over the entire
interface. Thus, it is necessary either to keep drop sizes
and size distribution as small as possible, which may not
always be practical, or to compare results for different
systems only within narrow embedment length ranges.

Microbond results reported in this paper have been obtained for embedded areas ranging from 3000 to 4500 square
microns. Other factors of importance such as composition,
stoichiometry and curing conditions for the epoxy resin
droplet have been kept constant and are specified below.

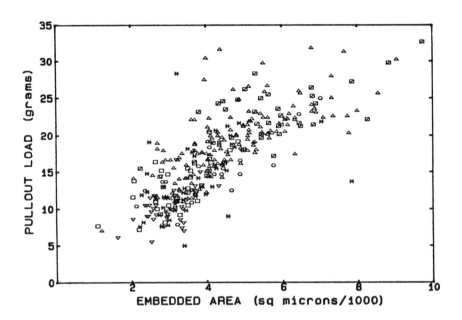

Figure 4. Effect of embedded length on pull-out force for Kevlar 49/Epon 828 microbonds.

Materials and Methods

Fibers. Kevlar 29 .1500 denier, finish free: average filament denier 1.43. Kevlar 49 .1140 denier, finish free: average filament denier 1.71, supplied by E. I. Du Pont and used without further treatment.

Nitration. Fiber (ca one gram) was wound around a cage-like support made of glass rods, and immersed in the nitrating solution which was stirred at the specified temperature. After a given time, the fiber assembly was removed, and

washed with water. The fiber was then unwound, soaked in
water for 24 hours and dried in a vacuum oven at 90°-100°C
for 24 hours. A summary of nitration conditions explored is
shown in Table 1. For <u>Method</u> 1 the nitrating mixture consisted of 40/2/370/100 parts of fuming HNO_3/conc. H_2SO_4/ace-
tic anhydride/acetic acid (by volume). For <u>Method</u> 6 the
nitrating mixture consisted of 5.5g NH_4NO_3 and 2g trichloroacetic acid in a 350/150 mixture of chloroform/acetic anhydride (by volume) [7].

<u>Table 1</u>. Summary of Nitration Methods

Method	Nitrating Agent	Solvent (Conditions)
M-1	HNO_3/H_2SO_4	Ac_2O/AcOH (10°-20°, 2-8 hrs)
M-2	HN_4NO_3/TFA	TFAA/$CHCl_3$ (Room temp, 3-24 hrs)
M-3	HNO_3/H_2SO_4	Ac_2O/TFAA/AcOH (0°-5°, 2 hrs)
M-4	HNO_3/H_2SO_4	Ac_2O/TCAA/AcOH (0°-5°),(10°-15°, 1-2 hrs)
M-5	NH_4NO_3/TCA	TCAA/$CHCl_3$ (Room temp, 2-8 hrs)
M-6	NH_4NO_3/TCA	Ac_2O/$CHCl_3$ (Room temp, 2-8 hrs)

TFA - Trifluoroacetic acid
TFAA - Trifluoroacetic anhydride
TCA - Trichloroacetic acid
TCAA - Trichloroacetic anhydride

Reduction. Nitrated fiber was wound around a cage-like glass support and immersed in the stirred medium under the conditions summarized in Table 2. After washing with water and soaking in water overnight, the reduced fiber was dried in a vacuum oven at 90°-100° for 24 hours. Under the conditions evaluated, results were comparable. Method (A) was employed for most experiments reported here.

Table 2. Summary of Reduction Methods

Reducing Agent (Ref)	Medium (Conditions)	Temp/Time
(A) $NaBH_4/Co(dipy)_3(ClO_4)_3$ (5)(8)(9)	THF/water (1/1) 400 ml 2.5gKH_2PO_3, 0.6gNa_2HPO_4	Room temp, 24 hrs
(B) Sulfrated $NaBH_4$ (10)	Dioxane (400 ml)	Room temp, 15 hrs + reflux, 30 min.
(C) Sodium Hydrosulfite (11)(12)	Water (400 ml) 2.5g Na hydrosulfite 5g NaOH	55°-60°C, 2 hrs

Characterization and evaluation. Ion exchange analysis was carried out with Ponceau 3R [2] [5], using optical density, Ponceau 3R calibration factor, and filament surface area:

$$NH_2/100 \ \overset{\circ}{A}^2 = \frac{OD - 0.0001}{\text{weight of fiber}} \times 1.453$$

Optical Density (OD) on UV Spectrophotometer - (Carry 2300)

FTIR spectra were obtained on a Bio Rad Spectrophotometer, Model FTS 60. Samples were mounted by hanging on the window of the sample holder; 512 scans with a resolution of 4 (nitrogen purge)

Scanning Electron Microscopy on gold coated fibers.

Single fiber tensile tests were run on an Instron tensile tester with gauge length of 1 inch, and crosshead speed of 0.1000 in./min.

Microbond tests – resin microdroplets were epoxy resin (Epon^R 828) with methylene dianiline (MDA) at 4/1 weight ratio, cured 2 hrs at 80°C, then 2 hrs at 150°C.

Results and Discussion

Based on the results of a study of nitration and reduction of methyl-substituted model diamides (reported elsewhere), a range of conditions was identified for the experimental investigation of surface reactions on fibers. Within the range, effects of reaction variables on concentration of functional groups, on surface topography (SEM), and on fiber mechanical properties were determined for Kevlar 49 and for Kevlar 29. The criteria for emphasizing specific methods and conditions in more extensive experiments were based primarily on an assessment of these effects, coupled with empirical observations relating to uniformity and appearance in modified fibers.

The summary of mechanical properties shown in Table 3 for fibers nitrated in different reaction environments (Table 1) is indicative of strength losses encountered under varying conditions of nitration. Additional evidence of surface topography obtained from SEM results on the nitrated fibers supported the decision to focus on nitration by Method 1 [5] and by Method 6, and on reduction with Na BH$_4$ (Co/dipy)$_3$ (ClO$_4$)$_3$ [8] [9] in subsequent work, primarily because of the absence of roughness and flaws in the surface of fibers modified in these reaction environments. The surface topography of Kevlar 49 modified by nitration (Method 1) and by subsequent reduction is illustrated in the micrographs of Figures 5A, B, C, D (X 3000). For nitrated fiber (5B), and for nitrated/reduced fiber at 0.4 NH$_2$/100 Å^2 (5C) and at 0.65 NH$_2$/100 Å ^{2} (5D), the smooth surface observed in the SEM is comparable to that of a control (5A).

The concentration of functional groups attained was clearly a primary consideration. For NO$_2$, this was not determined quantitatively, but inferred from NH$_2$ values measured after reduction assuming reduction to be complete. The NO$_2$ absorption peaks in FTIR spectra at 1518 cm^{-1}, 1341 cm^{-1}, and 838 cm^{-1} provided additional, albeit qualitative evidence for nitration of the aromatic ring in the fiber [13].

Figure 5. Scanning Electron Micrographs (3000 X) of Kevlar 49

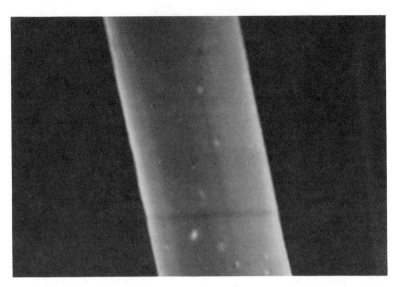

Figure 5A. Untreated Control

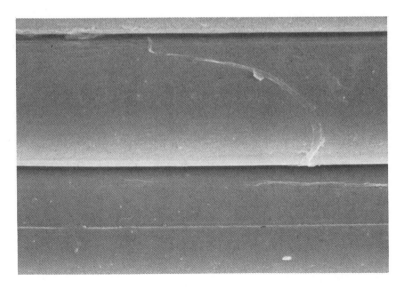

Figure 5B. Nitrated (Method 1)

Figure 5C. Nitrated and reduced (0.4 NH_2/100 $\overset{\circ}{A}{}^2$)

Figure 5D. Nitrated and reduced (0.65 NH_2/100 $\overset{\circ}{A}{}^2$)

Table 3. Effect of Nitration Method on Mechanical Properties
(Ranges)

Nitration Method (Table 1)	Average Tenacity(g/den)		Ave. Elongation %	
	Kevlar 49	Kevlar 29	Kevlar 49	Kevlar 29
M-1	17.6-27.2	20.3-25.6	3.9-5.2	3.8-5.1
M-2	17.8-25.4	15.8-23.3	2.8-3.8	3.3-4.5
M-3	22.7-29.1	19.2-27.7	3.8-4.5	3.8-5.4
M-4	26.7	22.8-25.2	4.3-	4.3-5.2
M-5	20.1-28.3	21.6-24.2	2.9-4.5	4.5-4.6
M-6	25.4-29.0	18.7-22.6	3.5-4.4	3.9-4.9
Control	26.0	26.2	4.1	5.1

For nitration with fuming HNO_3 (Method 1) effects of reaction time and of HNO_3 concentration on the level of modification are shown in Figures 6 and 7 respectively as $NH_2/100$ Å^2 on reduced fiber. Several trends believed to be important are apparent from these results. In spite of some scatter, the results provide clear evidence for somewhat higher reactivity of Kevlar 29 as compared to Kevlar 49, particularly as a function of HNO_3 concentration (Figure 7). This may reflect microstructural differences in the fiber surface. The maximum value of 0.8-0.9 $NH_2/100$ Å^2 reached in the surface-modified fibers is of particular interest. It coincides with maximum values of 0.75-1.0 $NH_2/100$ Å^2 reported by Allred for the modification of Kevlar 49 with ammonia plasma [2], and it suggests that higher values could be reached only under conditions causing disruption of surface structure and radial penetration of reagents into the fiber.

The value of 100 Å^2 area chosen to normalize amine concentration because of ease of visualization can be converted to moles NH_2/cm^2 by multiplying by 1.66×10^{-10}. Considering the average area of a PPTA (poly p-phenylene terephthalamide-Kevlar) repeat unit as 83.9 A^2 [14], a value of ca 1.0 functional group per unit obtained in the nitration/reduction sequence is consistent with the theoretical maximum for electrophilic substitution in the amide-flanked aromatic rings on

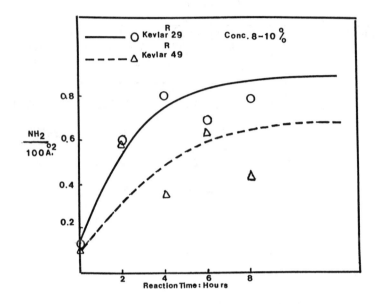

Figure 6. Effect of nitration time on modification
nitration (Method 1): 18°-20°C, Reduction (A)

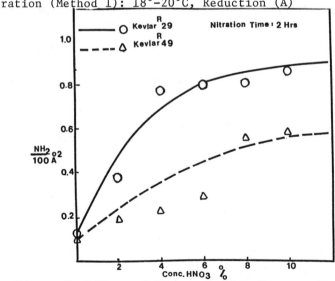

Figure 7. Effect of HNO$_3$ concentration on nitration
modification (Method 1): 18°-20°C, Reduction (A)

the fiber surface. The existence of this maximum value supports the hypothesis that reaction has been limited to the
surface, and that nitration does not involve the carbonyl-
deactivated rings in the PPTA.

The effect of reduction on the mechanical properties of
fibers nitrated by Method 1 or by Method 6 is essentially
nil, as shown by average tenacity values summarized in
Table 4. It is interesting to note that there is no apparent
correlation between surface concentration of NH_2 and average
tenacity for the reduced fibers, nor for their nitrated precursors. Interfacial bond strength is thus the ultimate
criterion for optimizing modification within the range of
functional group concentrations obtained in surface-limited
nitration and reduction, under the conditions investigated.

Table 4. Effect of Reduction on Tenacity (Value in
         Parenthesis Before Reduction)

Nitration Method	$NH_2/100 \, \mathring{A}^2$	Average Tenacity (g/den)
		Kevlar 29
M-1	0.38	22.5 (22.6)
	0.64	23.1 (19.2)
	0.78	22.8 (23.8)
	0.80	22.4 (22.6)
M-6	0.33	22.4 (18.7)
	0.56	23.3 (22.7)
	0.64	22.6 (21.4)
	1.06	20.4 (17.9)
		Kevlar 49
M-1	0.22	24.8 (28.7)
	0.40	20.2 (25.7)
	0.58	28.5 (23.7)
	0.63	22.5 (24.3)
M-6	0.19	27.7 (25.4)
	0.39	28.3 (27.6)

Measurements of shear strength at the fiber/resin interface were carried out by the microbond technique [6] for
modified Kevlar filaments in which nitro groups and amino
groups were present, in order to estimate the effect of the
polar surface groups on adhesion in epoxy composites. Results obtained for systematically varied functional group

content after nitration by Method 1 (Figure 5) and subsequent
reduction are summarized in Table 5 and in Figure 8. Values
calculated for a narrower range of embedded lengths are shown
in Table 6. The shear strength improvement for fiber modified with surface amine groups is evident, while values for
NO_2-modified fibers are comparable to those of the control.

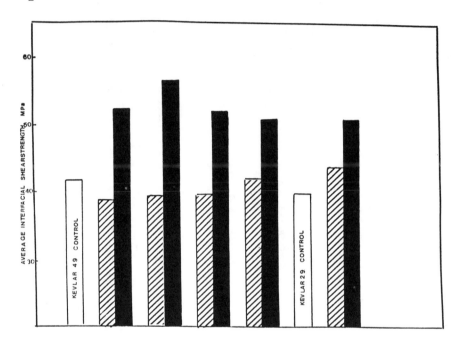

Figure 8. Summary of microbond results
           Shaded bars - nitrated only
           Black bars - nitrated and reduced

These results are consistent with the hypothesis [2] [5] that
covalent bonding of NH_2 on the Kevlar surface with epoxy
groups in the resin plays a significant role in fiber/matrix
adhesion. Some uncertainty remains concerning the possible
relationship of surface amine concentration to shear
strength, and also the effect which reaction environment may
have on the distribution of functional groups introduced.
The values of NH_2 content reported are averages of replicate
analyses which are generally in good agreement (eg 0.62, 0.57,
0.60, 0.47, 0.54), but visual observation of dye sorption
suggests that different reaction environment and conditions
can affect uniformity of the surface modification and the reproducibility of the analytical results.

Table 5. Interfacial Shear Strength (MPa) Averaged Over Complete Range of Experimental Embedded Area (2100-6600 square microns) ± 95% Confidence

Fiber	MPa	$NH_2/100\text{Å}^2$
Kevlar 49 Control	42.6±3.4	(0.10)
Nitrated-M-1-1	38.9±2.3	–
2	39.4±2.4	–
3	39.7±3.0	–
4	42.1±3.1	–
Nitrated/Reduced-1	52.4±2.7	0.58
2	56.7±3.1	0.35
3	52.1±4.0	0.63
4	50.8±4.2	0.43
Kevlar 29 Control	39.9±2.0	(0.13)
Nitrated M-1	43.8±2.6	–
Nitrated/Reduced	50.8±2.7	0.60

Table 6. Comparison of Treated Kevlar 49 Fibers Over Narrower Embedded Area Range (3000-4500 sq.microns)

Kevlar 49 Fiber	Average Embedded		Interfacial Shear Strength, MPa (±95% Confidence)
	Area (sq.μ)	Length (μ)	
Nitrated M1-1	3759	105	39.3±2.7
2	3724	104	39.4±3.2
3	3759	105	36.0±3.3
4	3759	105	38.4±4.3
Nitrated/Reduced 1	3494	97.6	46.6±2.5
2	3551	99.2	52.0±3.7
3	3795	106	48.2±3.9
4	3795	106	47.1±4.1

We have also noted that interfacial shear strengths for NO_2-modified and reduced fibers at comparable functional group content may depend on the specific nitration environment (eg Method 1 vs Method 6) and thus, possibly on reagent penetration during nitration and on the locus of modification (eg [15]). These uncertainties can be resolved only by

further study. Nevertheless, the results reported warrant
some conclusions and tentative generalizations regarding the
effects of functional groups introduced by surface-limited
reactions of Kevlar fibers on interfacial shear strength in
composites.

Summary and Conclusions

On the basis of results reported in this paper, the following conclusions are warranted:

1. Nitration of Kevlar filaments can be limited to fiber
surfaces and controlled to introduce NO_2 groups without apparent disruption of fiber topography, microstructure or
mechanical properties.

2. The $-NO_2$ substituents can be converted to $-NH_2$ by reduction without impairment of surface smoothness or strength
loss.

3. For nitration, the extent of reaction (determined by
ion exchange analysis of $-NH_2$ after reduction) depends on
the conditions used. For a given set of nitration conditions,
rate of reaction and maximum concentration of functional
groups are higher for Kevlar 29 than for Kevlar 49.

4. The microbond technique for determining shear
strength at the Kevlar/Epoxy interface provides an essential
tool for evaluating the effects of surface modification on
adhesion.

5. Microbond shear strength values for epoxy resin with
nitrated and nitrated/reduced fibers have shown that major
improvements can be attained after reduction for Kevlar 29
and for Kevlar 49 at values of $-NH_2$ ranging from 0.4 to 0.8
per 100 $\overset{\circ}{A}{}^2$.

Acknowledgements

The authors wish to acknowledge financial support from
E. I. du Pont de Nemours & Co. and cooperation by du Pont in
providing SEMs and measurements of mechanical properties for
the investigation. Advice and many helpful discussions with
Dr. B. Miller (TRI) and with Dr. T. Carney (du Pont) throughout the work are gratefully acknowledged.

References

[1] Wertheimer, M.R.; Schreiber, H.P. J. Appl. Polymer Sci. 1981, 26, 2087-2096.

[2] Allred, R.E.; Merrill, E.W.; Roylance, D.K. In "Molecular Characterization of Composite Interfaces"; Ishida, H. Ed., Plenum Press: New York, 1984; pp. 333-376.

[3] Keller, T.S.; Hoffmann, A.S.; Ratner, B.D.; McElroy, B.J. In "Physicochemical Aspects of Polymer Surfaces; Mittal, K.L., Ed., Plenum Press: New York, 1983; Volume 2, pp. 861-879.

[4] Penn, L.S.; Byerley, T.J.; Liao, T.K. J. Adhesion 1987, in press.

[5] Wu, Y.; Tesoro, G.C. J. Applied Polymer Sci. 1986, 31, 1041-1059.

[6] Miller, B.; Muri, P.; Rebenfeld, L. Composites Science and Technology 1987, 28, 17-32.

[7] Crivello, J.V. J. Org. Chem. 1981, 46, 3056-3060.

[8] Burstall, R.S.; Nyholm, R.S. J. Chem. Soc. 1952, 3570-3590.

[9] Antonin, A.V.; Rusina, A. Proc. Chem. Soc. 1961, 14, 161-162.

[10] Lalancette, J.M.; Brindle, J.R. Can. J. Chem. 1971, 49, 2990-2995.

[11] Joshi, G.G. Current Science India 1949, 18, 73-74.

[12] Hudlicky, M. In "Reduction in Organic Chemistry"; J. Wiley, New York, 1984; p. 216.

[13] Gibson, H.W.; Bailey, F.C.; Minur, J.L. J. Polymer Science (Chemistry Edition) 1979, 17, 2961-2974.

[14] Northolt, M.G. European Poly J. 1974, 10, 799-804.

[15] Penn, L.S., Tesoro, G.C.; Zhou, H.X. Polymer Composites 1987, in press.

ESTIMATION OF THE INTERFACIALLY BONDED POLYMER IN PP COMPOSITES

Béla Pukánszky and Ferenc Tüdős

Central Research Institute for Chemistry of the
Hungarian Academy of Sciences,
H-1525 Budapest, P.O.Box 17, Hungary

Introduction

It has been proved in the last 20 years that the interface plays a decisive role in the properties of polymer blends and composites [1-6]. There is, however, much controversy in the literature concerning the properties of the interface and the importance of interfacial adhesion, as well. Generally it is assumed that in filled or reinforced polymers a demobilized polymer layer covers the filler, but in some cases less ordered structures were observed [5] and according to Theocaris [7] soft interface can also develope under certain conditions. Kardos [2] states that a simple contact of the filler and the matrix is sufficient to increase stiffnes and excellent adhesion is not necessary, while Iisaka and Shibayama [8] have shown that with increased adhesion the amount of the interfacially bonded polymer increases resulting in an increase of the modulus. Dekker and Heikens [9] have proved that in an unidirectional tensile test the mechanism of failure changes with varying interfacial adhesion. Since no direct method is available to measure it, there are contradictions concerning the amount of the interfacially bonded material, as well.

The aim of our present work was to study the role of the interface in PP composites containing different anorganic fillers. An estimation of the interfacially bonded material was made and the factors influencing it were considered.

Experimental

Tipplen H 501 (TVK, Hungary) polypropylene was used as matrix in our study. The investigated fillers are listed in Table 1, they include six chalks, three talcs, a mica and a silica.

The composites were prepared in a Brabender W 50 EH mixing

Table 1. Designation, producer and characteristics of the investigated fillers

Trade name	Manufacturer	Filler type	Density ρ (g/cm^3)	Particle size at 50 w% (μm)	Oil No. F_0 (g/100 g)	Specific surface A_f (m^2/g)
Socal U1	Solvay	CaCO$_3$	2.65	0.08	50.2	16.5
Polcarb	English Clays	"	2.64	0.9	22.5	5.0
Calcilit 8	Alpha	"	2.71	7.2	22.0	2.4
Millicarb	Omya	"	2.65	4.0	21.4	2.2
Calcilit 100	Alpha	"	2.71	78.0	12.8	0.5
Mixture[a]			2.68	0.6	30.4	8.1
Finntalc M05	Finnminerals	talc	2.85	2.0	56.2	8.4
Finntalc M15	Finnminerals	"	2.80	5.1	40.0	5.9
Finntalc PF	Finnminerals	"	2.77	12.5	29.4	3.0
		mica[b]	2.77	75.0	65.3	4.9
		silica[b]	2.05	5.6	59.5	18.4

[a]Socal U1/Polcarb/Calcilit 4:33/33/34 v%
[b]Mica and fused silica were supplied by the State Research Institute of Materials, Prague, Czechoslovakia.

chamber attached to a Haake Rheocord EU 10 V plastograph. Compounding was carried out at 190 °C, 40 rpm for 15 min with a charge volume of 45 ml. During the compounding, torque and material temperature were recorded. The composites were compression moulded into 1 and 4 mm thick plates at 190 °C for further investigation.

Tensile tests were carried out at 5 mm/min crosshead speed. Melting and crystallization chracteristics were determined by a DuPont 990 DSC at 10 °C/min heating and cooling rate. Morphology was studied by SEM on fracture surfaces and by polarization light microscopy on thin sections.

Results

From the torque and temperature curves recorded during compounding the torque value at 185 °C was determined. This value is proportional to the apparent viscosity of the melt, thus giving an insight into the rheological properties of our composites. Ln M vs 1/T plots were used to determine the flow activation energy of the melt.

In Figure 1 the flow activation energy is depicted as a function of filler content for five different $CaCO_3$ fillers.

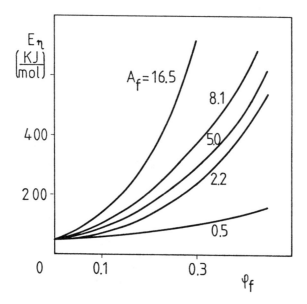

Figure 1. Dependence of flow activation energy of PP composites
  on filler content and specific surface of the fillers

With increasing filler content the activation energy increases, but significant increase can be observed as a function of the specific surface of the filler, as well. A similar tendency can be observed on the torque (M_{185}) vs filler content graph (Figure 2), i.e., increasing specific surface results in an increase

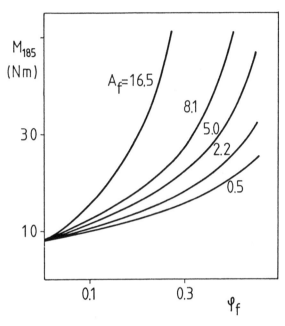

Figure 2. Torque (M_{185}) proportional to melt viscosity plotted against the composition. Running variable: specific surface of the fillers

of melt viscosity. Generally, the effect of a given amount of high specific surface filler corresponds to the effect of greater amount of filler with lower specific surface. In other words, the apparent amount of the filler in the composite increases with increasing specific surface.

As expected, Young's modulus of the composites also increases with increasing filler content (Figure 3). The specific surface of the filler significantly increases this mechanical property too. In fact the rheological properties (flow activation energy, torque at 185 °C) and Young's modulus show very similar behaviour.

The yield stress plotted against the filler content in Fig-

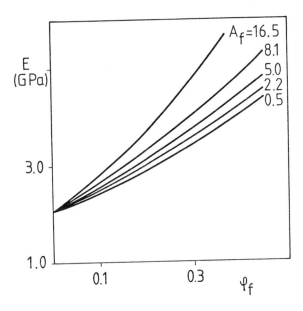

<u>Figure 3</u>. Dependence of Young's modulus of CaCO$_3$ filled PP composites on filler content and specific surface of the fillers

ure 4 shows a completely different dependence, it decreases with increasing filler content. However, concerning the effect of the specific surface it can be observed that with an increase of the latter the decrease of the yield stress is smaller, i.e., it increases relatively.

According to the above results the particle size, i.e., the specific surface of the fillers markedly influences the rheological and mechanical properties of the PP composites. To obtain an explanation we can turn to the theoretical models and equations, which describe the properties of the blends and composites as a function of the composition. Numerous models are available for the description of the rheological and mechanical properties, and especially for that of the viscosity and the elastic modulus.

For our study these equations can be divided into two groups. The first group contains those models and equations which include only the constants of the matrix and the filler, but no adjustable parameters. Such models are, for example, the Kerner equation [10], the Hashin-Strikhman bounds [11],

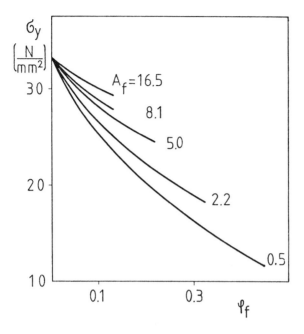

Figure 4. Dependence of tensile yield stress of CaCO$_3$ filled
PP composites on filler content and specific surface of the
filler

the Halpin-Tsai bounds [12], the Hill and Budiansky models
[13,14] etc. for the modulus, the Einstein euqation [15] and
the similar correlations derived thereof [16,17] for the viscosity, and the Nicolais-Narkis equation [18] for the yield
stress, respectively. Obiously these equations are incapable
of coping with the effect of the particle size or the specific
surface.

    The second group of the equations contains adjustable para-
meters, too. The most models of this category include the maximum packing fraction as adjustable parameter, i.e., the maximum
volume fraction of the filler which can be incorporated into
the continuous polymer matrix. Such equations are the Lewis-
Nielsen [19] and the McGee-McCullough equations [20] for the
modulus, the Mooney equation [21] for the viscosity, and the
generalized Nicolais-Narkis equation for the yield stress. The
maximum packing fraction, however, depends on the spatial distribution, but not of particle size.

The above mentioned equations, however, leave out of consideration the role of the interface in the composite properties. If we assume that the polymer adheres to the surface of the filler and a demobilized interface forms, the effect of the specific surface can be explained. At the same filling grade, higher specific surface results in a higher amount of demobilized polymer, i.e., increased stiffness and viscosity.

Somewhat more difficult is to explain the effect of the interface on the yield stress. If we take into consideration, however, the deformation mechanism of a PP composite under unidirectional tensile load, the observed behaviour can be easily understood. The basic deformation mode of PP is shear yielding, where no volume change is observed [22]. With increasing filler content a volume increase takes place during deformation caused by the dewetting on the polymer/filler interface [23]. With increasing contact surface the overall work of adhesion increases resulting in higher yield stress.

It is obvious from the results that in PP composites a hard interface forms, which consists of a demobilized polymer layer. The properties of this layer are between those of the polymer and the filler. The demobilized layer increases the apparent volume of the filler in the matrix. The question remains, however, how much polymer is bonded on the interface.

Estimation of the interfacially bonded polymer

Since the interface plays a decisive role in the determination of composite properties, it obviously would be advantageous to know its thickness and properties. Unfortunately, presently there are no means to determine these directly. We can, however, use the observed effect of specific surface to estimate its amount. From the measured composite properties an apparent volume fraction of the filler can be calculated. The difference of the actual and apparent volume fractions will give us a value proportional to the interfacially bonded polymer. To carry out this calculation, however, a reference state, i.e., the knowledge of the properties of a composite with zero interface is needed.

Similar calculations have already been carried out using always measured modulus values of the composites to estimate the amount of the interfacially bonded polymer. In these calculations different theoretical equations were used as reference functions. Iisaka and Shibayama [8] used the semiempirical equation of Ziegel and Romanov [24-26] and calculated the thickness of the interface using measured Young's moduli values.

Kolarik et al [27-28] utilized the Lewis-Nielsen equation
[19] and calculated the thickness of the interface in PU rub-
ber/crosslinked polymeric filler system on the ground of
Young's and shear moduli. Theocaris [29] used Lipatov's theory
[30] and determined the so called mesophase dimensions from
changes in the heat capacity jumps at the glass transition
temperature of the matrix and the composites, respectively.
The interface dimensions determined by these methods, however,
are biased by the assumptions of the reference functions and
in most cases do not represent a zero interface state.

To avoid this problem we fitted the formal $y = A \exp(B\varphi_f)$
equation to our experimental data, where $y = M_{185}$ and E, respectively. From a linearized plot of the data A and B can be determined and by extrapolating to zero specific surface the
reference function can be established. Figure 5 shows the
extrapolation procedure in the case of the torque (M_{185}), proportional to the viscosity. Apart from a few points, the fit
is excellent and the $A_f = 0$ parameters are easy to determine.

Unfortunately the above equation, which is easily applica-

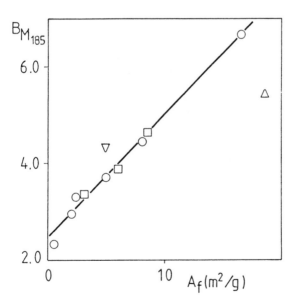

Figure 5. Determination of the parameter of zero interface
reference equation from the fitted viscosity curves. o: $CaCO_3$
□: talc; Δ: silica; ∇: mica

Béla Pukánszky and Ferenc Tüdos

ble in the case of M_{185} and E cannot be used in the case of the
yield stress, because the σ_a vs φ_f function has a quite different shape. In this case we chose the Nicolais-Narkis equation as a reference function. It is true that the basic assumptions of this model include also zero adhesion, which in this
case means zero interface as well, but by its derivation a
certain packing of the filler particles was also assumed. In
our composites, however, the packing can vary from filler to
filler, thus changing the reference function and prejudicing
our estimation. Still, we consider the thus calculated values
as appropriate for comparison with those derived from the
other properties.

In the following figures some measured curves are shown together with the reference functions and a theoretical equation
not containing any adjustable parameter. Figure 6 shows the

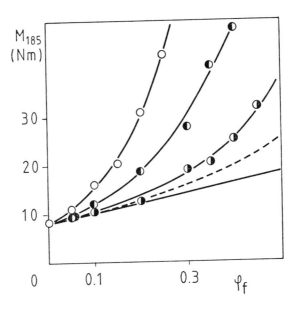

Figure 6. Zero interface reference function and model equation
compared to the M_{185} torque values of PP composites. o: Socal
U1 (A_f = 16.5 m^2/g); ◒: Mixture of $CaCO_3$ fillers (A_f = 8.1 m^2/g);
◐: Millicarb (A_f = 2.2 m^2/g); ---: reference function (A_f = 0);
——: Einstein equation

change of M_{185} as a function of filler content. The plotted theoretical equation is the Einstein model which is valid only up to a few percent of filler content and gives far lower values than even our reference function. Figure 7 depicts the change

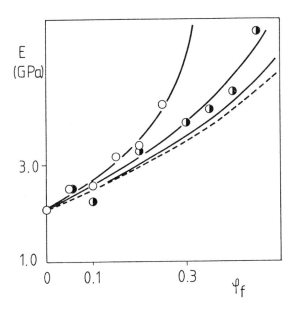

Figure 7. Zero interface reference function and theoretical model equation compared to the Young's modulus values of PP composites. o: Socal U1 (A_f = 16.5 m^2/g); ◑: Polcarb (A_f = 5.0 m^2/g); ---: reference function (A_f = 0); ——: Halpin-Tsai lower bound

of Young's modulus. Here the comparative theoretical model is the Halpin-Tsai lower bound, which is quite near to our reference function. In the case of the yield stress (Figure 8) the theoretical equation and the reference function is the same.

    In Figure 9 the amount of interfacially bonded polymer is plotted as a function of the specific surface of the filler. It is obvious from the graph that close correlation exists between the bonded material and the specific surface, but the amount of the interface strongly depends on the property from which we derive it. It is also worth to note that although the correlations are generally good, there are some deviations from the curves, which indicate that other factors than the specific surface are also effective in determining the amount of interfacial material.

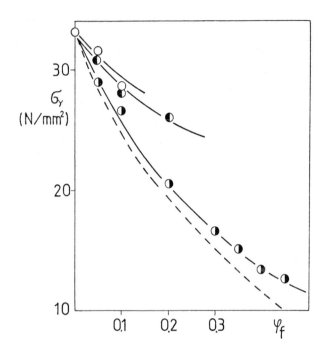

Figure 8. Reference function compared to the tensile yield
stress values of PP composites. o: Socal U1 (A_f = 16.5 m^2/g);
◑: Polcarb (A_f = 5.0 m^2/g); ◐: Calcilit 100 (A_f = 0.5 m^2/g);
---: Nicolais-Narkis equation

Discussion

The results of Figure 9 indicate that no unique thickness
of the interfacial layer exists - it highly depends on the conditions and the method of the measurement with which it is determined. Out of the three characteristics Young's modulus and
yield stress were determined in the solid state, while the
torque, proportional to viscosity, was measured in the melt.
In the two solid state measurements, however, the extent of deformation is quite different, the Young's modulus is determined
at zero deformation, while at yield the deformation is more
than 2 %. There is a large difference between the solid state
and the melt measurements, too. Interface properties are between those of the polymer and the filler. In the melt state
this difference is much larger than in the solid state. These
difference also may influence the thickness of the interface.

It further complicates the evaluation of the results, if we con-

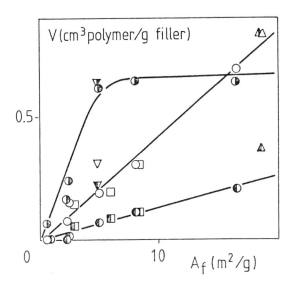

Figure 9. Calculated interfacially bonded polymer plotted
against the specific surface of the fillers. o, □, Δ, ∇; M_{185};
o, ◘, ▲, ▼: σ_y; o, ◘, ▲, ▼: E. The shape of symbols as in
Figure 5

sider the possible reasons for the deviations from the overall
correlations. The six different $CaCO_3$ fillers usually fit these
correlations, but the silica and the mica show considerable
deviation in most cases, they adsorb thicker interfacial layers
than $CaCO_3$. These differences indicate that the work of adhesion also play a role in the determination of interface thickness and thus in the determination of composite properties.
These two fillers have definitely higher surface free energies
than $CaCO_3$ resulting in higher work of adhesion. These observations are also corroborated by the results of Iisaka and Shibayama [8], who observed interface thicknesses between 0.06 - 1.4
μm and a correlation between the thickness and interaction
energy. Kolarik et al. [27,28] have found for a highly polar PU
elastomer apparent thicknesses for the demobilized layer between 0.1 and 1.5 μm depending on filler content and temperature. If we consider these values and the relatively low polarity
of PP the 0.01 - 0.13 μm interface thicknesses, which we have
calculated, seem to be realistic.

Although interfacial adhesion seems to be an important factor in the determination of interface and composite properties,
there are also other factors to be considered. In Figure 10 we
show the yield stress of PP composites as a function of filler

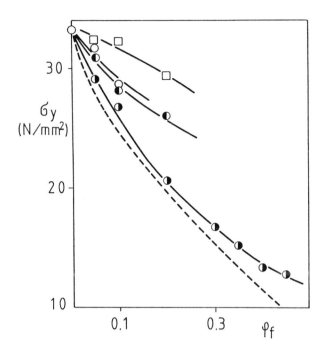

Figure 10. Anomalous tensile yield behaviour of talc-filled PP
composites. o: Socal U1 (A_f = 16.5 m^2/g); ●: Mixture of CaCO$_3$
fillers (A_f = 8.1 m^2/g); ◐: Calcilit 100 (A_f = 2.4 m^2/g); □: Finntalc PF (A_f = 3.0 m^2/g); ---: Nicolais-Narkis equation

content. Those composites containing CaCO$_3$ show the expected
tendency - similar shape of the curves was observed in the case
of mica and silica too - but the composites containing talc
behave quite differently. This kind of behaviour is usually
attributed to the plate like structure of the talc, but mica
has a similar anisotropic particle geometry. The anomaly may be
related to the stronger nucleation effect of the talc (Figure
11), but further study is needed to prove this correlation.

Crystallization on anisotropic filler surfaces may lead to
oriented transcrystalline structures, changing the properties
of the composite in a great extent. Further morphological organization of the filler also can change the properties, which
may lead to a changed interface thickness and/or to an erroneous
estimation of it. Such structural factors can be the aggregation of the filler having very low particle size (Figure 12) or
the orientation of anisotropic, plate-like fillers (Figure 13).

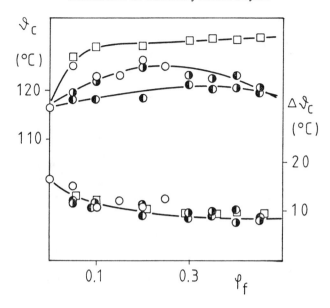

Figure 11. Crystallization temperature (ϑ_c) and crystallization rate ($\Delta\vartheta_c$) of PP composites as a function of filler content. o: Socal U1 ($A_f = 16.5$ m^2/g); ◓: Polcarb ($A_f = 5.0$ m^2/g); ◒: Calcilit 100 ($A_f = 0.5$ m^2/g): □: Finntalc PF ($A_f = 3.0$ m^2/g)

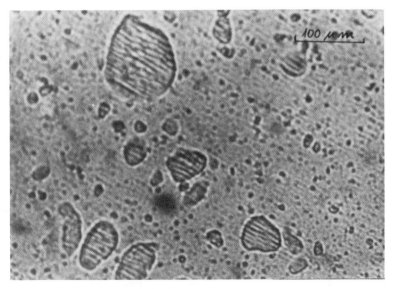

Figure 12. Aggregation of filler particles in a PP composite filled with Socal U1 ($A_f = 16.5$ m^2/g)

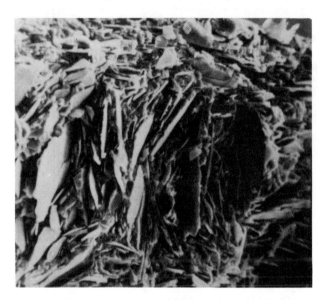

Figure 13. Orientation of mica particles in a 0.2 v% mica containing PP composites

Conclusions

In PP composites filled with inorganic fillers a hard interlayer forms on the polymer/filler interface, which consists of demobilized polymer molecules. This demobilized material increases the apparent volume percent of the filler. Although the thickness of the interface cannot be measured directly, it can be estimated by using this apparent increase of the filler content. To carry out the estimate, however, an appropriate, unbiased reference function is required. It has been established that with this method no universal interface thickness can be determined, it depends on the filler content, temperature, extent of deformation, loading conditions etc. Our results indicate that polymer/filler adhesion also plays an important role in the determination of interface properties, and special morphological formations must also be considered during the evaluation of composite and hence interface characteristics.

References

[1] Eirich, F.R. J. Appl. Polym. Sci., Appl. Polym. Symp. 1984, 39, 93
[2] Kardos, J.L. In "Molecular Characterization of Composite

Interfaces"; Ishida, H., Kumar, G., Eds.: Plenum: New York, 1981, 1

[3] Shanks, R.A. Polym. Prepr. 1984, 25, 132
[4] Kolarik, J.; Janacek, J., Nicolais, L. J. Appl. Polym. Sci. 1976, 20, 841
[5] Kosfeld, R.; Uhlenbroich, T., Maurer, F.H.J. Org. Coat. Appl. Polym. Sci. Proc. 1981, 46, 362
[6] Friedrich, K.; Karsch, U.A. Fibre Sci Technol. 1983, 18, 37
[7] Theocaris, P.S. Mat. Res. Soc. Symp. Proc. 1984, 21, 847
[8] Iisaka, K.; Shibayama, K. J. Appl. Polym. Sci. 1978, 22, 3135
[9] Dekkers, M.E.J.; Heikens, D. Ibid 1983, 28, 3809
[10] Kerner, E.H. Proc. Phys. Soc. 1956, 69B, 808
[11] Hashin, Z.; Shtrikhman, S. J. Mech. Phys. Solids 1963, 11, 127
[12] Halpin, J.C.; Kardos, J.L. Polym. Eng. Sci. 1976, 16, 344
[13] Hill, R. J. Mech. Phys. Solids 1965, 13, 213
[14] Budiansky, B. Ibid 1965, 13, 223
[15] Einstein, A. Ann. Physik, 1905, 17, 549; 1906, 19, 289
[16] Faulkner, D.L.; Schmidt, L.R. Polym. Eng. Sci. 1977, 17, 657
[17] Hoffman, R.L. NATO ASI Ser., Ser. E 1983, 68, 570
[18] Nicolais, L.; Narkis, M. Polym. Eng. Sci. 1971, 11, 194
[19] Lewis, T.B.; Nielsen, L.E. J. Appl. Polym. Sci. 1970, 14, 1449
[20] McGee, S.; McCullough, R.L. Polym. Compos. 1981, 2, 149
[21] Mooney, M. J. Colloid Sci. 1951, 6, 162
[22] Bucknall, C.B. "Toughened Plastics"; Applied Science: London, 1977
[23] Chacko, V.P.; Farris, R.J., Karasz, F.E. J. Appl. Polym. Sci. 1983, 28, 2701
[24] Ziegel, K.D. J. Colloid Interface Sci. 1969, 29, 72
[25] Ziegel, K.D.; Romanov, A. J. Appl. Polym. Sci. 1973, 17, 1119
[26] Idem, Ibid 1973, 17, 1133
[27] Kolarik, J.; Hudecek, S., Lednicky, F. Ibid 1979, 23, 1553
[28] Idem, Faserforsch. Textiltechn. 1978, 29, 51
[29] Theocaris, P.S. Adv. Polym. Sci. 1985, 66, 149
[30] Lipatov, Y. "Physical Chemistry of Filled Polymers"; Intern. Polym. Sci. Techn. Monograph, No. 2; Khimiya: Moscow, 1977

CHAIN LENGTH DEVELOPMENT IN COMPLEX EPOXY-AMINE MATRICES

I. Shapiro

Consultant, Los Angeles, California 90045-0124

Abstract

The role of complex polymer formulation for matrices for composite systems can be taken out of the realm of empiricism by studying mechanism and reaction kinetics of the polymer reactions. A first step in this direction is the measurement of the progressive change in the chain lengths of the polymer products during the course of reaction. Such measurements have been made by a Gel Permeation Chromatographic technique coupled with Gaussian curve-resolution methods. The study included two different types of epoxy resins and two primary aromatic amines, reacted both singly and in combination with one another. The extent of this current study covers epoxy-amine reaction with excess amine to the quarter point of stoichiometry. By varying the concentration of reactants, one can recognize the progressive changes in chain lengths of the polymeric products. The longest chain segment observed with certainty was one containing five epoxy and six amine units. With mixed amines both amine moieties can be discerned in the same chain length segment. Calibration charts are necessary to determine the composition of the chain segments; a method for preparing these calibration charts is described. The experimental techniques employed here also is amenable to the study of the kinetics of the reaction. There is a large contrast in reaction behavior of the two epoxy resins so that by using this combination one gets the effect of a pseudo "B-staging" which can be better controlled for longer periods of time before cross-linking takes place. Thus larger hardware pieces can be fabricated.

Introduction

The examination of chain length segments of epoxy-amine polymeric products gives a considerable insight into the mechanism and kinetics of the reaction. A convenient method to measure the distribution of such polymer chain lengths has been the Gel Permeation Chromatographic (GPC) technique in

conjunction with the Gaussian-resolution method. Here the
reaction can be followed in a step-wise manner to give a
panoramic view of the reaction during the course of the
polymerization process. The present resin system consists of
two epoxy resins, namely, diglycidyl ether of bisphenol A
(DGEBPA) and bis(2,3-epoxycyclopentyl) ether (BECPE), and two
primary aromatic amines, metaphenylene diamine (mPDA) and
4,4'-methylene dianiline (MDA). All combinations and
mixtures of the epoxies and amines have been studied under
condition of excess amine to the quarter point of
stoichiometry. An important facet of this study is to
construct a calibration chart relating the count number of
the individual peaks in the GPC curve to the molecular weight
of the corresponding chain species. This ability to locate
peak positions is especially important when several peaks
overlap each other.

Experimental

    In this section is described each of the materials used in
the experiments, the instrumental equipment for developing
the curves relating to the polymeric products and the
resolution of these curves (due to overlapping of curves),
the procedures for mixing the reactants, and the development
of the calibration chart for establishing peak positions.

Materials

    In a study of polymer reactions by the technique developed
here it is important to work with relatively pure materials,
otherwise even trace impurities could conceivably obscure the
true reaction products and lead to erroneous conclusions.
For this reason extreme care was taken to obtain "pure"
ingredients. Each of the ingredients used in this study are
described below.

    Diglycidyl ether of bisphenol A (DGEBPA). This epoxy
resin, generally known as the workhorse of the industry, is
made commercially by the reaction of phenol and acetone to
give bisphenol A, which in turn is reacted with
epichlorohydrin, followed by neutralization, to give a series
of compounds of various molecular weights, depending upon
conditions and concentrations of reactants. This series of
compounds can be generalized by a structural formula, shown
on the next page, in which the repeating section of the
formula is designated by the letter "n". The value of n can
vary from 0,1,2,3,.8 and perhaps even higher. Each species

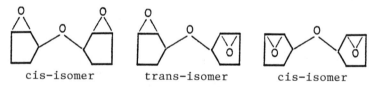

or compound with its various n values will have its own
distinct GPC peak. The commercial grade of this epoxy resin
is a liquid and consists of about 5 to 10 percent of the
structure of value n = 1, with the remaining material having
the value n = 0. However, there is available in limited
quantities from various chemical suppliers a purified resin
which is a solid with a waxy consistency and is essentually
pure n = 0. In all work described in this paper the DGEBPA
resin is solely the n = 0 variety. No impurity could be
detected by GPC analysis. The molecular weight of this resin
is 340 amu.

Bis(2,3-epoxycyclopentyl) ether (BECPE). This resin is
classed as one of the cycloaliphatic epoxy resins and has
been well described in the literature [1,2]. It has been
reported that theoretically this resin consists of ten liquid
and solid isomers, but no attempt had been made to separate
the individual isomers. The ratio of solid to liquid isomers
is approximately 65 to 35. The accuracy of this ratio is not
important since only the solid isomer was used in this study.
The structures of three of the isomers are shown below.

cis-isomer   trans-isomer   cis-isomer

The first and third structures would represent the solid
isomer; and the center one, the liquid isomer. Spatial
arrangements with respect to hydrogen bonding would give rise
to other isomers. At the time of the original work with
these cycloaliphatic resins the solid isomers were physically
separated from the liquid isomers by a distillation process
with the purity yield being 10 percent liquid isomer still
remaining in the solid isomer [1]. In the work described in
this paper the solid isomer was isolated by a recrystallizing
process to yield exceptionally high purity solid isomer. The
mixed isomers were dissolved at room temperature in a solvent
1,1,2-trichlorotrifluoroethane and the solution was chilled
in a refrigerator. A white crystalline material separated
out, and the supernatant liquid poured off the solid phase.
Second crops of crystals were obtained by repeated low
temperature fractional crystallization and the solids

repurified in a similar manner. Since it was not the purpose of this work to pursue further the separation of isomers, the solid material was used directly in the currently described reactions. The solid isomer has a low melting point and liquifies at slightly above ambient temperature.

Primary aromatic amines. In this paper the reaction of two primary diamines with the two epoxy resins were studied in some detail. The two primary amines are metaphenylene diamine (mPDA) and 1,4-methylene dianiline (MDA). Additionally, some experiments were conducted with aniline, which was used for modeling purposes. The structures of these amines are shown below.

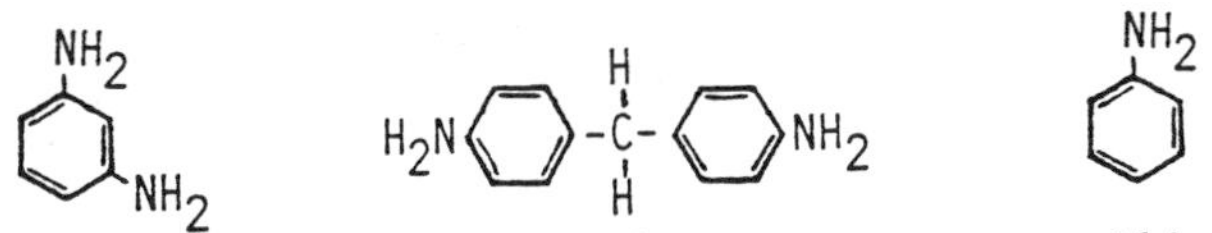

mPDA   MDA   aniline

Aniline is very similar to mPDA but has only one NH_2 group attached to the benzene nucleous. Thus aniline acts as a chain stopper since it cannot propagate the epoxy reaction beyond its two active hydrogen atoms on the single amine group. The two diamines are crystalline substances obtainable in very pure form. Titration analyses confirm the high purity of the materials and each of the amines show only one peak in the GPC spectrum. The molecular weight of mPDA is 108 amu; for MDA, 198 amu; and for aniline, 93 amu.

Solvents. Two solvents have been used extensively in this program. Tetrahydrofurane (THF) was used as the liquid carrier in the gel permeation chromatographic apparatus, and also was used as the medium for the epoxy-amine reactions. Large volumes of THF will quench the reactions. More details of this effect will be presented with the results on the specific reactions. The second solvent, 1,1,2-trichlorotri-fluoroethane, also known as Solvent-113, was used in the isomer separation and purification of the solid isomer of BECPE.

Apparatus

The experimental work in this study relied heavily on two special instruments, the Waters Gel Permeation Chromatograph and the Dupont 310 Curve Resolver. A working knowledge of these instruments is deemed necessary. Both of these instruments are discussed below in considerable detail because of their importance to the success of the program.

Waters Gel Permeation Chromatograph. This instrument is
capable of changing the flow time of compounds passing
through a packed column, and then detecting and recording
these changes. There are three major components to the
instrument: separator, detector, and recorder. The separator
contains a twelve foot column packed with poragel or styragel
granules through which a carrier liquid is pumped at constant
or programmable rates. The twelve foot column actually
consists of three columns in series, each four feet in
length, and packed separately with the porous material rated
for 80, 100, or 700 Å porosity. When a mixture of compounds
dissolved in a liquid is injected into the apparatus, the
liquid carrier sweeps the sample through the packed column.
Larger molecular weight (really "molar volume") material will
flow past the porous material if their molecular size is
greater than the porosity opening of the packing; however,
lower molecular weight material can penetrate the porous
packing material, the extent of which depends upon the pore
openings of the packed material, and thus follow a "longer
path" through the column. This change in path length causes
the simultaneously injected compounds to come out of the
column at different times. The liquid then flows through a
detector unit which recognizes the material by changes in
refractive index or ultraviolet absorption and this change is
transmitted electronically to a strip chart recorder. Both
means of detection are shown as separate curves on the strip
chart. Materials containing benzene rings in their
structures are sensitive to ultraviolet absorption; some
compounds are not sensitive and thus would not be detected
this way. However, the refractive index detector would pick
up the changes due to the material passing through the
detector. The chance of a compound not being detected by
either method is remote. The apparatus is complemented with
temperature controllers and flow-rate pumps, changes of which
will alter the times of exit or count time. Changes in
packing size granules also will have a profound impact on
count time.

Dupont 310 Curve Resolver. This instrument contains a bank
of ten individual generators for preparing distribution
curves following Gaussian or skewed probability laws. The
Gaussian curve, also known as the "normal distribution
curve", can be expressed by the following type exponential
equation.

$$F(d) = n = \frac{\Sigma n}{\sigma\sqrt{2\pi}} \exp\left(-\frac{(d - d_{av})^2}{2\sigma^2}\right)$$

where $\quad d_{av} = \dfrac{\Sigma(nd)}{\Sigma n}$

and $\quad \sigma = \sqrt{\dfrac{\Sigma[n(d - d_{av})^2]}{\Sigma n}}$

The Gaussian plot requires only two parameters to define the curve, one giving total height of peak; and the other, the standard deviation. Both of these parameters can be fed into any one curve generator. The position of the curve can be moved as desired. In practice the curves obtained on strip charts are projected onto a screen with the aid of a mirror. Then a curve generator superimposes a curve which is adjusted to fit the observed curve. When there is no overlap of curves, the problem of fit is relatively easy to do. However, for overlapping peaks, it is necessary to have a starting point which could be the edge or portion of a peak. After adjusting for a "good fit", a second generator is turned on. The equipment is designed such that curves can be positioned individually, or several curves can be made additive. In any situation, the minimum number of peaks needed to fit the observed curve is sought. By a series of reiterations and comparing of peaks corresponding to different concentrations of reactants, one can arrive at a reasonable fit of curves. Such data must be in agreement with a calibration curve, which in turn is prepared by examining a number of curves from the same chemical species. Such a calibration curve wii be described in a later section of this paper.

Procedure

This study involving the reaction products from various epoxies and amines was carried out in two different ways. In the first case a known weight of epoxy was added to an excess but known weight of amine. Reaction was normally carried out at 75°C and with vigorous mixing. The reaction was allowed to proceed to completion, i.e., all of the epoxy was consumed with the excess amine remaining. The addition of epoxy to amine is important since chain lengths could be established and measured by this procedure. In the event excess amine was added to epoxy, somewheres along the way the reaction would pass through the stoichiometric point and cross-linking would occur to obscure the chain segments. Once the reaction was completed, samples from the reaction mixture were dissolved in tetrahydrofurane (THF) solvent and injected into the GPC apparatus. In the second case the reaction was interrupted so samples could be taken during the course of the reaction.   Two ways were used to obtain samples before the reaction was completed.   One way was to take aliquot portions from the reaction mixture as a function of time, and then dilute with THF.   Sufficient dilution is capable of stopping the reaction from continuing.   A second way was to terminate the reaction also by the dilution technique after fixed periods of time.   This second way calls for considerably more material and longer periods of time of experimentation, but was deemed necessary to verify the aliquot portion type experiments.

Areas under the curves were measured by use of a polar planimeter.   Another convenient method is to trace the curve onto a good grade of paper, then cut out the curve and weigh the paper.   Actually both methods have been used to verify the accuracy of the area measurements.

Calibration

The calibration chart shown in Figure 1 represents the accumulation of considerable data for the epoxy-amine reaction products, and it relates the peak count number of the GPC curve to the molecular weight of the product.   In certain polymer systems the relationship of peak count to

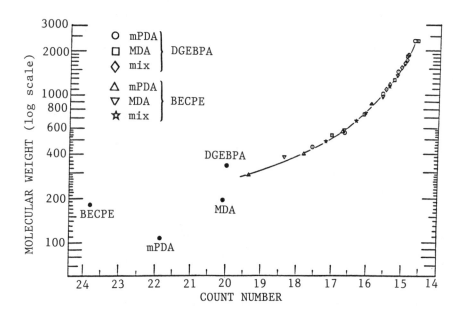

Figure 1. Molecular weight GPC calibration curve for epoxy-amine system.

molecular weight is linear on semi-log plot; however, in the epoxy-amine system the plot increases progressively with increasing molecular weight. Thus there is a practical maximum limit of molecular weights that can be distinguished from each other under the conditions of the experiment. Changing columns would affect the calibration curve so it would be wise for anyone starting such studies to first optimize the column (porosity packing size, length, temperature, pumping speeds, etc). Obviously, for any series of experiments the condition of the equipment and column should be maintained unchanged, otherwise the calibration curve would be incorrect. The count numbers should be considered only as relative numbers and should not be used interchangeably with other columns without first recalibrating.

In practice as the molecular weight of the products increases, the peaks will start to overlap one another, so curve resolving is necessary to separate out the products. In curve resolving one starts with the lower molecular weight products, and after getting a good fit of curves, the peak position for that product remains constant, so that the curve can be subtracted from the rest of the curves. The same process of curve resolving is applied to the balance of the curves. The peak position should fall on a smooth curve. Any deviation is a good indicator of an error, and the position of the peak must be adjusted until all data are consistent with each other.

Results

The results of the epoxy-amine reaction are presented by a series of GPC curves, covering the two epoxy systems, with each epoxy being reacted with the two amines individually as well as in combination of the two amines. Included here are the results of a kinetic experiment covering the combination of both epoxies and both amines simultaneously and at several different temperature levels. Finally, some experiments of a reaction of DGEBPA epoxy with aniline is presented for modeling purposes.

DGEBPA Epoxy System

mPDA amine. The reaction of mPDA with DGEBPA was the first experiment of the series, so consequently it is described in more detail. The technique of carrying out the reactions and analyses of GPC curves are similar throughout this study. A

weighed amount (4.38 gm) of DGEBPA epoxy was added to a known
weight (5.01 gm) of mPDA with stirring while maintaining a
constant temperature at 75°C, and the reaction was allowed to
go to completion. Afterwards samples from the reaction
mixture were taken and diluted with THF and run through the
gel permeation chromatographic column. Next, the resulting
curve on the strip recorder (maximum peak height of curve is
25 cm) was placed into the curve resolver window and the
curve was resolved by the trial and error method. In curve
resolving one must always try to find the minimum number of
peaks that would give a good fit. In this case in the 14-18
count region the minimum number of components was four peaks
as shown in Figure 2. The peak labelled "a" is identified as
unreacted (excess) mPDA. In resolving the other portion of
the curve one finds a peak c which fits a portion of the
original curve in the upper section of the curve. In
practice it is almost always necessary to find one peak that
coincides with a portion of the original curve to use as a
base for resolution. Once peak c is subtracted from the
original curve, the rest of the peaks fall into place.
Interestingly, if one tries to move, say, peak c, laterally
in either direction, the valley between peak c and peak d
gets grossly distorted. This change is very sensitive and
proved to be extremely useful in resolving curves. This
technique helped firm up the count numbers for calibration.

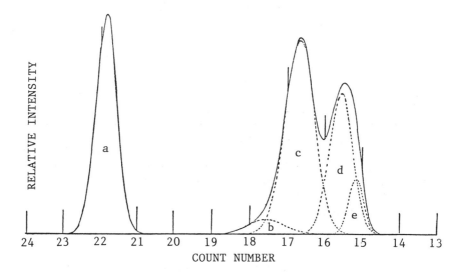

Figure 2. Gaussian resolution of gel permeation chromatograms
showing reaction products from DGEBPA epoxy with excess mPDA.
Areas of species peaks are given in Table 1.

Once the curve has been resolved, then the area under each peak can be measured with a planimeter. The value of the GPC area associated with each peak is shown in Table 1. Included in the table are the species and its corresponding molecular weight and count number. From this information one can separate the GPC area into component values corresponding to the epoxy content and amine content of the species. The significance of these values will be discussed later in the section under the heading of Discussion.

Another series of experiments studying the reaction of DGEBPA with mPDA involved the variation of the relative amounts of reactants, still having the amine in excess. The weights of the reactants were adjusted so as to yield molar ratios (MR) of 18, 4, and 2. Again these reactions were allowed to go to completion and then the resulting curves were resolved in a manner as described previously. These curves are shown in Figure 3, where the curves are arranged beneath each other to show the corresponding peaks in alignment. These curves give a panorama of the change in chain lengths that the species undergo as the reaction proceeds with increasing addition of epoxy to the amine. The changes in concentration of the various species were obtained by measuring the areas under the peaks. These values are shown in Table 2. While the interpretation of these changing values will be deferred to a later section, there are two series of species of interest. First, as the epoxy is added to the mixture, the concentration of unused or excess amine decreases. Such a change is expected, but the values in Table 2 gives a quantitative measurement of this change. The second series involve the peak b, interpreted as an unreacted epoxy group in the presence of excess amine, The product concentration increases with increasing addition of epoxy.

Table 1. Concentration of chain lengths of reaction products from the addition of DGEBPA epoxy to an excess of mPDA amine.

Peak	Species[a]	Mol wt	Count	GPC area	% Epoxy	% Amine
a	O	108	21.84	32.1		32.1
b	O—	448	17.58	2.2	1.7	0.5
c	O—O	556	16.69	37.8	23.1	14.7
d	O—O—O	1004	15.59	24.2	16.4	7.8
e	O—O—O—O	1452	15.16	3.7	2.6	1.1
				100.0	43.8	56.2

[a] O = mPDA
— = DGEBPA

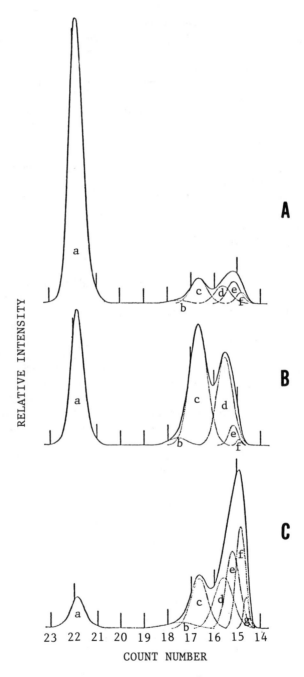

Figure 3. Effect of molar ratio mPDA/DGEBPA upon chain length of reaction products. A, 18 MR; B, 4 MR; C, 2 MR. Values of area of species peaks are given in Table 2.

Table 2. Concentration of polymer chain lengths as function
of molar ratio of mPDA and DGEBPA

Peak	Species[a]	Mol wt	Count	Molar ratio mPDA/DGEBPA 18	4	2
a	O	108	21.84	77.5[b]	32.1	9.9
b	O▬	448	17.58	0.9	2.2	3.4
c	O▬O	556	16.69	8.1	37.8	18.2
d	O▬O▬O	1004	15.59	5.6	24.2	23.0
e	O▬O▬O▬O	1452	15.16	5.7	3.2	20.9
f	O▬O▬O▬O▬O	1900	14.88	2.2	0.5	20.8
g	O▬O▬O▬O▬O▬O	2348	14.71			3.8

[a] O = mPDA
▬ = DGEBPA
[b] Values based on percent of GPC area.

The explanation of this phenomenon is straight forward if one
accepts the premise that a very low concentration of epoxy
has a defect on one end of the molecule, either in its
formation or subsequent loss of the epoxy moiety. Such an
explanation is entirely plausible as it will be explained in
the section dealing with the reaction of epoxy with aniline.

MDA amine. The reaction of DGEBPA with excess MDA followed
the same pattern as with mPDA. However, in this case 2.28 gm
of epoxy was added to 5.02 gm of MDA. In terms of mole ratio
the values are very close to each other; 3.8 for MDA/DGEBPA,
and 3.6 for mPDA/DGEBPA. The resulting GPC curve and its
resolved peaks are shown in Figure 4. This resolved curve
also has four peaks assigned to similar type species (see
Figure 2), but the relative concentration of the species
differ as illustrated by the data presented in Table 3 as in
contrast to the values cited in Table 1. It should be noted
that although the molar ratios of the two examples are rather
similar, on the weight basis there is a considerable
difference. The areas under the peaks of the curve are
really on a weight basis. However, before any comparison can
be made, one must apply a correction factor accounting for
the sensitivity of the species. This matter will be taken up
in a later section of this report.

As in the case of the mPDA amine cited earlier a series of
experiments were conducted in which the DGEBPA epoxy reaction
with the MDA amine was studied as a function of mole ratio of
reactants. Again, this technique give a panorama view of how
the reaction is proceeding upon further addition of epoxy to

I. Shapiro

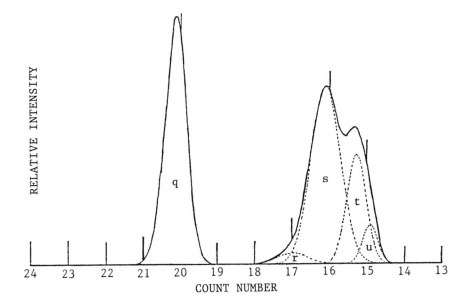

Figure 4. Gaussian resolution of gel permeation chromatograms showing reaction products from DGEBPA epoxy with excess MDA. Values of area of species peaks are given in Table 3.

the amine. Figure 5 shows these changes in chain lengths with the molar ratio of the reactants. The quantitative measure of the concentration of the various chain lengths is given in Table 4. Comparison of these data with the values given in Table 2 proves very enlightening. For the same molar ratio of reactants the mPDA tends to form longer chain lengths than does MDA. Thus, at the quarter-way point (molar

Table 3. Concentration of chain lengths of reaction products from the addition of DGEBPA epoxy to an excess of MDA amine.

Peak	Species[a]	Mol wt	Count	GPC area	% Epoxy	% Amine
q	□	198	20.11	39.8		39.8
r	□■	538	16.97	2.3	1.4	0.9
s	□■□	736	16.11	38.1	17.6	20.5
t	□■□■□	1274	15.27	15.6	8.3	7.3
u	□■□■□■□	1812	14.90	4.2	2.4	1.8
				100.0	29.7	70.3

[a] □ = MDA
■ = DGEBPA

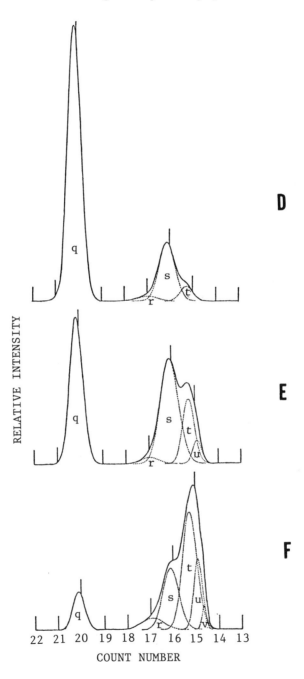

Figure 5. Effect of molar ratio MDA/DGEBPA upon chain length of reaction products. D, 12.5 MR; E, 4 MR; F, 2 MR. Values of area of species peaks are given in Table 4.

Table 4. Concentration of polymer chain lengths as function of molar ratio of MDA and DGEBPA

Peak	Species[a]	Mol wt	Count	Molar ratio MDA/DGEBPA		
				12.5	4	2
q	□	198	20.11	75.4[b]	39.8	11.2
r	□■	538	16.97	2.0	2.3	6.1
s	□■□	736	16.11	18.8	38.1	23.9
t	□■□■□	1274	15.27	3.8	15.6	41.3
u	□■□■□■□	1812	14.90		4.2	15.2
v	□■□■□■□■□	2350	14.62			2.3

[a] □ = MDA
 ■ = DGEBPA
[b] Values based on percent of GPC area.

ratio 2) the molar concentration of unreacted (unused) mPDA is greater than that of unreacted MDA by virtue that with the longer chain lengths for mPDA, less mPDA is required to react with a given quantity of DGEBPA than in the case with MDA. Obviously, with a difference in reaction rates of the two amines, the order of addition of ingredients and rate of mixing may play a dominant role in the chain structure of the polymerized resin.

Mixed amines. To determine the effect of mixed amines in the reaction with DGEBPA, an experiment was conducted to establish the above premise. For this purpose 5.69 gm of DGEBPA was added to a mixture of 5.01 gm of mPDA and 2.51 gm MDA. The resulting GPC curve and its resolved peaks are shown in Figure 6. It should be obvious that without a good calibration curve, the resolution shown here would not have been possible. The quantitative values of the peak areas and their associated species are listed in Table 5. Further, these peak areas are broken down into chemical components corresponding to the original ingredients, viz., epoxy content, mPDA content, and MDA content.

The addition of DGEBPA to a mixture of mPDA and MDA leads to new polymer species as shown in the table. Both amine moieties can be found in the same chain length species. For example, peaks 4, 5, and 6 each consist of two amines and one epoxy. Peak 5 is composed of one mPDA and one MDA group. Peaks 4 and 6 each contain two identical groups, i.e., either mPDA or MDA. From the values in Table 5 the relative concentrations of these three peaks are 2.1:1.0:0.6. Had these peaks been influenced solely by concentration of

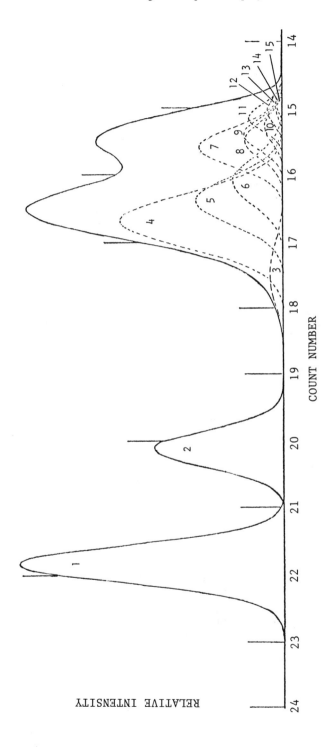

Figure 6. Gaussian resolution of gel permeation chromatographic curve showing reaction products of DGEBPA epoxy with a mixture of mPDA and MDA amines in which the amines are in excess of the stoichiometry of the reaction. Values of area of species peaks are given in Table 5.

Table 5. Concentration of chain lengths of reaction products from the addition of DGEBPA epoxy to a mixture of mPDA and MDA amines in which the amine concentration is in excess.

Peak	Species[a]	Mol wt	Count	% GPC	% Epoxy	% mPDA	% MDA
1	O	108	21.84	26.24		26.24	
2	□	198	20.11	12.50			12.50
3	O━	448	17.58	2.03	1.54	0.49	
4	O━O	556	16.69	22.40	13.70	8.70	
5	O━□	646	16.40	10.60	5.58	1.77	3.25
6	□━□	736	16.11	6.00	2.77		3.23
7	O━O━O	1004	15.59	8.20	5.55	2.65	
8	O━O━□	1094	15.50	3.48	2.16	0.69	0.63
9	O━□━□	1184	15.40	3.13	1.80	0.29	1.04
10	□━□━□	1274	15.27	1.44	0.77		0.67
11	O━O━O━O	1452	15.16	2.22	1.56	0.66	
12	O━O━O━□	1542	15.04	0.97	0.64	0.20	0.13
13	O━O━□━□	1632	14.95	0.39	0.24	0.05	0.10
14	O━□━□━□	1722	14.92	0.30	0.18	0.02	0.10
15	□━□━□━□	1812	14.90	0.10	0.06		0.04
				100.00	36.55	41.76	21.69

[a] O = mPDA
□ = MDA
━ = DGEBPA

ingredients, the ratio of these three peaks according to statistical distribution should be 4:4:1. The difference in the observed and calculated values is too great to be attributed to concentration effect. In the same way peaks 7, 8, 9, and 10, each containing three amine and two epoxy groups are present in the ratio 2.6:1.1:1.0:0.5, the first value referring to all three mPDA groups being present in the chain segment; and the last, all three MDA groups. Peaks 8 and 9 carry both amine moieties in the same chain segment. According to statistical distribution for the concentration of the two amines, the ratio would have to be 1.3:2.0:1.0:0.2 if concentration of ingredients were the controlling factor. Obviously, this disageement is too great even for marginal errors, and one must conclude that the distribution of groups within a chain segment is not controlled solely by concentration of ingredients. Peaks 11 through 15 contain four amine groups and three epoxy groups. While the concentration of these peaks are rather low, the distribution of observed values(2.2:1.0:0.4:0.3:0.1) and those calculated by statistical distribution (0.66:1.33:1.00§0.33:0.04) differ in the same manner as described for the previous example. It is believed that reaction kinetics holds the clue for these distributions, and this aspect will be explored later in the sections dealing with reaction kinetics and discussion.

BECPE Epoxy System

The reaction of amines with BECPE epoxy was followed in the same manner as in the case of DGEBPA epoxy. The notable difference was concerned with the rate of reaction; higher temperatures were required for the BECPE system in order to get sufficient concentration of reaction products for analysis. In the case of GPC peak areas it was mentioned previously that the resolved peaks are not directly proportional to the weight fraction, but that sensitivity factors must be taken into account, especially when BECPE is involved. This subject will be considered next.

mPDA amine. The reaction of mPDA with BECPE at various values of molar ratio is illustrated by the GPC-resolved curves shown in Figure 7. The magnitude of the resolved peaks and corresponding species are given in Table 6. In dealing with resolved peaks the subject of sensitivity comes to the forefront. A convenient way of measuring sensitivity factors is as follows. It had been observed that upon mixing dilute solutions of amines and epoxides (in tetrahydrofurane solvent) no reaction occurs. Consequently, known quantities of amines and epoxides in various combinations are mixed under these conditions and the areas of the corresponding GPC curves are measured. For the basic reactants there are no overlapping peaks. Selecting mPDA for reference (sensitivity factor = 1), one calculates the following factors: MDA,

Table 6. Concentration of polymer chain lengths as function of molar ratio of mPDA and BECPE.

Peak	Species[a]	Mol wt	Count	Molar ratio mPDA/BECPE		
				12.1	3.4	1.7
z	ᴧ	182	23.80	1.7[b]	3.0	9.3
a	O	108	21.84	74.8	57.3	38.3
b	Oᴧ	290	19.37	2.8	9.0	22.2
c	OᴧO	398	17.80	11.2	16.0	20.7
d	OᴧOᴧ	580	16.68	5.4	10.2	8.2
e	OᴧOᴧOᴧ	870	±5.87	4.1	4.5	1.3

[a]O = mPDA
[b]ᴧ = BECPE
[b]Values based on percent of GPC area.

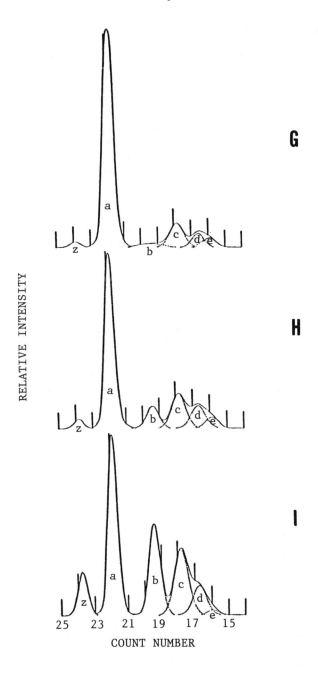

Figure 7. Effect of molar ratio mPDA/BECPE upon chain length of reaction products. G, 12.1 MR; H, 3.4 MR; and I, 1.7 MR. Values of area of species peaks are given in Table 6.

0.947; DGEBPA, 1,409; and BECPE, 2.936 (factor x GPC area = relative weight). Sensitivity factors for reaction products were not measured because sufficient quantities of the products were not isolated. In all resolution of GPC curves the sensitivity factor for the reaction products was assumed to be unity, but the ones corresponding to the original reactants were corrected in the tables according to the figures cited above. The low sensitivity of BECPE is attributed to its aliphatic nature, whereas all other materials, products as well as reactants, contain aromatic benzene rings which respond well to the ultraviolet detector in the GPC apparatus. The refractive index detector curves were always examined carefully to make certain no peaks were omitted.

Comparison of the reaction of mPDA and BECPE (Table 6) with that of mPDA and DGEBPA (Table 2) reveals significant differences. First, all of the BECPE epoxy reactant was not completely consumed in spite of the presence of excess amine. Further, the species in the reaction products take on a completely different makeup. A number of species contain a terminal epoxy group, again pointing to the low reactivity of the BECPE. Further comparisons will be made in later sections.

MDA amine. The reaction of MDA with BECPE at various molar ratio values is shown in Figure 8, and the corresponding data given in Table 7. In contrast to the reaction of mPDA with BECPE, MDA reacts more vigorously as evidenced by the small amount of epoxy remaining in the presence of excess amine, and the relatively low concentration of species with terminal

Table 7. Concentration of polymer chain lengths as function of molar ratio of MDA and BECPE.

Peak	Species[a]	Mol wt	Count	Molar ratio MDA/BECPE		
				13.8	3.7	1.8
p	∿	182	23.80		1.1	2.4
q	□	198	20.11	82.1	62.2	38.3
r	□∿	380	18.37	2.8	6.3	9.2
s	□∿□	578	16.68	11.3	13.0	19.0
t	□∿□∿□	958	15.60	3.1	13.8	22.0
u	□∿□∿□∿□	1338	15.15	0.7	3.5	9.1

[a] □ = MDA
∿ = BECPE
[b] Values based on percent of GPC area.

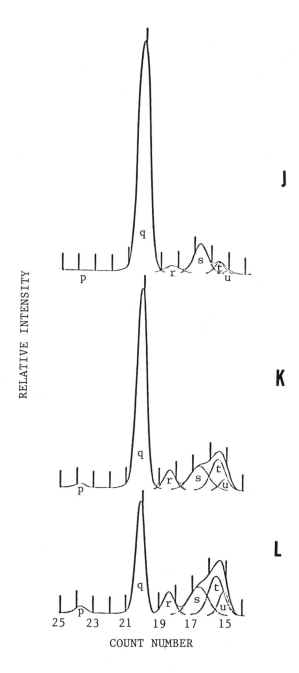

Figure 8. Effect of molar ratio MDA/BECPE upon chain length of reaction products. J, 13.8 MR; K, 3.7 MR; and L, 1.8 MR. Values of area of species peaks are given in Table 7.

epoxy groups. Further, MDA reaction products tend to form longer chain segments as compared to the mPDA products. This difference is just the reverse of the reaction products of amines with DGEBPA epoxy. In the latter case mPDA favors longer chain segments. A startling revelation is the absence of the chain segment composed of three mPDA groups and two BECPE groups. The counterpart in the MDA system is present. All of these differences are related to the relative rates of reaction or competition among the ingredients. The reaction of BECPE with a mixture of the two amines should prove enlightening.

Mixed amines. The reaction of BECPE with a mixture of the two amines at two different concentrations of reactants is shown in Figure 9. The peak areas of the resolved curves with corresponding species are given in Table 8. In terms of weight quantities of reactants curve M represents 3.009 gm BECPE added to a mixture of 5.009 gm mPDA and 2.508 gm MDA; curve N, 3.002 gm BECPE to 2.506 gm mPDA and 2.536 gm MDA.

Table 8. Concentration of chain length of reaction products from the addition of BECPE epoxy to a mixture of mPDA and MDA amines at two different levels of amine concentration.

Peak	Species[a]	Mol wt	Count	GPC areas, Figure 9	
				Curve M	Curve N
1	⌁	182	23.80		0.31
2	o	108	21.84	38.53	11.16
3	□	198	20.11	17.50	11.25
4	o⌁	290	19.37	9.48	1.48
5	□⌁	380	18.37	5.19	4.15
6	o⌁o	398	17.80	9.74	8.15
7	o⌁□	488	17.15	11.00	15.75
8	□⌁□ / o⌁o⌁	578 / 580	16.68	4.42	9.56
9	□⌁o⌁	670	16.31	1.66	5.64
10	o⌁o⌁□	778	16.05	1.44	7.20
11	o⌁□⌁□ / o⌁o⌁o⌁	868 / 870	15.87	0.59	3.89
12	□⌁□⌁□ / □⌁o⌁o⌁	958 / 960	15.60	0.41	11.53
13	□⌁□⌁o⌁	1050	15.38	0.04	5.93
14	□⌁o⌁□⌁□	1338	15.15		4.00

[a] o = mPDA
□ = MDA
⌁ = BECPE

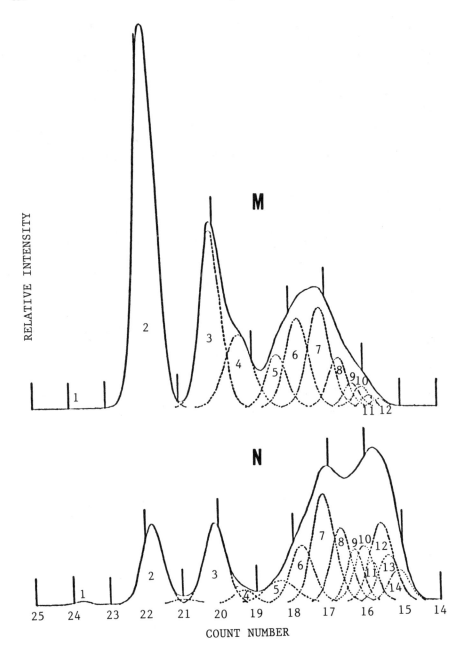

Figure 9. Effect of mole percent of mixed amines with BECPE
upon chain lengths of reaction products. M: mPDA, 61.7;
MDA, 16,7; BECPE, 21.7. N: mPDA, 43,9; MDA, 24.4; BECPE,
31.7 mol %. Values of area of peaks are given in Table 8.

As in the case of the other epoxy system new species have
become evident. In some instances two species of differing
composition have very similar values of molecular weights so
that one peak represents both species (case of overlapping).
In these cases one of the pair of species has an active epoxy
group present. The rational for the existence of such
species can be found in the compositions of peaks 4 and 5
where one amine group is attached to one epoxy group. The
concentration of these species is rather appreciable.
Likewise peak 13 is another example that points to the
existence of an active epoxy group in the presence of excess
amines.

Total Epoxy System

Stoichiometric reaction. Some experiments involving both
epoxies and both amines in the same mixture were carried out
at several temperature levels. The epoxies were premixed
with the BECPE constituting 43 percent by weight; and DGEBPA,
57 percent by weight. In terms of epoxy equivalents, BECPE
has 58.5 percent; and DGEBPA, 41.5 percent. The two amines
also were premixed with MDA constituting one-fourth of the
amine equivalent and mPDA containing the balance. Next, an
amount of the mixed amines was added to a stoichiometric
amount of the mixed epoxy resin with vigorous stirring at
ambient (23°C) temperature. Specimens were taken at various
time intervals and the reaction quenched by dilution of the
samples in THF to 0.5 percent concentration. Each sample was
run in a GPC apparatus (Note: This instrument was a different
one used in previous experiments so the count numbers will
differ. For simplicity the "count number" of the present
experiment is labelled "GPC Elutriation Number"). The first
sample for the series was taken after an elapsed time of 20
minutes. To get a true zero-time reading, separate samples
of resin and hardener (amines) were made up to 0.5 percent
concentration in THF and then mixed in the proper ratio and
analyzed by GPC. The GPC curves taken at different intervals
of time are shown on the left hand side of Figure 10. The
progressive changes can be readily followed in the sequence
of curves. The BECPE peak remains unchanged after 54 hours,
signifying that this material did not react during this time
interval. The disappearance of the mPDA peak indicates that
the mPDA entered completely into the reaction products. The
same can be said about DGEBPA; however, there is a small
amount of unreacted MDA. The progressive increase in
reaction products and shift to higher molecular weight
products attests to the extent of reaction. Since BECPE did
not partake so far in the reaction, these reaction products
are probably linear with very little cross-linking.

I. Shapiro

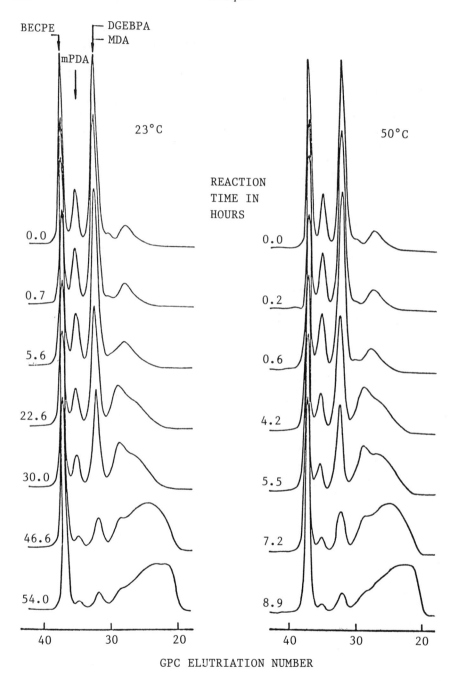

Figure 10. Change in GPC curves for mixed epoxy resin system as a function of reaction time at two temperature levels.

Reaction Kinetics. The stoichiometric reaction of the two epoxies and two amines was studied at four selected temperature levels: 23°C, 50°C, 65°C, and 85°C. The reaction at 23°C was already described in the previous section. The same techniques were used for all temperature reactions. At 50°C the set of reaction curves yields similar shaped GPC curves except there was a six-fold change in reaction times at the 50°C level. The comparison of the two sets of curves is shown in Figure 10. The progressive change in shape of the GPC curves for 65°C and 85°C was surprisingly similar to those for the lower temperatures, except, of course, the time to reach certain contour profiles decreased as the temperature increased. From an inspection of the curves one concludes that the BECPE did not participate in the reaction with the amine hardener up to the gel temperature of 85°C.

To be able to reduce the cure cycle without resorting to empiricism, a knowledge of the kinetics of cure (reaction) is necessary. In this connection the GPC data generated above readily lends itself for calculating the energy of activation, E, of the reaction as defined by the Arrhenius equation.

$$\text{Rate constant} = Ae^{-E/RT}.$$

Since the rate constant is a measure of the amount (percent or extent) of reaction per unit time, the rate constant would be proportional to the reciprocal of time necessary to reach a certain level or degree of reaction. Such values of time can be obtained by comparison of GPC profiles at the different temperatures (similar profile, similar degrees of reaction). Incidentally, a number of curves were obtained but only those with similarity of shapes at all temperatures were retained for the purpose of comparison. A plot of logarithm of reciprocal of time vs the reciprocal of absolute temperature is shown in Figure 11. The four lines in Figure 11 correspond to the last four GPC profiles of Figure 10. The same slope for the four lines lends credulity to accuracy of the data. The value of E calculated from the slope is 13 Kcal/mole. This value is the energy of activation for the reaction of DGEBPA with the two mixed aromatic amines. It is anticipated that at higher temperatures the slope of these curves will increase to correspond to the activation of BECPE. The temperature at the point of intersection of the two slopes should represent the highest temperature for gelation, i.e., the end-point of linear polymerization of DGEBPA and the start of cross-linking (cure) of BECPE. Too rapid a heat-up time at a fixed temperature could cause the BECPE to distill out of system.

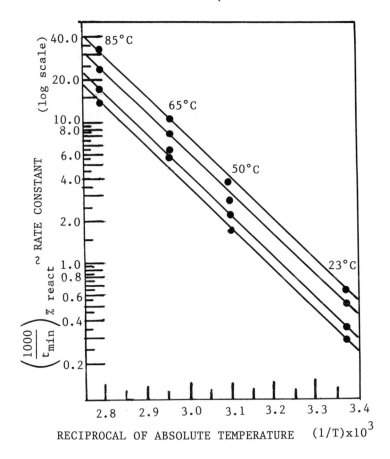

Figure 11. Arrhenius plot of different degrees of reaction
of stoichiometric mixture of two epoxies and two amines based
on GPC data given in Figure 10.

Aniline. Exploratory experiments with the reaction of
DGEBPA with aniline were undertaken to help develop a model
for the epoxy-amine reaction. Aniline was selected as a
simplified amine because it has only one amine group instead
of two amine groups as are present in both mPDA and MDA. It
was necessary to raise the temperature of the DGEBPA-aniline
mixture inasmuch as no evidence of reaction occured at room
temperature over a reasonable period of time.

With a four-fold molar ratio of aniline to epoxy, samples
of the mixture were heated to 75°C over a period of 4 hours
and then quenched in THF for GPC analyses. Figure 12 shows
the resolved curves in a sequence of thermal aging. Four

distinct reaction products were recognized, and from a
molecular weight calibration chart, the four products were
derived in terms of ratio of molecules of aniline and epoxy
as indicated in the figure. The changing intensity of curves
shown in Figure 12 illustrates that the 1:1 ratio product
first builds up in concentration and then decreases. The
other three reaction products continue to build up in
concentration with time. Reaction stops after all epoxy and
the 1:1 product have been consumed.

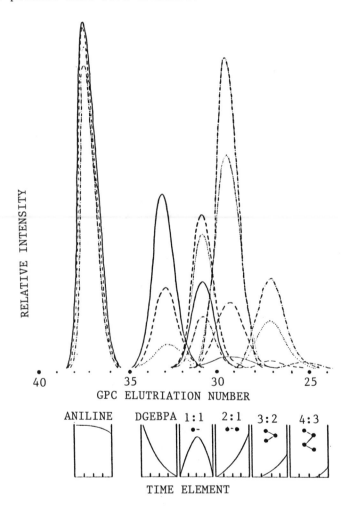

Figure 12. Resolved GPC curves showing reaction products from
reaction of DGEBPA epoxy with aniline in excess as a function
of thermal aging. Sequence of age is ———,— — —,····,—·—·—,
Changes in concentration of species is illustrative.

Discussion

The results of the current study of the reaction of an epoxy with an amine under different conditions of reactants and their concentrations to form polymer chain length segments have lead to a better understanding of the mechanism of the reaction, and ultimately, to the means of improving the processing parameters for composite matrices. All experimental results point to the fact that the mechanism of the polymer reaction is a stepwise process with the initial reaction product being the 1:1 addition. Then depending upon the reactivity of the remaining "active" group in the 1:1 product, further addition to the initial 1:1 product can take place. Apparently there are a series of competing reactions that can occur in the system. In the case of the DGEBPA epoxy with the mixed amines, mPDA and MDA, the remaining epoxy group of the 1:1 product is still very reactive so that that group reacts further to give amine groups on both ends of the DGEBPA molecule. Actually the 1:1 product is not isolated in this case because the reaction occurs too rapidly at high amine content. It should be pointed out that there is a very small quantity of an apparent 1:1 product (see Tables 1-5) but its existence is attributed to a defective end group which cannot react further.

In the competition of the amines for the DGEBPA epoxy, the mPDA is the more reactive of the two amines, so the concentration of chain length segments will contain a preponderance of mPDA groups in the reaction products as indicated in Table 5. Relative concentration of the two amines may have some influence, but it has already been established that from statistical considerations concentration alone is not the controlling factor.

In the reaction of the BECPE epoxy with the mixed amines, the roles of the two amines appear to be reversed in that the MDA is the more dominant in its reaction. A plausible reason for the observed differences in reaction ability or rate may be the difference in distance between active sites in the epoxy molecule. The reported distance of active sites in DGEBPA is 16.4 Å; whereas in BECPE, 7.0 Å [3]. Likewise, from the structure of the amines, the distance between active sites in MDA would be greater than in mPDA. It is well known that after the first hydrogen of a primary amine is reacted, the second hydrogen has a much lower reactivity in that it has the character of a secondary amine. In a similar fashion in the case of the BECPE epoxy once one of the active groups has reacted, the reactivity of the other active site is relatively less reactive; however, in the case of DGEBPA the

two epoxy groups are sufficiently separated that the
reactivity of either does not influence the other. This
possible difference in reactivity due to distance separation
of active sites merits further experimentation. .

   In practice the relative reactivity of the BECPE towards
the two amines is academic when both epoxies are involved in
a common mixture. The great difference between reactivity of
the two epoxies is such that at low (ambient) temperatures
the DGEBPA takes preference in reaction with the amines, and
all of the DGEBPA is consumed before the BECPE partakes in
the reaction. This situation actually is ideal for an epoxy
system in that the system has a built-in "B-stage" operation.
The fabric or fiber of a composite system can be impregnated
or coated with the mixed epoxy system at ambient temperature,
and then the entire system can be heated for the matrix to
cure (cross-linking). The role of the BECPE is to act as an
active diluent until further reaction is desired. The
processing parameters involve choice of relative ratio of
epoxies and of amines (stoichiometry or slight excess of
amine must be maintained for optimum physical and mechanical
properties), and manner of mixing of ingredients, i.e., order
of adding ingredients and degree of stirring.

Summary

   The reaction of epoxies and primary aromatic amines has
been followed by a technique of measuring the chain length
segments of the polymeric reaction products. The
concentration of chains has been determined by analyzing
reaction mixture samples in a GPC followed by curve-
resolving. The mechanism of reaction is the formation of the
1:1 addition of epoxy to amine, which in turn can be reacted
further. In the mixed epoxied - mixed amine system at low
temperatures the more reactive DGEBPA epoxy reacts completely
before the BECPE epoxy starts to react. Higher temperatures
are needed for the second epoxy to react, which, in this
case, is the cross-linking of the linear chain segments.

References

[1] Soldatos, A,C.; Burhans, A.S.   Ind. Eng. Chem. Prod. Res.
    Develop. 1967, 6, 205.
[2[Soldatos, A.C.; Burhans, A.S.; Cole, L.F.   SPE Journal.
    1969, 25,61.
[3] Busso, C.J.; Newey, H.A.; Holler, H.V.   25th Conference
    SPI Reinforced Plastics/Composites Div. 1970.

POLYURETHANE ADHESIVES WITH SILANE COUPLING AGENTS

H. Dodiuk, A. Buchman, S. Kenig

A.D.A., Haifa, Israel

The effect of silane coupling agents incorporated into the bulk of two polyurethane adhesives or applied on aluminum adherends prior to bonding was studied. The physical and mechanical properties of the bulk and corresponding aluminum bonded joints were characterized in ambient and humid-hot environments. Experimental results have demonstrated the advantage of silane addition to the performance of polyurethane adhesives especially under humid atmosphere. For polyol cured polyurethane best results were obtained with 0.5% vinyl terminated silane, while amine cured polyurethane showed best performance with up to 2% amine or epoxy terminated silane. Property enhancement of polyurethane with silane addition has been attributed to crosslinking and reaction with absorbed humidity in the bulk and the interface. This main mechanism was substantiated using FTIR analysis. It could be concluded that by proper selection of the silane reactive species and its appropriate matching with a given polyurethane, enhancement of both bulk and joint properties as well as durability in humid environment could be realized.

Introduction

Application of coupling agents for surface modification of fillers and reinforcements incorporated in polymers composites have generally been directed towards improved mechanical and chemical properties (1-5). The resulting chemical bond occurs by virtue of the coupling agent functional groups which react both with the filler and the polymer. One of the most common cases, comprises silanes and glass-fibers. In this case the mechanical properties of the polymer composite as well as its chemical characteristics are enhanced by a large extent. It is well accepted that water primarily degrades the fiberglass polymer interface via hydrolysis of the silane coupling agent in addition to its adverse effect on the polymer matrix and filler.

Silane coupling agents are also used as additives for adhesion promotion of polymers to adherend at level of 1% or less (3). In this case they function in the same way as coupling agents or primers provided there is a driving force for their migration to the interface.

Selection of the appropriate coupling agent for a given polymer substrate combination is difficult. Factors like wetting, surface energies, polar absorption, acid-base interaction, interpenetrating network formation and chemical reactions, have to be taken into consideration.

Even the best coupling agent for a given system may perform poorly if it is not applied properly. Orientation of the molecules at the inteface and their physical properties can be controlled by the method of application, and may be as important as the chemistry of the selected silane.

Silane coupling agents have been used to improve adhesion and durability of polyuerthane adhesive exposed to aggressive environment. F. Liang and P. Dreifuss (6) have studied the durability of polybutadiene-polyurethane joints to glass or metals. The effects of silane pretreatment surface morphology on adhesion were examined. K. E. Creed (7) has studied the effect of added silane to polyurethane on its durability. F. D. Swanson and S. J. Price (8) have analysed the chemistry of urethan adhesives including silane coupling agent. A. Kaul has used silane coupling agent to improve polyurethane durability, either as substrate coating or as additive (9) The general conclusion of the above studies is that silanes are very effective when incorporated into polyurethanes, especially as it relates to durability.

The present study is aimed at evaluating the effect of silane coupling agents; either incorporated into two polyurethane adhesives or applied on the substrate surface prior to bonding.

The characterization included long term durability of the bulk urethane adhesives with and without silane and the effect of silane on the strength of polyurethane-aluminium joint when added to the adhesives or applied on the adherends.

Experimental

Materials and processes. Two adhesives were used in the
present study, a two component urethane based on a polyol
cured methylene diphenyl diisocyanate MDI-[PU-A], and a two
component urethane based on an aromatic amine cured MDI-
[PU-B]. Both adhesives are cured at room temperature.

A series of silanes coupling agents were used:

(a) Two epoxy terminated silanes:
     A-187 supplied by Union Carbide, U.S.A., and
     Z-6040 supplied by Dow Corning, U.S.A., CH_2-CH-
     $CH_2O(CH_2)_3SiO(CH_3)_3$

(b) An amine terminated silane, A-1100 supplied by Union
     Carbide, U.S.A., $NH_2(CH_3)_3Si(OC_2H_5)_3$.

(c) A methacryl terminated silane, Z-6030, supplied by Dow
     Corning, U.S.A., $CH_2 = CH(CH_3) - COO(CH_2)_3Si(OCH_3)_3$.

(d) A vinyl terminated silane, Z-6032 supplied by Dow
     Corning, U.S.A.,
     $CH_2 = CH(C_6H_4)CH_2NH(CH_2)_2NH(CH_2)_3Si(OCH_3)_3 \cdot HC\ell$.

The silanes were incorporated into the adhesives or
applied on the adherends by brushing as described later. The
adherends used for the present study were Al-2024-T6, chromic
acid anodized without sealing in accordance with MIL-B-8625.

In the course of the study a two component primer for
polyurethanes adhesives AD-6 manufactured by Conap, U.S.A.,
was used. The primer was applied by brushing on the
adherends and curing was carried out at room temperature for
at least half an hour.

Testing. The bulk properties, as well as thin bond line
joint properties, were characterized using the following
testing methods:

(a) Bulk Tests

     Hardness: Hardness level (shore D scale) was measured
     on cast specimen (50 × 50 × 10mm) according to ASTM D-
     2240.

Swelling: Cast specimens were swollen until equilibrium
in toluene (21-25 days). Weight increase of the swollen
material and of the extracts were determined. The same
procedure was applied following aging.

(b) Adhesive Joint Tests

Shear and T-peel strengths: Single lap shear (SLS) and
T-Peel tests were performed by polymerizing the adhesive
between two aluminum plates according to ASTM D-1002-73,
and ASTM D-1876, respectively. Five specimens were
fabricated for each test. Bondline thickness for all
specimens was 0.04 - 0.1mm.

Testing was carried out by means of an Instron
Mechanical Tester (crosshead speed was 2mm/min for lap
shear test strength and 200mm/min for T-peel test) at
room temperature. The mode of failure (adhesive or
cohesive) was evaluated by visual inspection expressed
as the percentage of adhesive coverage on both adherends
(if 100% cohesive both sides are completely covered).

Fourier Transform Infra Red (FTIR): FTIR was used to
analyze the fracture surfaces of the SLS specimen after
fracture. The analysis was performed using a Nicolet
5DX spectrophotomether in a horizontal specular
reflectance mode. Screening range was 400-4600cm^{-1}. A
gold coated mirror was used as a reference.

Sample preparation and aging: The neat silanes were
incorporated in part A of the adhesives by direct
addition. When applied on the adherend directly, a
solution of 2% silane in 80:20 V/V ethanol/water was
stirred for 30 minutes in RT, a thin layer was applied
on the adherend by brushing, the specimens were dried
for 30 minutes at ambient temperature and 30 minutes at
100° C. The samples were cured at room temperture for 48
hours under pressure of 0.5atm. Postcure was performed
for PU-A at 50°C during 20 days and for PU-B at 50°C
during 10 days. Non-treated specimens were prepared
under the same conditions.

Aging conditions for all specimens were 50°C and 95%
relative humidity (RH) for various periods of time
(Associated Environmental System Humidity Chamber Model BHK
4105). Table 1 and Table 2 summarize the type and
concentration of the silanes added for bulk, SLS and T-Peel
specimens, respectively.

Adhesive Aging Time (Days)	PU-A Primer Type and Concentration in the Adhesive	PU-B Primer Type and Concentration in the Adhesive
0, 20, 40, 60, 100, 130	0	0
0, 20, 30, 40, 60, 100, 130	A-187 0.5%, 1% A-1100 0.5%, 1%	A187 0.5%, 1% A1100 0.5%, 1%
0, 60, 100, 130	-	A187 0.5%, 2% A1100 0.5%, 2%
0, 10, 30, 60, 100, 130	Z-6030 0.5% Z-6032 0.5% Z-6040 0.5%	Z-6030 0.5% Z-6032 0.5% Z-6040 0.5%

Table 1:   A List of Samples Prepared for Bulk
Characterization by SLS Experiments

Adhesive Aging Time (Days)	PU-A Primer Type and Concentration (Application Mode)	PU-B Primer Type and Concentration (Application Mode)
0, 60, 130	0	0
0, 60, 130	0.5% A-1100 (in the adhesive)	0.5% Z-6032 (in the adhesive)
0, 60, 130	A-1100 (on the adherend)	Z-6032 (on the adherend)
0, 60, 130	AD-6	AD-6

Table 2:   A List of Samples Prepared for T-Peel Experiments

Results and Discussion

The Bulk Adhesive

Hardness tests. Figures 1a and 1c describe the changes of hardness with A-187 silane contents and aging time at 50° C/95 RH of PU-A and PU-B, respectively.

Figures 1b and 1d depict the changes of hardness with A-1100 silane contents and aging time at 50° C/95 RH of PU-A and PU-B, respectively. Results indicate that incorporation of silane into the polyurethane adhesives results in significant changes in hardness for all aging periods, but the direction of change is dependent on the type of the polyurethane, the chemical nature and the amount of silane added, and on the aging time.

As distinct from PU-A, addition of A-187 and A-1100 silanes to PU-B gives rise to an increase in hardness in unaged and aged samples, the maximum increase is obtained after environmental exposure for 60-100 days. Longer exposure times and high silane concentration result in a decrease of hardness values (Figure 1c and 1d).

It can be seen clearly, that initially hardness values increase to a maximum followed by a decrease at longer aging duration. The increase in hardness is a result of higher crosslinking of the polyurethane network induced by heat and humidity, especially in the presence of silanes. At longer exposures, aging dominates due to the effect of humidity on the silane, a well known phenomena (1-5). As the silane concentration exceeds some optimal value (depending on its chemical nature and the type of polyurthane) it functions as a plasticizing agent which both softens the polymer and accelerates water diffusion into the material. Consequently, it may be concluded from the hardness tests, that the incorporation of silanes into the polyurethane is advantageous provided an optimal concentration is used. Best results are obtained using 1.5% of A-187 or 0.5% of A-1100 silanes for PU-B and none or 0.5% of A-1100 for PU-A.

Generally, it seems that the incorporation of the silane into PU-B is more effective than for PU-A in terms of hardness, probably because PU-B has more reactive groups from its amine curing agents and therefore could react with the functional terminated silane groups, leaving the alkoxy group to react with water. PU-A reacts more slowly with the

H. Dodiuk et al.

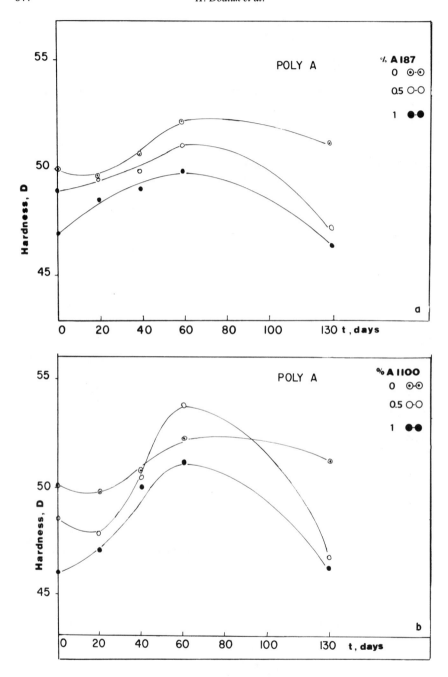

Figure 1 : Hardness vs. aging time (days) for:
            (a) PU-A containing A-187 silane (0, 0.5%, 1.0%).
            (b) PU-A containing A-1100 silane (0, 0.5%, 1.0%).

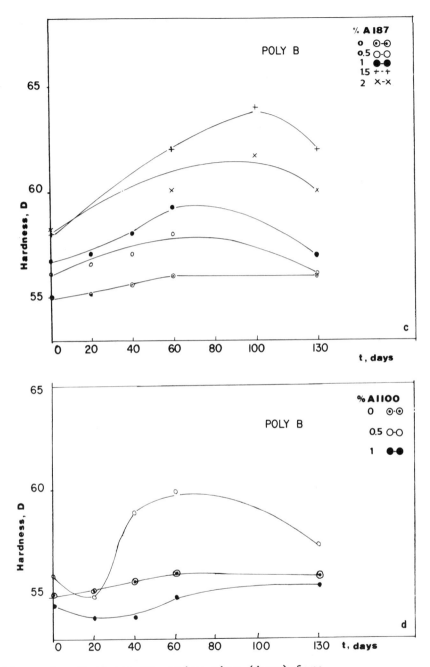

Figure 1 : Hardness vs. aging time (days) for:
      (c) PU-B containing A-187 silane (0, 0.5%, 1.0%,
          1.5%, 2.0%).
      (d) PU-B containing A-1100 silane (0, 0.5%, 1.0%).

silanes and, therefore, they turn to be more as a
plasticizing agent, which softens the material.

Swelling tests.   Swelling curves (weight changes vs
immersion time in toluene) are depicted in Figure 2a for PU-A
and Figure 2b for PU-B.   An increase in swelling for samples
of PU-B containing 0.5% and 1% A-1100 and for PU-A containing
all silanes types and concentrations compared to non-treated
polyurethanes indicates a lower degree of crosslinking of the
treated resins.   The swelling increase depends on the
chemical nature and the amount of the sil‑.ue. The higher the
silane level the larger is the swelling and with A-187 more
than with A-1100.   These results are supported by the
hardness values (Figures 1a and 1b).   In the case of PU-B
containing 0.5-2% A-187 a significant decrease in swelling is
observed as a result of an increase in crosslinking density
compard to the nontreated samples. This phenomenon was also
observed in the hardness measurements (Figure 1c). Figure 3
describes the weight change (ΔW%) of fresh samples after
exposure to a combination of heat and humidity (95RH/50^{0}C).
It seems that in all PU-A samples there is a weight gain
(Figure 3a) during aging, the highest gain occurs for the
untreated samples and the lowest one when 0.5% Z-6032 is
added.   In all PU-B samples there is a weight loss during
aging, the highest occurs for the untreated sample (Figure
3b).   These results indicate that PU-B, which has an amine
curing agent, behaves differently during humidity exposure
than PU-A which has a polyol curing agent.   The amine curing
agent reacts probably with water causing some solubulization
in water and, consequently, weight loss.

PU-A, on the other hand, absorbs water during aging,
resulting in weight gain.   The presence of the silanes in
both polyurethanes modifies to some extent their behaviour by
either reacting with water and reducing weight gain (PU-A) or
by reacting with the amine-curing agent and reducing its
solubilization in water (PU-B). However, it does not change
the general aging pattern.   It is also noticeable from the
swelling and hardness results during aging that the treated
and non treated PU-A samples are less durable than the
corresponding PU-B samples, and that the effectiveness of
incorporating silanes into PU-A is not advantageous, compared
to PU-B.

The extracts weight of the swelling solution (ΔW%) as a
function of exposure time for 50^{0}C/95RH are shown in Figure
4a for PU-A, and Figure 4b for PU-B.   After 20 days of
exposure to these aging conditions there is a dramatic change

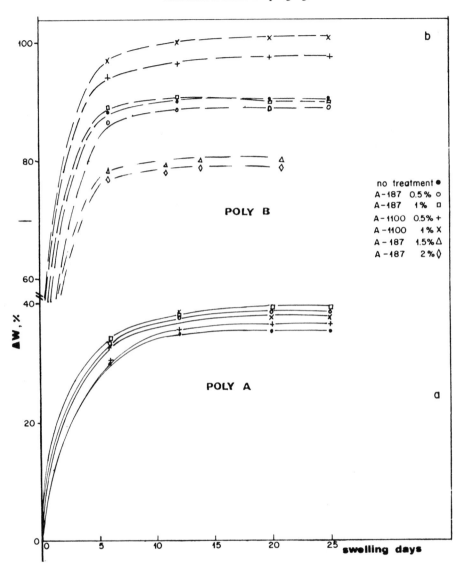

Figure 2 : Swelling curves (weight changes vs. time of
immersion) of:
(a) PU-A containig A-187 silane (0.5%, 1.0%) and
A-1100 silane (0.5%, 1.0%).
(b) PU-B containig A-187 silane (0.5%, 1.0%, 1.5%,
2.0%) and A-1100 silane (0.5%, 1.0%).

H. Dodiuk et al.

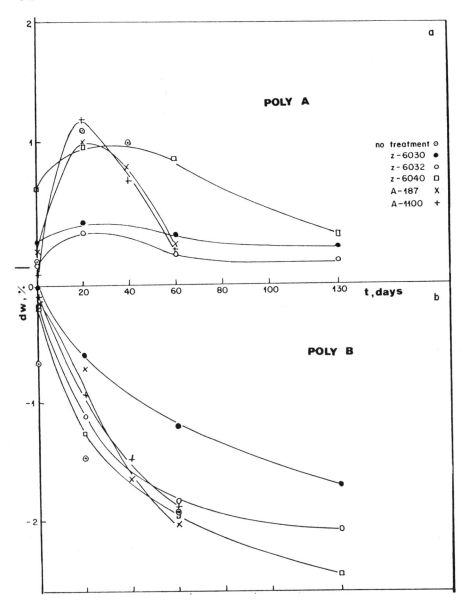

Figure 3 : Weight change vs. aging time (days) for:
   (a) PU-A containing 0.5% of various silanes.
   (b) PU-B containing 0.5% of various silanes.

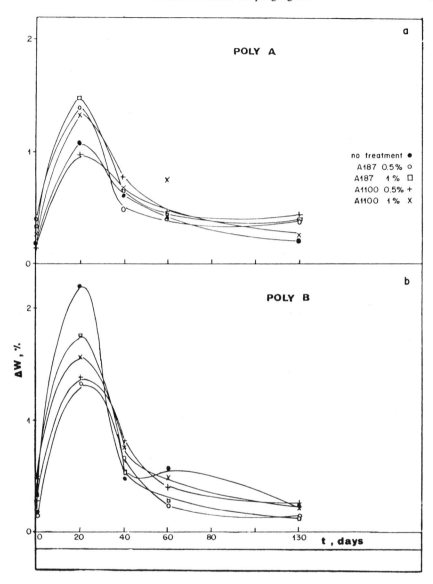

Figure 4 : Extract weight vs. aging time (days) for:
(a) PU-A containing A-1100 (0.5%, 1.0%) or A-187 (0.5%, 1.0%).
(b) PU-B containing A-1100 (0.5%, 1.0%) or A-187 (0.5%, 1.0%).

in the weight of all samples tested. This result may be due
to some unreacted materials or other additives extraction
from the 20 days aged samples. It appears again, that
treated PU-A loses more material than the untreated ones,
except for 0.5% A-1100, while the treated PU-B loses less.
This result may indicate that silanes react chemically with
PU-B and, therefore, cannot be extracted out in the toluene
(swelling solution) while they do not fully react with PU-A,
and thus can be extractd with toluene.

Swelling curves (weight change in equilibrium in toluene
of aged sample vs. aging time in 95RH/50°C) are shown in
Figure 5a for PU-A and Figure 5b for PU-B containing various
concentrations of A-1100 and A-187. Results show that for
treated PU-A after 130 days ΔW% is higher than for the
untreated one and some treated PU-B samples show smaller ΔW%
than the untreated ones. This behaviour depends on the type
and amount of the silane used.

Swelling curves (weight change in equilibrium in toluene
of aged sample vs. aging time in 95RH/50°C) are depicted in
Figure 6a for PU-A and in Figure 6b for PU-B containing 0.5%
of various silanes. An increase in swelling for all treated
PU-A aged samples compared to the non-treated ones indicates
a lower degree of water absorption during aging. This
demonstrates the contribution of the coupling agents to the
prevention of degradation processes as a result of water
diffusion into the polyurethanes. The most pronounced effect
is obtained for Z-6032. In the case of PU-B (Figure 6b), the
silane contributes also to an improved durability as
concluded from the decrease in ΔW of swollen aged treated
samples compared to the non-treated ones.

To summarize it could be concluded from Figures 5a, 6a,
compard to Figures 5b, 6b that the aging behavior of treated
PU-A is very different from that of treated PU-B, as found in
other properties studied. This is due to the different
chemical nature of the curing agent (aromatic amine vs.
polyol) which modifies the interaction of the polyurethane
with the incorporated silanes. In the case of PU-A, it seems
that Z-6032 (vinyl terminated silane) improves its stability
during aging, and A-1100 (amine terminated silane) improves
the stability of PU-B.

Adhesive joint tests. The effect of incorporating silanes
into polyurethanes on the strength of aluminum joints (peel
and shear) and its durability in combined environments
(50°C/95RH) was evaluated.

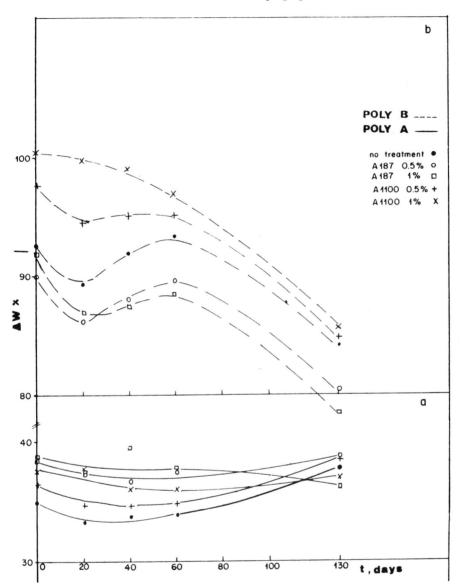

Figure 5 : Swelling curves (weight changes of swelled
           specimens in equilibrium in toluene vs.
           aging time)of:
           (a) PU-A containing A-187 silane (0.5%, 1.0%) or
               A-1100 silane (0.5%, 1.0%).
           (b) PU-B containing A-187 silane (0.5%, 1.0%) or
               A-1100 silane (0.5%, 1.0%).

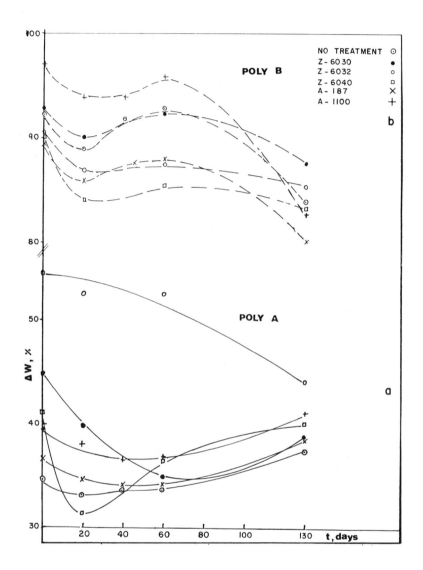

Figure 6 : Swelling curves (weight changes of swelled
specimens in equilibrium in toluene vs. aging time)
of:
(a)   PU–A containing 0.5% of various silanes.
(b)   PU–B containing 0.5% of various silanes.

Table 3 summarizes shear strength data for untreated and treated polyurethane adhesive before and after exposure to a combination of heat and humidity. It is clear that after long aging periods of time (130 days), the incorporation of silanes enahnce the lap shear strength of both PU-A and PU-B; Z-6032 silane shows the largest improvement in durability for PU-A, A-187 and A-1100 for PU-B. Generally, the silane treatment is more effective for PU-A than for PU-B in terms of lap shear strength of aged and unaged joints.

The effect of silane addition on the peel strength of polyurethanes is given in Table 4. As can be seen, the peel strength is doubled in most cases when silane has been added. This outstanding increase may be attributed to the combined effects of the two mechanisms previously mentioned: (a) the softening of polyurethane by silanes which act as a plasticizing agent; and (b) the reaction with humidity which contributes to enhanced durability in humid environment. Since for peel properties those two mechanisms act in the same direction, the effect of incorporating silane to polyurethane is the most striking one compared to the other properties studied (shear, swelling and hardness).

Best peel properties are obtained by A-1100 added to PU-B either directly to the polymer or applied on the adherend. As can be concluded from Table 4, the commercial primer (AD-6) for polyurethanes, is less effective than silanes applied on the adherends. In the case of PU-A best results are obtained when Z-6032 was applied on the aluminum adherend. In addition, the combination of AD-6 and Z-6032 results in satisfying results compared to the untreated PU-A aluminum joint.

Adhesive	Lap Shear Strength [b,c] (kg/cm^2)	
	Unaged	Aged
PU–A	65 ± 11	48 ± 11
PU–A + 0.5% A–187	77 ± 13	53 ± 2
PU–A + 0.5% A–1100	101 ± 30	52 ± 8
PU–A + 0.5% Z–6030	76 ± 21	83 ± 32
PU–A + 0.5% Z–6032	106 ± 31	123 ± 35
PU–A + 0.5% Z–6040	123 ± 35	111 ± 22
PU–B	105 ± 10	119 ± 16
PU–B + 0.5% A–187	127 ± 11	150 ± 11
PU–B + 0.5% A–1100	110 ± 11	127 ± 25
PU–B + 0.5% Z–6030	116 ± 7	121 ± 12
PU–B + 0.5% Z–6032	126 ± 5	125 ± 11
PU–B + 0.5% Z–6040	129 ± 12	125 ± 36

[a] 130 days at 50°C/95RH
[b] Adhesive failure
[c] 5 specimen for standard deviation

Table 3:   Lap Shear Strengths of Polyurethanes –
Aluminum Joints Before and After Aging[a]

Adhesive	T-Peel Strength [b,c] (lb/inch)		
	Unaged	60 Days	130 Days
PU-A	9.6 ± 2.0	14.0 ± 2.5	2.4 ± 2
PU-A + 0.5% Z-6032	21.5 ± 3.4	21.0 ± 2.0	5.4 ± 0.5
PU-A, Z-6032 Applied on the Adherend	35.5 ± 1.9	32.1 ± 3.6	39.3 ± 0.2
PU-A, AD-6 Applied on the Adherend	26.0 ± 2.1	30.0 ± 2.0	24.7 ± 2.0
PU-A + 0.5% Z-6032 and AD-6 Applied on the Adherend	33.2 ± 3.7	27.8 ± 2.5	24.3 ± 2.5
PU-B	13.0 ± 2.2	9.9 ± 1.2	12.7 ± 2.2
PU-B + 0.5% A-1100	22.0 ± 1.3	25.1 ± 2.9	40.0 ± 3.0
PU-B + A-1100 Applied on the Adherend	22.0 ± 3.0	19.3 ± 1.2	29.9 ± 2.0
PU-B, AD-6 Applied on the Adherend	25.0 ± 3.0	17.3 ± 1.2	9.3 ± 0.3
PU-B + 0.5% A-1100 and AD-6 Applied on the Adherend	26.0 ± 2.7	4.5 ± 0.6	4.5 ± 0.6

[a] 50° C/95RH
[b] Adhesive Failure
[c] 5 specimens for standard deviation

Table 4: T-Peel Strengths of Polyurethanes –
Aluminum Joints Before and After Aging[a]

FTIR analysis of failure surfaces. Figure 7a depicts the
FTIR spectra of PU-A containing A-187 (0.5% and 1.0%)
compared to the neat resin. Only the FTIR regions were
significant changes occur are shown. It can be seen that
adding A-187 to PU-A does not change the main absorption but
adds to the spectra its typical bands, i.e., epoxy functional
group (866cm^{-1}), Si-0-Si, (1070cm^{-1}). The carbodiimide which
is a known additive to polyurethanes for protection from
water attack (10), has its typical absorption at 2140 (cm^{-1}).
After exposure to heat/humidity (for 30 days) (Figure 7b) the
carbodiimide absorption disappears, probably as a result of
reaction with absorbed humidity, therefore no additional
water peaks are observed in the spectra.

Figure 8a represents FTIR spectra for PU-A with addition
of A-1100 silane. Additional amine absorption appears from
the amine terminated silane functional group. After exposure
to heat/humidity no additional water absorption is observed
a n d the carbodiimide band disappears (Figure 8b).

Figure 9a depcits the IR spectra of PU-B containing A-187
silane (0.5%, 1.0%) compared to the untreated ones. It is
clear that the added silane causes an increase in the
intensities of Si-0-Si (1070cm^{-1}), C=0 (1734cm^{1}), N=C=0
(2290cm^{-1}), OH (3300cm^{-1}) absorptions. Those observed
changes might be due to reaction of the silane with MDI or
with the aromatic amine curing agent (8) leaving unreacted
isocyanate group or amine groups and free hydroxyl and
carbonyl groups.

Figure 10a shows the same analysis for PU-B with A-1100
silane. Before aging it seems that the N=C=0 and the
carbonyl bands appear when silane is added to polyurethane,
in the same manner as with A-187. But the aged samples
(Figures 9b and 10b) represent a different behaviour. While
the N=C=0 absorption increases when A-187 is added to PU-B,
it disappears in the aged sample, probably as a result of
reaction with humidity.

It seems from the FTIR that in the PU-B the presence of
silanes has a significant effect on the polyurethane both
aged and unaged; while in the PU-A its effect is minor.
Part of it might be due to the presence of carbodiimide in

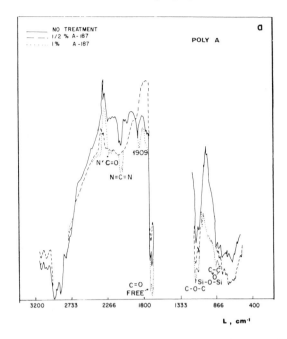

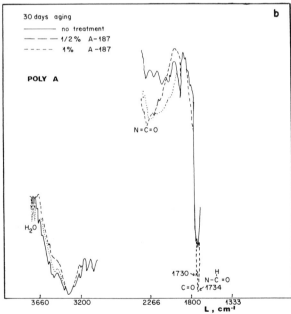

Figure 7 : FTIR spectra of fractured surfaces of:
   (a)   PU–A containing A–187 (0.5%, 1.0%).
   (b)   PU–A containing A–187 (0.5%, 1.0%) after 30
         days of aging.

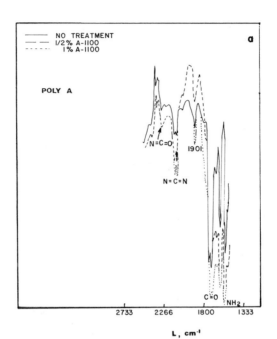

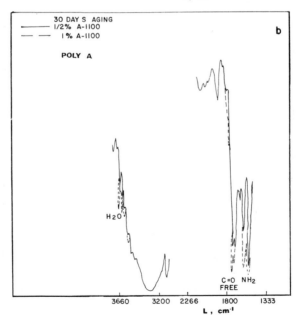

Figure 8 : (a) PU–A containing A-1100 (0.5%, 1.0%).
            (b) PU–A containing A-1100 (0.5%, 1.0%) after 30
                 days of aging.

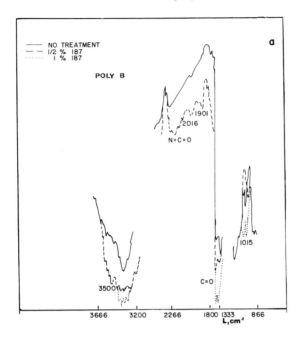

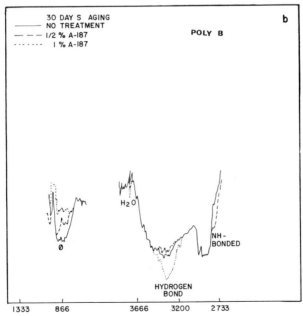

Figure 9 : FTIR spectra of fractured surfaces:
(a) PU–B containing A-187 (0.5%, 1.0%).
(b) PU–B containing A-187 (0.5, 1.0%) after 30 days of aging.

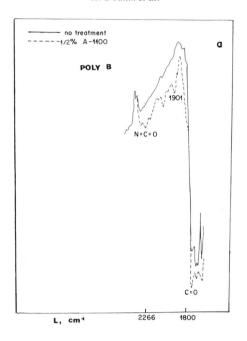

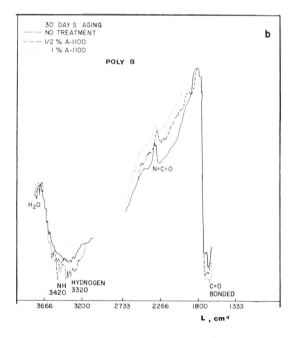

Figure 10: (a) PU–B containing A-1100 (0.5%, 1.0%).
            (b) PU–B containing A-1100 (0.5%, 1.0%) after 30
                 days of aging.

PU-A which is more reactive toward water than the functional terminated groups of the silanes.

Summary and Conclusions

In the present study, changes in properties as a result of incorporating silanes into two polyurethane adhesives using polyol (PU-A) and amine (PU-B) curing agents were characterized in terms of physical and mechanical properties of the bulk and in terms of shear and peel strengths of aluminum joints, before and after exposure to aging (heat/humidity) environment.

The two studied polyurethanes exhibit different behaviour as a result of silane incorporation, the observed changes depend on the type and the amount of the added silane. While silane treated PU-A demonstrates a small reduction in hardness both before and after aging, PU-B shows a substantial increase in hardness.

From the swelling results it might be concluded that Z-6032 is the most effective for PU-A durability and A-1100 for PU-B durability.

Since during aging PU-B, treated or non-treated, loses weight and PU-A gains weight, it is evident that their aging mechanisms are different. This distinct behaviour is due to their chemical structure. While PU-A is a polyol, PU-B is amine cured polyurethane. It seems that this difference in chemical nature modifies its chemical interaction with the added silanes and with humidity. It may be concluded that the added silane reacts chemically with PU-B components either with the isocyanate or with the amine curing agent. PU-A,   seems not to react with silanes which acts more as a plastisizing agent. It is also clear from the lap shear results that Z-6032 causes a dramatic improvement in shear strength for PU-A aluminum joints, while for PU-B all the silanes have minor efects. In terms of T-peel strength, their effect is of primary importance either on PU-A or on PU-B, either added to the polymer or applied on the adherent both aged and unaged joints, when compared to commercial polyurethane primer. FTIR analysis of the fractured surfaces substantiates our previous conclusions.

The durability improvement depends on the physical and chemical nature of the polyurethane resin, on the type and the amount of the silane added, and each case has to be studied separately. However, it is recommended that the silanes will be incorporated into the polyurethane resin, in addition or without its application on the adherend, in order to enhance their bulk and adhesive joint durability.

Acknowledgement

The authors would like to thank Mrs. A. Heller and Mrs. I. Liran for the technical assistance and Mr. L. Drori for helpful discussions.

References

[1] D. Leyden, W. Collins, "Symposium on Silylated Surfaces", Gordon and Breach, 1980.

[2] B. Arkles, "Tailwing Surfaces with Silanes", Chemtech. 7, 766 1977.

[3] E. P. Plueddemann: "Silane Coupling Agents", Plenum N.Y., 1982, and Poly. Mat. Sci. Eng., 1984, 50, 430.

[4] I. Ishida, J. Coll. Interface. Sci., 1980, 73, 936.

[5] J. L. Koenig, Advances in Polym. Sci., 1983, 54, 89.

[6] F. Liang, P. Dreyfuss, J. Appl. Polym. Sci., 1984, 29, 3147.

[7] K. E. Creed, Rep. of G.E. Co., No. 274A/5973A (1981), and No. DE-82012868, 1982.

[8] F. D. Swanson, Sampe Tech. Conf., 1972, 339.

[9] A. Kaul, N. Sung, Poly. Eng. and Sci., 1983, 24, 493,

[10] S. L. Cooper, G. Strate, T. A. Spechhard, Polym. Eng. Ed. Sci., 1983, 23, 337.

Part IX
Special Topics

STRUCTURE AND PROPERTIES OF FLUORINATED ELASTOMERS

C. Garbuglio - G. Ajroldi - M. Pianca - G. Moggi

Montefluos Research and Development Center
Via Bonfadini, 148 MILANO (Italy)

The developments of fluorine chemistry, especially those starting from the fifties, have made available more and more sophisticated materials that, because of their intrinsic properties, match the increasing needs of:
- high performance
- continuous service temperature
- safety

Fluoroelastomers are mainly employed in the chemical, oil drilling, automotive and aerospace industries: development trends are towards higher and low temperature resistance, higher stability to more aggressive fuels and oils and better processability, related to technological developments like injection molding, fuel-hoses, expansion joint production and so on.

Today the most common fluorocarbon elastomers are copolymers of vinyldenefluoride (VDF) with comonomers such as hexafluoropropene (HFP) and tetrafluoroethylene (TFE).

The usual curing system of these elastomers is based on nucleophilic agents (diamines, bisphenols in the presence of inorganic acid-acceptors).

Modifying the ratio of monomers and more specifically introducing TFE as a third monomer, the fluorine content may be varied from 53 to 70%. The higher is the fluorine content of these elastomers the higher is the resistance towards fuels, oils and corrosive chemicals.

New developments are towards peroxide curable elastomers in order to increase oil resistance, in particular. [1]

Hereafter a review is given of a collaborative work carried out at the Research Center of Montefluos on the structure and order of copolymers of VDF and HFP and terpolymers of VDF, HFP and TFE.

This is a part of a more comprehensive research program on structure and properties of fluoroelastomers, also comprising investigations on curing mechanism, processability, physico-mechanical properties of end products.

Rheological and viscoelasticity studies are here considered as the connection point beween the polymer structure and the processability investigations, which play a relevant role in the development of fabrication technologies.

EXPERIMENTAL

VDF-HFP copolymers and VDF-HFP-TFE terpolymers have been prepared by semicontinuous emulsion polymerization at 80-90°C and 10-20 atm [2] . The polymerization was done using ammonium persulphate as free-radical initiator and continuously feeding a gaseous monomer mixture with the same composition as the desired polymer.

Temperature, pressure and amount of catalyst were properly regulated for obtaining the requested values of intrinsic viscosity and molecular weight distribution.

The initial gas composition in the reactor was adjusted to yield the desired polymer composition, according to the reactivity ratios. The polymers were coagulated by pouring the latexes into an equal volume of a stirred 6 gl^{-1} aluminium sulphate solution, washed with demineralized water and dried at 70°C for 16 h. Owing to the impossibility of homopolymerizing HFP under normal free radical conditions, the composition range that can be experimentally investigated is limited; therefore no copolymer or terpolymer can contain more than 50 mol% HFP.

Intrinsic viscosity $\sqrt{\eta}$ measurements have been carried out at 30°C in methyl-ethyl-ketone.

Molecular weight distributions (MWD) have been determined by gel permeation chromatography (G.P.C.), at 30°C, using tetrahydrofurane (THF) as solvent.

Data on physico-chemical characteristics and Mooney viscosity data for copolymers and terpolymers here investigated are collected in table 1.

TAB.1 CHARACTERISTIC OF COPOLYMERS AT CONSTANT COMPOSITION
        AND TERPOLYMERS

SAMPLE	MOLAR COMPOSITION			INTRINSIC VISCOSITY (1)	MOONEY VISCOSITY (2)	Mw/Mn
	VDF	HFP	TFE			
COPOLYMER						
A	80	20	–	65	50	4,7
B	80	20	–	75	60	5,0
C	80	20	–	90	80	5,2
D	80	20	–	82	60	7,0
TERPOLYMER						
E	65.5	19,0	16,5	108	120	4,7
F	65	20	15,0	63	68	3,7
G	64.5	19.5	16,0	67	75	2,4

(1) Intrinsic viscosity at 30°C in metylethylketone
(2) Mooney viscosity at 100°C,(1+4)minutes

The ^{19}F NMR spectra have been recorded on a VARIAN XL–200 spectrometer at 188 MHz.

I.R. spectra have been recorded on a Nicolet 20 SXB FT–IR instrument (Fourier Transform I.R.). Accumulation technique is used, if necessary. Cast films, annealed at 150 degrees, were used as specimens.

Glass transition temperatures (Tg) were measured by differential scanning calorimetry (DSC) using a Perkin Elmer DSC 2C calorimeter, with a heating rate of 10°C/min. The intercept of the line of maximum slope with the base-line was assumed to be the Tg.

A free oscillation, torsion pendulum (Torsion Automat/Brabender) has been used for dynamic–mechanical measurements. The oscillation frequency was in the range of 5 Hz at low temperature and about 0.5 Hz above Tg.

Shear and elongation viscosities have been determined by means of a capillary Göttfert Rheometer for shear rates higher than 1 s $^{-1}$. Elongational viscosity was obtained following the Cogswell method of convergent flow.

Dynamic moduli G' and G" and dynamic viscosity η have been determined by means of a Weissenberg Rheogoniometer, with parallel plates geometry.

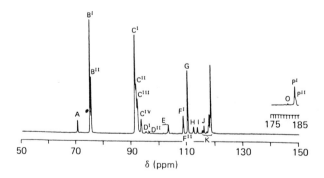

Fig.1. Copolymer VDF/HFP 80/20 ; ^{19}F NMR Spectrum

CHAIN COMPOSITION AND VDF SEQUENCES

The chemical microstructures of the VDF/HFP and VDF/HFP/TFE copolymers have been determined by ^{19}F NMR spectroscopy. A typical ^{19}F NMR spectrum of VFD/HFP copolymer (4/1 molar ratio) is reported in fig. 1.

The elastomer composition can be easily determined from 19F NMR spectra and, for VDF/HFP copolymers, the concentration of all the possible VDF-HFP triads can be measured $\mathcal{L}$ 3 $\mathcal{J}$.
This task is simplified by the fact that HFP cannot homopolymerize under the reaction conditions used in the synthesis.

The probabilities of finding every triad have been calculated by means of the equations developed by Price for a first-order Markovian statistical model $\mathcal{L}$ 4 $\mathcal{J}$: the experimentally determined concentrations and the calculated ones have been matched and an excellent agreement between experimental data and the statistical model has been found (fig. 2).

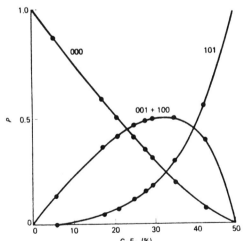

Fig.2. Frequences of monomer triads as a function of composition of VDF/HFP copolymers.
• Experimental results from ^{19}F NMR spectra.
── Computed values according to the Bernoullian model (0= VDF; 1 = HFP).

The agreement with the model leads to a complete structural resolution of the copolymers and allows to evaluate the probability of finding VDF sequences of a given lenght.

It has to be pointed out that also in copolymers having 20% mole of HFP, the long sequences (more than 10 units) have concentrations presumably high enough to allow some ordered structures to be formed (fig. 3)

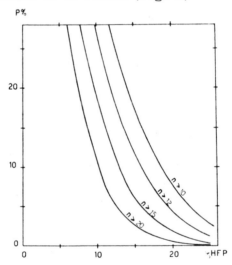

Fig. 3. Calculated probabilities (P%) of finding sequences of VDF of a given length reported as a function of the composition of the VDF/HFP copolymers.

CRISTALLINITY

The crystal structure of VDF/HFP copolymers has been investigated by means of X-Ray diffractometry and IR spectroscopy.

It is well known that poly(vinyldenefluoride) (PVDF) may exist in three crystalline modifications in which the polymer chains assume a planar zig-zag structure (I or beta), a TGTG' chain conformation (II or alpha), and a nearly-planar zig-zag structure (III or gamma) $\lfloor$ 5, 6 $\rfloor$. The addition of small percentages of HFP comonomer does not affect the polymorphism of PVDF. As a matter of fact,

it is well known   that the introduction of
perfluorinated units in the PVDF chain favours
crystallization in phase β for steric and electronic
reasons. This tendency is more pronounced in case of
introduction of tetrafluoroetylene units in PVDF and
apparently also holds for hexafluoropropene units.

According to our result ⌐ 7 ⌐ the crystallinity of the
copolymers decreases with increasing of the HFP fraction,
according to the linear relationship of Fig. 4.

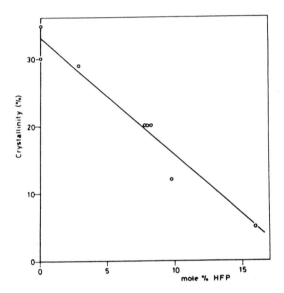

Fig. 4. Crystallinity index of VDF/HFP copolymers
as a function of the composition.

This is expected on the basis of the perturbation of the
constitutional regularity due to the trifluoromethyl
groups.
The effect is accompanied by a gradual shift in position
of the maximum of the amorphous fraction (from 18.5° to
17.0°, 2 θ Cu Kα), which is readily interpreted on the
basis of the substituent effect of the HFP units. This
trend sets trough also for higher HFP percentages (max. at

16.3° for a 41 mole % HFP content – tab. 2.).

TAB. 2 CRYSTALLINITY INDEX (C.I.) AND X-RAY DIFFRACTION
MAXIMUM (2 ϑ) OF AMORPHOUS FRACTION AGAINST COMPOSITION
FOR VDF/HFP COPOLYMERS

MOLE % HFP	C.I. %	2ϑam
0	33	18,5
2,9	29	18,5
7,8	21	18,4
7,9	20	18,4
8,2	20	18,3
9,7	12	18,0
15,9	5	17,0
16,2	–	17,0

A great deal of investigation was performed by means of
vibrational spectroscopy and particular absorption bands
were attributed to the three different crystalline forms
$\angle$ 8, 9, 10 $\rfloor$.

We observed that some typical IR bands of PVDF are also
present in VDF/HFP copolymers by cast solution of PVDF and
VDF/HFP copolymers in methylethylketone on KBr windows; we
obtained films whose spectra show bands mainly due to form
II, both for PVDF and low HFP content copolymers.

The bands corresponding to these crystalline modifications
are easily observed in copolymers containing up to 10%
HFP. For higher HFP concentrations, FT-IR spectroscopy has
been used to detect bands of very low intensity. Even in
some copolymers containing 20% mole of HFP we observed,
after repeated annealing, weak but sharp IR bands, that
are attributed to the crystalline form II of PVDF.

This is clearly shown in fig. $\lceil 5 \rfloor$ where the intensities of the crystallinity bands are reported for different compositions and in fig. $\lceil 6 \rfloor$ and $\lceil 7 \rfloor$ where amorphous and annealed samples of copolymers with high contents of HFP are shown. The difference spectrum is self-explaining.

FIG. 5.

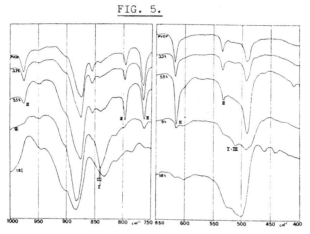

Residual crystallinity of polyvinylidenefluoride in copolymers with different content of HFP (I, II and III refer to the different crystalline forms of PVDF).

FIG. 6.   FIG. 7

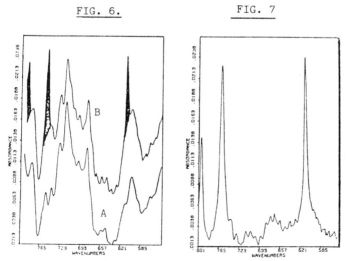

Fig.6. FT-IR of VDF/HFP copolymers with 80/20 molar ratio (A = untreated sample, B = annealed sample)

Fig.7. FT-IR difference spectrum referred to Fig. 6 spectra (B-A).

TRANSITIONS AND CHAIN COMPOSITION

a) Copolymer. The glass transition temperature measured
either from DSC scans, or by DMS, increases gradually by
increasing HFP content (fig. 8.).

However the law of variation cannot be straightforwardly
described by well-known equations, for example by the very
simple Fox equation Γ 11 $\overline{J}$, on the basis of Tg of PVDF and
PHFP (-45 and + 160°C, respectively - fig. 8).

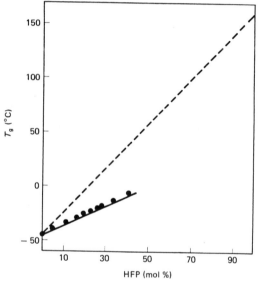

Fig. 8. Glass transition temperatures of VDF/HFP copolymers
versus molar HFP content: (- - - -) as predicted by Fox
equation; (———) as predicted by Johnston equation; (●)
experimental values.

This is due to the fact that, by increasing the HFP
content, the copolymer becomes progressively less random,
up to the limiting case of 1:1 molar ratio, where the
copolymer has a totally alternated structure. If the
sequence distribution is taken into account, by using the
Johnston equation a good agreement is obtained between
experimental and calculated data Γ 2 $\overline{J}$.

In fig. 9 and 10 the influence of composition on dynamic
mechanical spectra is shown.

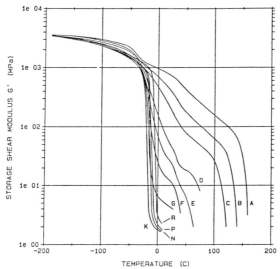

Fig.9. Dynamic modulus for VDF/HFP copolymers as a function
of molar composition. (HFP: A = 0, B= 2.5, C= 5.2, D=10.2,
E= 12.9, F=15.5, G= 17.5, K= 19.2, N= 26.0, P= 30.3,R=39,0).

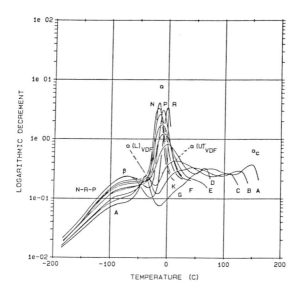

Fig. 10. Loss factor for VDF/HFP copolymers as a function
of molar composition (key as in Fig. 9).

VDF-HFP copolymers with 80/20 molar composition show two
relaxation processes: the first (α), related to the glass
transition temperature, and a secondary one (β) lying at
lower temperature. For copolymers with a molar ratio
higher than 80/20 T_α increases with HFP content and
approaches an asymptotic value of about 0°C.

For compositions lower than 80/20 the relaxation
behaviour becomes complex: the α process is split in two
parts Fig. 11.

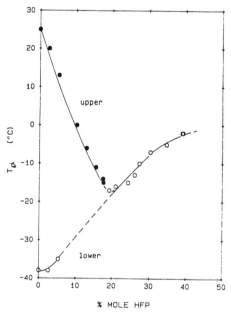

Fig. 11. α- Relaxation of VDF/HFP copolymers : T_α versus
composition.

The upper α process is better revealed by DMS, while the
lower one is normally observed by DSC $\angle$ 2,12 $\underline{7}$. The
difference in temperature between the two α relaxations
increases with lowering the HFP content and also the
relaxation strength depends on the HFP content. Such a
behaviour is certainly related to the double glass
transition shown by the pure poly-vinylidenefluoride
$\angle$ 13, 14 $\underline{7}$.

The temperature of the β transition, which appears as a shoulder of the α process in a Log Δ - T plot, has been determined from a Log G"-1/T plot, where the resolution is better.

T_β (about 85°C) seems to be practically independent on the composition so that the β process could be tentatively related to local motions of VDF units. Its relaxation strength, or its intensity (fig.12),increases with HFP content.

Since the cristallinity or the order decrease with HFP this relaxation could be related to the amorphous part of the polymer.

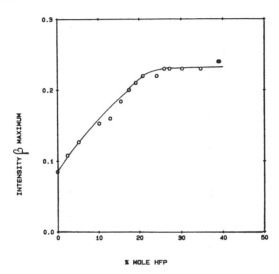

Fig. 12. Intensity of β relaxation as a function of composition for VDF/HFP copolymers.

The G' modulus above Tg decreases with HFP (fig. 13) and reaches a lowest plateau value.

At constant composition, no influence on dynamic-mechanical spectra has been found, as expected, by changing molecular weight or molecular weight distribution.

It has to be pointed out that the 20% composition is a critical one; small fluctuations in composition due to the polymerization technique can bring about a few percent of

crystallinity, as revealed also from FT-IR measurements.
The behavior of dynamic modulus above Tg can be attributed
to the presence of crystallites.
In fact, as a first approximation, crystalline polymer can
be considered as a two phase system (according to the
fringed micelle model).
Crystallites act as physical crosslinks which are
destroyed by increasing temperature and stress. These
results have a relevant interest as fas as rheological
properties at high temperature are concerned (as it will
be further discussed).

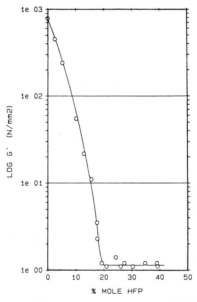

<u>Fig.13.</u> Dynamic modulus above T$_\alpha$ for VDF/HFP copolymers
as a function of composition.

In fig. 14 the crystallinity indexes and Tg's are reported
against composition.

It is clearly evident that the 80/20 molar composition of
the VDF/HFP copolymers is the best one in order to balance
the needs of a low Tg and that of a completely amorphous
material.

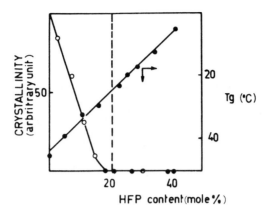

Fig. 14. Crystallinity index and Tg as a function of
composition.

b) <u>Terpolymers</u>. The influence of chain composition at
constant fluorine content on dynamic-mechanical spectra of
terpolymers is reported in fig. 15 and 16 and in tab.3
∠ 15 ⌋.

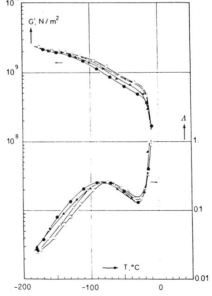

Fig. 15. Dynamic-mechanical spectra for terpolymers:
AT (O), BT (Δ), CT (▲), DT (●) (for compositions
see Table 3).

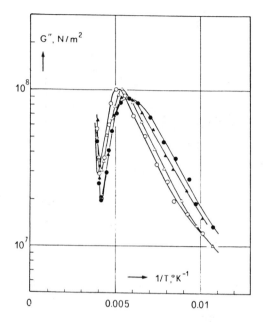

Fig. 16. The maximum of G" for terpolymers (key as in
Fig.15).

TAB.3   COMPOSITION MOLECULAR AND DYNAMIC-MECHANICAL
                    PROPERTIES OF TERPOLYMERS

SAMPLE	100 M 1+4 (1)	(η)	COMPOSITION Mol% VDF	HFP	TFE	T_g (°C) from G'	T_β (°C) from G"
AT	74	0.67	63	37	0	− 10	− 84
BT	70	0.60	61	33	6	− 10	− 86
CT	72	0.70	58	26	16	− 15	− 91
DT	78	0.72	55	20	25	− 10	− 98

While the glass transition temperature Tg is nearly indipendent on the composition, the content of TFE monomer affects both shear modulus and secundary dissipation peak (β peak). In fact, in the temperature range between T_β and Tg, G' decreases as TFE increases; in other words the relaxation strength of the process increases. As far as the relaxation maximum is concerned, one can observe that T_β is shifted towards lower temperatures and that the maximum is broadened by increasing the amount of TFE units.

These results are not unexpected since the β process, due to the local motion of few monomeric units, lies at lower temperature for PTFE than for VDF-HFP copolymers having the composition here considered. Moreover it is reasonable that the introduction of a third monomeric unit must bring about a broadening of the β relaxation process; in fact the kinds of kinetic units responsible of the β dissipation are increased.

SEQUENCE DISTRIBUTION AND RHEOLOGICAL PROPERTIES

It was said before that small fluctuations in composition due to the polymerization technique can induce crystalline structures in 80:20 molar ratio copolymers.

In fig. 17 shear viscosity data, measured at constant shear stress, are reported, as a function of the reciprocal absolute temperature for two samples having the same gross composition.

A stress dependent rheological transition, i.e. a change in the apparent activation energy of flow, is observed for only one of the two samples, at about 165°C, at the two lowest stress levels. The other one shows a constant activation energy of 15.3 kcal/mole, in the overall explored temperature range.

Below 165°C, the apparent activation energy of the first

sample increases by decreasing the stress level. The asymptotic value at high stresses is practically the same for all the samples. Such a behavior resembles that of polyvinylchloride (PVC) found several years ago by Pezzin $\lceil$ 16 $\rfloor$. PVC shows several stress dependent rheological transitions: their location on the temperature axis depends in turn on the content of crystalline or, more generally, ordered structures.

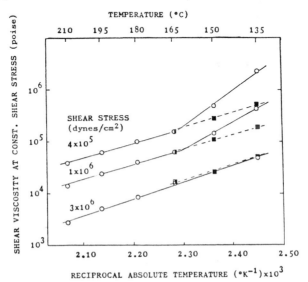

Fig. 17. Shear viscosities as a function of the reciprocal absolute temperature for two samples having the same gross composition (20% HFP by mole) but with random structure (■) or with VDF sequences (o). A stress dependent rheological transition is evident.

Also in the case of the copolymers here studied, the transition is shifted towards lower temperatures or disappears by decreasing of the ordered structures.

It is worthy to observe that the activation energy of the 80:20 copolymers is not far from that of pure polyvinylidenefluoride (13.5 Kcal/mole) in spite of the bulkiness of the CF_3 side group. This could point out that the polar interactions of VDF units are dominating the flow mechanism.

FLOW PROPERTIES

<u>Copolymers.</u>   The   shear   viscosity   functions   of   three specimens having increasing Mooney viscosities and similar MWD are shown in fig. 18 .

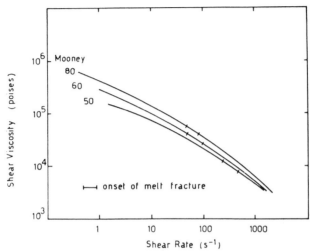

<u>Fig. 18.</u>   Influence of molecular weight on shear viscosity (key as in Table 1 of the text). Capillary data at 150°C.

This   figure   shows   the   well-known   fact   that   the   melt viscosity increases with molecular weight.

Larger differences are found in uniaxial tensile viscosity functions   fig. 19.

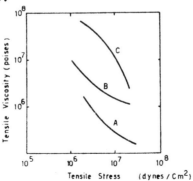

<u>Fig. 19.</u> Influence of molecular weight on tensile viscosity (key as in Table 1 of the text). Capillary data at 150°C.

Especially for the highest molecular weights a decreasing
slope with **ϐ** towards a troutonian behaviour is not found.
Such a result should imply a large maximum in μ (**ϐ**) as
found by Münstedt and coworkers [20] for low and
high-density polyethylenes.

It is interesting to compare two samples having the same
Mooney viscosities but quite different molecular weight
distributions.
The shear viscosity function in the capillary range seems
to be only scarcely affected by the broadening of MWD, as
shown in fig. 20 .

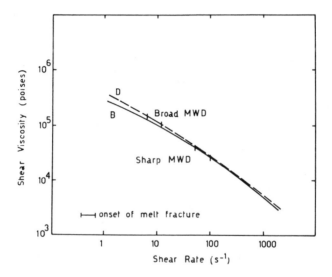

Fig. 20. Influence of molecular weight distribution on shear
viscosity (key as in Table 1 of the text).Capillary data at
150°C.

However the tensile viscosity function shows a very
dramatic increase in the μ curve for the sample with the
broader MWD fig. 21 .

In fig. 22 the dynamic moduli G' (ω) and G" (ω) are
reported against circular frequency ω .

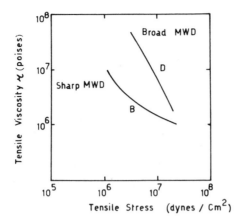

Fig.21. Influence of molecular weight distribution on tensile viscosity (key as in Table 1 of the text). Capillary data at 150°C.

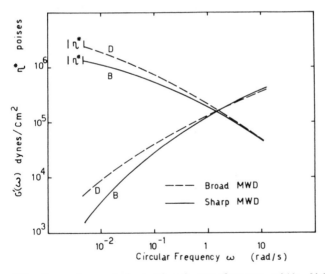

Fig. 22. Dynamic modulus G' of copolymers with different molecular weight distribution. Rheogoniometer at 150°C.

The relaxation spectrum for the polymer having the broader MWD extends over a larger relaxation time scale.
It has to be noted that differences in MWD from GPC are evident but not dramatic, while differences in some rheological functions are. This derives from the well known fact that they are very sensitive to small fractions

of high molecular weight components.

Information given by dynamic moduli allows to explain the
differences in μ (**σ**) function.
Large differences in relaxation spectra imply first of all
differences in η_o, but as data show, it was practically
impossible to reach the newtonian behavior also at low
shear rates. Moreover according to the molecular theories
of rheology and experimental data on PE and PS ⌈ 17 ⌉ it
appears that, by increasing MW and broadening MWD, becomes
more probable to find larger and larger maxima in the
μ (**σ**) function.

Terpolymers. The influence of the molecular weight and
molecular weight distribution on the rheological
properties of terpolymers is shown in fig. 23 to 25 .
As expected the trend is the same as observed for the
copolymers even if there is some uncertainty on molecular
data because of the complexity for the GPC calibration
curve for terpolymers.

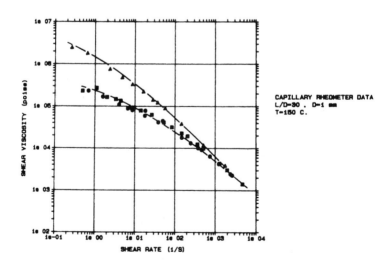

Fig. 23. Shear viscosity of terpolymers : ▲=E, ●= F, ■=G
(key as in table 1 of the text). Capillary data at 150°C.

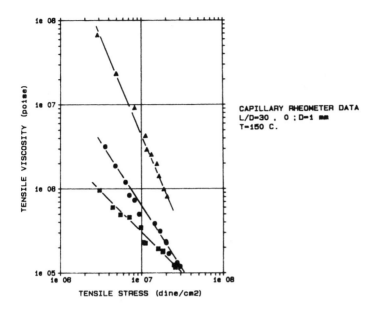

Fig. 24. Tensile viscosity of terpolymers (key as in Fig.23)
Capillary data at 150°C.

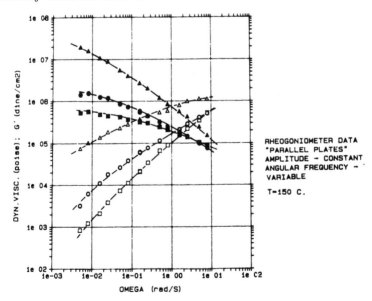

Fig. 25. Dynamic viscosity (full symbols) and storage
modulus (open symbols) of terpolymers (key as in Fig.23).
Rheogoniometer at 150°C.

Flow Anomalies. Molecular weight and molecular weight distribution influence the elastic behavior and the flow instability phenomena for either copolymers and terpolymers. In the case of convergent flow for an L/D= 0 die an inverse logarithmic relationship has been found between the μ_c/E_c ratio (which is the ratio between tensile viscosity and tensile modulus at the onset of the flow instability) and the apparent critical shear rate at the flow instability onset. Experimental findings are shown in fig. 26 and 27 for copolymers and terpolymers respectively.

The μ_c/E_c ratio has the dimensions of a relaxation time, a quantity dependent on both M and MWD.
Such a relationship has been found also for PMMA samples tested at different temperatures $\lfloor 18 \rfloor$.
These results, even if not totaly unexpected, are however quite new. It is believed that more investigation is needed to have a deeper insight on the flow instability problem.

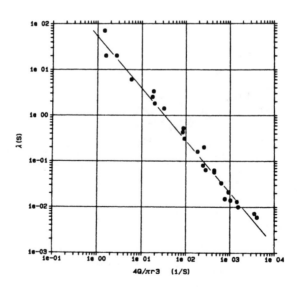

Fig. 26. $\lambda = \mu_c/E_c$ against apparent critical shear rate for copolymers.

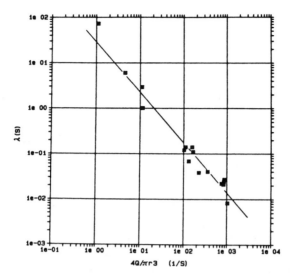

__Fig. 27.__ $\lambda = \mu_c / E_c$ against apparent critical shear rate for terpolymers.

CONCLUDING REMARKS

The results here presented on the relationships between the structure and order of macromolecular chains and rheological properties of fluoroelastomers provide the main guidelines for tailoring materials.

These results together with those coming from a thorough investigation on vulcanization mechanism and on phisico-mechanical behavior of elastomeric compounds, now under development in our laboratories, should provide relevant information to develop fabrication, technologies and applications.

REFERENCES

(1) Caporiccio G., Garbuglio C., Moggi G. Proceedings of A.I.M., invited lectures, Pisa 1983 Symposium, pag. 141.

(2) Bonardelli P., Moggi G., Turturo A. Polymer, 1986, 27,905.

(3) Pianca M.,Bonardelli P., Tatò M., Cirillo G.,Moggi G. Polymer, 1987, 28, 224.

(4) Price F.P. J. Chem. Phys., 1964, 36, 1.

(5) Hasegow R. et al. Polym. S., 1972, 3, 1591.

(6) Bachmann M.A. et al. J. Appl. Phys., 1979, 50, 10.

(7) Moggi G., Bonardelli P., Bart J. Polymer Bulletin, 1982, 7, 115.

(8) Cortili G., Zerbi G. Spectrochim. Acta, 1967, 23A, 285.

(9) Zerbi G., Cortili G. Spectrochim. Acta, 1970, 26A, 733.

(10) D'Alessio A., Del Fanti N., Benedetti E. Proceeding of Symposium "Future Trends in Polymer Science and Technology", Capri, 1984, pag. 54.

(11) Fox T.G. Bull. Am. Phys. Soc., 1956, 1, 123.

(12) Ajroldi G. et al. International Rubber Conference - Göteborg 1986.

(13) Sasabe H. et al. J. Polym. Sci., 1969, A2, 7, 1105.

(14) Boyer R.F. J. Macromol. Sci. Phys., 1973, B8 (3-4),503.

(15) Moggi G. et al. Proceeding Int. Rubber Conference, Venice, 1979.

(16) Pezzin G. et al. Proceeding of 23th Meeting of It. Soc. Rheology, Siena, 1973.

(17) Münstedt, Laun H.M. Rheol. Acta, 1981, 20, 211.

(18) Ajroldi G., Fumagalli M. unpublished results.

Figures 1, 2, 8 are reproduced with permission $\int 3 \rfloor$.
Figure 4 is reproduced with permission $\int 7 \rfloor$.
Figures 9, 10, 11, 12, 13, are reproduced with permission $\int 12 \rfloor$.

NEW DEVELOPMENTS IN RAMAN SPECTROSCOPY OF POLYMERS

Bernard J. Bulkin[*], Menachem Lewin[†]

Polytechnic Institute of New York, Brooklyn, New York, USA
* Present Address: The Standard Oil Company, Cleveland, Ohio 44128, USA.
† Present Address: School of Applied Science and Technology, Hebrew University and Israel Fiber Institute, Jerusalem, Israel.

The vibrational (infrared and Raman) spectra of polymers have been investigated for decades, and have long been known to contain information about the crystallinity of polymers. Indeed, crystallinity in polymers such as polyethylene has been quantitatively measured by infrared spectroscopy, and there are many other cases in more complex polymers where 'crystallinity bands' are known.

In recent years, however, the capability of instrumentation has improved, as has our ability to theoretically describe vibrational frequencies and intensities of polymers. These parallel developments make it worthwhile to re-examine what is actually involved in crystallinity bands in polymers, and to extend our understanding of what polymer properties these bands are really reporting on.

There is more reason to do this than just an extension of our fundamental understanding. The quality of both infrared and Raman instrumentation now makes it quite feasible to use vibrational spectroscopy as on-line instrumentation in a variety of polymer processing applications. To be sure, such applications can be built up empirically and individually, and be rugged for that individual use. But it is the fundamental understanding of the information content of polymer spectra that should determine the scope of potential applications, allowing us to use our development time most effectively. It is in that context that this paper is set.

The results to be presented will deal with aromatic
polyesters, specifically with poly (ethylene terephthalate)
(PET). Poly (propylene terephthalate) (PPT), and poly
(butylene terephthalate) (PBT) have also been the subject of
related studies from our laboratory [1-3] that are not
described herein. But the principles described are
sufficiently general that these need only be viewed as
examples. Indeed, nearly parallel studies of other polymer
systems are going on in various laboratories, and while this
paper will not attempt to review these, they are easily
identified by reference to regular reviews that are
published [4].

What We Expect to Find in Vibrational Spectra

There are so many analytical techniques for the study
of polymers, that it is necessary to briefly explore what
our expectations should be from the vibrational spectrum.
In this and the subsequent section we are, perhaps, going
'back to basics' in some respects, but that is useful from
the scientific and as well as pedgogical points of view.

Most of the vibrational spectrum (vibrations from say
200-4000 cm^{-1}, consists of vibrations fairly well localized
to functional groups. Thus we are generally quite correct
to describe these as 'carbonyl stretching', 'methylene
deformation', etc. The frequencies of these modes are
primarily determined by the bond force constants, but have
second order influences from the environment (solvent,
conformation, crystallinity, etc.). Fortunately, these
second order influences hare readily measured with existing
instrumentation. Even more important than the influences
of environment on vibrational frequency is the influence on
intensity. These effects are generally larger than the
frequency shifts and tend to dominate the changes seen in
the spectrum. Until the early 1970's, there were few
infrared or Raman spectrometers capable of measuring spectra
at signal-to-noise ratios of greater than 20, within a
reasonable period of time (say minutes). Today, signal to
noise ratios of 500 or more are routinely obtainable in
seconds using commercial instrumentation. This means that a
variety of diagnostics, making use of intensity changes, are
now possible. In infrared spectra, the accuracy and
precision of frequency measurement has also greatly improved
through such instruments.

We do not expect vibrational modes of functional groups to be very sensitive to long range order. The accumulated experience of many years says that environmental effects on these modes are mainly from nearest neighbors. This is a caution to be taken seriously, because crystallinity is long range order, and vibrational modes may sometimes confuse shorter range effects, for example conformational ordering, with crystallinity.

However, the vibrational spectrum also includes the lower frequency vibrations, those below 200 cm^{-1}. These include both delocalized intra-chain modes and inter-chain modes of polymers. These low frequency vibrations have not been much investigated in the past, mostly because of instrumentation or perceived instrumentation problems. In the infrared spectrum there are definite difficulties with making far infrared measurements, as the region is known, on polymers. Absorption bands are often weak, and are distorted by crystallite size in semi-crystalline materials, because the wavelengths are large. In the Raman spectrum it was the case with older instruments that accessing the low frequency vibrations was made difficult by the intense Rayleigh scattering near the exciting frequency. This is no longer the case, and the field has become very active, particularly since the exciting work from a number of laboratories on the longitudinal acoustic modes (LAM's) of simple polymers, and their connection to polymer mechanical properties.

The low frequency region of the spectrum can thus be expected to be more genuinely reflective of crystallinity in polymers, containing as it does the modes that are arising from long range order. The spectroscopist's task is to sort our inter- from intra-chain modes. In the case of intra-chain modes, such as LAM's, we expect to see propagation once several monomer units are present, and continued change in frequency and intensity as order or chain length increases over a considerable range, depending on the particular case involved. For inter-chain modes, we can only begin to see these appear once several unit cells are built up. Thus both cases present an attractive (from the information point of view) contrast to the internal functional group modes seen in the rest of the spectrum.

The low frequency region also has another important piece of information. For semi-crystalline polymers, undergoing various processing steps, the only important changes to the specific heat are in the vibrational portion. The low frequency region of the spectrum, because of the exponential dependence relative to kT, dominates these contributions. It is possible to calculate at least the relative specific heats of a polymer sample before and after a processing step (such as drawing or annealing) from the low frequency Raman spectrum, with rather good relative accuracy.

What Are the Measurements We Can Make on the Spectra

Vibrational spectra produce several parameters on any band: the wave number maximum, the absolute or relative intensity, the half-band width, and the band asymmetry. In studying crystallization and other processing changes in polyesters, we make use of all of these.

Measurement of frequency maximum position needs little explanation. As already mentioned, it can be done to about .3 cm^{-1} in the Raman spectrum, and to about .05 cm^{-1} in the infrared spectrum if laser interferometric schemes are used, as they are routinely in commercial FT-IR instruments. In fact, instrumentation is now rarely the limitation in this measurement, as the band maxima tend to be too flat for more accurate measure.

Absolute intensity is rarely measured, and is never a routine measurement on a vibrational spectrum. But relative intensity is straightforward, whether measured as area or peak height.

Band width, generally measured as the width at half the intensity maximum or half band width is also a wave number measurement. However, because of the band slope at the half width points it is possible to measure half band widths more precisely than wave number maxima. Even in Raman spectra these can be easily measured to .1 cm^{-1} if desired.

Finally, though rarely used, band asymmetry is a parameter of interest. It is most simply characterized as the ratio of the half band widths on either side of a normal dropped from the band center.

In actual practice, with complex polymer systems, almost all the changes seen as a function of processing variables are intensity changes, and for various reasons

these are characterized as relative intensity, half width,
asymmetry, or even wave number change. This is because in
complex polymer spectra almost all bands are composed of
overlapping, unresolved modes. These may arise from a
distribution of conformations, orientations, or
crystallinity, causing the second order changes in frequency
referred to earlier. Or bands may appear with different
selection rules on intensity, the rules changing as symmetry
changes with crystallization, etc.

When the relative intensity of unresolved overlapping
bands changes, the position of the frequency maximum may
change, the band may narrow or broaden, and it can become
more or less symmetric. But virtually all of these changes
are still relative intensity changes, no matter how they are
measured.

Poly (Ethylene Terephthalate)- Thermal Annealing and Take Up Speed as Variables

Bulkin, Lewin and McKelvy [5] studied the annealing of
bulk, amorphous PET using Raman spectroscopy. This study
focussed on an examination of the crystallization kinetics
of the bulk material, with two variables being examined -
temperature and intrinsic viscosity.

In this study, and in many of the ones described
herein, the spectroscopic approach used was rapid scanning
Raman spectroscopy. In that technique, a region of the
spectrum is repeatedly scanned over a time scale of a few
seconds, and the changes resulting from crystallization in
situ are monitored.

In the study of bulk PET, crystallization was monitored
using the half band width of the carbonyl stretching
vibration (1725 cm^{-1}) and the ratio of intensity
I_{1096}/I_{1117}. The 1096 band is one that appears and grows as
PET is annealed. In a separate study, to be described in
more detail below, we have demonstrated rather convincingly
that both the observed decrease in the carbonyl stretching
mode half width and the increase in the 1096 band relative
intensity are associated with gauche-trans conformational
change in the glycol unit, rather than true crystallinity.
This had been understood by previous workers in this area as
well.

The results of the study indicated that bulk PET crystallization appears to occur in stages. There is an initial induction period, followed by a relatively rapid change, then a second induction period, and again a relatively rapid change.

The first level description of our results is that stage 1 is the activation of conformational change by nucleus formation and spherulite growth, proceeding as described in the traditional literature of polymer crystallization. Stage 2 is the growth of ordered regions, representing short chain lengths, which as a result of statistics can proceed without significant chain disentanglement. In this stage, the effects of the thermal and entropic driving forces may lead to higher conformational order for the carbonyl groups than for the glycol linkages.

In stage 3, the crystallization process stops. Our results provide no direct evidence for what is happening at this stage, only that conformational change as monitored by our experiment is not happening. This is the chain disentanglement barrier.

Stage 4, that follows this activation period, is very similar in mechanism to stage 2. The spectroscopic data indicates that conformational reorganization is proceeding smoothly again during this stage. This is the period in which crystallites grow in size, and phase propagation occurs.

All of the kinetic data in [5] were analyzed using the Avrami equation, and detailed results are presented in that reference. This equation has been heavily criticized in the literature, primarily because too many studies have attempted to use values of $\underline{n}$ derived from plots of the Avrami equation to determine the mechanism of nucleation and the dimensionality of growth of the crystalline phase.

It is now generally agreed that deriving such information from Avrami equation plots is not reasonable, because too many factors influence the values of $\underline{n}$ obtained. However, it is our view that the Avrami equation nonetheless plays a useful role in the study of crystallization kinetics. It allows for the facile comparison of data on the same or similar systems between a wide variety of different techniques. In the case of a new technique, such as rapid scanning Raman spectroscopy, the use of the Avrami equation allows us to test whether our data are in any way comparable to those obtained by other techniques.

When we turn to fibers, the situation at once becomes more complex, yet simplifies in certain ways. First, kinetic data on fibers by DeBlase [6] indicate that only a single stage of crystallization is seen, or, occasionally, a short induction time followed by the rapid crystallization step. Crystallization also occurs at considerably lower temperatures than in bulk samples.

This is not surprising. Essentially, we can understand these results as saying that fiber spinning puts in the activation energy needed for chain disentanglement, and even may build in some of the trans content in conformational change needed as a precursor to formation of crystals.

The complication in the fiber samples is that many more bands in the spectrum change with thermal annealing. Bulkin, Lewin and DeBlase [7] have shown that, in fact, almost every band in the Raman spectrum of a fiber sample changes with thermal annealing.

When a set of five such fibers were studied, it was found that the bands divided into three distinct groups. The first group was bands already referred to, such as the carbonyl stretch that are sensitive to conformational change. The second group is bands that are sensitive to orientation, these are primarily phenyl ring modes. Finally, group three is bands that are true crystallinity bands. These three sets of changes correlate with density, birefringence, and wide angle X-ray scattering respectively. Thus there are true crystallinity bands in the spectrum, but they need to be carefully distinguished, in the case of fibers, from conformationally sensitive modes.

This point was further emphasized in a subsequent study by Bulkin, Lewin, and DeBlase [8]. In that work the effect of take-up speed on the spectra was analyzed. Again, the same sorts of changes as those seen in thermal annealing was observed. The two studies thus reinforced the notion that the Raman spectrum contains all of the information about conformation, orientation, and crystallinity that was previously obtained from a wide variety of techniques. This lends considerable support, in our view, to the use of Raman spectroscopy as an on-line quality control technique for fiber spinning.

Low Frequency Data

DeBlase, McKelvy, Lewin, and Bulkin [9] have presented Raman data on PET bulk and fiber samples in the region from 20-300 cm^{-1}. The data for the crystalline sample identifies two bands, at 73 and 129 cm^{-1}, that are clearly associated with crystallinity. Bulkin, Lewin, and DeBlase have correlated the 73 cm^{-1} band with X-ray scattering.

The study [9] also showed that the Raman spectrum of amorphous PET fibers was different from bulk amorphous material. Fibers show a broad scattering in the low frequency region, peaked near 35 cm^{-1}. It has been suggested that this band is of intermolecular origin, involving loosely coupled chain segments. Such scattering is reminiscent of the low frequency scattering of liquid crystalline phases.

By comparing the spectra of thermally annealed fibers, it was possible to see how thermal annealing transformed this order, as manifested by the 35 cm^{-1} ordered amorphous phase band, into either disorder (in the case of annealing at low temperatures with shrinkage) or crystallinity (for higher temperature annealing with fiber ends fixed). The low frequency region is an excellent probe of these transitions.

Bulkin, DeBlase, and Lewin [10] also used these data in a quantitative way to derive vibrational heat capacities for bulk PET and 11 fiber samples. Background on how to do this calculation is presented in the reference. The results clearly show the increase in heat capacity from bulk amorphous to amorphous fiber. As crystallinity increases by annealing of fibers, the calculated values of the vibrational heat capacity actually decreases.

Summary

The data presented briefly in this paper, and described in more detail in the references, indicates just how much information there is in the Raman spectra of polymers, particularly in the spectra of polymer fibers. Modern instrumentation combined with our understanding of the spectroscopy make this area ripe for exploitation in characterizing and controlling the quality of polymers for advanced technologies.

Acknowledgments

The support of the National Science Foundation, Polymers Program grant DMR-8304220, is gratefully acknowledged. We thank our students, Drs. M. McKelvy, F. DeBlase, and J. Kim for their many contributions to this work.

References

[1] Kim, J. S.; Lewin, M.,; Bulkin, B. J. J. Polym. Sci. Polym. Phys. Ed. 1986, 24, 1783.
[2] Bulkin, B. J.; Lewin, M.; Kim, J. S. Macromolecules, 1987, 20, 830.
[3] Kim, J. S. Ph.D. Thesis, Polytechnic University, 1986.
[4] See, for example, the annual review issue of Analytical Chemistry.
[5] Bulkin, B. J.; Lewin, M.; McKelvy, M. L. Spectrochim. Acta 1985, 41A, 251.
[6] DeBlase, F. J. Ph.D. Thesis, Polytechnic Institute of New York, 1986.
[7] Bulkin, B. J.; Lewin, M.; DeBlase, F. J. Macromolecules, 1985, 18, 2587.
[8] Bulkin, B. J.; Lewin, M.; DeBlase, F. J. to be published.
[9] DeBlase, F. J.; McKelvy, M. L.; Lewin, M.; Bulkin, B.J. J. Polym. Sci. Polym. Lett. Ed. 1985, 23, 109.
[10] Bulkin, B. J.; DeBlase, F.; Lewin, M. SPIE 1986, 665, 234.

MICROPARACRYSTALS IN POLYMERS AND THEIR EQUILIBRIUM STATE

R. Hosemann (Gruppe Parakristallforschung, c/o BAM, Unter den Eichen 44-46, 1ooo Berlin 45, F.R.G.)

The positions of atoms in materials can be explored most directly by diffraction of X-rays. This was demonstrated 1912 for crystals by M. Laue, F. Friedrich and P. Knipping (1). The positions of atoms in polymers could be analysed in a similar way by using an adequate theory which transforms the intensity of X- or neutron rays to detailed information of the structure. Instead of this, words as small range order, middle range order, clusters, micells, mesophases or amorphous phases, semicrystalline or liquid-crystalline materials are used. New definitions as alignments, nodules, inclinations and meander-cubes appear. Another kind of research uses the mathematical language: Starting from the results of M.v. Laue (2), the notations of the reciprocal space by P.P. Ewald (3) and the idea of the same a-priori probability function for all atoms in noncrystalline state by P. Debye (4), the theory of the "ideal paracrystal" was published in 1950 (5). M.v. Laue's idea of lattices is blended with P. Debye's a-priori distance function: The three lattice cell vectors $\underline{a}_k$ each have a-priorily the same statistical 9 relative variances g_{ki} in three directions $\underline{a}_i$ and are given by

$$g_{ki} = \left[\overline{(\underline{a}_k - \bar{\underline{a}}_k, \bar{\underline{a}}_i)^2} \right]^{1/2} / \bar{\underline{a}}_i .$$

Then the lattice factor no longer is P.P. Ewald's lattice peak function, but the three-dimensional generalization of L.D. Landau's convolution polynom (6). Three a-priori correlation functions $H_k(\underline{x})$ of the vectors $\underline{a}_k$ define the lattice factor. M.v. Laue early recognized the importance of the theory of paracrystals for all materials inbetween crystalline and liquid state. Nevertheless, the new theory till now only is applied in the USA and U.K. preferentially on promoted catalysts (7,8). The caution of most scientists may be due to the new mathematical language and to the fact that the ideal paracrystal consists of lattice cells with the shape of parallelo epipeds. Its lattice therefore contains certain irrational statistical correlations. Recently

it was proved that in all paracrystals only certain partial
correlations exist which, on account of the so called α^*-
relation, affect the intensity function of the ideal paracrystal
by less than 1,5 % and hence can be neglected (9). The theory of
paracrystals offers fundamental equations by which most substances can be deciphered as microparacrystals (10) by measuring
their integral widths δb of reflections. These widths increase
linearly with h^2 (h order of reflection). From the slope of this
straight line and from its intersection with the ordinate the g_{ik}
can be calculated as well as the number N of netplanes. All
materials investigated by us until now are submitted to the
α^*-relation which connects N with the g-value

$$g\sqrt{N} = \alpha^* ; \qquad \alpha^* = 0,15 \pm 0,05 \ .$$

Figure 1 gives some examples of organic and inorganic materials
in a plot $\sqrt{N}$ against $1/g$. There is no doubt that this α^*-
relation conceals a law in nature which is of fundamental importance for most noncrystalline materials. The model of Figure 2
shows all essentials of a twodimensional paracrystal in equilibrium state. Some larger coins are statistically mixed with smaller ones and represent atoms with four orthogonal valences.

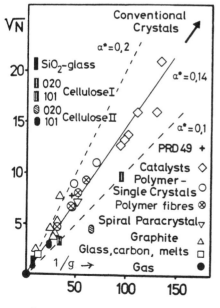

Figure 1

$$\sqrt{N}g = \alpha^* \sim 0.15$$
$$\Delta_N = \sqrt{N}\Delta_1 \longrightarrow \alpha^*\bar{d} \qquad g = \Delta_1/\bar{d}$$

Figure 2

Their radial potentials are responsible for the size of adjacent
distances from center to center. Their tangential potentials on
the other hand are demanded,because the coins lie in statistical
curved lines and rows. These deviations, resulting from straight
lines, reach maximal $360° \alpha^*/2\pi$ and influence the diffraction by
less than 1,5 %. Those rows where atoms directly touch each other
are called "bearing netplanes", because the tangential strengths
are transformed along them. The consequently arising distortions
of the valence angles require a new kind of energy which in
crystalline lattices is unknown. This leads to a new term of free
enthalpy ΔG_p which for cubic paracrystals with N^3-lattice cells
is given by (11) :

$$\Delta G_p = \frac{3}{2} N^4 A_o g^2$$

A_o depends on the course of the tangential potentials and determines the value of α^*. Figure 3 shows a characteristic example
with a particular value of the surface free energy σ and volume
enthalpy ΔG_v, g = 0 for crystals, g = 4,2 % for polymer blends
and g = 7,5 % for melts. After attaining the critical size the
paracrystals grow till to a minimum of enthalpy where $N = (\alpha^*/g)^2$.

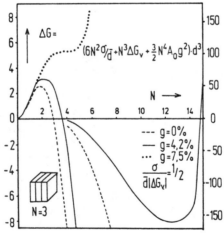

Figure 3

Figure 4 shows the cross section through a PE-microparacrystal
in equilibrium state. The so called "bearing netplanes" then
form the surface of the lattice. In Figure 2 the (10)- and (01)-

netplanes are the "bearing netplanes" and in Figure 4 the (110)-
and (1Ī0)-netplanes of the orthorhombic face-centred lattice.

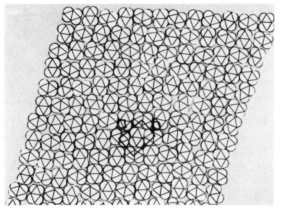

Figure 4

Figure 5 finally confirms this structure obtained by electron
microscopy of linear PE precipitated from high diluted perchlor-
ethylene solution (12). An exceptionally convincing evidence for
the reality of this new kind of equilibrium state of microparacrystals may be recognized in the fact that the size distribution of microparacrystals can be calculated by statistical mechanics (13) which could be fully established lateron by high precision TEM-diagrams of M.G. Dobb et al. (14), where lattice planes
of an ensemble of microparacrystals in the PPT-fibre (Kevlar
DuPont) with N̄ = 12 could be detected (15). The agreement with the
theoretical results is quite satisfactory and may support to
understand better the properties of microparacrystals.

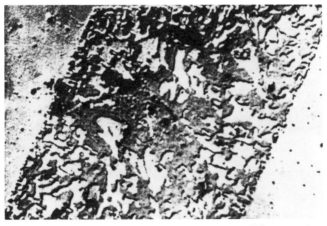

Figure 5

References:

(1) M. Laue, F. Friedrich and P. Knipping (1912), Sitzungs-Ber. Bayer. Akad. Wiss. Math. Phys., Kl. 303, 363-373.

(2) M.v. Laue (1960), Röntgenstrahl-Interferenzen, 3. Aufl., Akad. Verl. Ges., Frankfurt/Main.

(3) P.P. Ewald (1940), Proc. Phys. Soc. (London), 52. 167-178.

(4) P. Debye (1927), Z. Phys., 28, 135-147.

(5) R. Hosemann (1950), Z. Phys., 128, 1-35 and 465-492.

(6) L.D. Landau and E. Lifshitz (1938), Statistical Physics, Oxford.

(7) W.S. Borghard and M. Boudart (1983), J. Catal., 80, 194-206.

(8) D.C. Puxley, C.J. Wright and C.G. Windsor (1982), J. Catal., 78, 257-261.

(9) R. Hosemann, W. Schmidt, A. Fischer and F.J. Baltá-Calleja (1983), Anales de Fisica, A79, 145-154.

(10) R. Hosemann (1970), Ber.Bunsen-Ges.phys.Chem., 74, 755-767.

(11) R. Hosemann (1982), Colloid & Polymer Sci., 260, 864-870.

(12) P.H. Geil, F.R. Anderson, B. Wunderlich, T. Avakawa (1964), J. Polymer Sci., A2, 3707-3712.

(13) R. Hosemann, W. Schmidt, A. Lange and M. Hentschel (1981), Colloid & Polymer Sci., 259, 1161-1169.

(14) M.G. Dobb, A.M. Hindeleh, D.J. Johnson, B.P. Saville (1975), Nature, 253, 189-195.

(15) R. Hosemann and A.M. Hindeleh (1982), Polymer Commun., 23, 1101-1103.

Statistical-Thermodynamic Theory of Polymer Precipitation Fractionation

H. Craubner

Max-Planck-Institut für Festkörperforschung

Heisenbergstr. 1, D-7000 Stuttgart 80, West Germany

A statistical-thermodynamic theory of polymer precipitation fractionation is developed. It is based on the statistical-thermodynamic perturbation theory of phase separation in multicomponent macromolecular solutions. The fractionation efficiency $\bar{\Phi}$ is well founded as probability of confidence which, in this context, serves as reliability measure of the molecular weight distribution function. The theory is scrutinized and corroborated experimentally at, for example, poly(methyl methacrylate). For the prefractionated polymer sample investigated it resulted a bimodal molecular weight distribution ($\bar{M}_W = 44,000$) with a reliability corresponding to $\bar{\Phi} > 99\ \%$.

I. Introduction

Fractionation precipitation from solution is, besides successive solution fractionation /1/, one of the most widely used methods for analytical or/and preparative separation of heterogeneous macromolecular substances. In this context, the precipitation of the fractions building up the polymer ensemble is carried out by a stepwise or continuous decrease in the solvent power of the system which may be achieved mainly by any of the following three methods: (i) addition of nonsolvent; (ii) elimination of solvent by evaporation; (iii) lowering of the temperature of the system. The first member of such a set of polymer fractions has the highest molecular wt. (M) and the succeeding ones progressively decreasing mol. wts. In this way, fractional precipitation may be used to solve one of the most important problems in Polymer Science, namely, the characterization of the distribution of mol. wts. (MWD) or/and any other properties throughout a polymer sample. Solely in this fashion, one can also achieve full advantage of the characteristic qualities being inherent in polymeric materials /2/. Even though new techniques are rapidly being developed, it is likely that those above will continue to be popular, because they are usually based upon simple operations and may, furthermore, offer a first access to quantitative theoretical treatment not only in regard to such complex measuring processes as e.g. HPLC and GPC, but also regarding advanced multiphase polymers and polymer alloys/1-3/. Here, there is only little progress in regard to the needed

coordination of molecular structure with the physical and
chemical properties.

Up to now, there does not exist a quantitative theory of
polymer fractionation by precipitation. Therefore, this paper is concerned with a statistical-thermodynamic   aproach
to the fractional precipitation of polymers. The treatment is
based on concepts that were developed in previous papers /1,
3,4/, i.e. above all the statistical-thermodynamic perturbation theory of phase separation in multicomponent macromolecular solutions /3/ and the molecular distribution theory of
macromolecules /4/.

II. Probability Theory of Reliability

According to Huggins et al. /5/ and Flory /6/ (cf. also
ref. 7), it is evident that each determination of the MWD of
a polymer by fractional precipitation has to answer the question of the extent to which higher components are selectively transferred   under given thermodynamic equilibrium conditions to the precipitated phase, i.e., to the more concentrated gel phase. This means, however, that the principal issue
involved in the fractional precipitation of polymers consists
in the efficiency of the sorting of species between the two
phases, i.e., the polymer-rich gel phase (superscript β) and
the highly diluted polymer-lean sol phase (superscript α).

In this context, it is assumed that both phases which separate in equlibrium are amorphous and that the possibility
that the polymer in the precipitated gel phase may be semi-
crystalline, can be disregarded.

If we try to determine the MWD of an array of homologs
present in an ordinary polymer ensemble as described, we must
be prepared that the result is subject to error /4/. Therefore, we should know the confidence probability Φ that the
experimentally recovered MWD function

$$I(\hat{P}) = {}_1\!\int^{\hat{P}} H(P') \, dP' \; ; \quad I\big|_1^\infty = {}_1\!\int^\infty H(P) \, dP = 1 \; , \qquad (1)$$

represents the real but actually unknown MWD function

$$\mathcal{J}(P) = {}_1\!\int^P H(P') \, dP' \; ; \quad \mathcal{J}\big|_1^\infty = {}_1\!\int^\infty H(P) \, dP = 1 \; , \qquad (2)$$

of the original flake which may also be estimated by theory
to

$$\tilde{I}(P) = {}_1\!\int^P \tilde{H}(P') \, dP' \; ; \quad \tilde{I}\big|_1^\infty = {}_1\!\int^\infty \tilde{H}(P) \, dP = 1 \; . \qquad (3)$$

In the situation analysis given by equations (1) - (3), the
real statistical polymer component /4/ is given within the
probability range (P, P+dP) of the degree of polymerization
P (DP) by the expression $d\mathcal{J} = H(P) \, dP$, whereas $d\tilde{I}$ and dI
designate the corresponding theoretical estimation and the
experimental approximation, respectively. Moreover, the different symbols for the MWD $H(P)$, $\tilde{H}(P)$, and $H(\hat{P})$ represent the
attached probability densities to the corresponding MWD func-

tions $\mathcal{I}(P)$, $\tilde{I}(P)$, $I(\hat{P})$, and $1 \leqq P < \infty$ mean the real DP, as well as $1 \leqq \hat{P} < \infty$ the measured DP.

Therefrom, it follows that $\bar{\Phi}$ means a statistical-thermodynamic reliability measure of the fractionation efficiency (FE). In this context, the total FE corresponds to the product of the partial efficiencies $\bar{\Phi}_k$ ($k = m, p$) in regard to mass (m) and DP (P),

$$\bar{\Phi} = \Pi_k \bar{\Phi}_k = 1 - \xi^2 \simeq \exp(-\xi^2) = \exp(-\Sigma_k \xi_k^2), \quad (0 \leqq \xi^2 << 1). \quad (4)$$

Here,

$$\xi^2 = \sum_k \xi_k^2 = 1 - \bar{\Phi}, \qquad (0 \leqq \xi_k^2 << 1), \quad (5)$$

designates the inefficiency as the compliment to $\bar{\Phi}$, given by the sum of the partial inefficiencies where the inequalities in equations (4) and (5) define the acceptance range of a fractionation.

III. Thermodynamic Computation of Fractionation Efficiency

In order to calculate the FE, let us remember that fractional precipitation is carried out in multicomponent macromolecular solutions consisting of solvent ($\mathbb{L}$), nonsolvent or precipitant ($\mathbb{F}$), and the well-ordered subset of the statistical polymer components (v.s.) $\mathbb{P} = \{\mathcal{P}_j \mid j \in \mathbb{N}_{r_j}\}$. Such complex systems may be described briefly by the set-topological structure-theory, $\mathbb{K} = \mathbb{P} \cup \mathbb{L} \cup \mathbb{F} = \{L, F, \mathcal{P}_j \mid j \in \mathbb{N}_{r_j}\}$, $\mathbb{N}_{r_j} = \{1, 2, \ldots, r_j\}$ (cf. ref. 3,4). Here, the united set $\mathbb{K}$ generally describes such a quasiternary macromolecular solution. After mapping of $\mathbb{K}$ to the thermodynamic $\{\gamma\}$-space, according to the statistical-thermodynamic perturbation theory of polymer phase separation /3/, microscopic (quasi-)ternary thermodynamic phase equilibria defined by $(dG_j)_{T,p} = 0$ (G = free enthalpy) are established after any perturbation of the system, e. g. by addition of nonsolvent, change of temperature or pressure, etc. During the relaxation of such a constraint, the well-ordered statistical polymer components $\mathcal{P}_j$ are successively adjusted to the coordinated microscopic phase equilibria before they are separated by partial precipitation. In this fashion, unequivocal functional relationships are established between the precipitant fraction (volume fraction of nonsolvent at the precipitation threshold, γ_F^*) and the different statistical-thermodynamic variables of state, such as e.g. T, M (mol. wt.), initial concentration c_o, and pressure p. In differential notation it is (cf. ref. 3)

$$d\gamma_F^* = -\sigma_c^* \, d \ln c_o - \frac{1}{2} \sigma_M^* \, d \ln M + \frac{\Delta S'_{sep}}{R} \, d \ln T \qquad (6)$$

($\sigma_{(M,c)}^* = $ const; R = gas constant; $\Delta S'_{sep} = $ apparent molar entropy of phase separation).

In the jth (quasistationary) microscopic (quasi-)ternary phase equilibrium the infinitesimal weight fraction of the jth statistical polymer component ($dm_j = dm(P)$) is distributed between the highly diluted sol phase $(dm_j^\alpha = dm^\alpha(P))$ and the polymer-rich gel phase $(dm_j^\beta = dm^\beta(P))$, $dm_j = dm_j^\alpha + dm_j^\beta$. Therefrom, the original flake corresponds to $(m_g = \text{original total weight})$

$$m_g = {}_0\!\int^{m_g} d\,m(P) = {}_0\!\int^{m_g}(d\,m^\alpha(P) + d\,m^\beta(P)), \qquad (7)$$

and the total recovery after precipitation is given by

$${}_0\!\int^{m^\beta} d\,m^\beta(P) = m^\beta \implies {}_0\!\int^{m_g^\beta} d\,m^\beta(P) = m_g^\beta \leq m_g. \qquad (8)$$

Normalization to the total weight of the original flake m_g yields the experimental MWD function via its complement

$$I^*(\hat{P}) = \frac{1}{m_g} {}_0\!\int^{m^\beta} d\,m^\beta(\hat{P}) = m^\beta(\hat{P})/m_g = 1 - I(\hat{P}). \qquad (8a)$$

The differential mass ratios $\boldsymbol{\varphi}^\beta = 1 - \boldsymbol{\varphi}^\alpha = d\,m_j^\beta/d\,m_j$ designate the partial separation coefficients for gel $(\boldsymbol{\varphi}^\beta)$ and sol $(\boldsymbol{\varphi}^\alpha)$ phases. They are closely related to the corresponding partial molar phase volumes and the corresponding volume fractions of the statistical polymer components in question. From another viewpoint, $\boldsymbol{\varphi}^{(\alpha,\beta)}$ may also be interpreted as the respective probabilities to find macromolecules of the species $\mathcal{P}_j$ in the gel and sol phases, i.e., however, to find macromolecules in the probability P-range $(P_j, P_j + dP_j)$. Finally, the averages of the separation coefficients are given by the expression

$$\langle \boldsymbol{\varphi}^\beta \rangle = 1 - \langle \boldsymbol{\varphi}^\alpha \rangle = {}_1\!\int^\infty \boldsymbol{\varphi}^\beta(P)\,H(P)\,dP. \qquad (9)$$

The theoretical basis described shortly, enables us to calculate the FE in regard to DP and weight fraction. According to equation (4), one finally obtains

$$\eth = \eth_p \eth_m \simeq \exp[-(\xi_p^2 + \xi_m^2)] = \exp\left\{-\frac{1}{\boldsymbol{\varphi}^\beta}\left[\frac{\overline{(P-\hat{P})^2}}{\overline{P_0}^2} + \overline{w^2}(1-\overline{\boldsymbol{\varphi}^\beta})^2\right]\right\} (10)$$

$(w = m^\beta/m_g$ wt. ratio; $P = $ optimum estimate of DP; $\hat{P} = $ experimental DP; $\overline{P_0} = $ mean wt. average DP of the original flake).

According to eqation (10), the FE depends exponentially on the sum of the partial inefficiencies. It turns out to be a function of the relative dispersions of DP and the wt. ratios of the fractions, as well as of the mean separation coefficients. To achieve optimum FE $(\eth \to 1)$, i.e., to separate a polymer ensemble by fractional precipitation in the best possible way, one has to take care of the 2nd moments of the DP dispersion and the wt. ratios which should be as small as possible, and, finally, $\boldsymbol{\varphi}^\beta \to 1$.

IV. Experimental

To corroborate the statistical-thermodynamic reliability theory of FE derived, we need the quantitative analysis of the MWD functions of polymer ensembles by fractional precipitation. This means, however, that we require the numerical data of the perturbation-theoretical relations represented by the differential equation (6) for each type of polymeric material , in order to compute the different fractions in dependence on mass, mol. wt., temperature, concentrations, and pressure. In this context, all our measurements were performed by fast turbidimetric titrations /3,8/. In regard to the determination of MWD , the partial turbidity ratio τ/τ_g ($\tau_g =$ turbidity of total precipitation of the mass of the original flake m_g) is equal to the weight fraction of the precipitated polymer $w = m^\beta/m_g$ and, therefore, according to equation (8), equal to the complementary MWD function $I^* = 1-I(P)$,

$$I(P) = 1-I^*(P) = \int_1^P H(P')dP' = 1-m^\beta/m_g = 1-\tau(P)/\tau_g \ . \qquad (11)$$

For the MWD experiments, I used double-fractionated poly (methyl methacrylate) (PME). To introduce an intrinsic preparative control test, the sample, prepared by triangle double-fractionation, was taken from that part of the original flake with the lower mol. wt. average. Therefore, the precision of any fractionation procedure may be checked: (i) by the condition of small nonuniformity $U = (\bar{P}_w/\bar{P}_n)-1 < 0.2$; (ii) by the cutting out of higher mol. weights. The experimental conditions for fast turbidimetric titration of PME were as follows: $\bar{M}_n$ = 43,000; T = 298 K; initial polymer conc. c_o= 0.75 mg/cm^3; solvent: benzene; precipitant: cyclohexane; injection rate $\dot{v}$ = 20 cm^3/min.

V. Results

The results of the MWD analysis of the triangle prefractionated PME original sample are shown in figures 1 and 2. In figure 1, the MWD function I(P) is plotted vs. P. It is relatively narrow and shows a small transition region (marked by dotted lines). This is evidently illustrated by the plot of the number frequency distribution h(P) = H(P)/P vs. P. The curve of h(P) is rather narrow, showing a bimodal character which demonstrates evidently the preparative cutting out of higher mol. wts. The mathematical formulation corresponds to a two-branch MWD function with a rather complex curved transition region between the two peaks of the bimodal MWD ($f_p =$ transition probability),

$$I(P) = {}_1\!\int^P H(P')dP' = 1-f_p\exp(-b_1 P^{a_1})- (1-f_p)\exp(-b_2 P^{a_2}); \quad (12)$$

$$f_p = \begin{cases} 1 & \text{if } P < P_1 ; \\ \exp\{ -[(P-P_1)/(P-P_2)]^2\} & \text{if } P_1 \leq P \leq P_2; \\ 0 & \text{if } P_2 < P \ . \end{cases}$$

The coordinated numerical values of the constants of equation (12) are given in table 1.

The statistical characterization of the prefractionated

original PME flake is given by the following data:

$\bar{M}_n$ = 37,000

$\bar{M}_w$ = {44,100 from MWD;
              43,900 by direct turbidimetric determination;

$\bar{M}_z$ = 50,100

$\bar{M}_\eta$ = 43,300 from viscosity measurements;

$U_W = (\bar{M}_w/\bar{M}_n)-1 = 0.19$ nonuniformity;

$U_z = (\bar{M}_z/\bar{M}_w)-1 = 0.13$ polydispersion (cf. ref.4);

Φ = 99.99×10^{-2} fractionation efficiency.

The complex situation in regard to the prefractionated PME sample under consideration is made evident by the relative small polydispersion U < 0.2, on the one side, even though the MWD is, on the other side, of bimodal character which was made safe by the value of FE Φ > 99 %. This demonstrates that reliable characterization of macromolecular material demands generally the determination of the MWD function I(P) and, moreover, the corroboration of the latter by the determination of the fractionation efficiency Φ. Only in this way, it can be corroborated within the confidence probability Φ as reliability measure that the experimentally or/and also theoretically determined MWD represent actually the true MWD of the original flake.

References

1. H. Craubner, Makromol. Chem. 175, 2171, 2461(1974);
   Z. Phys. Chem. NF 103, 45(1976).
2. H. Craubner, 23rd IUPAC Congress, Boston 1971, Macromol.
   Preprint Vol.I, p.513.
3. H. Craubner, Ber. Bunsenges. Phys. Chem. 81, 1060(1977);
   Macromolecules 11, 1161(1978); Coll. & Polym. Sci. 257,103
   (1979); Polymer Preprints (ACS) 21/2, 243(1980); 27/1, 435
   (1986); IUPAC Internat. Symposium on Macromolecules, Mainz
   1979, Preprints, Vol. II, p. 962.
4. H. Craubner, Z. Phys. Chem. NF 122, 21(1980).
5. M. L. Huggins and H. Okamoto, Polymer Fractionation, M.J.R.
   Cantow, Ed., Academic Press, New York, 1967, p. 1.
6. P. J. Flory, Principles of Polymer Chemistry, Cornell University Press, Ithaca, NY, 1953, p. 559.
7. L. H. Tung, Ed., Fractionation of Synthetic Polymers, Dekker, New York, 1977.
8. H. Craubner, Ber. Bunsenges. Phys. Chem. 74, 962, 1262
   (1970); 75, 326(1971).

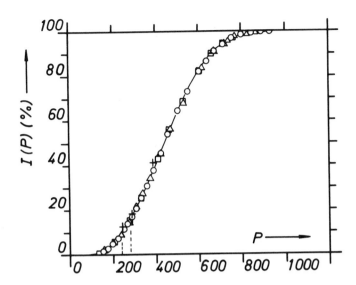

Figure 1. Molecular weight distribution of prefractionated poly(methyl methacrylate) (the sample was prepared by tri-angle double-fractionation and corresponds to the lower mol. wt. cut of the original PME flake): mol. wt. distribution function I(P) plotted vs. degree of polymerization P (dotted lines mark transition region, P_1=240; P_2=280).

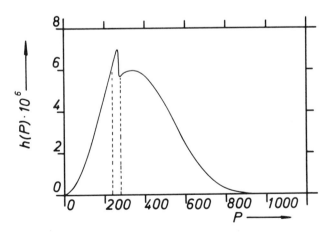

Figure 2. The same experiment as in Figure 1 but here the differential number frequency distribution h(P) is plotted vs. degree of polymerization P (dotted lines mark transition region, P_1=240; P_2=280).

Table 1. Constants of the bimodal MWD function eq (12) for the prefractionated poly(methyl methacrylate)

Constant	Branch 1	Branch 2
a	3.762 ± 0.004	2.960 ± 0.001
b	$(1.179 \pm 0.027) \times 10^{-10}$	$(1.043 \pm 0.007) \times 10^{-8}$
P	240	280

THE SYNTHESIS OF BLOCK AND GRAFT COPOLYMERS
FROM MACRORADICALS

Raymond B. Seymour

Department of Polymer Science
University of Southern Mississippi
Hattiesburg, Mississippi 39406-0076

Introduction

Macroradicals have been present in all organisms since
the beginning of life on earth. The existence of radicals
was proposed in the early part of the 19th Century by
Berzelius and this unsound hypothesis was revived in the mid-
1800's. Berzelius' contemporaries, Wohler and Liebig used
the term benzoyl radical and suggested that stable radicals
could exist. However, stable free radicals were not
isolated until 1900 when Gomberg prepared triphenylmethyl
$(C_6H_5)_3C\cdot$. After much controversy, his results were accepted
and he became recognized as the discovered of organic free
radicals.

Early Experiments with Macroradicals

Simon (1) and Blyth (2) unknowingly, worked with
macroradicals in 1839 and 1849 when they thermally
depolymerized polystyrene. Likewise, Regnault produced
macroradicals during the free radical polymerization of
vinyl chloride in 1838 (3). However, the first systematic
study in which the existence of intermediates
(macroradicals) was recognized in the polymerization of
vinyl monomers was suggested by Kronstein in 1902 (4).
Another equally significant study in the polymerization of
unsaturated hydrocarbons was reported by Lautenschlager in
1913 and amplified by Staudinger in 1931 (5).

Early Block Copolymers

One of the first commercial block copolymers was a
block copolymers of ethylene oxide and propylene oxide made
by ionic copolymerization and used as a surfactant
(Pluronics) (6). Block copolymers have also been produced
by mechanochemical synthesis in which vinyl monomers have
been added to macromolecules which were cleaved by
mastication, milling or ultrasonic irradation (7). A
commercial block copolymer of vea rubber and methyl

methacrylate, called Hevea Plus, is made by mechanochemical
synthesis or by the copolymerization of the vinyl monomers
in rubber latex (8).

Commercial block copolymers of styrene and isoprene,
called Kraton,are produced by anionic polymerization
techniques (9). Segmented block copolymers of polyesters-
polyurethanes, polyester-polyethers and polyamides-polyethers
are produced by condensation polymerization (10).

Experiments with "Trapped" Free Radicals

The existence of "trapped" free radicals, occluded
macroradicals or microgels has been recognized for many
years but few commercial block copolymers have been produced
from these macroradicals (11). These occluded macroradicals
which have been called "pop corn" polymers are produced when
polymers, such as polyacrylonitrile or polyvinyl chloride,
precipitate in their monomers as formed (12). Nevertheless,
it is difficult to produce block copolymers from these
"trapped free radicals" because of the inability of many
monomers to penetrate the occluded macroradicals.

Synthesis of Stable Macroradicals

Shapiro (13) and Hiemeleers (14) controlled the
production of these occluded macroradicals by polymerization
in poor solvents and Bamford (15) characterized these free
radicals by electron spin resonance. Seymour and coworkers
showed that the criterion for precipitation of macroradicals
was a difference in Hildebrand solubility parameters of
macroradical and solvent of at least 1.8 (H) (16).
Atkinson (17), Tsuchida (18), and Seymour (19), showed that
the rate of polymerization increased as the one approached
the 1.8H solubility parameter limit.

Application of Solubility Parameter Concepts

Because of the relatively large differences in
solubility parameters, it is difficult to form block
copolymers by adding styrene or methyl methacrylate to
acrylonitrile "trapped" free radicals. However, some
block copolymers can be obtained when small amounts
of dimethylformamide are added with the monomers (20). It
has been shown that the criterion for block copolymerization
between a vinyl monomer and occluded macroradical is that
the difference in Hildebrand solubility parameters is less
than 3.1H (21).

Thus, it is possible to produce block copolymers of acrylonitrile and styrene (22) or methyl methacrylate (23). It is also possible to produce block copolymers by the addition of equimolar quantities of styrene and maleic anhydride (24) or styrene and acrylonitrile (25) to acrylonitrile macroradicals. Styrene blocks may also be formed on these blocks with styrene terminal groups.

Macroradicals of alkyl methacrylates may be used to produce block copolymers (26). However, macroradicals of acrylic esters tend to terminate by coupling because of their low glass transition temperatures. However, block copolymers were readily produced from vinyl acetate macroradicals (27).

Synthesis of Block Copolymers In Viscous Media

Stable macroradicals, which could be used for block copolymer synthesis, were also produced by the polymerization of the monomers in a viscous medium. Extremely fast polymerization was noted when the difference between solubility parameters of the monomer and macroradicals in the viscous solvent approached 1.8H (28).

Because of similarities in solubility parameters and glass transition temperatures, phase separation of the blocks is not well defined (29). However, better phase separation was noted when p-butylstyrene was used as the blocking monomer (30). Macroradicals of acrylamides have also been produced by polymerization in poor solvents and block copolymers have been produced by the addition of vinyl monomers to these occluded radicals (31).

Synthesis of Block Copolymers in Emulsion Systems

Macroradicals are also present in the micelles in aqueous emulsion in polymerization systems but few attempts have been made to use these systems for the synthesis of block copolymers. Nozaki of Shell Oil Company patented several emulsion block copolymer systems (32), but these were not commercialized.

Block copolymers of styrene and methyl methacrylate were prepared in the presence of a removable heterogeneous initiator, by Hori (33). Seymour and Tinnerman (34) also produced and characterized block copolymers and other monomers in an emulsion polymerization system using an organic soluble initiator (tert. butyl peroxy pivalate). The

second monomer was added 5 days after the original
polymerization was complete, i.e., after complete decay
of the initiator.

Conclusions

Stable macroradicals can be readily prepared in
solution and emulsion polymerization systems. Providing
the primary initiator is no longer active, these
macroradicals can be used to produce good yields of block
copolymers.

References

1. E. Simon,Ann 31 265 (1839).
2. J. Blyth, A. W. Hofmann Ann 53 289 (1945).
3. V. Regnault, Ann Chim Phys 2 69 152 (1838).
4. A. Kronstein, Ber 35 4150 (1902).
5. H. Staudinger, L. Lautenschlager, Ann 488 1 (1931).
6. T. H. Vaughn, D. R. Jackson, L. C. Lunsted J. Am. Oil
    Chem. Soc. 29 240 (1952).
7. R. J. Ceresa,"Block and Graft Copolymers" Buttersworth,
    Washington, D.C., 1962.
8. L. C. Bateman, Ind. Eng. Chem 49 704 (1957).
9. B. M. Walker,Elastomerics 39 (April 1981).
10. I. Goodman, "Developments in Block Copolymers" Applied
    Science Publishers, London, 1982.
11. V. E. Shoashua, K. E. Van Holde, J. Polym. Sci 28 398
    (1958).
12. G. S. Whitby,Ind Eng Chem 47 806 (1955).
13. A. J. Chapiro, Chem. Phys 47 747 (1950).
14. J. Hiemeleers, G. Smets Makromolecular Chem 47 7 (1961).
15. C. H. Bamford,J. Polym Sci 48 37 (1968).
16. Seymour, R. B. et al Adv. Chem Ser. 99 418 (1971).
17. B. Atkinson, G. R. Cotton,Trans Faraday Soc 54 877 (1958).
18. E. V. Tsuchida et al, J. Chem. Soc. Japan 70 573 (1967).
19. R. B. Seymour et al, The Texas J. of Sci 21 (1) 13 (1968).
20. Y. Minoura, Y. Ogata, J. Polym Sci A-1 7 2547 (1969).
21. R. B. Seymour, P. D. Kincaid, D. R. Owen, J. Paint
    Technol 45 (580) 33 (1973).
22. R. B. Seymour et al, Polymer Preprints 17 (1) 216 (1976).
23. R. B. Seymour et al, Advan Chem Ser 142 309 (1975).
24. R. B. Seymour, D. R. Owen, P. D. Kincaid Chem Tech 3 (9)
    549 (1973).
25. R. B. Seymour, G. A. Stahl, H. Wood, W. N. Tinnerman,
    Appl. Polym. Sci. Symp 25 (69) (1974).
26. R. B. Seymour, D. R. Owen, H. Wood, Polymer Preprints
    25 69 (1974).

27. R. B. Seymour, G. A Stahl, J. Polym Sci 14 2545 (1976).
28. R. B. Seymour, G. A, Stahl, Polym Sci Technol 10 217 (1977).
29. H. S. Makowski, R. D. Lundberg, U.S. Pat 3,821,148 (1974).
30. H. Makowski, D. G. Buckley, Ger. Pat, 2,443,875 (1973)
31. T. Sato, T. Iwaki, O takayuku, J. Polym Sci, Polym Chem Ed., 21, 943, (1983).
32. K. Nozaki, U.S. Pat 2,666,042 (1954), 3,069,380 1 (1962).
33. K. Hori, D. Mikulasova Die Makromol Chem 175 2091 (1977).
34. R. B. Seymour, W. Tinnerman Coatings and Plastics Preprints 36 (2) 417 (1976).

A NOVEL APPROACH TO THERMOPLASTIC ELASTOMERS WITH ENHANCED TEMPERATURE RESISTANCE

by

B. C. Auman *, H.-J. Cantow

Wu Jishan **, V. Percec * and H. A. Schneider

Institut für Makromolekulare Chemie der Universität Freiburg, Hermann-Staudinger-Haus D-7800 Freiburg - Federal Republic of Germany

* Department of Polymer Science, Case Western Reserve University, Cleveland, Ohio 44106, USA

** Chenguang Research Institute of Chemical Technology, Fushun 101-209, Sichuan, PR China

Introduction

Functional polyaromatics which contain terminal or pendant groups which undergo thermally initiated crosslinking reactions without evolving volatile by-products are considered good candidates for matrix resins in high performance composites. This is due to their good thermal stability and high glass transition temperatures. We have prepared fastly curing polysulfone (PSU) oligomers containing terminal or pendant vinylbenzyl(styrene type) groups [1,2] as a viable alternative in the reactive polymer area. Often, higly crosslinked networks tend to be brittle, so we extended our work with α,ω-di(vinylbenzyl)PSU and sought to incorporate rubbery, thermally stable poly(dimethylsiloxane) (PDMS) segments into our PSU reactive oligomers. Due to the drastic incompatibility of these two polymers, phase separation into rubbery and glassy domains could be realized which would lend the desired properties of both components in the network. Ideally, the high T_g and the modulus of PSU would be maintained while the impact resistance of rubbery PMDS domains would be added.

Synthesis

α,ω-di(vinylbenzyl)PSU has been prepared from α,ω-di(hydroxyphenyl)PSU and p-chloromethylstyrene, using tetrabutylammonium hydrogen sulfate as phase transfer catalyst. These bifunctional telechelics contain terminal vinylbenzyl (styrene-type) groups which enable the oligomer to be thermally cross-linked. By virtue of their end-standing vinyl functions, these styrene terminated PSU's can be coupled with PDMS containing end-standing ($\equiv$Si-H) functions via hydrosilation.

The synthesis of PSU with incorporated PDMS segments has been performed in chlorobenzene solution by reaction of the α,ω-di(vinylbenzyl)PSU with α,ω-di(silane)PDMS with hexachloroplatinic acid as the catalyst. A 2/1 molar excess of the PSU to the PDMS component has been choosen with the consequence that, statistically, "triblock" copolymers have been obtained. It was generally sought to maintain as high a crosslink density as possible, as well as the desired properties of the blocks. To this end, it was desired to keep the molecular weight of both blocks low enough to minimize the distance between the crosslinks, yet high enough so good phase separation could be attained.

All synthesis steps have been followed by 200 MHz proton NMR. GPC has been applied to check the "triblock" formation [3].

Results and Discussion

The thermal characterization of "triblocks" - all of them with a ~ 50 % vinyl conversion as expected - shows that the data split into different groups. Samples 1, 5, 7, and 9 present a PSU series with the lowest molecular weight PDMS (M_n = 540). Because the segment has an average lenght of 7 repeat units only, phase separation is very poor. Consequently, the PDMS serves to lower the T_g of the PSU and of the resulting network. No separate T_g for the PDMS is observed in DSC. However, dynamic-mechanical measurements of "triblock" 9 revealed a small gradual step in G', the storage modulus, in the temperature range -120 to -75°C, indicating minor phase separation.

Another group (Samples 1-4) represents a series of different PDMS molecular weights with the lowest molecular weight PSU, M_n = 1200. Due to its low M_n, phase separation is very poor, and therefore not only is the PSU T_g somewhat lowered, but the PSU T_g* is also drastically reduced with incorporation of PDMS. The deterioration of the PSU T_g

B.C. Auman et al.

Table 1: PSU-PDMS triblock glass transition °C

Sample	2PSU/1PDMS M_n	T_g before cure PDMS	PU	T_g* after cure PDMS	PU	T_g* $-T_g$
1	1200/ 540	–	60	–	102	42
2	1200/1200	–	49	–	93	44
3	1200/**3500**	-122	44	-123	82	38
4	1200/**5500**	-123*	38	-123*	70	32
	1200- PSU		75		213	138
5	2100/ 540	–	104	–	134	30
6	2100/**3400**	-124	97	-122	128	31
	2100- PSU		113		201	88
7	**3200**/ 540	–	127	–	153	26
8	**3200/3400**	-126	137	-122*	**178**	**41**
	3200- PSU		**139**		**190**	51
9	**3800**/ 540	–	139	–	162	23
10	**3800**/1200	–	136	–	166	30
11	**3800/3400**	-126	150	-126*	**180**	30
	3800- PSU		**147**		**190**	**43**

* Samples 8 and 11 exhibit additionally a melting transition after curing, -56 and - 47°C, resp., sample 4 melts at -43°C before and after curing

is more pronounced with PDMS molecular weight, a result of the poor phase separation and increasing distance between crosslinks. Samples 3 and 4 from this group show clearly a separate PDMS phase due to the higher PDMS molecular weight. Sample 4 even exhibits a melting transition.

The next group (samples 9-11), like the foregoing, represents a series of different molecular weight PDMS; however, in this case the PSU block has higher M_n = 3800. As the PDMS molecular weight is raised, phase separation improves and the PSU T_g's approach those of the parent homopolymer. "Triblocks" 9 and 10 show no DSC transition for the PDMS segment due to its low M_n, but "triblock" 11 does exhibit a sharp well defined PDMS T_g both before and after curing due to the higher PDMS M_n = 3400. In fact after curing, this PDMS segment even shows melting transition [4].

Based on these results, it was realized that good phase separation was obtained with PDMS M_n ~ 3400. The samples 3, 6, 8, 11, therefore, were prepared from this molecular weight PDMS with different molecular weight PSU. The thermal data indicate improved PSU phase separation with molecular weight as expected. It also indicates that for this molecular weight of PDMS, a PSU M_n of only ca. 3000 is necessary for good phase separation. Indeed, for "triblocks" 8 and 11, PSU T_g's before cure are essentially equivalent to those of the corresponding homopolymers, and post-cure T_g's approach those of the PSU networks without PDMS. For the PDMS segment, a weel-defined T_g is always seen before and after cure. The crystallization of the PDMS that occurs in samples 8 and 11 may be somewhat undesirable for low temperature network properties because of its stiffening effect. It may be possible, therefore, to avoid this through incorporation of a small percent of ether or phenyl groups in place of the methyl groups of PDMS.

In conclusion, the thermal data indicate that good phase separation is obtained is obtained with M_n of both components ~ 3000 only.

An additional approach for network manipulation is blending of the block copolymers with varying amounts of PSU (M_n= 1200). E. g., it was noted that although purified "triblocks" 3 and 4 yielded clear films from solution, films of blends showed certain degrees of translucence or opacity due to phase separation into relatively large domains.

A study of the mechanical properties of the two phase PSU-PDNS-PSU "triblock networks" has been performed by torsional pendulum as well as by forced oszillation [5]. Mesurements on samples 9-11 may be discussed shortly. The forced oscillation studies were executed on an INSTRON 3250 Rheometer using torsion bars of 35x3x3 mm. In the temperature range of the PSU T_g, the temperature was varied in steps of 5°. The rheological data were then shifted to isotherm mastercurves using both the WLF relation [6] and the temperature dependent activation energy method suggested by Brekner et al. [7]. The latter method was also applied for constructing isochrone mastercurves.

The complex temperature dependence of the PSU glass transition on the lenght of the incorporated

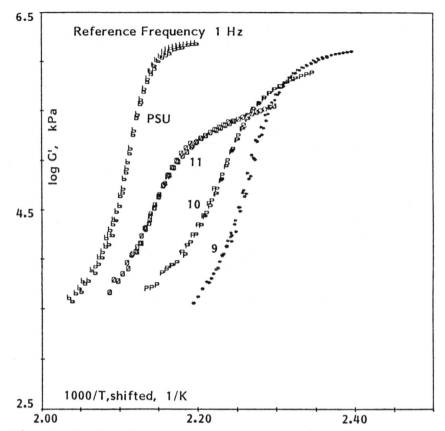

<u>Figure 1:</u> Isochrone mastercurves of G' in the PSU T_g range of PSU M_n=3800 and of the triblock net-works 9-11 (M_n PDMS = 540, 1200 and 3400, resp.)

PDMS block as suggested by DSC is confirmed by the isochrone mastercurves of the storage modulus, G', and of the loss modulus, G", of the cured "tri-block" networks 9-11. The PSU T_g shift to lower temperatures is maximum for the triblock with the shortest PDMS middle block. With increasing M_n of PDMS the PSU glass transition approaches gradually the T_g of the parent homo-PSU network (M_n = 3800). The broadening of the loss modulus peak increases with increasing molecular weight of the PDMS.

    In conclusion, by proper choice of the block-lenght of the components of the "triblock" network, the machanical properties are conveniently adjust-able. The conclusions drawn by DSC concerning the optimum block length are are confirmed.

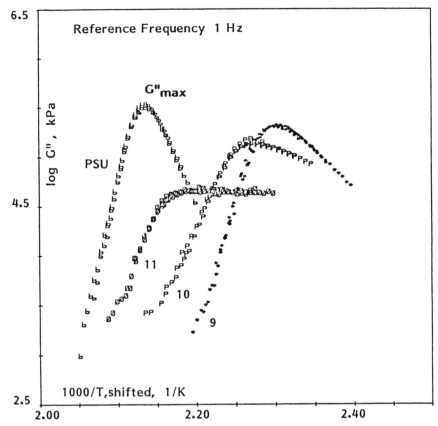

Figure 2: Isochrone mastercurves for G" (comp. F 1)

For better insight into the rheological behaviour, the obtained data were applied for the evaluation of the constants of the WLF relation, the related "Vogel" temperature, and the temperature dependent activation energy of flow. A recent approach of Schneider et al. [8,9] has been used. It was concluded that, concerning the rheological behaviour, entropic and energetic contributions are distinguishable. The contributions are related with the WLF constants. The changes in these contributions, as well as of the WLF constants, are block specific and seem to obey a "compensation effect" law [5].

Generous financial support from the SFB 60, DFG and a NATO traveling grant are gratefully acknowledged.

References

1. V. Percec and B. C. Auman, Makromol. Chem. 19, 1 (1984)
2. V. Percec and B. C. Auman, Polymer Preprints 25, 122 (1984)
3. B. C. Auman, V. Percec, H. A. Schneider, Wu Jishan and H.-J. Cantow, Polymer 28, 119(1987)
4. Wu Jishan, B. C. Auman, H. A. Schneider and H.-J. Cantow, Makromol. Chem., RC 7, 303 (1986)
5. H. A. Schneider, Wu Jishan, H.-J. Cantow, B. C. Auman and V. Percec, Polymer 28, 132 (1987)
6. M. L. Williams, R. F. Landel and J. D. Ferry J. Amer. Chem. Soc. 77, 3701 (1955)
7. M.-J. Brekner, H.-J. Cantow and H. A. Schneider Polymer Bull. 10, 328 (1983)
8. H. A. Schneider, M.-J. Brekner and H.-J. Cantow Polymer Bull. 14, 479 (1985)
9. M.-J. Brekner, H.-J. Cantow and H. A. Schneider Polymer Bull. 13, 51 (1985)

925